Probiotics against Antimicrobial Resistance

This book provides an overview of antimicrobial resistance (AMR), including its underlying causes, mechanisms, global ramifications, and the potential of probiotics as a viable approach in combating AMR. It explores the potential of probiotics to counteract resistance by competing with pathogenic bacteria, modulating the immune system, and producing antimicrobial compounds. It also examines the applications of probiotics in biofilm-related infections, respiratory health, urinary tract infections, and oral health. Additionally, this book covers important considerations such as safety, regulatory frameworks, challenges, and risks associated with the use of probiotics in the context of AMR. It provides insights into the development of novel approaches, including engineering probiotics for enhanced antimicrobial activity and exploring synergistic approaches by combining probiotics with traditional antibiotics. Toward the end, this book explores future directions and potential applications of probiotics, including their roles in neonatal care and personalized probiotic interventions. This book is primarily intended for healthcare professionals, researchers, and individuals working in the fields of microbiology, infectious diseases, and public health.

Probiotics against Antimicrobial Resistance

Opportunities and Challenges

Edited by

Mohmmad Younus Wani, Irfan Ahmad Rather, and Aijaz Ahmad

CRC Press
Taylor & Francis Group
Boca Raton London New York

CRC Press is an imprint of the
Taylor & Francis Group, an **informa** business

Designed cover image: Shutterstock

First edition published 2026
by CRC Press
2385 NW Executive Center Drive, Suite 320, Boca Raton FL 33431

and by CRC Press
4 Park Square, Milton Park, Abingdon, Oxon, OX14 4RN

CRC Press is an imprint of Taylor & Francis Group, LLC

ISBN: 9781032676920 (hbk)
ISBN: 9781032676944 (pbk)
ISBN: 9781032676951 (ebk)

DOI: 10.1201/9781032676951

Typeset in Times
by codeMantra

Dedication

We dedicate this book to the worldwide network of scientists, clinicians, and innovators working to curb antimicrobial resistance. Your collaborative spirit and dedication are vital in safeguarding the health of future generations.

With enduring respect and heartfelt appreciation.

The Editors

Contents

Preface

The escalating challenge of antimicrobial resistance (AMR) has emerged as one of the most critical global health issues of our time. Misuse and overuse of antibiotics have accelerated the emergence of resistant bacterial strains, rendering many conventional treatments ineffective. This alarming trend demands innovative and sustainable solutions to mitigate its impact.

Among the promising strategies to combat AMR, probiotics have gained significant attention. These live microorganisms, known primarily for their positive effects on gut health, have shown remarkable potential in addressing AMR through mechanisms such as enhancing host immunity, modulating gut microbiota, and inhibiting pathogenic bacteria. Although the research on probiotics in this context is still evolving, their potential role as complementary or alternative tools to combat AMR is an exciting area of exploration.

Probiotics against Antimicrobial Resistance: Challenges and Opportunities offers a comprehensive insight into the intersection of probiotics and AMR. The chapters, authored by experts from diverse disciplines, address critical topics, including the mechanisms behind AMR, the role of probiotics in combating multidrug-resistant bacteria, and their application in both human and animal health. Emerging themes such as encapsulation and delivery strategies, safety and regulatory frameworks, and synergistic approaches combining probiotics with traditional antibiotics are also explored. The latter sections focus on clinical trials, evidence-based research, and the potential for probiotics to be integrated into personalized medicine. This book concludes with a forward-looking discussion on the opportunities and challenges that lie ahead, emphasizing the importance of interdisciplinary collaboration in addressing AMR.

As editors and contributors, we believe this book serves as a valuable resource for researchers, healthcare professionals, policymakers, and educators alike. It aims to inspire innovation, drive policy changes, and encourage further research to harness the full potential of probiotics in combating AMR. While the path ahead is fraught with challenges, we are optimistic that the insights and strategies discussed here will contribute meaningfully to global efforts against this critical health crisis.

We extend our gratitude to the authors, reviewers, and publishers who have contributed their expertise to this volume. We hope this book serves as a catalyst for meaningful progress in the fight against AMR and inspires a new wave of innovation in probiotic science.

Editors

Mohmmad Younus Wani

Irfan Ahmad Rather

Aijaz Ahmad

About the Editors

Dr. Mohmmad Younus Wani is Associate Professor of Organic and Medicinal Chemistry in the College of Science, University of Jeddah, Jeddah, Saudi Arabia. He received his Ph.D. in Organic and Medicinal Chemistry from Jamia Millia Islamia, New Delhi, India, and Postdoctoral training from the University of Coimbra, Portugal, and Texas Therapeutics Institute, UTHealth, Houston, Texas, USA, on the development of new treatment strategies and non-traditional approaches to combat antimicrobial drug resistance (AMR). With extensive experience in the field of medicinal chemistry and drug discovery, he has authored/co-authored over 80 peer-reviewed research articles, 6 books, and several US patents. He has received numerous prestigious honors, awards, and research grants. A professional member of several scientific societies, he also serves on the editorial boards of various esteemed scientific journals. Dr. Wani is working at the interface of chemistry and biology to develop novel small molecule inhibitors, new treatment strategies, and novel approaches that could circumvent AMR, which is within the ambit of World Health Organizations global action plan on antimicrobial resistance.

Dr. Irfan Ahmad Rather is an accomplished Associate Professor at the Department of Biological Sciences, King Abdulaziz University. Prior to his current position, he served as an assistant professor at the School of Biotechnology, Yeungnam University in Korea. With a strong academic background and extensive research experience, Dr. Irfan specializes in the development of novel probiotics for metabolic diseases and zoonotic viruses. Throughout his career, Dr. Irfan has made significant contributions to the field of biomedical research. He has authored over 100 research articles, reflecting his expertise in his chosen areas of study. Additionally, his innovative work has been recognized with the granting of 10 US and Korean patents, demonstrating his commitment to translating scientific discoveries into practical applications. Dr. Irfan's dedication to advancing knowledge and finding solutions to pressing health challenges has earned him recognition as a respected researcher in the scientific community. His work continues to contribute to the development of novel approaches in the fields of probiotics and viral diseases.

Dr. Aijaz Ahmad is a senior medical scientist in the Division of Infection Control of the National Health Laboratory Service at Charlotte Maxeke Johannesburg Academic Hospital and a professor in the Department of Clinical Microbiology and Infectious Diseases, School of Pathology, University of the Witwatersrand. His core research interest is understanding fungal pathogenesis and the development of antifungal vaccines and novel antifungal drugs. He obtained his Ph.D. in Medical Mycology. During his training as a Postdoctoral Scientist from different institutions, he gained lot of experience to be a medical microbiologist. He is an HPCSA-registered medical scientist in Microbiology and is also a National Research Foundation (NRF)-rated researcher. With an extensive research experience in infectious diseases, microbial pathogenesis, and drug development, he has authored/co-authored over 90 peer-reviewed publications, supervised more than 25 postgraduate students, and is the recipient of several research grants and awards.

Foreword

It is a distinct pleasure and honor to write the foreword for this important work that brings together timely insights into the intersection of probiotics, microbiome therapeutics, and antimicrobial resistance.

Over the past decades, we have witnessed remarkable advances in our understanding of the human microbiome and its vital role in maintaining health and preventing disease. Probiotics and microbiome-based interventions have emerged as promising alternatives and adjuncts to conventional therapies, offering new hope for conditions ranging from inflammatory bowel disease to metabolic and dermatological disorders.

At the same time, we are facing one of the gravest threats to global health in the form of **antimicrobial resistance (AMR)**. The overuse and misuse of antimicrobials in human medicine, agriculture, and animal husbandry have accelerated the emergence of multidrug-resistant pathogens. The World Health Organization has recognized antimicrobial resistance as one of the top 10 global public health threats facing humanity. If current trends continue, we risk entering a post-antibiotic era where common infections and minor injuries could once again become deadly.

In this context, the relevance of this book cannot be overstated. It reflects the urgent need to explore **sustainable, microbiome-friendly approaches** to infection control, gut health, and immune modulation. Probiotics, postbiotics, and microbiome therapeutics offer not only a preventative paradigm but also potential tools for reducing our dependency on antimicrobials, restoring microbial balance, and slowing the development of resistance.

The editors have done an admirable job in curating this volume, bringing together scientific rigor, innovation, and a multidisciplinary perspective. It is a valuable contribution that will serve both seasoned researchers and young scientists entering this critical field.

As someone who has spent my life researching probiotics and functional microbiology, I am heartened to see the next generation taking bold steps to address one of the most pressing challenges of our time. I hope this book will inspire deeper inquiry, cross-disciplinary collaboration, and a renewed commitment to developing safer, smarter, and more sustainable health solutions.

Professor Yong-ha Park
Founder & Chairman
ProBionic Institute & Probiotics House
Former Professor, Yeungnam University
South Korea

1 The Intricate Relationship between Probiotics and Antimicrobial Resistance

Lamar Al-Muslem[1], *Ayman M. Al-Qaaneh*[2,3], *Zainab Al-Salem*[4], *Irfan A. Rather*[5,6], *and Suriya Rehman*[4*]
[1]Mawhiba Research Enrichment Program, Institute for Research and Medical Consultations (IRMC), Imam Abdulrahman Bin Faisal University, Dammam, Saudi Arabia
[2]Faculty of Allied Medical Sciences, Al-Balqa Applied University (BAU), Al-Salt, Jordan
[3]Department of Pharmaceutical Technology, Faculty of Pharmacy, Jordan University of Science and Technology (JUST), Irbid, Jordan
[4]Department of Epidemic Diseases Research, Institute for Research and Medical Consultations (IRMC), Imam Abdulrahman Bin Faisal University, Dammam, Saudi Arabia
[5]Faculty of Science, Department of Biological Sciences, King Abdulaziz University, Jeddah, Saudi Arabia
[6]Center of Excellence in Bionanoscience Research, King Abdulaziz University, Jeddah, Saudi Arabia
*Corresponding author: surrehman@iau.edu.sa

1.1 INTRODUCTION

More than 33,000 deaths annually in the European Union and the European Economic Area are attributed to antibiotic-resistant bacteria, which also account for approximately 900,000 disability-adjusted life years (DALYs) (Cassini et al., 2019). Antimicrobial resistance (AMR) is proving to be a global health crisis, driven primarily by the misuse and overuse of antibiotics in human medicine, animal husbandry, and agriculture. AMR undermines the efficacy of treatments for numerous infections, significantly increasing morbidity and mortality risks. Additionally, it threatens food security and imposes substantial economic burdens. Vulnerable populations, particularly those in resource-limited settings, are disproportionately impacted by the consequences of AMR (Murray et al., 2022) (Figure 1.1).

The history of antimicrobials dates back to 1928 when Alexander Fleming discovered penicillin. Fleming foresaw that bacterial resistance to this "miracle drug" would eventually emerge a prediction that has proven accurate as the overuse and misuse of antibiotics pose a significant threat to global health (Khardori et al., 2020; Tan & Tatsumura, 2015). As Fleming anticipated, the effectiveness of penicillin has diminished over time as bacteria have evolved various mechanisms of resistance due to its excessive use. AMR takes place when pathogens that include bacteria, fungi, viruses, and parasites become unresponsive to antimicrobial agents, rendering these agents ineffective (Rehman, 2023).

DOI: 10.1201/9781032676951-1

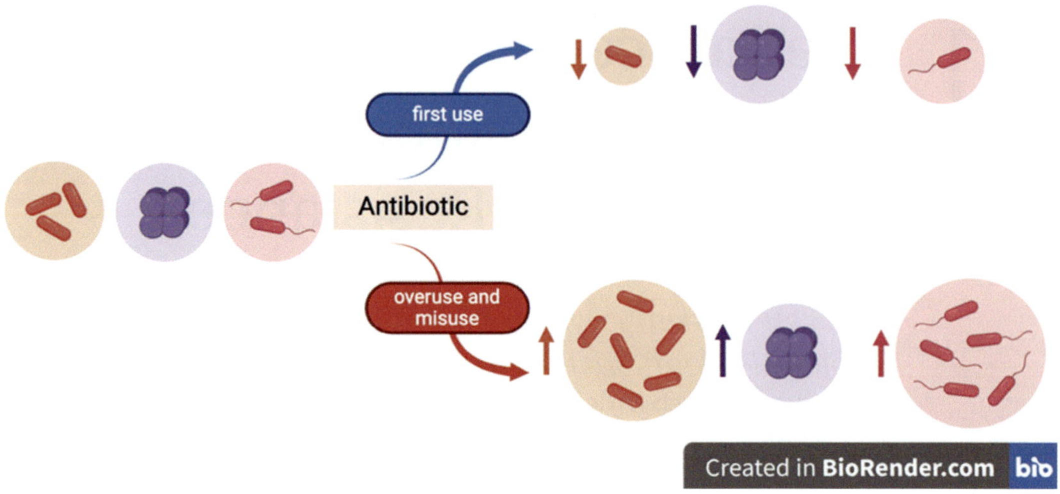

FIGURE 1.1 Figure illustrates the impact of antibiotic use on bacterial populations. The inappropriate use of antibiotics results in the survival of some bacteria and subsequent development of resistance.

Antimicrobial agents target various aspects of microbial physiology through distinct mechanisms and can be classified as natural, semisynthetic, or synthetic. The first mechanism involves cell wall synthesis inhibition, where agents such as β-lactams and glycopeptides disrupt the microbial cell wall structure, leading to cell death (Rehman et al., 2024). The second mechanism is protein synthesis inhibition, where macrolides and tetracyclines interfere with protein synthesis, preventing microorganisms from producing essential proteins. Another mechanism is metabolic pathway inhibition, in which sulfonamides block vital metabolic pathways necessary for microbial growth and survival (El-Nashar et al., 2024; Ullah et al., 2024). Additionally, DNA replication and transcription interference prevent microorganisms from reproducing, as fluoroquinolones inhibit DNA replication and transcription (Al-Ghamdi et al., 2023; Zhou et al., 2015).

This chapter provides a comprehensive overview of the issue of AMR and explores potential solutions, including the use of probiotics.

1.2 CURRENT LANDSCAPE OF AMR AND ITS UNDERLYING CAUSES

AMR has been jeopardizing the ability to treat common infections that were once easily managed. This crisis is primarily driven by the overuse and misuse of antibiotics in both human medicine and agriculture (Church & McKillip, 2021). Poor-quality antimicrobial products further exacerbate AMR by creating suboptimal drug concentrations that fail to eliminate infections, fostering an environment where resistant bacteria thrive. This enables the survival and proliferation of resistant strains, ultimately leading to the emergence of drug-resistant infections. Many developing countries, particularly in Africa, lack robust systems to monitor and control the quality of antibiotics. The absence of stringent regulatory frameworks facilitates the widespread availability of counterfeit or substandard drugs. Additionally, weak enforcement mechanisms in regulating the manufacturing and distribution of drugs significantly contribute to escalate AMR crisis (Figure 1.2).

Excessive and inappropriate prescribing practices by healthcare providers significantly contribute to the development of AMR. This includes prescribing antibiotics for viral infections, using broad-spectrum antibiotics when narrow-spectrum options would suffice, and failing to adhere to recommended dosages, all of which promote AMR. Additionally, drug vendors or dispensers often lack adequate knowledge and training, leading to incorrect dosages or inappropriate medication recommendations, further driving the emergence of resistant strains. The misuse of antibiotics for non-bacterial infections also exacerbates resistance. Moreover, the widespread use of antibiotics in animal

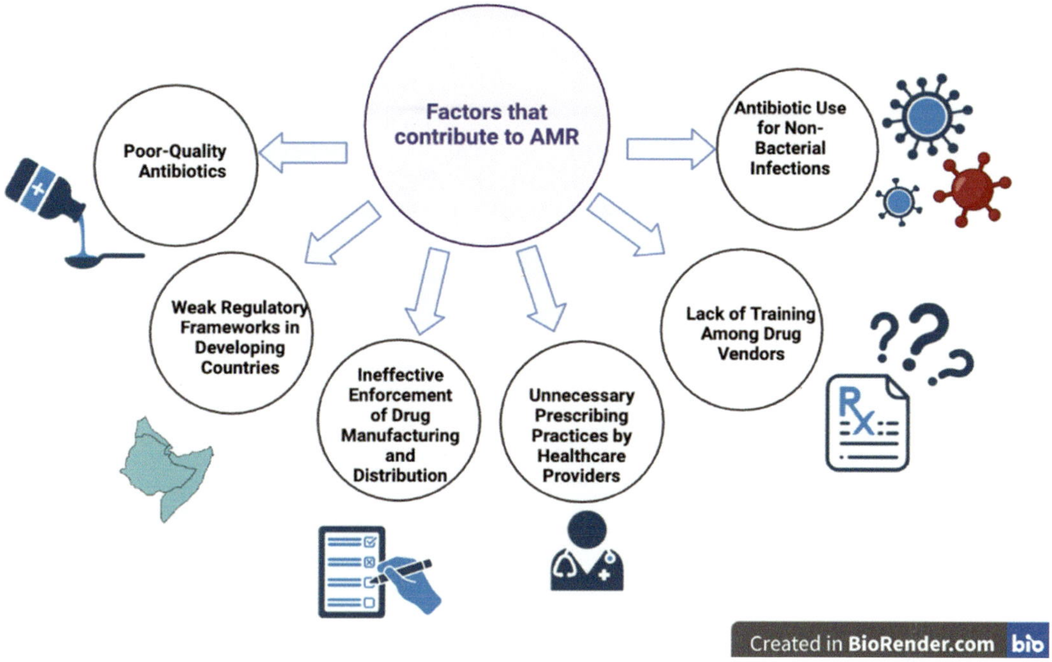

FIGURE 1.2 This figure summarizes the various contributing factors to AMR.

agriculture (both as growth promoters and for treating infections) creates a strong selective pressure that encourages bacterial survival and adaptation. Resistant strains arising from these practices can spread to humans through the food chain or direct contact with animals (Ayukekbong et al., 2017).

1.3 EMERGING ALTERNATIVES TO CONVENTIONAL ANTIBIOTICS

Phage therapy, which involves using bacteriophages to target and eliminate bacterial infections, dates back to the early 20th century. This approach made significant progress in the 1920s, spearheaded by Félix d'Hérelle in Brazil, where it demonstrated promising results in treating various infections. However, the advent of antibiotics, with their ease of manufacturing and use, overshadowed the development and application of phage therapy (Ayukekbong et al., 2017). Antimicrobial peptides (AMPs) are short chains of amino acids with broad-spectrum antimicrobial activity. They occur naturally in various organisms, including plants, fungi, bacteria, and animals (Figure 1.3).

AMPs serve as a primary defense mechanism against infections, targeting a wide range of pathogens. Discovered in the early 20th century, the first identified AMPs included gramicidin and tyrocidin. Since then, thousands of AMPs have been discovered or synthesized. AMPs directly kill pathogens by disrupting their cell membranes or interfering with intracellular processes. Additionally, they modulate the host's immune response, promoting healing while reducing inflammation. AMPs are often regarded as a promising alternative to traditional antibiotics due to their broad-spectrum activity, lower risk of resistance, and potential immunomodulatory properties (Bahar & Ren, 2013).

Vaccines present another promising alternative to combat AMR. They offer direct prevention of infections by stimulating the immune system to recognize and fight specific pathogens, thereby reducing the need for antibiotics. By preventing infections, vaccines decrease the selective pressure that drives the emergence of antibiotic-resistant bacteria. Vaccines also indirectly protect unvaccinated individuals, particularly vulnerable populations such as infants and the elderly, further reducing antibiotic dependence (Jansen & Anderson, 2018).

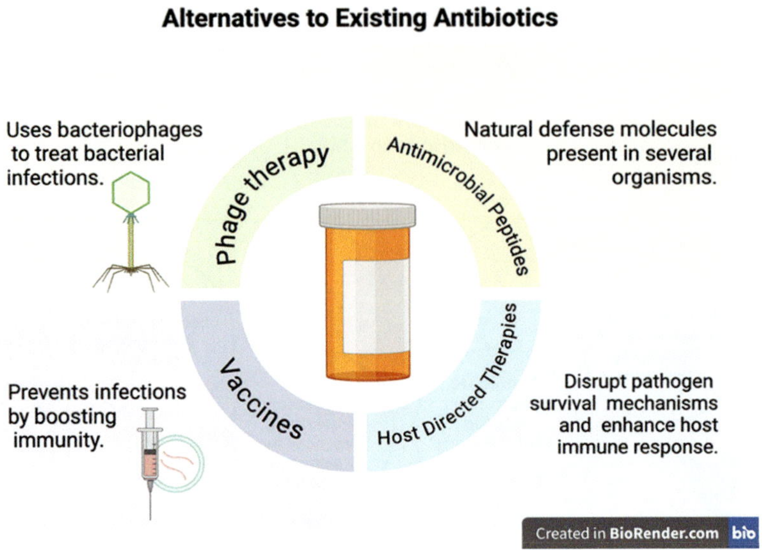

FIGURE 1.3 This illustration presents the potential alternative strategies to the existing antibiotic therapies.

Host-directed therapies (HDTs) represent a novel approach to combating infections by targeting the host's immune response rather than directly attacking the pathogen. HDTs stimulate the host's immune system to fight infections more effectively and regulate inflammatory responses, preventing excessive tissue damage. Additionally, HDTs can interfere with pathogen survival mechanisms, disrupting their ability to replicate and persist within host cells (Shapira et al., 2024).

1.4 PROBIOTICS: A VIABLE ALTERNATIVE TO CONVENTIONAL ANTIBIOTICS

Probiotics trace their origins to 1908, when Nobel Prize-winning scientist Élie Metchnikoff proposed that "ingested lactobacilli can displace toxin-producing bacteria, promoting health and prolonging life." Probiotics are defined as microorganisms that, when administered in adequate amounts, confer health benefits to the host (Roberfroid, 2000). Historically, probiotics have been safely used in fermented dairy products like yogurt, with their health benefits recognized and utilized for generations (Otieno, 2011).

For probiotics to be effective, they must survive and thrive under the harsh conditions of the stomach and small intestine, adhere to the intestinal lining to colonize the gut, and effectively compete with harmful bacteria to reduce the risk of infections (Collins et al., 1998; Ouwehand et al., 2002). One key mechanism of probiotics is competitive exclusion, where probiotics compete with pathogenic bacteria for binding sites on the intestinal epithelium, preventing the colonization of harmful microorganisms (Callaway et al., 2008).

Probiotics also produce antimicrobial substances, such as bacteriocins (an AMP that inhibits the growth of other bacteria, including pathogens). Additionally, probiotics ferment dietary carbohydrates, producing organic acids like lactic acid and acetic acid. This metabolic activity creates an acidic environment in the gut that suppresses the growth of harmful bacteria, thereby promoting a healthier gut microbiome. Some probiotic strains also produce hydrogen peroxide, a potent antimicrobial agent effective against bacteria, fungi, and viruses, further enhancing the gut's defense mechanisms (Mishra & Lambert, 1996) (Figure 1.4).

Another significant mechanism is gut barrier strengthening, where probiotics upregulate the expression of tight junction proteins such as occludin, claudin-1, and ZO-1. These proteins form

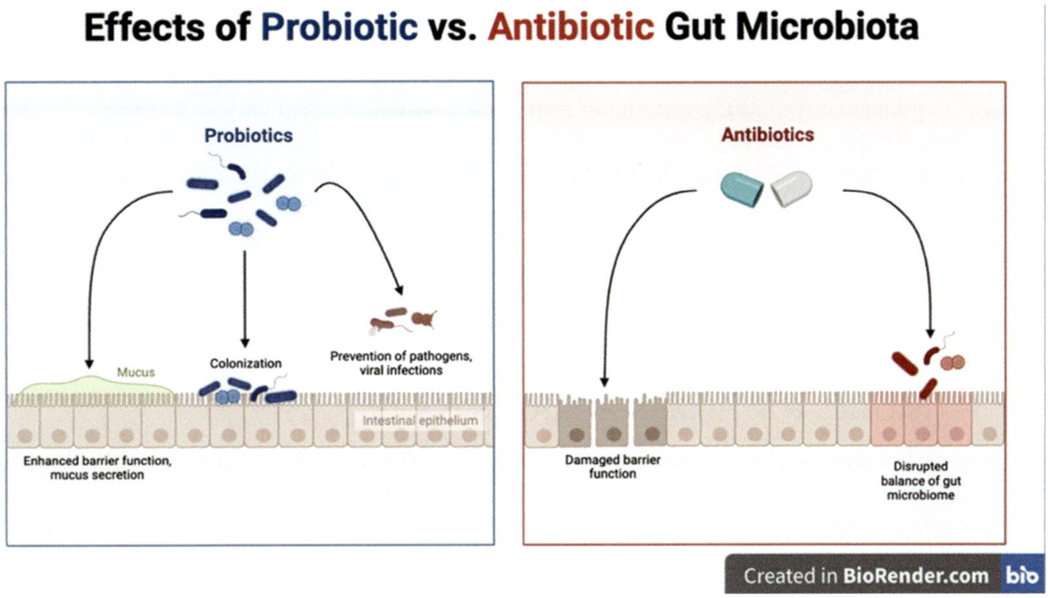

FIGURE 1.4 Presentation of the effect of consumption of probiotic and antibiotic on the normal flora of the gut.

a tight seal between intestinal epithelial cells, preventing the leakage of harmful substances into the bloodstream (O'Callaghan & van Sinderen, 2016). Probiotics also stimulate goblet cells in the intestinal epithelium to generate more mucus. This mucus layer acts as a protective barrier, trapping pathogens and toxins and reducing the risk of infections (Gou et al., 2022).

Furthermore, probiotics modulate immune responses by stimulating the production of beneficial immune cells and cytokines, such as regulatory T-cells and interleukin-10. They also help suppress the production of inflammatory cytokines, stabilizing the immune response. This regulation is crucial for maintaining gut health and reducing the risk of inflammatory bowel diseases and allergies (Kang & Im, 2015).

1.5 PROSPECTUS AND LIMITATION OF PROBIOTICS AS ANTIBIOTIC ALTERNATIVES

Probiotics present a promising strategy to address the growing threat of antibiotic resistance. By modulating the gut microbiota, probiotics can reduce the need for antibiotics and limit the emergence of antibiotic-resistant bacteria. These benefits contribute to decreased antibiotic use, improved gut health, enhanced immune function, and a potential reduction in the risk of antibiotic-resistant infections (Kechagia et al., 2013).

Probiotics offer several health benefits beyond AMR prevention. They can help alleviate constipation, diarrhea, and irritable bowel syndrome (IBS) by restoring gut microbiota balance. Probiotics also support immune function, and some studies suggest they may improve mood and reduce symptoms of anxiety and depression (Johnson et al., 2021). Furthermore, probiotics can prevent and even treat vaginal and urinary tract infections by maintaining and restoring bacterial balance in these areas (Abdelhamid et al., 2019).

Although many studies suggest a potential role for probiotics in preventing AMR, the evidence remains inconclusive. The Centers for Disease Control and Prevention (CDC) continues to research this area, but specific recommendations have not yet been issued. Currently, probiotics are used as an adjunct therapy alongside antibiotics (Ouwehand et al., 2016). Probiotics are generally regarded as safe (GRAS); however, rare cases of serious side effects, such as septicemia and fungemia, have been reported in vulnerable individuals (Vandenplas et al., 2007).

The effectiveness of probiotics depends on the specific strains, as not all strains exhibit the same efficacy (Crittenden et al., 2005). Individual responses to probiotics also vary, influenced by factors such as diet, lifestyle, and gut microbiota composition. Additionally, the lack of standardization in probiotic formulations can impact their quality and effectiveness. More rigorous clinical trials are needed to evaluate their long-term safety and benefits.

Probiotics can interact with certain medications, particularly antibiotics and immunosuppressants. Antibiotics may disrupt gut bacterial balance, and while probiotics can help restore this balance, they must be timed appropriately to avoid interfering with the natural microbiome restoration process (Razavi et al., 2021).

1.6 PROBIOTICS FOR HUMAN APPLICATIONS

The potential of probiotics as an alternative to antibiotics has been demonstrated through numerous in vivo and ex vivo experiments, as well as various clinical trials. Probiotics, including lactic acid bacteria and soil-based bacteria, have exhibited bacteriostatic and bactericidal activities against several pathogens, as shown in Table 1.1. These include *Staphylococcus aureus*, *Listeria monocytogenes*, *Pseudomonas aeruginosa*, and *Candida albicans.*

Beyond antimicrobial properties, probiotics offer a range of health benefits. They improve nutritional status by enhancing the bioavailability of vitamins and minerals and aiding in the secretion of digestive enzymes. Probiotics are effective in treating various types of diarrhea, including those caused by infections, antibiotics, and radiation exposure. Additionally, they have been shown to lower cholesterol levels, boost the immune system, and increase intestinal motility, thereby alleviating constipation. Probiotics also support mucosal integrity, balance intestinal microbiota, and reduce symptoms of lactose intolerance.

TABLE 1.1
Summary of Findings from Previous Studies on the In Vivo Antimicrobial Activity of Probiotics

Probiotics	Pathogens	Observation	Reference
B. subtilis MB40	*S. aureus*	A potential decrease in the *Staphylococcus aureus* colonization was detected in the human body; this reduction did not cause any modifications to the overall microbiome and did not disrupt the natural microbial diversity and balance within the human body.	Piewngam et al. (2023)
L. casei	*C. albicans*	Antifungal effect was noticed in a murine model of vulvovaginal candidiasis (VVC), indicating effectiveness in reducing fungal infection.	Liao et al. (2019)
Pediococcus acidilactici HW01	*P. aeruginosa*	The biofilm formation was suppressed by bacteriocin, and virulence factors reduction was observed such as pyocyanin, protease, and rhamnolipid.	Lee et al. (2020)
	L. monocytogenes	Bacterial lysate suppressed the formation of biofilm, later adhesion, and penetration of HT-29 cells (human intestinal epithelial cells), to prevent bacterial interaction with the cells without influencing their function.	Lee et al. (2022)

Furthermore, probiotics exhibit anticarcinogenic properties and can help prevent food-borne allergies and atopic dermatitis. Regular consumption of probiotic-rich foods such as yogurt and kefir has been linked with minimized risk of cardiovascular diseases, relief from lactose malabsorption, and improvement in symptoms of IBS (Yadav & Shukla, 2017).

1.7 PROBIOTICS FOR ANIMAL APPLICATIONS

The gut microflora of animals is vital in regulating their overall health and supporting uninterrupted digestive events. Although probiotics have been predominantly used as an addition to human nutrition, their use is now expanding to include animal nutrition. Supplementing animal feed with probiotics derived from intestinal microbiota has been shown to improve nutrient utilization, thereby benefiting the animal production industry. Research indicates that probiotics can enhance intestinal microbial balance, leading to better health and productivity of the host animal (Valdes et al., 2018).

Several factors, including genetics, age, diet, and sex, influence animal growth, but providing optimal nutrition remains critical for achieving optimal development. Probiotics improve nutrient digestibility, leading to enhanced growth rates and feed efficiency. Recent studies have reported improved growth performance in pigs, broiler chickens, and calves when their diets were supplemented with probiotics (Al-Shawi et al., 2020).

As global food production demands continue to rise, the use of probiotics in animal agriculture is expected to increase. This trend is driven by their demonstrated positive impact on production indices and their potential to prevent certain diseases, benefiting both human and animal health (Evangelista et al., 2023).

1.8 CONCLUSION

This chapter explored the potential of probiotics to mitigate the consequences of AMR, an ongoing and significant global health challenge. While probiotics demonstrate substantial promise and may serve as alternatives to traditional antibiotics, further research is crucial to fully elucidate the mechanisms underlying their benefits. A deeper understanding of these processes is indispensable to promote the clinical application of probiotics and ensure their effective use in combating AMR in the future.

CONFLICT OF INTEREST

None to declare.

REFERENCES

Abdelhamid, A. G., El-Masry, S. S., & El-Dougdoug, N. K. (2019). Probiotic Lactobacillus and Bifidobacterium strains possess safety characteristics, antiviral activities and host adherence factors revealed by genome mining. *EPMA J*, *10*(4), 337–350. https://doi.org/10.1007/s13167-019-00184-z

Al-Ghamdi, F., Al-Qaaneh, A., & Al-Shehri, B. (2023). Effect of antibiotics' electromagnetic spectrums on the bacterial growths: Beyond drug-receptor interaction. *J Med Pharm Allied Sci*, *12*, 6241–6246. https://doi.org/10.55522/jmpas.V12I6.5904

Al-Shawi, S. G., Dang, D. S., Yousif, A. Y., Al-Younis, Z. K., Najm, T. A., & Matarneh, S. K. (2020). The potential use of probiotics to improve animal health, efficiency, and meat quality: A review. *Agriculture*, *10*(10), 452. https://www.mdpi.com/2077-0472/10/10/452

Ayukekbong, J. A., Ntemgwa, M., & Atabe, A. N. (2017). The threat of antimicrobial resistance in developing countries: Causes and control strategies. *Antimicrob Resist Infect Control*, *6*(1), 47. https://doi.org/10.1186/s13756-017-0208-x

Bahar, A. A., & Ren, D. (2013). Antimicrobial peptides. *Pharmaceuticals*, *6*(12), 1543–1575. https://www.mdpi.com/1424-8247/6/12/1543; https://mdpi-res.com/d_attachment/pharmaceuticals/pharmaceuticals-06-01543/article_deploy/pharmaceuticals-06-01543.pdf?version=1385649896

Callaway, T. R., Edrington, T. S., Anderson, R. C., Harvey, R. B., Genovese, K. J., Kennedy, C. N., Venn, D. W., & Nisbet, D. J. (2008). Probiotics, prebiotics and competitive exclusion for prophylaxis against bacterial disease. *Anim Health Res Rev*, *9*(2), 217–225. https://doi.org/10.1017/s1466252308001540

Cassini, A., Högberg, L. D., Plachouras, D., Quattrocchi, A., Hoxha, A., Simonsen, G. S., Colomb-Cotinat, M., Kretzschmar, M. E., Devleesschauwer, B., Cecchini, M., Ouakrim, D. A., Oliveira, T. C., Struelens, M. J., Suetens, C., & Monnet, D. L. (2019). Attributable deaths and disability-adjusted life-years caused by infections with antibiotic-resistant bacteria in the EU and the European Economic Area in 2015: A population-level modelling analysis. *Lancet Infect Dis*, *19*(1), 56–66. https://doi.org/10.1016/s1473-3099(18)30605-4

Church, N. A., & McKillip, J. L. (2021). Antibiotic resistance crisis: Challenges and imperatives. *Biologia*, *76*(5), 1535–1550. https://doi.org/10.1007/s11756-021-00697-x

Collins, J. K., Thornton, G., & Sullivan, G. O. (1998). Selection of probiotic strains for human applications. *Int Dairy J*, *8*(5), 487–490. https://doi.org/10.1016/S0958-6946(98)00073-9

Crittenden, R., Bird, A. R., Gopal, P., Henriksson, A., Lee, Y. K., & Playne, M. J. (2005). Probiotic research in Australia, New Zealand and the Asia-Pacific region. *Curr Pharm Des*, *11*(1), 37–53. https://doi.org/10.2174/1381612053382304

El-Nashar, H. A. S., Abbas, A. M., Al-Qaaneh, A. M., Abdel-Maksoud, M. A., El-Tayeb, M. A., Almutairi, S., Hafeez Kiani, B., & Elhawary, E. A. (2024). Chemical composition and antimicrobial investigation of the essential oil isolated from *Montanoa bipinnatifida* (Kunth) K. Koch leaves. *J Essent Oil Bear Plants*, *27*(5), 1428–1438. https://doi.org/10.1080/0972060X.2024.2411398

Evangelista, A. G., Corrêa, J. A. F., Pinto, A., Gonçalves, F. D. R., & Luciano, F. B. (2023). Recent advances in the use of bacterial probiotics in animal production. *Anim Health Res Rev*, 1–13. https://doi.org/10.1017/s1466252323000063

Gou, H. Z., Zhang, Y. L., Ren, L. F., Li, Z. J., & Zhang, L. (2022). How do intestinal probiotics restore the intestinal barrier? *Front Microbiol*, *13*, 929346. https://doi.org/10.3389/fmicb.2022.929346

Jansen, K. U., & Anderson, A. S. (2018). The role of vaccines in fighting antimicrobial resistance (AMR). *Hum Vaccine Immunother*, *14*(9), 2142–2149. https://doi.org/10.1080/21645515.2018.1476814

Johnson, D., Thurairajasingam, S., Letchumanan, V., Chan, K. G., & Lee, L. H. (2021). Exploring the role and potential of probiotics in the field of mental health: Major depressive disorder. *Nutrients*, *13*(5). https://doi.org/10.3390/nu13051728

Kang, H. J., & Im, S. H. (2015). Probiotics as an immune modulator. *J Nutr Sci Vitaminol (Tokyo)*, *61 Suppl*, S103–105. https://doi.org/10.3177/jnsv.61.S103

Kechagia, M., Basoulis, D., Konstantopoulou, S., Dimitriadi, D., Gyftopoulou, K., Skarmoutsou, N., & Fakiri, E. M. (2013). Health benefits of probiotics: A review. *Int Sch Res Notices*, *2013*(1), 481651. https://doi.org/10.5402/2013/481651

Khardori, N., Stevaux, C., & Ripley, K. (2020). Antibiotics: From the beginning to the future: Part 1. *Indian J Pediatr*, *87*(1), 39–42. https://doi.org/10.1007/s12098-019-03087-z

Lee, D. H., Kim, B. S., & Kang, S. S. (2020). Bacteriocin of *Pediococcus acidilactici* HW01 inhibits biofilm formation and virulence factor production by *Pseudomonas aeruginosa*. *Probiotics Antimicrob Proteins*, *12*(1), 73–81. https://doi.org/10.1007/s12602-019-09623-9

Lee, H. B., Kim, K. H., Kang, G. A., Lee, K.-G., & Kang, S.-S. (2022). Antibiofilm, antiadhesive and anti-invasive activities of bacterial lysates extracted from *Pediococcus acidilactici* against *Listeria monocytogenes*. *Foods*, *11*(19), 2948. https://www.mdpi.com/2304-8158/11/19/2948; https://mdpi-res.com/d_attachment/foods/foods-11-02948/article_deploy/foods-11-02948.pdf?version=1663751839

Liao, H., Liu, S., Wang, H., Su, H., & Liu, Z. (2019). Enhanced antifungal activity of bovine lactoferrin-producing probiotic *Lactobacillus casei* in the murine model of vulvovaginal candidiasis. *BMC Microbiol*, *19*(1), 7. https://doi.org/10.1186/s12866-018-1370-x

Mishra, C., & Lambert, J. (1996). Production of anti-microbial substances by probiotics. *Asia Pac J Clin Nutr*, *5*(1), 20–24.

Murray, C. J. L., Ikuta, K. S., Sharara, F., Swetschinski, L., Robles Aguilar, G., Gray, A., Han, C., Bisignano, C., Rao, P., Wool, E., Johnson, S. C., Browne, A. J., Chipeta, M. G., Fell, F., Hackett, S., Haines-Woodhouse, G., Kashef Hamadani, B. H., Kumaran, E. A. P., McManigal, B.,... Naghavi, M. (2022). Global burden of bacterial antimicrobial resistance in 2019: A systematic analysis. *The Lancet*, *399*(10325), 629–655. https://doi.org/10.1016/S0140-6736(21)02724-0

O'Callaghan, A., & van Sinderen, D. (2016). Bifidobacteria and their role as members of the human gut microbiota. *Front Microbiol*, *7*, 925. https://doi.org/10.3389/fmicb.2016.00925

Otieno, D. O. (2011). Biology of prokaryotic probiotics. In M.-T. Liong (Ed.), *Probiotics: Biology, Genetics and Health Aspects* (pp. 1–28). Springer Berlin Heidelberg. https://doi.org/10.1007/978-3-642-20838-6_1

Ouwehand, A. C., Forssten, S., Hibberd, A. A., Lyra, A., & Stahl, B. (2016). Probiotic approach to prevent antibiotic resistance. *Ann Med*, *48*(4), 246–255. https://doi.org/10.3109/07853890.2016.1161232

Ouwehand, A. C., Salminen, S., & Isolauri, E. (2002). Probiotics: An overview of beneficial effects. *Antonie Van Leeuwenhoek*, *82*(1–4), 279–289.

Piewngam, P., Khongthong, S., Roekngam, N., Theapparat, Y., Sunpaweravong, S., Faroongsarng, D., & Otto, M. (2023). Probiotic for pathogen-specific *Staphylococcus aureus* decolonisation in Thailand: A phase 2, double-blind, randomised, placebo-controlled trial. *Lancet Microbe*, *4*(2), e75–e83. https://doi.org/10.1016/s2666-5247(22)00322-6

Razavi, S., Janfaza, S., Tasnim, N., Gibson, D. L., & Hoorfar, M. (2021). Nanomaterial-based encapsulation for controlled gastrointestinal delivery of viable probiotic bacteria. *Nanoscale Adv*, *3*(10), 2699–2709. https://doi.org/10.1039/d0na00952k

Rehman, S. (2023). A parallel and silent emerging pandemic: Antimicrobial resistance (AMR) amid COVID-19 pandemic. *Journal of Infection and Public Health*, *16*(4), 611–617.

Rehman, S., Albhishiri, G., Alsalem, Z., AlJameel, S. S., Al-Qaaneh, A., Shah, A. H., Akhtar, S., Hameed, S., & Jermy, B. R. (2024). Bionanocomposites comprising mesoporous metal organic framework (ZIF-8) phytofabricated with *Allium sativum* as alternative nanomaterials to combat antimicrobial drug resistance. *Bioprocess Biosyst Eng*, *47*(8), 1335–1344. https://doi.org/10.1007/s00449-024-03027-y

Roberfroid, M. B. (2000). Prebiotics and probiotics: Are they functional foods? *Am J Clin Nutr*, *71*(6 Suppl), 1682S–1687S; discussion 1688S–1690S. https://doi.org/10.1093/ajcn/71.6.1682S

Shapira, T., Christofferson, M., & Av-Gay, Y. (2024). The antimicrobial activity of innate host-directed therapies: A systematic review. *Int J Antimicrob Agents*, *63*(5), 107138. https://doi.org/10.1016/j.ijantimicag.2024.107138

Tan, S. Y., & Tatsumura, Y. (2015). Alexander Fleming (1881–1955): Discoverer of penicillin. *Singapore Med J*, *56*(7), 366–367. https://doi.org/10.11622/smedj.2015105

Ullah, I., Ullah, A., Al-Qaaneh, A. M., Aman, K., Bekhit, M. M., Ullah, I., Zengin, G., & Farid, A. (2024). Green synthesis of silver and copper nanoparticles using *Parthenium hysterophorus*: Antibacterial, antioxidant, and anticancer properties. *ChemistrySelect*, *9*(40), e202401573. https://doi.org/https://doi.org/10.1002/slct.202401573

Valdes, A. M., Walter, J., Segal, E., & Spector, T. D. (2018). Role of the gut microbiota in nutrition and health. *BMJ*, *361*, k2179. https://doi.org/10.1136/bmj.k2179

Vandenplas, Y., Salvatore, S., Vieira, M., Devreker, T., & Hauser, B. (2007). Probiotics in infectious diarrhoea in children: Are they indicated? *Eur J Pediatr*, *166*(12), 1211–1218. https://doi.org/10.1007/s00431-007-0497-9

Yadav, R., & Shukla, P. (2017). Probiotics for human health: Current progress and applications. In P. Shukla (Ed.), *Recent advances in Applied Microbiology* (pp. 133–147). Springer Singapore. https://doi.org/10.1007/978-981-10-5275-0_6

Zhou, G., Shi, Q.-S., Huang, X.-M., & Xie, X.-B. (2015). The three bacterial lines of defense against antimicrobial agents. *Int J Mol Sci*, *16*(9), 21711–21733. https://www.mdpi.com/1422-0067/16/9/21711; https://mdpi-res.com/d_attachment/ijms/ijms-16-21711/article_deploy/ijms-16-21711.pdf?version=1441788905

2 Understanding Antimicrobial Resistance

Mechanisms, Impacts, and Global Concerns

Iram Saba[1,2], *Kaiser Wani*[3,4], *Khalid Mujassam Batoo*[5], *and Saif Hameed*[2]

[1]Research and Scientific Center, Sultan Bin Abdulaziz Humanitarian City, Riyadh, Saudi Arabia

[2]Amity Institute of Biotechnology, Amity University Haryana, Gurugram (Manesar), India

[3]Department of Biochemistry, College of Science, King Saud University, Riyadh, Saudi Arabia

[4]University Institute of Biotechnology, Chandigarh University, Mohali, India

[5]King Abdullah Institute for Nanotechnology, King Saud University, Riyadh, Saudi Arabia

2.1 INTRODUCTION

Antibiotic resistance, though not new, continues to evolve into an increasingly complex global issue. Factors such as excessive antibiotic use in both humans and animals, unrestricted access to antibiotics in the market, increased international travel, poor sanitation and hygiene, and the environmental release of non-metabolized antibiotics or their residues through manure and waste are key contributors to the rise of antimicrobial resistance (AMR). These factors drive the emergence of multidrug-resistant (MDR) diseases through genetic adaptations (Laxminarayan et al., 2016). In recent decades, the use of antimicrobial in livestock has become a major concern, particularly in regions more susceptible to the adverse effects of antibiotics. The consequences of this growing trend will have long-lasting economic and public health impacts (Davies & Davies, 2010). The routine use of antibiotics in farm animals raised for human consumption is standard practice, and with the increasing global population, antibiotic consumption is projected to rise by as much as 67% by 2030 (Aslam et al., 2018).

According to a statement made by the Centers for Disease Control and Prevention (CDC) in 2019, we are currently living in a "postantibiotic" environment (CDC, 2019). In May 2014, the World Health Organization (WHO) issued a warning about the potential catastrophe caused by AMR. Comprehending the scientific foundations of AMR is critical in the fight against this imminent danger to public health. This awareness should extend from the mechanisms of resistance, which facilitate the development of novel therapeutics and diagnostic approaches, to the societal and environmental drivers of AMR, which are crucial for the formulation of effective intervention policies (Dar et al., 2016). Antibiotic resistance around the world doesn't seem to be declining, but it might be moving differently. There are many causes of drug resistance, and its effects can be seen all over the world. A great deal of effort has been devoted to understanding the different aspects

DOI: 10.1201/9781032676951-2

of drug resistance and finding potential solutions to this global problem. However, there isn't a well-coordinated effort, especially at the government level around the world (Prestinaci, Pezzotti, & Pantosti, 2015). Without effective antibiotics to prevent bacterial infections, organ transplants, cancer treatment, preterm newborn care, and advanced major surgeries would not have been possible. Implementing timely global action plans could have severe social, medical, and economic repercussions. This chapter aims to provide a comprehensive overview of the global scale, major causes, effects, and key areas that require immediate attention (Laxminarayan et al., 2013). This chapter discusses the fundamental reasons behind the AMR challenge, along with the ongoing and future attempts to find a solution to the problem. In addition, we consider the likelihood that there will be no conclusive answer to this catastrophe and that continuing management will be required throughout the whole of human civilization. To conclude, we provide some recommendations for further actions that should be taken to assist for managing and settling this situation.

2.2 IMPACT OF THE AMR CATASTROPHE ON HUMAN SOCIETY

In 2019, an unprecedented 4.95 million people were expected to lose their lives because of illnesses that were resistant to medicines. Sub-Saharan Africa and other low- and middle-income nations bore the brunt of this clinical load. The number of deaths caused by this figure is significantly higher than the annual global fatalities from tuberculosis, malaria, and HIV/AIDS. Estimates suggest that without action, yearly worldwide deaths related to AMR might rise to 10 million by 2050 (O'Neill, 2016; Ranjbar & Alam, 2023). The issue of AMR presents a promising chance to enhance public health and requires collaboration from various sectors, including the natural, economic, and human domains. It is not limited to hospitals or nations with poor and moderate incomes; improper antibiotic prescriptions are also widespread. Antoine Ardillon and his research team used data from a community-based mother-and-child cohort study (BIRDY) that was conducted in Cambodia, Madagascar, and Senegal between 2012 and 2018. Their investigation uncovered a concerning trend of overprescribing antibiotics to pediatric outpatients, with 76.5% of the prescriptions found to be unwarranted. The results of their investigation uncovered a concerning and troubling pattern of antibiotic prescription that was both excessive and inappropriate among pediatric outpatients, with 76.5% of the prescriptions found to be unwarranted. The most common causes of improper prescribing were rhinopharyngitis, gastroenteritis, and uncomplicated bronchiolitis. It was observed that rural children over 3 months received a higher number of prescriptions. These findings emphasize the need for community-level programs to optimize antibiotic prescription in low- and middle-income countries (LMICs) (Ardillon et al., 2023).

Charlie Luchen and his colleagues showed that antibiotics used in LMICs affect kids' gut microbiomes and even antibiotic-resistant genes. AMR bacteria from stomachs move to humans, animals, and the environment in LMICs with poor water and sanitation, creating a cycle of resistant microorganisms (Luchen et al., 2023). In high-income settings, antibiotic abuse and misuse have created well-known "super-bugs" like methicillin-resistant *Staphylococcus aureus*. Antibiotic stewardship programs often shorten treatment duration to reduce the occurrence of antimicrobial-resistant microbes. Yin Mo et al. investigated the relationship between the duration of antibiotic treatment and the risk of antimicrobial-resistant bacteria colonization in hospitalized patients. The study focused on examining this impact at both the individual and population levels, providing valuable insights into the potential outcomes. Researchers found that reducing antibiotic treatment time may affect the colonization of resistant bacteria, depending on individual and combination bacterial and antibiotic properties (Mo, Oonsivilai, Lim, Niehus, & Cooper, 2023).

The United Nations (UN) conducted a high-level summit in September 2024 to examine AMR and proposed an improved framework from 2016. It is essential for world leaders and policymakers to fully understand the importance of this issue and dedicate resources to the development of affordable vaccinations, antimicrobials, diagnostics, and reporting systems. The COVID-19 epidemic has taught global society a valuable lesson about our interconnectedness, both economically and socially. The impact of one nation's actions can have far-reaching consequences for others. The global impact of COVID-19 has been devastating, with millions of lives and trillions of dollars lost from the world economy since December 2019. Currently, AMR may cause 10 million deaths yearly and cost the world economy over $100 trillion by 2050, surpassing COVID-19 and highlighting its severe concern (Walsh, Gales, Laxminarayan, & Dodd, 2023).

2.3 ORIGIN OF ANTIMICROBIAL RESISTANCE

A specific antimicrobial agent does not necessarily induce uniform susceptibility or resistance in bacteria or species. Resistance levels can vary significantly among related bacterial species. The resistance or response of microbial strains to transgene expression can be transferred between bacteria through various mechanisms such as plasmids, transgenes, genetic elements, and phages. Additionally, chromatin instability can also lead to cross-resistance (Davies & Davies, 2010). Rapid proliferation of resistant microorganisms can occur when resistance factors are attached to plasmids. The bacterial cell wall is protected from various threats by employing biochemical mechanisms that bring about modifications to the target, enzymatic degradation, and alterations in the uptake of efflux pump proteins (Figure 2.1). Antibiotics of the first generation have faced resistance in various clinical settings because of natural processes (Bartlett, Gilbert, & Spellberg, 2013). Additionally, it is important to consider that bacteria have the capability of acquiring resistance genes from other species that have a close contact with both. It is possible for resistance to vary depending on the species as well as the genes that are acquired (Martinez, 2014).

The first antibiotic resistance was discovered in 1930 when sulfonamides were introduced; it anticipated preantibiotic resistance in the natural environment without supporting lethal resistant organisms. The antibiotic era began with antibiotic development and use in various areas of life. Due to human activities, excessive amounts of harmful antibiotics were utilized, altering their intrinsic importance, and accelerating antibiotic-resistant bacteria development (Muteeb, Rehman, Shahwan, & Aatif, 2023). Drug compounds that inactivate cell membranes have made antibiotic resistance widespread since their clinical usage. Novel therapeutic techniques and broad-spectrum antimicrobials are required for developing a few new antibacterial medications in clinical trials. In patients with bacteremia and other critical infections, pathogenic microorganisms and antibiotic

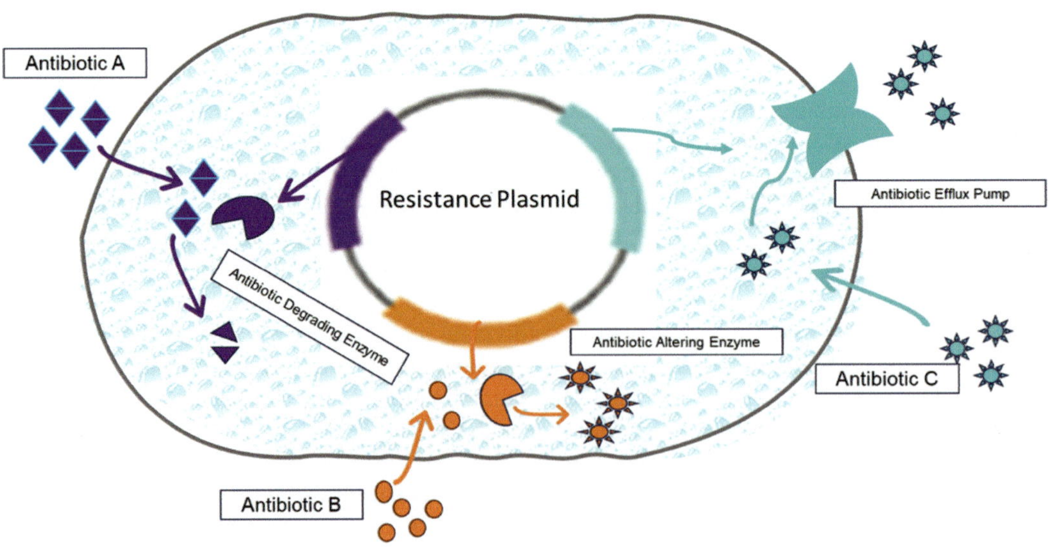

FIGURE 2.1 Different molecular components of drug resistance inside a bacterial cell.

sensitivity trends are not quickly identified. The researchers discovered that by comprehending the biochemical mechanisms and virulence strategies of infectious diseases, they can effectively target and engage with crucial factors that contribute to the bacteria's pathogenicity and infectivity, without inducing resistance (D'Costa, McGrann, Hughes, & Wright, 2006; Muteeb et al., 2023).

It's quite fascinating how the use of antimicrobial medications contributes to the development of resistance, creating a significant contradiction. This chapter highlights the significant impact that subinhibitory concentrations of antimicrobials can have on bacteria. Even at low or extremely low levels, these concentrations have the potential to contribute to the development of high-level resistance in the future generations of bacteria. They also have the potential to promote the growth of hypermutable strains, leading to an increased mutation rate. In addition, subinhibitory concentrations have the potential to enhance the development of resistance to other antimicrobial agents and assist in the transfer of mobile genetic elements (Blázquez, Couce, Rodríguez-Beltrán, & Rodríguez-Rojas, 2012).

2.4 MECHANISMS OF RESISTANCE AGAINST MICROBIAL AGENTS

Bacterial resistance mechanisms evolved rapidly in response to competition for selection. The antibiotic defense mechanisms include the production of deactivating enzymes like β-lactamases and aminoglycoside-modifying enzymes, as well as the modification of antibiotic targets. Additionally, these mechanisms work to reduce intracellular antibiotic concentrations by either limiting or facilitating expulsion. The Ambler classification categorizes each β-lactamase based on its main protein structure (Figure 2.2). In enzymes belonging to classes A, C, and D, the presence of a serine residue is necessary for the hydrolysis of the beta-lactam ring located in the active sites of these enzymes. However, it is essential to mention that metal o β-lactamases, which are classified as class B enzymes, need the presence of a zinc ion cofactor to perform their functions (Ambler, 1980; Coronell-Rodríguez, Arteta-Acosta, & Dueñas-Castell, 2018).

The main way antibiotics work is by disrupting or killing microorganisms by attacking their biochemistry and physiology. Certain antibiotics break down the -lactam and glycopeptide components of bacterial cell walls and membranes, while other antibiotics target the machinery that makes proteins by attaching themselves to ribosomal units, which inhibits the antibacterial action (Darby et al., 2023). Aminoglycosides, tetracycline, linezolid, chloramphenicol, and macrolides are effective in targeting cell walls. Rifampin and fluoroquinolones have a positive impact on cell machinery and nucleic acid synthesis. Some medications, such as folic acid analogues, daptomycin,

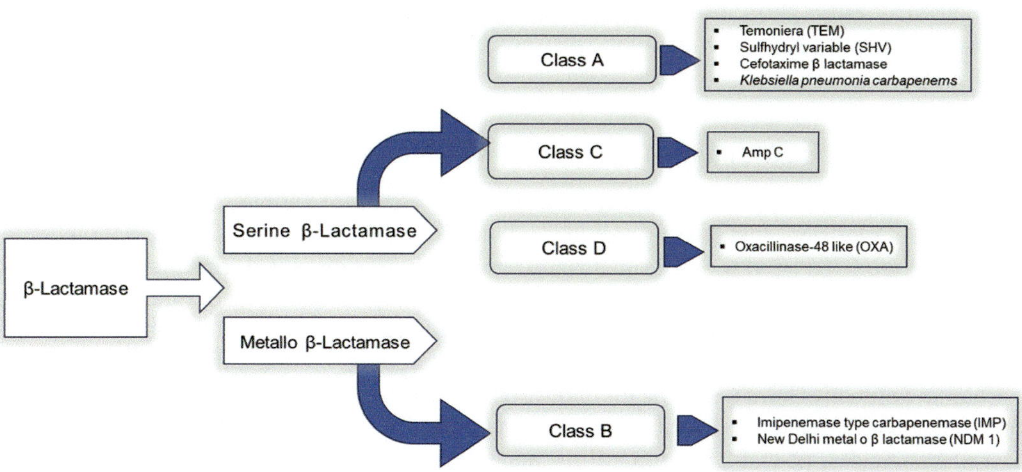

FIGURE 2.2 The β-lactamases according to the Ambler classification.

polymyxins, and sulfonamides, have the potential to disrupt metabolic processes and damage the membrane matrix (Chellat, Raguž, & Riedl, 2016). The researchers have made an important discovery regarding antibiotic resistance mechanisms in bacteria, identifying several mechanisms, including bacterial cell inactivity, target-protecting protein secretion, metabolism regulation, and self-repair systems, which collectively form the bacterial defense system against antibiotics. Below are some of the mechanisms of antibiotic resistance in microorganisms.

2.4.1 Efflux Pump

Efflux pumps are complex bacterial systems that are situated on the membrane that surrounds the cytoplasm. They play a crucial role in expelling toxic molecules from the cell and require energy to function. In 1980, the first plasmid-encoded efflux pump was discovered. This pump was able to expel tetracycline from *Escherichia coli.* Antibiotic resistance, achieved by efflux mechanism, has been generated by a considerable number of bacteria over time by both Gram-positive and Gram-negative. Although drug-specific efflux pumps exist, most efflux systems may transport many unrelated compounds causing multidrug resistance (Christaki, Marcou, & Tofarides, 2020; Nikaido & Pagès, 2012). Some bacteria's inherent antibiotic resistance may be determined by their chromosomal multidrug efflux systems. Despite the persistence of multidrug efflux systems in bacteria, few provide clinically relevant antibiotic resistance (Salam et al., 2023). Mutations that increase pump activity or expression often cause clinical resistance. Genes that encode substrate-specific efflux pumps are usually on free genetic elements in the bacterial cell (Fernández & Hancock, 2012).

2.4.2 Reduced Permeability

Gram-negative bacteria's outer membrane is vital for both protecting the organism and controlling the movement of vital materials. This barrier exhibits a dual capacity and seems to be an extraordinarily complex assemblage of macromolecules, the intricacies of which have only been investigated in recent times (Winterhalter & Ceccarelli, 2015). The outer membrane's hydrophobic lipid bilayer contains permeability-forming proteins (Omps) with size-exclusion capabilities to build a selective barrier (Ghai, 2017). The main route of entry for hydrophilic antibiotics, such as β-lactams, fluoroquinolones, tetracyclines, and chloramphenicol, is via holes in the bacterial outer membrane. The amount and kind of porins produced on the outer membrane affect the bacterial cell's sensitivity to hydrophilic antibiotics by determining how easily the antibiotics may enter the cell (Vergalli et al., 2017). Mutations in prion function or expression may cause acquired antibiotic resistance. Mutations may cause porin loss, size or conductance changes, or reduced expression. Most porin expression changes cause minor antibiotic resistance (Pinet, Franceschi, Davin-Regli, Zambardi, & Pagès, 2015).

2.4.3 Enzymatic or Mutational Target Site Modification

Bacteria's protein synthesis can nullify the bacteriostatic effects of antibiotic combinations on specific targets. TRPP-ribosome binding provides protection for type I targets. The interaction between 16S rRNA base C1054 and the deformed ribosome structure is disrupted, resulting in a modification of its shape. The ribosome is protected by 13 TRPP classes that cannot bind or disassociate from the binding site's 30S subunit (Zampieri, 2021). Changes in target conformation can lead to the indirect elimination of antibiotics through Type II target protection. ABC-F proteins that are resistant to antibiotics have a significant impact on the development of clinical resistance to different antimicrobials that target the ribosomal 50S subunit. These antimicrobials include lincomycin, macrolides, Azides, phenols, pleuromutilins, and streptogramins of groups A and B (Sharkey & O'Neill, 2018; Su et al., 2018). Target protection proteins of Type III alter the conformation of targets, enabling the proper functioning of antibiotic-bound targets. The acquisition of the Fus B-type protein has resulted in a notable boost in resistance to fusidic acid in clinically isolated *S. aureus* and other *staphylococci*. Fusidic acid does not affect FusB resistance proteins because they attach to EF-G and help it dissociate from ribosomes. Fusidic acid may dissociate from EF-G after leaving the ribosome because of its low affinity (Castanheira, Watters, Mendes, Farrell, & Jones, 2010).

2.4.4 Drug Inactivation

For bacteria to acquire antibiotic resistance, one of two things may happen: either the antibiotic is destroyed, or it is chemically altered by adding a new chemical group. The two primary methods by which bacteria inactivate medications are via real drug breakdown or by chemical group transfer to the medicine (Salam et al., 2023). Members of the *Enterobacter ales* family generate hydrolyzing enzymes called β-lactamases, which are very good at inactivating β-lactam antibiotics. One particularly vast class of drug hydrolyzing enzymes is called β-lactamases. Another drug that, via the tetX gene, may be inactivated by hydrolysis is tetracycline. This enzyme, initially referred to as penicillinases and cephalosporinases, could selectively target and deactivate the β-lactam ring structure. Due to this, the ring structure loses its ability to effectively interact with the target, known as penicillin-binding proteins (Santos & Lamers, 2020; Triggiano et al., 2020). It is well known that β-lactamase genes can be passed from one type of Gram-positive bacteria to another. These genes can be found in *S. aureus*, *Enterococcus faecalis*, and *Enterococcus faecium*, among others. Also, some bacteria can break down tetracycline by making an enzyme from the text gene (Blair, Webber, Baylay, Ogbolu, & Piddock, 2015).

2.5 MOLECULAR INSIGHTS OF ANTIBIOTIC RESISTANCE

In bacteria, the characteristic of resistance is either acquired or natural. A naturally occurring characteristic resulting from an organism's biology is intrinsic resistance; vancomycin resistance in *E. coli* is one such characteristic. Acquired resistance results from the development of resistance by a bacteria susceptible to antibiotics either by mutation or by acquisition of additional DNA. Not having antibiotics around doesn't stop mutations from happening on their own. An antibiotic quickly kills the sensitive cells, leaving behind a group of bacteria that are immune. This gives the bacteria a huge edge. In 1959, it was first known that resistance genes could be spread when they were found in *Shigella* and then moved to *E. coli* via plasmids. (Ventola, 2015). Plasmids are round pieces of DNA that can copy themselves. They are smaller than a bacterium's genome and direct another bacterial strain or species how to copy them. They can carry and pass on many resistance genes. These genes may be on a transposon, which is a piece of DNA that can move from one plasmid to another or to the genome (Tao, Chen, Li, Wang, & Liang, 2022). Sometimes, plasmids can only spread to a certain type of bacteria. This is why transposons are important for spreading resistant genes across these types of borders (Jian et al., 2021). There is a possibility that the mecA gene that was discovered in Methicillin-Resistant *Staphylococcus aureus* (MRSA) was acquired by transposition (Hiramatsu et al., 2013).

The evolution of plasmids may be a complicated process; nevertheless, contemporary molecular methods can provide insight into this process (as is the case with the plasmids that carry the tetM gene and are present in *Neisseria gonorrhoeae* all over the globe) (Muteeb et al., 2023). Bacteriophages, which are viruses that infect bacteria, can transfer resistance, and staphylococci are typically the organisms that exhibit this phenomenon. In the process known as transformation, bacteria that have died release their DNA, which may then be picked up by bacteria that can pass it on. When it comes to the transmission of penicillin resistance in *Streptococcus pneumoniae*, this mechanism is arguably the most significant pathway since it results in the production of "mosaic penicillin-binding protein genes." This process is recognized as crucial in the environment (Michaelis & Grohmann, 2023). Multiple antibiotic resistance pathways are depicted in Figure 2.3. Bacteria may horizontally transmit genes. Transformable receivers directly absorb ambient genetic

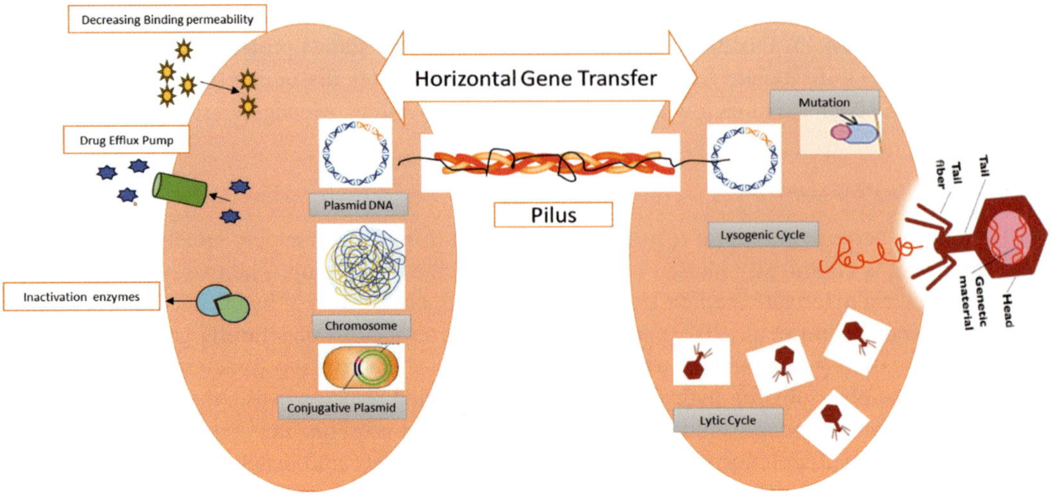

FIGURE 2.3 Multiple antibiotic resistance pathways.

information via chromosomal proteins. Transduction inserts prophage DNA into chromosomes, which duplicate and package host DNA. This passage offers a precise and detailed explanation of the mechanism by which conjugative plasmids establish a connection with recipient cells and facilitate the transfer of genetic material through a protein structure known as a pilus. The plasmids can transfer a full bacterial chromosome, a plasmid with multiple copies, or a small segment of genetic material. If compatible with the resident plasmid, they can either insert the genetic material or replicate it independently. The dissemination of gene cassettes is achieved through site-specific recombination mediated by integrins. Not homologous recombination copies genes and causes mutations, deletions, and rearrangements by inserting transposons and insertion sequences into new chromosomal or plasmid regions (Sultan et al., 2018).

2.6 TRANSMISSION OF ANTIMICROBIAL RESISTANCE

Under the "One Health" paradigm, antibiotic resistance is a multifarious ecological issue endangering the interdependent people, animals, and environmental integrity (White & Hughes, 2019). Direct contact or the food chain and surrounding environment allow resistant bacteria to move from one place to another, thus spreading worldwide over many different areas (McEwen & Collignon, 2017). Diseases can be transmitted to humans through the consumption of vegetables or contaminated water. Similarly, the transmission of diseases from animals to humans is prone to happen either through direct interaction with animals or by consuming food that has been contaminated (EFSA, 2018). The propagation of resistance genes occurs through ecological niches between individuals in various habitats, such as communities, hospitals, domestic or wildlife animals, and the environment, wherever antibiotics are administered. There is ongoing debate about whether antibiotic usage leads to resistance (Chatterjee et al., 2018).

There are numerous reports of bacterial lineages that span across different species or niches (Randad et al., 2021). Nevertheless, a recent study that aimed to compare the lineages of animals and humans has uncovered an unexpectedly low number of shared clones. There is a possibility that the apprehensions regarding the transmission of resistance among humans, animals, and the environment may be overstated (Thorpe et al., 2021). While it is generally uncommon for bacteria to transfer across species, there are instances where this can become more prevalent and noticeable. This typically occurs when there is a selective advantage or a high number of crossover opportunities (Liu et al., 2018). On the other hand, certain species can carry resistant organisms temporarily, without causing any noticeable effects. This can make it challenging to detect them in general surveys until further development occurs (Ingham et al., 2021). The research on the spread of resistant bacteria to humans often assumes short transfer pathways, which can introduce bias in the research design. On the other hand, it is possible that open multidirectional or circular transmission schemes with intermediate hosts could be feasible. The transmission of bacterial lineages may be less significant compared with the migration of mobile elements and genes, which might be difficult to identify and demonstrate (Hickman et al., 2021). Despite the little study undertaken outside Europe and North America, it is essential for studies in low-income countries to use approaches that can accurately detect transmission, even in uncommon instances. These situations are probably more likely to occur in low-income countries' environments compared to other places. The use of whole-genome sequencing-based approaches allows for a detailed analysis that can help determine whether inter-host transmission or other niche-entry events have occurred (Jørgensen et al., 2019). As the costs of sequencing continue to rise, conducting research with limited resources remains a significant challenge. Therefore, it is difficult to determine when and how genes, elements, or resistant organisms transfer from one niche to another (Okeke, 2016).

2.7 GLOBAL STRATEGY TO CONTROL THE MENACE OF ANTIBIOTIC RESISTANCE

AMR continues to pose a substantial threat to the well-being of human and animal health, and it doesn't look like it will go away any time soon. AMR is a serious environmental problem that doesn't seem to have a clear answer, just like global warming. Before penicillin was discovered, scientists and doctors didn't know much about AMR, even though germs that were immune to it had already been found (Graham et al., 2019). Some nations have reported controlling resistant diseases, such as MRSA, however, the causes for the fall in their resistance are debated. The United States' apparent effectiveness in reducing MRSA via sanitation and screens has been questioned due to other variables such as a countrywide drop in MRSA prevalence (Jain et al., 2011). Similar behavior has also been quite prevalent at the national level in Europe. The scientific community should appreciate and prioritize these debates, as they are crucial for identifying the key strategies needed to manage AMR. While it is not feasible to eliminate AMR, there are measures that can be taken to slow down its development (Gurieva, Bootsma, & Bonten, 2012). The Global Antimicrobial Resistance Surveillance System, in partnership with WHO, Food and Agriculture Organization, CDC, and Office International des Epizooties, join forces to tackle antibiotic resistance. There are other initiatives, like the Global Health Security Agenda (GHSA) and Antimicrobial Resistance Action Package (GHSA Action Package Prevent-1), that are also actively addressing the global challenge of antibiotic resistance (Sirijatuphat et al., 2018). Diagnostics promote incorrect antibiotic use by clinicians, contributing to AMR. Microbiological tools that are commonly used for germ identification can pose challenges in developing countries, hindering effective diagnostics. Customized medicine can address these gaps by utilizing advanced molecular diagnostics to identify patients who require antibiotics.

The One Health Approach has the potential to assist researchers in developing screening procedures and investigating the dynamics of human–animal interaction. This problem is important since AMR transmission between people, animals, and the environment affects one's health (Cantón et al., 2012). The development of new technologies, as well as innovations in the fields of antibiotics and combination treatment, are crucial. An additional facet that is of utmost significance in this respect is the incorporation of antibiotic stewardship into the healthcare system. Programs devoted to antimicrobial stewardship (AMS) aim to reduce AMR and maximize clinical outcomes in infection prevention and treatment while minimizing or eliminating adverse drug reactions (EDRs) in patients. An improvement in AMU and patient treatment results has been shown with the proper implementation of AMS programs, which are crucial in addressing AMR. Public health campaigns, awareness-sensitization initiatives, NAPs, and the GAP are all examples of AMS programs (Mudenda et al., 2023).

2.8 CONCLUSION

In conclusion, this chapter provides an in-depth exploration of AMR, covering its origins, mechanisms, and far-reaching impacts. By analyzing the molecular foundations of resistance and its historical progression, we highlight the sophisticated survival strategies of microorganisms. The discussion underscores the urgent need for innovative therapeutic solutions, robust policy frameworks, and coordinated global action to combat the escalating threat of AMR. Addressing this critical issue requires a unified, informed response to safeguard public health. The insights and strategies presented here serve as a valuable resource for tackling this pressing global challenge effectively.

REFERENCES

Ambler, R. P. (1980). The structure of β-lactamases. *Philosophical Transactions of the Royal Society of London. B, Biological Sciences, 289*(1036), 321–331.

Ardillon, A., Ramblière, L., Kermorvant-Duchemin, E., Sok, T., Zo, A. Z., Diouf, J.-B., ... Borand, L. (2023). Inappropriate antibiotic prescribing and its determinants among outpatient children in 3 low-and middle-income countries: A multicentric community-based cohort study. *PLoS Medicine, 20*(6), e1004211.

Aslam, B., Wang, W., Arshad, M. I., Khurshid, M., Muzammil, S., Rasool, M. H., ... Qamar, M. U. (2018). Antibiotic resistance: A rundown of a global crisis. *Infection and Drug Resistance*, 1645–1658.

Bartlett, J. G., Gilbert, D. N., & Spellberg, B. (2013). Seven ways to preserve the miracle of antibiotics. *Clinical Infectious Diseases, 56*(10), 1445–1450.

Blair, J. M., Webber, M. A., Baylay, A. J., Ogbolu, D. O., & Piddock, L. J. (2015). Molecular mechanisms of antibiotic resistance. *Nature Reviews Microbiology, 13*(1), 42–51.

Blázquez, J., Couce, A., Rodríguez-Beltrán, J., & Rodríguez-Rojas, A. (2012). Antimicrobials as promoters of genetic variation. *Current Opinion in Microbiology, 15*(5), 561–569.

Cantón, R., Akóva, M., Carmeli, Y., Giske, C. G., Glupczynski, Y., Gniadkowski, M., ... Rossolini, G. (2012). Rapid evolution and spread of carbapenemases among Enterobacteriaceae in Europe. *Clinical Microbiology and Infection, 18*(5), 413–431.

Castanheira, M., Watters, A. A., Mendes, R. E., Farrell, D. J., & Jones, R. N. (2010). Occurrence and molecular characterization of fusidic acid resistance mechanisms among *Staphylococcus* spp. from European countries (2008). *Journal of Antimicrobial Chemotherapy, 65*(7), 1353–1358.

Centers for Disease Control and Prevention (CDC). (2019). Antibiotic Resistance Threat in the United States. Available from: https://ndc.services.cdc.gov/wp-content/uploads/Antibiotic-Resistance-Threats-in-the-United-States-2019.pdf

Chatterjee, A., Modarai, M., Naylor, N. R., Boyd, S. E., Atun, R., Barlow, J., ... Robotham, J. V. (2018). Quantifying drivers of antibiotic resistance in humans: A systematic review. *The Lancet Infectious Diseases, 18*(12), e368–e378.

Chellat, M. F., Raguž, L., & Riedl, R. (2016). Targeting antibiotic resistance. *Angewandte Chemie International Edition, 55*(23), 6600–6626.

Christaki, E., Marcou, M., & Tofarides, A. (2020). Antimicrobial resistance in bacteria: Mechanisms, evolution, and persistence. *Journal of Molecular Evolution, 88*(1), 26–40.

Coronell-Rodríguez, W., Arteta-Acosta, C., & Dueñas-Castell, C. (2018). Interpretive reading of the antibiogram: A tool for clinical practice. In G. Ortiz-Ruiz & C. Dueñas-Castell (Eds.), *Sepsis*, pp. 95–115. Springer New York, New York.

D'Costa, V. M., McGrann, K. M., Hughes, D. W., & Wright, G. D. (2006). Sampling the antibiotic resistome. *Science, 311*(5759), 374–377.

Dar, O. A., Hasan, R., Schlundt, J., Harbarth, S., Caleo, G., Dar, F. K., ... Shahid, M. (2016). Exploring the evidence base for national and regional policy interventions to combat resistance. *The Lancet, 387*(10015), 285–295.

Darby, E. M., Trampari, E., Siasat, P., Gaya, M. S., Alav, I., Webber, M. A., & Blair, J. M. (2023). Molecular mechanisms of antibiotic resistance revisited. *Nature Reviews Microbiology, 21*(5), 280–295.

Davies, J., & Davies, D. (2010). Origins and evolution of antibiotic resistance. *Microbiology and Molecular Biology Reviews, 74*(3), 417–433.

European Food Safety Authority (EFSA). (2018). The European Union summary report on antimicrobial resistance in zoonotic and indicator bacteria from humans, animals and food in 2016. *EFSA Journal, 16*(2), e05182.

Fernández, L., & Hancock, R. E. (2012). Adaptive and mutational resistance: Role of porins and efflux pumps in drug resistance. *Clinical Microbiology Reviews, 25*(4), 661–681.

Ghai, I. (2017). *Quantifying the Flux of Charged Molecules through Bacterial Membrane Proteins*. Jacobs University, Bremen.

Graham, D. W., Bergeron, G., Bourassa, M. W., Dickson, J., Gomes, F., Howe, A., ... Simjee, S. (2019). Complexities in understanding antimicrobial resistance across domesticated animal, human, and environmental systems. *Annals of the New York Academy of Sciences, 1441*(1), 17–30.

Gurieva, T., Bootsma, M. C., & Bonten, M. J. (2012). Successful Veterans Affairs initiative to prevent methicillin-resistant *Staphylococcus aureus* infections revisited. *Clinical Infectious Diseases, 54*(11), 1618–1620.

Hickman, R. A., Leangapichart, T., Lunha, K., Jiwakanon, J., Angkititrakul, S., Magnusson, U., ... Järhult, J. D. (2021). Exploring the antibiotic resistance burden in livestock, livestock handlers and their non-livestock handling contacts: A one health perspective. *Frontiers in Microbiology, 12*, 651461.

Hiramatsu, K., Ito, T., Tsubakishita, S., Sasaki, T., Takeuchi, F., Morimoto, Y., ... Hishinuma, T. (2013). Genomic basis for methicillin resistance in *Staphylococcus aureus*. *Infection & Chemotherapy, 45*(2), 117–136.

Ingham, A. C., Urth, T. R., Sieber, R. N., Stegger, M., Edslev, S. M., Angen, Ø., & Larsen, A. R. (2021). Dynamics of the human nasal microbiota and *Staphylococcus aureus* CC398 carriage in pig truck drivers across one workweek. *Applied and Environmental Microbiology, 87*(18), e0122521.

Jain, R., Kralovic, S. M., Evans, M. E., Ambrose, M., Simbartl, L. A., Obrosky, D. S., ... Muder, R. R. (2011). Veterans affairs initiative to prevent methicillin-resistant *Staphylococcus aureus* infections. *New England Journal of Medicine, 364*(15), 1419–1430.

Jian, Z., Zeng, L., Xu, T., Sun, S., Yan, S., Yang, L., ... Dou, T. (2021). Antibiotic resistance genes in bacteria: Occurrence, spread, and control. *Journal of Basic Microbiology, 61*(12), 1049–1070.

Jørgensen, S. L., Stegger, M., Kudirkiene, E., Lilje, B., Poulsen, L. L., Ronco, T., ... Pedersen, K. (2019). Diversity and population overlap between avian and human *Escherichia coli* belonging to sequence type 95. *MSphere, 4*(1). DOI: 10.1128/msphere.00333-00318.

Laxminarayan, R., Duse, A., Wattal, C., Zaidi, A. K., Wertheim, H. F., Sumpradit, N., ... Goossens, H. (2013). Antibiotic resistance—The need for global solutions. *The Lancet Infectious Diseases, 13*(12), 1057–1098.

Laxminarayan, R., Matsoso, P., Pant, S., Brower, C., Røttingen, J.-A., Klugman, K., & Davies, S. (2016). Access to effective antimicrobials: A worldwide challenge. *The Lancet, 387*(10014), 168–175.

Liu, C. M., Stegger, M., Aziz, M., Johnson, T. J., Waits, K., Nordstrom, L., ... Statham, S. (2018). *Escherichia coli* ST131-H 22 as a foodborne uropathogen. *MBio, 9*(4). DOI: 10.1128/mbio.00470-00418.

Luchen, C. C., Chibuye, M., Spijker, R., Simuyandi, M., Chisenga, C., Bosomprah, S., ... Harris, V. C. (2023). Impact of antibiotics on gut microbiome composition and resistome in the first years of life in low- to middle-income countries: A systematic review. *PLoS Medicine, 20*(6), e1004235.

Martinez, J. L. (2014). General principles of antibiotic resistance in bacteria. *Drug Discovery Today: Technologies, 11*, 33–39.

McEwen, S. A., & Collignon, P. J. (2017). Antimicrobial resistance: A one health perspective. *Microbiol Spectrum, 6*(2), ARBA-0009-2017. DOI: 10.1128/microbiolspec.ARBA-0009-2017.

Michaelis, C., & Grohmann, E. (2023). Horizontal gene transfer of antibiotic resistance genes in biofilms. *Antibiotics, 12*(2), 328.

Mo, Y., Oonsivilai, M., Lim, C., Niehus, R., & Cooper, B. S. (2023). Implications of reducing antibiotic treatment duration for antimicrobial resistance in hospital settings: A modelling study and meta-analysis. *PLoS Medicine, 20*(6), e1004013.

Mudenda, S., Chabalenge, B., Daka, V., Mfune, R. L., Salachi, K. I., Mohamed, S., ... Matafwali, S. K. (2023). Global strategies to combat antimicrobial resistance: A one health perspective. *Pharmacology & Pharmacy, 14*(8), 271–328.

Muteeb, G., Rehman, M. T., Shahwan, M., & Aatif, M. (2023). Origin of antibiotics and antibiotic resistance, and their impacts on drug development: A narrative review. *Pharmaceuticals, 16*(11), 1615.

Nikaido, H., & Pagès, J.-M. (2012). Broad-specificity efflux pumps and their role in multidrug resistance of Gram-negative bacteria. *FEMS Microbiology Reviews, 36*(2), 340–363.

Okeke, I. N. (2016). African biomedical scientists and the promises of "big science". *Canadian Journal of African Studies/Revue canadienne des études africaines, 50*(3), 455–478.

O'Neill, J. (2016). Tackling drug-resistant infections globally: Final report and recommendations. Government of the UK. https://apo.org.au/sites/default/files/resource-files/2016-05/apo-nid63983.pdf.

Pinet, E., Franceschi, C., Davin-Regli, A., Zambardi, G., & Pagès, J.-M. (2015). Role of the culture medium in porin expression and piperacillin-tazobactam susceptibility in *Escherichia coli*. *Journal of Medical Microbiology, 64*(11), 1305–1314.

Prestinaci, F., Pezzotti, P., & Pantosti, A. (2015). Antimicrobial resistance: A global multifaceted phenomenon. *Pathogens and Global Health, 109*(7), 309–318.

Randad, P. R., Larsen, J., Kaya, H., Pisanic, N., Ordak, C., Price, L. B., … Stewart, J. R. (2021). Transmission of antimicrobial-resistant *Staphylococcus aureus* clonal complex 9 between pigs and humans, United States. *Emerging Infectious Diseases, 27*(3), 740.

Ranjbar, R., & Alam, M. (2023). Antimicrobial Resistance Collaborators (2022). Global burden of bacterial antimicrobial resistance in 2019: A systematic analysis. *Lancet, 399*(10325), 629–655. DOI: 10.1016/S0140-6736(21)02724-0.

Salam, M. A., Al-Amin, M. Y., Salam, M. T., Pawar, J. S., Akhter, N., Rabaan, A. A., & Alqumber, M. A. (2023). Antimicrobial resistance: A growing serious threat for global public health. *Healthcare, 11*(13), 1946.

Santos, J. A., & Lamers, M. H. (2020). Novel antibiotics targeting bacterial replicative DNA polymerases. *Antibiotics, 9*(11), 776.

Sharkey, L. K., & O'Neill, A. J. (2018). Antibiotic resistance ABC-F proteins: Bringing target protection into the limelight. *ACS Infectious Diseases, 4*(3), 239–246.

Sirijatuphat, R., Sripanidkulchai, K., Boonyasiri, A., Rattanaumpawan, P., Supapueng, O., Kiratisin, P., & Thamlikitkul, V. (2018). Implementation of global antimicrobial resistance surveillance system (GLASS) in patients with bacteremia. *PLoS ONE, 13*(1), e0190132.

Su, W., Kumar, V., Ding, Y., Ero, R., Serra, A., Lee, B. S. T., … Yang, L. (2018). Ribosome protection by antibiotic resistance ATP-binding cassette protein. *Proceedings of the National Academy of Sciences, 115*(20), 5157–5162.

Sultan, I., Rahman, S., Jan, A. T., Siddiqui, M. T., Mondal, A. H., & Haq, Q. M. R. (2018). Antibiotics, resistome and resistance mechanisms: A bacterial perspective. *Frontiers in Microbiology, 9*, 2066.

Tao, S., Chen, H., Li, N., Wang, T., & Liang, W. (2022). The spread of antibiotic resistance genes in vivo model. *Canadian Journal of Infectious Diseases and Medical Microbiology, 2022*(1), 3348695.

Thorpe, H., Booton, R., Kallonen, T., Gibbon, M. J., Couto, N., Passet, V., … Mitchell, S. (2021). One Health or Three? Transmission modelling of Klebsiella isolates reveals ecological barriers to transmission between humans, animals and the environment. *BioRxiv*, 2021–2008.

Triggiano, F., Calia, C., Diella, G., Montagna, M. T., De Giglio, O., & Caggiano, G. (2020). The role of urban wastewater in the environmental transmission of antimicrobial resistance: The current situation in Italy (2010–2019). *Microorganisms, 8*(10), 1567.

Ventola, C. L. (2015). The antibiotic resistance crisis: Part 1: Causes and threats. *Pharmacy and Therapeutics, 40*(4), 277.

Vergalli, J., Dumont, E., Cinquin, B., Maigre, L., Pajovic, J., Bacqué, E., … Pagès, J.-M. (2017). Fluoroquinolone structure and translocation flux across bacterial membrane. *Scientific Reports, 7*(1), 9821.

Walsh, T. R., Gales, A. C., Laxminarayan, R., & Dodd, P. C. (2023). Antimicrobial resistance: Addressing a global threat to humanity. *Public Library of Science*, *20*, e1004264.

White, A., & Hughes, J. M. (2019). Critical importance of a one health approach to antimicrobial resistance. *EcoHealth, 16*, 404–409.

Winterhalter, M., & Ceccarelli, M. (2015). Physical methods to quantify small antibiotic molecules uptake into Gram-negative bacteria. *European Journal of Pharmaceutics and Biopharmaceutics, 95*, 63–67.

World Health Organization (WHO). (2014). Antimicrobial Resistance: Global Report on Surveillance. https://www.who.int/publications/i/item/9789241564748.

Zampieri, M. (2021). The genetic underground of antibiotic resistance. *Science, 371*(6531), 783–784.

3 The Role of Probiotics in Combating Multidrug-Resistant Bacteria

Pumza Magangane[1], *Phuti Mphaki*[2], *Sanelisiwe Thinasonke Duze*[3], *and Musa Marimani*[1]

[1]Anatomical Pathology, School of Pathology, Health Sciences, University of the Witwatersrand, Johannesburg, South Africa

[2]Oral Biological Sciences, School of Oral Health Sciences, Health Sciences, University of the Witwatersrand, Johannesburg, South Africa

[3]Clinical Microbiology and Infectious Diseases, School of Pathology, Health Sciences, University of the Witwatersrand, Johannesburg, South Africa

3.1 INTRODUCTION

3.1.1 Antimicrobial Resistance

Antimicrobials are compounds that work against bacteria, viruses, parasites, and fungi, among other microorganisms. One type of antimicrobial is an antibiotic, which is a drug intended to either eradicate or inhibit the growth of bacterial pathogens and improve the prognosis (Ferri et al., 2017). The World Health Organization (WHO) has classified eight pathogens as a matter of concern. These include *Mycobacterium tuberculosis* (*Mtb*), *Plasmodium* parasite, HIV/AIDS, Influenza viruses, *Escherichia coli*, *Klebsiella pneumoniae* (*K. pneumoniae*), *Staphylococcus aureus*, and *Neisseria gonorrhoeae* (*N. gonorrhoeae*). The use of antibiotics and other antimicrobials, particularly in veterinary medicine, cannot be completely eradicated, given the prevalence of infectious diseases and the need to treat them. This protects human health and animal welfare (Schmerold et al., 2023). In the 20 years that followed their discovery and widespread manufacturing, the use of antimicrobials, particularly antibiotics, increased rapidly.

Antimicrobial usage has also grown over time because of new product development, scientific advancements, and accessibility. Today, both the food industry and the medical profession extensively use antibiotics (Freeland et al., 2023). Alcohol-based hand sanitisers are widely used and are available at various outlets (Ahmad et al., 2022). Chlorine is frequently added as a disinfectant to public drinking water sources. Heavy metals are still employed as disinfectants; aluminium, for instance, is frequently utilised in antiperspirants to aid in the reduction of bacteria (Freeland et al., 2023). Indiscriminate antibiotic use is a major contributor to the introduction of selective pressures in our natural environments that have significantly contributed to the rapid emergence of antibiotic-resistant microbial strains (Imperial & Ibana, 2016). The use of probiotics instead of antibiotic therapy may alleviate these antibiotic-mediated selective pressures (Imperial & Ibana, 2016). This chapter discusses how probiotics may help combat antimicrobial resistance.

DOI: 10.1201/9781032676951-3

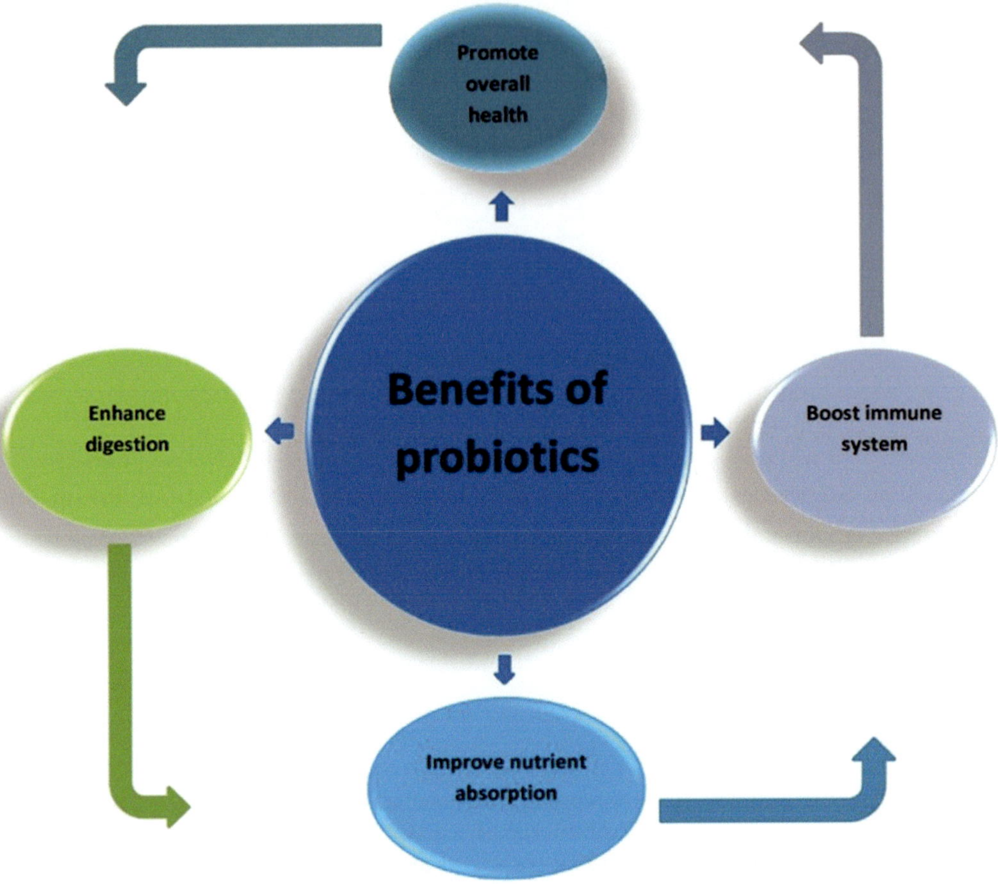

FIGURE 3.1 Benefits of probiotics.

3.1.2 Probiotics and Their Role in Health

The human gut is a complex ecosystem teeming with trillions of microorganisms, collectively known as the gut microbiome (Imperial & Ibana, 2016). This diverse community of bacteria, fungi, and other microbes is essential in digestion, nutrient absorption, immune function, and overall health (Figure 3.1) (Bull & Plummer, 2014). Disruptions in this delicate balance, known as dysbiosis, have been linked to various health problems, including inflammatory bowel diseases (IBD) (Li et al., 2023), irritable bowel syndrome (IBS) (Sharma et al., 2023), and even neurodegenerative disorders (Alshehri et al., 2021). The following sections explore the mechanisms by which probiotics work to maintain a healthy gut microbiome.

3.1.2.1 Competition and Exclusion

One of the primary ways probiotics promote gut health is by competing with harmful bacteria for space and resources (Stecchini et al., 2001). Beneficial bacteria occupy adhesion sites on the intestinal lining, thereby preventing colonisation by pathogenic bacteria like *E. coli* and *Salmonella* (Wilkins & Sequoia, 2017). Studies suggest that probiotics like *Lactobacillus* and *Bifidobacterium* effectively outcompete pathogens, reducing the risk of infections (Vlasova et al., 2016; Maftei et al., 2024). In addition, the gut microflora also aids in regulating the host pH and is also critical for food digestion and absorption.

Probiotics aid in digestion, nutrient absorption, immune function, and the overall health of an individual. These beneficial microorganisms also disrupt the growth of pathogenic microbial strains and maintain a favourable environment for the host's normal flora.

3.1.2.2 Short-Chain Fatty Acid Production

Probiotics ferment dietary fibres, producing short-chain fatty acids (SCFAs) like butyrate, acetate, and propionate. These SCFAs serve as a vital energy source for gut epithelial cells, promoting gut barrier function and reducing inflammation (Xiao et al., 2024; Liu et al., 2023; Ney et al., 2023). Additionally, SCFAs can modulate the immune response, thus contributing to a balanced gut environment (McFarland, 2007).

3.1.2.3 Modulation of the Immune System

Probiotics interact with the immune system at various levels, promoting immune tolerance and reducing inflammation. They can stimulate the production of immunoglobulins (antibodies) and activate immune cells like dendritic cells, which help regulate immune responses (Vlasova et al., 2016; Gupta & Garg, 2009). This immunomodulatory effect can be beneficial in managing IBD and allergies (Bermudez-Brito et al., 2018; Schultz et al., 2002).

3.1.2.4 Microbial Balance and Diversity

A healthy gut microbiome is characterised by a rich diversity of bacterial species, each contributing a unique function. Probiotics play a crucial role in restoring and maintaining this delicate balance. Probiotics can directly compete with pathogenic and opportunistic bacteria for limited resources and adhesion sites within the gut (Bezkorovainy, 2001). This competition prevents the overgrowth of harmful microbes, promoting a more balanced gut environment (Stecchini et al., 2001; Arasu et al., 2016; Wei et al., 2022). Certain probiotic strains produce antimicrobial compounds like bacteriocins, which can inhibit the growth of pathogenic bacteria. This helps maintain a balanced gut microbiota by suppressing harmful species (Kwun et al., 2024). Probiotics can influence the immune system to promote tolerance towards beneficial bacteria and suppress inflammatory responses against commensal microbes (Ivec et al., 2007). This targeted modulation helps maintain a balanced gut environment and prevents excessive immune activation. A diverse gut microbiome is more resilient to environmental disturbances and infections. Probiotics may help increase microbial diversity by introducing new beneficial strains and creating a favourable environment for their growth (Hemarajata & Versalovic, 2013). This increased diversity strengthens the gut ecosystem and improves its ability to resist colonisation by pathogens (Alshehri et al., 2021; Khan et al., 2019).

3.1.3 Bacterial Probiotics

Live microorganisms that confer health benefits when consumed have emerged as a promising strategy for restoring gut health (Rodrigues et al., 2022). The intestine is mainly colonised by four phyla of bacteria: *Firmicutes*, *Bacteroidetes*, *Proteobacteria*, and *Actinobacteria*. Several factors, such as antibiotics, diet, and pH can interfere with the gut microbiota (Hasan & Yang, 2019). A change from a healthy to a pathogenic microbiota (i.e. dysbiosis) can induce abnormal immune responses in the gut-associated lymphatic tissue. Subsequently, these alterations can compromise the systemic immune response (Rodrigues et al., 2022). The gut microbiota regulates the host immune response through two main mechanisms: activating the innate immune response via the toll-like receptor (TLR) and/or activating free fatty acid receptors (FFAR) via microbial metabolites such as SCFAs, including acetate, propionate, and butyrate (Yoo et al., 2020). A joint Food and Agriculture Organization of the United Nations/World Health Organization (FAO/WHO) expert consultation redefined probiotics as "live microorganisms which, when administered in adequate amounts, confer a health benefit on the host population" (Córdoba, 2001). In 2002, the FAO/WHO proposed criteria for selecting probiotic microorganisms, including antimicrobial activity, ability to adhere to

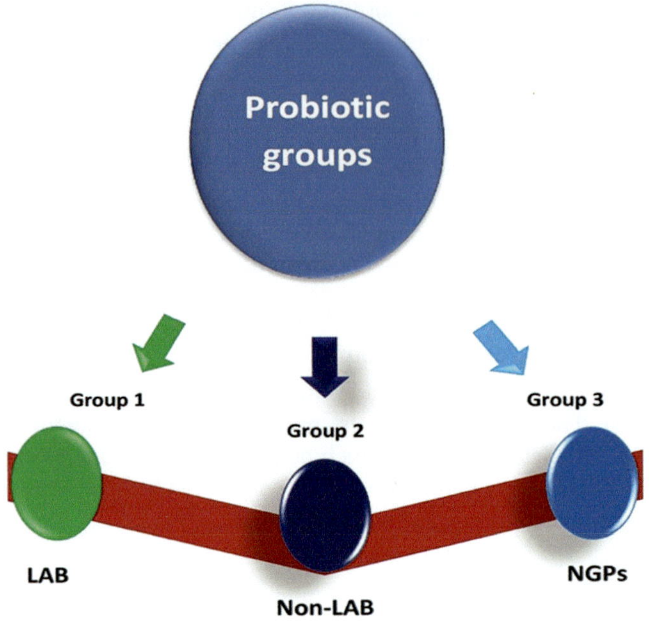

FIGURE 3.2 Categories of probiotic microorganisms.

epithelial tissues, resistance to harsh conditions inside the host and safety (El-Sohaimy & Hussain, 2023). Probiotics can be classified into three groups: traditional lactic acid bacteria (LAB) probiotics, non-LAB probiotics, and next-generation probiotics (NGP) (Figure 3.2) (Gao et al., 2021).

3.1.3.1 LAB Probiotics

LAB probiotics comprise the following genera: *Lactobacillus* (the largest genus of LAB probiotics), *Aerococcus*, *Carnobacterium*, *Enterococcus*, *Lactococcus*, *Leuconostoc*, *Oenococcus*, *Pediococcus*, *Sporolactobacillus*, *Streptococcus*, *Tetragenococcus*, *Vagococcus*, and *Weissella* bacteria (Thokchom & Joshi, 2012). Several species such as *Bifidobacterium* and *Lactobacillus* are also employed on an industrial scale for fermenting dairy and other food products (Papizadeh et al., 2017). Furthermore, various members of these genera are naturally associated with mucosal surfaces in the vagina, the oral cavity, and the gastrointestinal tract (GIT) (Shah, 2007). Also, it is commonly known that these microorganisms are native to environments associated with food, such as cereal grains, fruits, meat, milk, plants, vegetables, and vinegar (Wells & Mercenier, 2008). Numerous strains of *Bifidobacterium* and *Lactobacillus* are even available on the market as probiotic supplements, medical devices, or cosmetics. The consumption of food products containing probiotics has shown positive effects on human health (Million & Raoult, 2013).

In particular, the consumption of LAB-fermented dairy products may elicit antitumor effects. These effects are attributed to the inhibition of mutagenic activity and a decrease in several enzymes implicated in the generation of carcinogens, mutagens or tumour-promoting agents (Kumar et al., 2010). Specific cellular components in LAB strains seem to induce strong adjuvant effects, including the modulation of cell-mediated immune responses, the activation of the reticuloendothelial system, the augmentation of cytokine pathways, and the regulation of interleukins and tumour necrosis factors (Coelho et al., 2022). Oral administration of LAB is well tolerated and proven safe in 143 human clinical trials, and no adverse effects were reported in any of the 7,526 subjects studied during 1961–1998 (Naidu et al., 1999). To decrease the reliance on synthetic antimicrobials and control the emerging immunocompromised host population, it is crucial to carefully explore the prophylactic and therapeutic applications of LAB probiotics (Al-Fakhrany & Elekhnawy, 2024). However, traditional probiotics, primarily consisting of LAB strains like *Lactobacillus* and *Bifidobacterium*, have limitations in terms of strain viability, delivery, and efficacy. This has paved the way for exploring NGPs, a new frontier in gut health management.

3.1.3.2 Non-LAB Probiotics and NGPs

The non-LAB group contains the following genera: *Bifidobacterium*, *Enterococcus*, *Escherichia*, *Bacillus*, *Clostridium*, and the probiotic yeasts (*Saccharomyces* species) (Gao et al., 2021). Conventionally, LAB bacteria such as *Lactobacilli*, *Bifidobacteria*, and others have been used as probiotics. These microorganisms were primarily isolated from fermented dairy products and faecal microbiota. However, most microorganisms inhabiting the human gut have remained unidentified because they are mostly anaerobic bacteria that are difficult to cultivate. Recent advances in complete genome sequencing and culture techniques have facilitated the isolation and characterisation of various novel microbes from the human microbiome. These microbes were demonstrated to have promising health benefits and probiotic potential. NGPs are defined as "living microbes identified on the base of comparative microbiome investigations which confer health advantages to their host when taken to suitable extents." (Al-Fakhrany & Elekhnawy, 2024). Moreover, NGPs encompass a diverse range of microbes beyond traditional LAB strains, including *Akkermansia muciniphila* (*A. muciniphila*), *Faecalibacterium prausnitzii* (*F. prausnitzii*), *Bacteroides fragilis* (*B. fragilis*), *Eubacterium hallii* (*E. hallii*), and Roseburia species (Zhang et al., 2022). *A. muciniphila* is a Gram-negative anaerobic mucus-layer-degrading bacterium that colonises the intestinal mucosa of humans and rodents (Rodrigues et al., 2022). This emerging probiotic candidate has garnered significant interest due to its ability to promote mucus layer thickness and gut barrier function, thereby protecting against intestinal inflammation (Di Vincenzo et al., 2024; Wang et al., 2023).

The bacterium *F. prausnitzii* belongs to the Firmicutes phylum and the *Ruminococcaceae* family. Currently, this microbe is the only representative probiotic characterised within the genus. *F. prausnitzii*, is a Gram-positive, anaerobic, non-spore-forming, and non-motile rod that is exceedingly oxygen-sensitive (He et al., 2021). *Bacteroides* species are important clinical pathogens present in most anaerobic infections, with an associated mortality of approximately 19%. *B. fragilis* is an obligate anaerobic Gram-negative *Bacillus* (Elshaghabee et al., 2017; Elsaghir & Reddivari, 2024). The human colon has the greatest population of bacteria in the body (over ten organisms per gram of wet weight), and the largest part of these organisms is anaerobes. Of these, approximately 25% are species belonging to the *Bacteroides* genus. *B. fragilis* is part of the normal microbiota of the human colon. Disruption of the mucosal surface by inflammation, trauma, or surgery, and spread of *B. fragilis* to the bloodstream or surrounding tissues results in clinically significant infection. The mammalian intestine is densely populated by trillions of bacteria belonging to an estimated 500–1,000 species that are predominantly classified as *Firmicutes* and *Bacteroidetes* (Engels et al., 2016). Due to its ability to utilise the intermediate lactate, *E. hallii* is considered an important microbe with regard to intestinal metabolic balance, as lactate accumulation has been associated with several intestinal diseases and malabsorption (Engels et al., 2016). The genus *Roseburia* consists of obligate Gram-positive anaerobic bacteria that are slightly curved, rod-shaped, and motile using multiple subterminal flagella. It includes five species: *Roseburia intestinalis* (*R. intestinalis*), *Roseburia hominis*, *Roseburia inulinivorans*, *Roseburia faecis*, and *Roseburia cecicola*. Gut *Roseburia* spp. metabolise dietary components that stimulate their proliferation and metabolic activities. They are part of commensal bacteria producing SCFAs, especially butyrate. This affects colonic motility, immunity, and anti-inflammatory properties.

A significant change in *Roseburia* spp. representation may affect various metabolic pathways and is associated with several diseases including IBS, obesity, type 2 diabetes, nervous system conditions, and certain allergies. *Roseburia* spp. could also serve as biomarkers for symptomatic pathologies (e.g. gallstone formation) or as probiotics for restoration of beneficial flora (Tamanai-Shacoori et al., 2017). However, there is a big challenge with viable intestinal delivery as they are oxygen-sensitive and require new cultivation and storage methods. In addition, the need for viable intestinal delivery of probiotics has encouraged scientists to consider the use of *Bacillus* spp. and *Clostridium* spp. Both species are naturally encapsulated members of the gut microbiome and are capable of spore formation. Spores are highly resistant to acidity, heat, and chemicals. As such, they can survive in harsh environments like different host gastrointestinal fluids and heat processing in bakery products. They also play crucial roles in human health through immune modulation and secreting beneficial metabolites (Khomeiri et al., 2023). Yeast can also serve as probiotics, as discussed in Section 3.1.4.

3.1.4 Yeast as Probiotic

Saccharomyces boulardii (*S. boulardii*) is the most prevalent yeast and exhibits probiotic benefits. Another name for it is *Saccharomyces cerevisiae* var. Over-the-counter preparations of *Saccharomyces cerevisiae* (*S. cerevisiae*) Hansen CBS 5926, also known as boulardii, are commonly advised for the treatment of both chronic conditions like IBD and acute gastrointestinal diseases like viral and bacterial diarrhoea (Kelesidis & Pothoulakis, 2012). *S. boulardii*, a non-pathogenic yeast, effectively manages antibiotic-associated diarrhoea and IBD due to its ability to modulate the immune response and restore gut barrier function (McFarland, 2007; Szajewska et al., 2020). Recent mycobiome studies have found that *Saccharomyces* strains are present in up to 96.8% of samples derived from the gut (Dollive et al., 2012; Nash et al., 2017). Since humans have been intentionally consuming live *S. cerevisiae* and related species for thousands of years in bread, beer, and other fermented foods and beverages, the prevalence of *Saccharomyces* strains in the human GIT is not surprising. Currently, the strain can be found in over-the-counter probiotic supplements all over the world. Its non-pathogenic nature and relative safety for use properties are demonstrated by the fact that *S. boulardii* preparations have been produced, marketed, and used in clinical studies since 1977 (McFarland, 2010). A thorough analysis of *S. boulardii*'s probiotic properties was released in 2012 (Kelesidis & Pothoulakis, 2012). This work aimed to explore the role of *S. boulardii* in treating non-GIT disorders and to provide an update on clinical progress and new developments that have occurred since then.

3.1.4.1 Studies on Animals

The majority of *S. boulardii's* preclinical trials have used rodents as hosts, with variable degrees of success. The GIT is the primary target of these studies; however, some have also investigated the effect of *S. boulardii* on infections that result in hepatic disorders (Duman et al., 2013; Amandine et al., 2014). A proprietary strain called CNCM I-745, which is sold commercially worldwide under the brands Florastor, Reflor, Bioflor, Codex, Ultra-leveur, Floratil, and several others, is used in many studies. Some research indicates that prophylactic administration of *S. boulardii* does not always completely prevent infections caused by different pathogens such as *Salmonella* and *Helicobacter pylori*. On the other hand, other studies highlight the potential of this organism in treating GIT disorders (Hudson et al., 2016). The observed disparity implies that the possible advantages of delivering this probiotic could differ significantly depending on the type of infection or the host. Corthier and colleagues investigated the advantages of administering *S. boulardii* in gnotobiotic mice in 1986. This was one of the earliest investigations to show that *S. boulardii* administration was effective in treating *Clostridium difficile* (*C. difficile*) infections (Corthier et al., 1986).

3.1.4.2 Clinical Trials

The safety and effectiveness of using *S. boulardii* in a range of different organisms have led to numerous clinical trials on humans, with differing degrees of success (Kelesidis & Pothoulakis, 2012). Some of the systemic and GIT-specific effects of *S. boulardii* have been established by prior human clinical trials. For instance, *S. boulardii* can improve the gut microbiota's composition and boost innate and humoral immunity in healthy people (Vanhoutte et al., 2006), providing foundational evidence for the application of probiotics in managing diverse illnesses. Earlier research that focused on prevalent GIT conditions like *S. boulardii* as a component of the treatment regimen for Crohn's disease and *C. difficile* infections gave vital early data on how the probiotic can reduce the recurrence of these diseases (Guslandi et al., 2000).

3.2 THE APPLICATION OF PROBIOTICS

3.2.1 The Use of Probiotics in the Food Industry

There is currently a global surge in the application of probiotics as functional ingredients in food, animal feed, and pharmaceutical products. In the food industry, probiotics are used in dairy, beverage, and baking, with the dairy industry being the largest sector using probiotics in products such as milk, yogurt, cheese, and infant formula. In dairy products, probiotics are used as either starter cultures alone or in combination with traditional starter cultures to carry out probiotic-based fermentation. Additionally, they may also be incorporated into fermented and non-fermented dairy products as functional ingredients (Gao et al., 2021). The most commonly used probiotics are *Lactobacillus* spp. and *Bifidobacterium* spp. as they are "generally recognised as safe" by several regulatory agencies (Min et al., 2019). LAB help increase the shelf life of fermented products and also act as antimicrobial agents against many pathogens. Prezzi et al. (2020) reported that the addition of *Lactobacillus rhamnosus* (*L. rhamnosus*) GG inhibited the growth of *Listeria monocytogenes* (*L. monocytogenes*) in Minas Frescal cheese *with* no negative effect on the physicochemical (i.e. pH, moisture, fat, and protein) and textural (e.g. hardness, elasticity, cohesiveness, and chewiness) properties of the cheese.

Therefore, probiotics-based dairy products have been recommended as safe and healthy due to their beneficial effects such as aiding mineral absorption in the body, being efficient against *H. pylori* infection, and preventing diarrhoea and constipation (Gao et al., 2021). Non-dairy food products such as cereals, soy, meat, fruits, and vegetables supplemented with probiotics are an emerging group of functional foods (Min et al., 2019). In the beverage industry, the use of fermented or unfermented fruit juices, fermented vegetables, and minimally processed fruits as vehicles to deliver probiotic bacteria has been extensively studied. Mantzourani et al. (2019) showed that the application of probiotic *Lactobacillus plantarum* (*L. plantarum*) ATCC 14917 in the fermentation of pomegranate juice leads to the production of a functional fruit beverage. This beverage had a low alcohol concentration, higher levels of total phenolic content, and antioxidant activity compared to non-fermented pomegranate juice (Mantzourani et al., 2019). Similarly, Giri et al. (2018) found that probiotic fermentation of coconut water using *Lactobacillus casei* (*L. casei*) L4 may facilitate the development of an improved functional beverage with probiotic benefits. The beverage contained essential minerals, vitamin B12, and antioxidants and displayed antimicrobial activity against foodborne pathogens such as *Bacillus cereus* (*B. cereus*), *L. monocytogenes*, *S. aureus*, and *Salmonella* Typhi (*S.* Typhi) (Giri et al., 2018). In another study, a beverage prepared from whey, germinated millet flour, and barley extract treated with *Lactobacillus acidophilus* (*L. acidophilus*) NCDC 13 was found to be effective in controlling the pathogenicity induced by *Shigella* in mice models (Ganguly et al., 2019).

In food safety, probiotics are used to prevent biofilm formation as well as to remediate heavy metal contamination in food processing industries. Biofilms are complex aggregates of bacteria established in a three-dimensional structure and embedded in a matrix synthesised by the microbial community (Azeredo et al., 2017). They pose a significant challenge in many food processing industries, including dairy processing, seafood processing, meat processing, food brewing, and fresh produce, as they serve as reservoirs for bacterial contamination of food. In addition, biofilms act as a protective shield against stressful conditions, enabling bacteria to tolerate chemical sanitisers and disinfectants (Srey et al., 2013; Olanbiwoninu & Popoola, 2023; Duze et al., 2021). Probiotics are a promising alternative strategy for controlling biofilm formation, thereby preventing antimicrobial resistance associated with foodborne microorganisms (Tazehabadi et al., 2021;

Tomé et al., 2023). They inhibit the growth of microorganisms and biofilm formation by competing for nutrients, adhesion sites, or the release of antimicrobial metabolites such as bacteriocins, biosurfactants, organic acids, hydrogen peroxide, and inhibitory exopolysaccharides (Tomé et al., 2023). Tazehabadi et al. (2021) showed that *Bacillus subtilis* (*B. subtilis*) KATMIRA1933 and *Bacillus amyloliquefaciens* B-1895 can inhibit biofilm formation of several *Salmonella* serovars, pointing to the high potential of using these strains as effective antimicrobial producers and prophylactic agents against *Salmonella*. Similarly, in a study by Cisneros et al. (2021), *Lactiplantibacillus plantarum* (*L. plantarum*) CRL 1075 inhibited the growth and surface colonisation of Enterohaemorrhagic *E. coli* (EHEC) O157:H7, thus a good candidate for application in food processing environments to mitigate undesirable surface contaminations by pathogenic *E. coli* strains.

Heavy metal contamination is another growing concern in food processing industries as it facilitates the spread of antimicrobial-resistant genes (ARGs) and can induce antibiotic resistance in microorganisms by selecting the resistance determinants for both antibiotic and metal resistance through mechanisms of co-resistance, cross-resistance and co-regulation (Vats et al., 2022; Anedda et al., 2023). As a food-safe class of probiotics, LAB can be used as a biosorbent for the removal of heavy metals in soil, wastewater, and most importantly in food (Ma, 2024). LAB remove heavy metals from contaminated matrices through adsorption and bioaccumulation. They can also utilise metabolites such as lactic and acetic acids to increase the bioavailability and migration of heavy metals, thus making metal ions more easily captured (Ma, 2024).

3.2.2 The Application of Probiotics in Agriculture

In the agricultural sector, production, well-being, and welfare of the livestock are the ultimate concerns for a profitable business and have led to the increased use of antibiotics as growth promoters and antimicrobials to reduce bacterial infections. However, with the excessive use of antibiotics, bacterial resistance has emerged as a global threat (Xu et al., 2022). Probiotics are amongst the most promising approaches developed to replace the use of antibiotics as growth promoters in agriculture (Rahman et al., 2022). The use of probiotics in animal feed improves gut microbiota composition, immune response, nutrient digestibility and absorption, animal growth, and meat quality (Al-Shawi et al., 2020; Mirzaei et al., 2022; Nalla et al., 2022).

Similarly, probiotics in feed have a significant effect on the absorption and utilisation of feed, daily increase of body weight, and the total body weight of various animals, including turkeys, chickens, piglets, sheep, goats, cattle, and horses (Markowiak & Ślizewska, 2018). Studies on the use of probiotics in animal feed have reported improved yield and quality of milk, meat, and eggs (Al-Shawi et al., 2020; Saha et al., 2023; Yuniarti et al., 2023; El Jeni et al., 2024). Animal production also influences the nutrition and health of the consumer and animal intestinal pathogens such as *Campylobacter*, *Salmonella*, *Listeria*, and *Yersinia*. These serve as a direct source of food contamination (Markowiak & Ślizewska, 2018). The use of probiotics in animal feed increases the population of beneficial microorganisms such as *Lactobacillus* and *Bifidobacterium* species. This is achieved through the production of lactic acid and SCFAs and the reduction of pH. Ultimately, this inhibits the growth of harmful microorganisms such as *Campylobacter jejuni* (*C. jejuni*) or *Salmonella enteritidis* through the release of inhibiting substances like organic acids or bacteriocins (Rahman et al., 2022). Deng et al. (2020) reported that the addition of probiotics to animal feeds is effective in reducing and eliminating the colonisation of *Campylobacter* in the GIT of poultry. Probiotic use in livestock agriculture has been demonstrated to provide increased resistance against *Salmonella* spp. infections in chickens and turkeys through the accelerated establishment of normal and healthy microbiota. Moreover, decreased shedding of the *E. coli* O157:H7 strain was noted in cattle and sheep following the application of probiotics (Imperial & Ibana, 2016). The antimicrobial activity of probiotics against enteric pathogens further suggests their utility as effective biotherapeutics to prevent related GIT infections (Nalla et al., 2022). Probiotics can also be administered to animals that have been treated therapeutically with antibiotics or other medications to allow recolonisation or reinforcement of the gut flora that may have been weakened or depopulated during therapy (Rahman et al., 2022).

The application of probiotics in the aquaculture industry for the welfare of aquatic animals as well as the conservation of the aquatic environment is another emerging field (Bhakta et al., 2012). In aquaculture, probiotics have been shown to enhance growth and reproduction, provide protection against pathogens, have positive effects on immunity, optimise digestion, and increase water quality, which boosts fish production and safeguards human health (Mirzaei et al., 2022; El-Saadony et al., 2021). Research on the use of probiotics for the removal of heavy metals from aquatic environments has also gained interest as heavy metal contamination in the aquatic environment not only threatens aquatic and human health but also induces the development of resistance in microbial communities in these environments (Bhakta et al., 2012).

3.2.3 The Utility of Probiotics in the Health Sector

Probiotics are known to improve metabolic parameters, modulate the composition of the intestinal microbiota, and regulate the immune system making them vital in the health sector (Figures 3.3 and 3.4). They exhibit antioxidant, anti-inflammatory, and antimicrobial properties which can inhibit the growth of other pathogenic bacteria through the production of antimicrobial molecules, such as SCFAs, bacteriocins, and microcines (Doron & Gorbach, 2006). Several studies have shown the role of probiotics in gut health and their potential in the treatment of acute gastroenteritis, *H. pylori* infection, IBD, nosocomial diarrhoea, and bacterial vaginosis (BV), amongst other medical

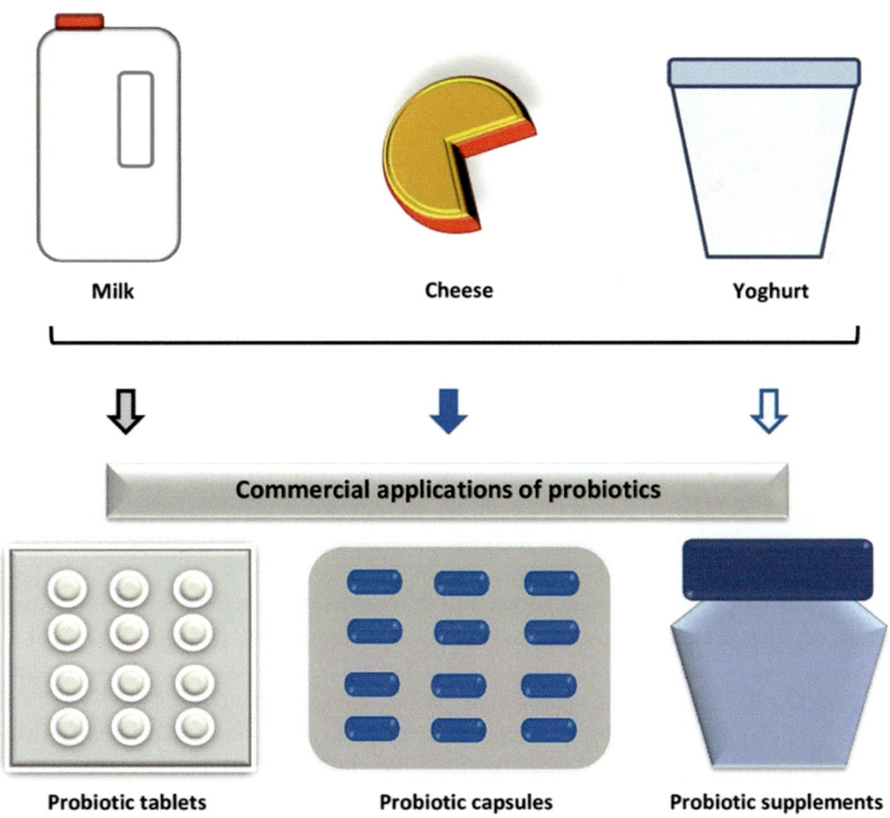

FIGURE 3.3 The source and commercial applications of probiotics.

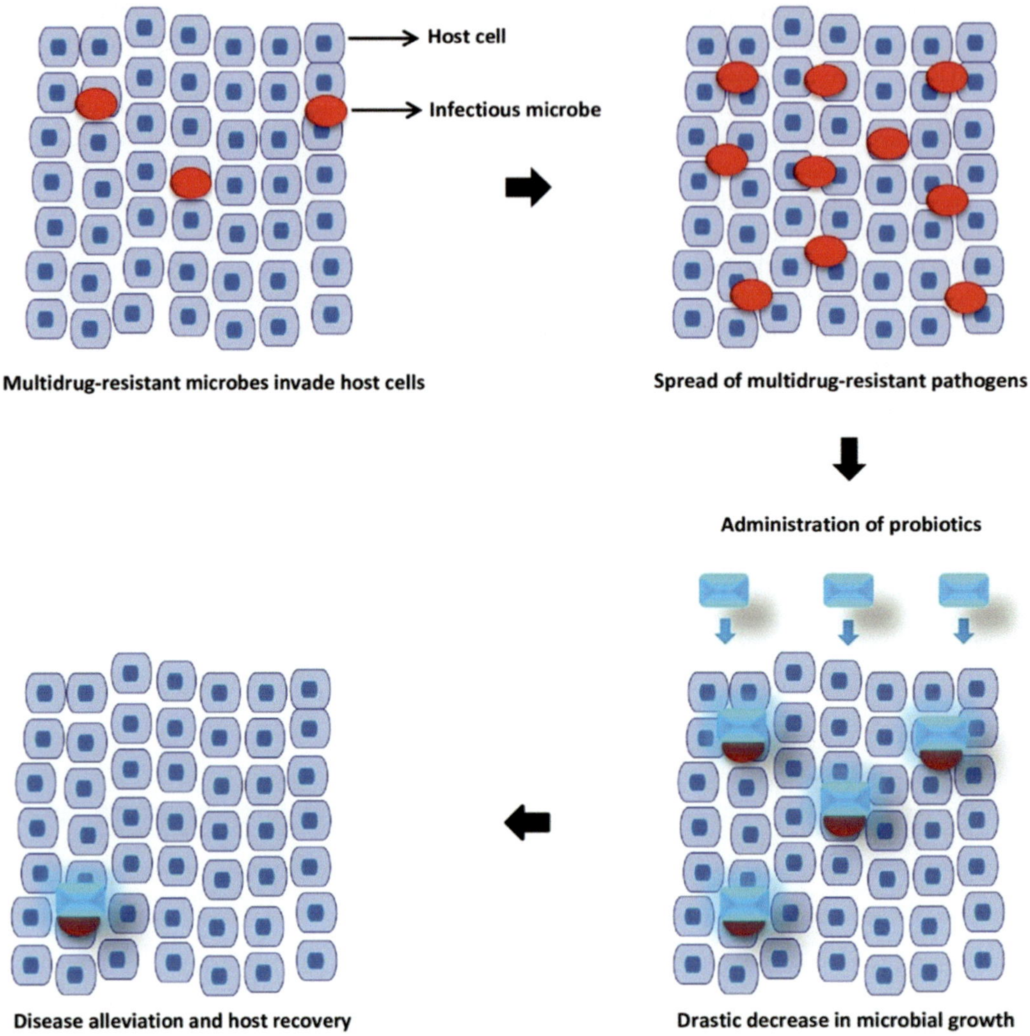

FIGURE 3.4 Mechanism of action of probiotics against multidrug-resistant pathogens.

conditions (Maftei et al., 2024; Mitra et al., 2023). Huang et al. (2021) alluded that probiotics can shorten the duration of diarrhoea, increase treatment efficacy, and reduce the length of hospital stay. Evidence suggests that probiotics used in conjunction with antibiotic therapy can improve the eradication of *H. pylori* (Penumetcha et al., 2021) and consumption of multispecies probiotics may substantially reduce the gut transit time, increase stool frequency, and improve stool consistency in adults with functional constipation (Zhang et al., 2020).

In hospital settings, probiotics may be used to prevent healthcare-associated infections, such as ventilator-associated pneumonia (VAP), antibiotic-associated diarrhoea (AAD), and *C. difficile*-associated diarrhoea (Manzanares et al., 2016; Sharif et al., 2022; Sun et al., 2022). A large systemic review and meta-analysis study revealed that probiotics significantly reduced the incidence of infectious complications, including new episodes of VAP in critically ill patients (Manzanares et al., 2016). Similarly, Sun et al. (2022) reported that prophylactic administration of probiotics might be utilised as a preventive method for VAP in neonates, children, and adult patients who require mechanical ventilation. Probiotics can also be used to prevent and treat AAD by providing the gut barrier and restoring the gut microflora (Yang et al., 2024). AAD results from dysbiosis of the intestinal flora due to long-term use of antimicrobial drugs (Gao et al., 2021). Probiotics have an immunomodulatory effect that can counteract the proinflammatory effect of pathogenic bacteria. They also offer a promise to patients in the intensive care unit (ICU) for the prevention of AAD, *C. difficile* infections, multiple organ dysfunction syndrome, and VAP (Morrow et al., 2012). Yang et al. (2023) reported that probiotics alone or in combination with conventional Western medical treatment could effectively prevent the incidence of AAD, *C. difficile*-associated diarrhoea in children. They shorten the duration of diarrhoea and the average hospitalisation time.

Lukasik et al. (2022) showed that administering a multi-strain probiotic product containing eight strains, including *Bifidobacterium bifidum* (*B. bifidum*) W23 and *L. acidophilus* W37, during antibiotic therapy reduced AAD risk in children and adolescents aged 3 months to 18 years. The European Society for Paediatric Gastroenterology, Hepatology and Nutrition (ESPGHAN) recommends high doses (≥5 billion CFU/day) of *L. rhamnosus* or *S. boulardii* started simultaneously with antibiotic treatment to prevent AAD (Szajewska et al., 2023). Probiotic intervention is widely practiced in BV to restructure the vaginal microbiome as they confer anti-adhesion properties via competition for the binding site. This displaces the other pathogens from the adhesion site on the vaginal epithelium (Joseph et al., 2021). The mechanistic action of probiotics against BV pathogens is multi-directional and includes antibacterial, anti-biofilm, anti-colonisation or anti-adhesion, co-aggregation, and host immunomodulation (Basavaprabhu et al., 2020). Clinical trials have investigated the probiotic potential against BV administered as a single strain or multi-strain through the oral route or intravaginally. Reid et al. (2023) explored the benefits of *L. rhamnosus* GR-1 and *Limosilactobacillus reuteri* RC14 as probiotics in vaginal health and demonstrated that this probiotic mixture was not only safe for daily use in healthy women but also prevented vaginal colonisation by potentially pathogenic bacteria and yeast. Similarly, Hantoushzadeh et al. (2012) reported a significant reduction in vaginal pH in pregnant women after receiving oral probiotic yoghurt (100 g, twice daily for 1 week) compared to orally administered clindamycin (300 mg, twice a daily for 1 week). In this chapter, the authors concluded that probiotic and antibiotic treatments were equally effective (Hantoushzadeh et al., 2012). Probiotics can also increase the colonisation rate, prevent biofilm formation, and restore the normal flora in the vagina. McMillan et al. (2011) demonstrated that *Lactobacillus* spp. can infiltrate BV biofilms and cause cell death by producing acid, bacteriocins, or biosurfactant-like substances with anti-biofilm activities.

The phenotypic surface properties of *Lactobacillus* spp. such as aggregation, adhesion, and biofilm formation as well as their ability to produce lactic acid and hydrogen peroxide makes them ideal protective strains (Sousa et al., 2023). An increasing amount of research supports the role of probiotics in promoting human health. Moreover, probiotics can also help slow down the spread of antibiotic-resistant bacteria. With increasing antibacterial resistance rates and fewer new antibiotics in the research pipeline, the focus has been shifted to non-antibiotic approaches such as the use of probiotics for the prevention and treatment of nosocomial infections. As such, probiotics have been widely employed to suppress the virulence and spread of several multidrug-resistant pathogens. The diversity of probiotics has enabled industries to obtain and process them from various sources such as milk, cheese, yoghurt, and other dairy products. Fruits such as almonds, bananas, custard apples, watermelon, and grapefruit are rich sources of probiotics. In addition, vegetables like cucumbers, carrots, green beans, radishes, and pickled bell peppers serve as an essential probiotic reservoir. A number of methods have been developed to extract, test, and analyse these probiotics. The source and application of probiotics are portrayed in Figure 3.3, while their mechanism of action against multidrug-resistant microbes is depicted in Figure 3.4.

Probiotics can be sourced from fermented foods and dairy products such as milk, yoghurt, and cheese. These foods are a rich source of acetic acid and LAB, which are essential for regulating the host microbiota and corresponding microbiome. In host cells, these bacteria act as probiotic agents by attacking, disrupting, and suppressing the proliferation of wild-type, drug-resistant, and multidrug-resistant pathogens. The bacterial cultures are also used commercially for producing probiotic tablets, capsules, and supplements for providing essential immune, nutritional, and physiological benefits.

The susceptible host cells are infected with multidrug-resistant pathogens, leading to exponential microbial spread. The application of probiotics impairs the proliferation of multidrug-resistant microorganisms. A dramatic decline in microbial growth is accomplished after the administration of probiotics, resulting in disease clearance and host recovery.

3.3 ADMINISTRATION OF PROBIOTICS TO COUNTER ANTIMICROBIAL RESISTANCE

3.3.1 Application of Probiotics to Inhibit Drug-Resistant *Pseudomonas aeruginosa*

In recent years, genetic mutations leading to the proliferation of multidrug-resistant bacterial pathogens, coupled with extensive biofilm formation among pathogenic microorganisms, have become major public health problems. Due to the general failure of conventional antibacterial agents to treat diseases caused by multidrug-resistant strains, combination therapy has been applied to disrupt microbial replication and alleviate disease symptoms. Unfortunately, the long-term application of this strategy often results in the development of unintended side effects. For example, although the administration of combination therapy markedly reduces microbial load in patients infected with tuberculosis (TB) and HIV, adverse effects such as hepatotoxicity, stomach cramps, hypersensitivity, heart complications, and kidney failure usually accompany their long-term use. Accordingly, alternative therapeutic approaches such as the use of probiotics have been applied to treat diseases caused by drug-resistant microbial strains. Importantly, some of these probiotics also have a protective effect as they prevent the further emergence of drug-resistant pathogens.

Díaz et al. (2020) examined the use of chloroform extracts recovered from culture supernatants of *L. casei* CRL 431 and *L. acidophilus* CRL730. These *Lactobacillus* probiotic species were employed to counter the replication of three drug-resistant *Pseudomonas aeruginosa* (*P. aeruginosa*) strains by specifically regulating the quorum sensing mechanism. The application of these probiotics led to a considerable decline in markers of microbial replication such as elastase enzyme production (63%), peptide synthesis (77%), the rate of biofilm metabolism (39%), and the formation of biofilm biomass (50%) (Díaz et al., 2020). In addition, a significant decrease in biofilm formation of up to 40% was observed in all three studied *P. aeruginosa* strains (Díaz et al., 2020). The analysis of chloroform extracts by gas chromatography/mass spectrometry (GC–MS) indicated that the primary compounds responsible for antibacterial were diketopiperazines. Results from this study highlighted that the metabolites of *L. casei* and *L. acidophilus* may be employed as probiotics to disrupt the growth of different drug-resistant *P. aeruginosa* strains (Díaz et al., 2020). The spread of multidrug-resistant bacterial pathogens due to drug misuse impairs global public health systems.

It has previously been demonstrated that fermented foods are a rich source of probiotics that are essential and beneficial for the human immune system. The traditional Korean fermented food called Kimchi was tested to regulate the proliferation of multidrug-resistant *P. aeruginosa* (Yi & Kim, 2023). In particular, the cell-free extracts from LAB found in Kimchi were examined for antimicrobial activity. Compared to untreated controls, the administration of cell-free extracts derived from LAB present in Kimchi led to substantial growth inhibition of *P. aeruginosa* (Yi & Kim, 2023). Moreover, a marked decline in biofilm formation was noted after applying the cell-free extracts to different *P. aeruginosa* cultures. DNA sequencing revealed that the predominant lactic acid bacterium present in Kimchi was *Pediococcus inopinatus* (*P. inopinatus*). The use of ultra-high performance liquid chromatography (HPLC) with quadrupole time-of-flight mass spectrometry (UPLC–QTOF–MS) identified curacin A and pediocin A as substances that confer anti-*P. aeruginosa* activity (Yi & Kim, 2023). This investigation demonstrated that *P. inopinatus* isolated from Kimchi may be utilised to impair the growth and biofilm-forming capability of multidrug-resistant *P. aeruginosa*. These data support the potential use of *P. inopinatus* to regulate diseases induced by multidrug-resistant *P. aeruginosa* pathogens (Yi & Kim, 2023). In an independent investigation, 100 carbapenem-resistant *P. aeruginosa* isolates were obtained from patients located at three separate hospitals in the Isfahan province in Iran (Ghiaei et al., 2023). Of these isolates, 61% displayed multidrug resistance and were tested against 20 probiotic bacteria extracted from local dairy products.

Phenotypic screening for the presence of metallo-β-lactamase enzymes responsible for the hydrolysis of β-lactam drugs including carbapenems indicated that 74.5% of the *P. aeruginosa* isolates were positive (Ghiaei et al., 2023). Administration of the two most effective *Lactobacillus* probiotic bacteria triggered exceptional suppression of *P. aeruginosa* proliferation, which was documented after performing the minimum inhibitory concentration (MIC), minimum bactericidal concentration (MBC), combined disc test, and biofilm assays. HPLC revealed that both acetic acid and lactic acid present in the *Lactobacillus* probiotic isolated from local dairy products induced antimicrobial efficacy against the multidrug-resistant *P. aeruginosa* clinical isolates (Ghiaei et al., 2023). These encouraging results showed the utility of the two *Lactobacillus* probiotics in exerting antimicrobial properties against all the tested carbapenem-resistant clinical isolates. This also reinforces their potential application as biocontrol agents against carbapenem-resistant *P. aeruginosa* isolates (Ghiaei et al., 2023).

3.3.2 Administration of Probiotics to Suppress Methicillin-Resistant *S. aureus*

The use of therapeutics that target the quorum sensing system to modulate the pathogenicity of methicillin-resistant *S. aureus* (MRSA) may serve as an effective approach to mitigate the deficiencies associated with monotherapy or combination therapy. Quorum sensing plays a crucial role in modulating toxin production, the production of antioxidants, the evasion of host immune surveillance, and drug resistance. For these reasons, quorum sensing is an attractive therapeutic mechanism to suppress microbial proliferation and reduce drug resistance. The advantage of utilising bioactive peptides extracted from LAB was evaluated against two MRSA clinical isolates. The introduction of bioactive peptides caused an exceptional decrease in markers of bacterial virulence such as biofilm formation, biofilm structural integrity, the rate of oxidative killing, and carotenoid production (Cella et al., 2023). The fractional-inhibitory concentration tests containing semi-purified and bulk bioactive peptides displayed synergy with β-lactam antibiotics and markedly enhanced cefoxitin-mediated killing of clinical MRSA isolates. This was accompanied by a marked decline in staphyloxanthin production and a sharp increase in oxidative stress-mediated killing (Cella et al., 2023).

Reverse transcription quantitative polymerase chain reaction (RT-qPCR) assays performed using the two clinical strains confirmed differential expression in the *agrA* and *luxS* genes involved in quorum sensing and modulated the *hla* gene which plays an important role in the host immune response. In addition, scanning electron microscopy and light microscopy showed attenuated biofilm-forming capacity and clustering behaviour in the MRSA clinical isolates treated with bioactive peptides (Cella et al., 2023). Data documented in this study demonstrated that co-administration of bioactive peptides and β-lactam drugs may be used as an effective strategy to attenuate the virulence, growth, and persistence of MRSA clinical isolates. The results also revealed that the use of LAB as probiotics to target important genes involved in quorum sensing is a feasible approach to mitigating the virulence of MRSA (Cella et al., 2023).

A group of antibacterial metabolites extracted from *Bacillus licheniformis* (*B. licheniformis*) have been used as probiotics to counter the spread of *Acinetobacter baumannii* (*A. baumannii*), MRSA, and other pathogenic Gram-negative and Gram-positive bacteria (Ham et al., 2023). The novel active metabolite called bacillimidazole G produced low MIC values ranging from 0.7 to 2.6 μg/mL upon testing against various drug-resistant bacterial pathogens of clinical importance. Impressively, bacillimidazole G also had synergistic properties following co-application with the drug colistin. This subsequently triggered a dramatic decrease in the overall MIC and exerted robust antibacterial activity against MRSA, *A. baumannii*, and *E. coli* MG1655 (Ham et al., 2023). These useful features underscore the essential role of probiotics in repressing microbial pathogenicity and virulence of wild-type, drug-resistant, and multidrug-resistant bacterial strains (Ham et al., 2023). The antimicrobial efficacy of bacteriocin secreted by LAB has been unravelled in biofilms formed by MRSA (Ibraheim et al., 2023). Bacteriocin was isolated from 40 white soft cheeses collected from sheep and 40 similar cheeses derived from cows. The treatment of bacterial cultures with the probiotic bacteriocin robustly repressed MRSA replication, spread, and biofilm formation (Ibraheim et al., 2023). The DNA sequencing data indicated that 2 of the 36 MRSA isolates possessed the *mecA* gene, which confers resistance to the methicillin antibiotics. This investigation highlighted that the bacteriocin metabolite secreted by LAB can disrupt the proliferation and persistence of various MRSA isolates. Thus, this probiotic may potentially be employed to regulate the spread of MRSA and other pathogenic microbes (Ibraheim et al., 2023).

3.3.3 Harnessing Probiotics to Suppress Drug-Resistant *E. coli*

Previously, a mixture of microcin C7-producing *E. coli* probiotic strain and bacteriophages was used to inhibit colonisation of the gut by the drug-resistant *E. coli* ST131-H30R strain (Porter et al., 2022). The treatment of the *E. coli* ST131-H30R strain with bacteriophages remarkably reduced subcutaneous sepsis in cell culture and mouse models. However, this notable antibacterial efficacy did not culminate in a marked suppression of gut colonisation (Porter et al., 2022). Similarly, the introduction of *E. coli* probiotic strain induced a sharp decline in indicators of bacterial virulence *in vitro* and *in vivo* but was ineffective in disrupting gut colonisation. Conversely, the simultaneous delivery of the *E. coli* probiotic strain and bacteriophages in cultured cells and mice exerted a pronounced decrease in indicators of bacterial virulence such as bacterial load, subcutaneous sepsis, and gut colonisation (Porter et al., 2022). This outcome disclosed the benefit of co-administration of the *E. coli* probiotic strain and bacteriophages to mitigate bacterial growth, subcutaneous sepsis, and gut colonisation in patients infected with the drug-resistant *E. coli* ST131-H30R pathogenic strain (Porter et al., 2022). In the past, the probiotic *Lactobacillus paracasei* (*L. paracasei*) was proven to decrease cholesterol levels *in vitro*. Subsequently, this probiotic was probed for its antimicrobial properties in cultured mammalian cells and Wistar rats (Shahverdi et al., 2023). Relative to negative controls, a noticeable decrease in the replication of drug-resistant *E. coli* and *S. aureus* was observed following treatment with *L. paracasei*. This remarkable antimicrobial efficacy was observed *in vitro* and *in vivo* (Shahverdi et al., 2023). Notably, the antibacterial activity observed in cultured cells and rats was achieved without a marked elevation in markers of cytotoxicity such as aspartate transaminase, alanine transaminase, and alkaline phosphatase. Overall, the probiotic *L. paracasei* has useful properties and was able to decrease the amount of cholesterol and suppress the replication of drug-resistant *E. coli* and *S. aureus* pathogenic strains (Shahverdi et al., 2023).

The cell-free supernatant of *Lactobacillus* species was surveyed against ampicillin and tetracycline-resistant *E. coli* isolates collected from fresh fish fillets (Fernandes et al., 2023). In particular, these multidrug-resistant isolates were identified as *E. coli* P12, P25, P35, and P36 strains. The outcome of this interrogation indicated that the application of the *Lactobacillus* species drastically reduced markers of bacterial virulence in both planktonic and biofilm *E. coli* cells (Fernandes et al., 2023). Additionally, scanning electron microscopy indicated a vast variation in cell growth, biofilm formation capacity, surface adhesion, surface topography, and surface compatibility between untreated and bacterial cultures treated with the *Lactobacillus* species (Fernandes et al., 2023). The versatility of this probiotic agent in effecting potent antibacterial activity in both biofilm and planktonic multidrug-resistant *E. coli* pathogenic cells may be an ideal therapeutic mechanism for impairing the proliferation and pathogenicity of mutant microbes of clinical significance (Fernandes et al., 2023).

3.3.4 The Use of Probiotics to Counter Drug-Resistant *Enterobacterales*

A 2021 study investigated the impact of *Lactobacillus* species on the replication and survival of *Enterobacterales* that can produce carbapenemase enzyme, known for hydrolyzing carbapenem antibiotics (Chen et al., 2021). For this reason, 18 *Enterobacterales* clinical isolates and 5 *Lactobacillus* species were utilised for microbiological experimentations. Potent anti-*Enterobacterales* efficacy was reported following treatment with the 5 *Lactobacillus* species (Chen et al., 2021). Moreover, this antimicrobial efficacy was not hindered by treatment of *Enterobacterales* with proteinase, catalase, or heating. Interestingly, the proliferation of carbapenem-resistant *K. pneumoniae* CPE0011 was completely abrogated at a concentration of 50% LUC0180 cell-free supernatant. Lactic acid and mix acid were responsible for the observed microbial inhibition, while acetic acid triggered 67.8% antibacterial activity (Chen et al., 2021). Conversely, both citric acid and succinic acid cannot reduce the growth of carbapenemase-producing *Enterobacterales* (CPE). At a concentration of 25%, extensive bacterial inhibition was achieved by lactic acid and mix acid. However, acetic acid, citric acid, and succinic acid displayed insignificant inhibitory effects (Chen et al., 2021). This important observation highlighted that *Lactobacillus* species display potent activity against CPE and that lactic acid is the primary antimicrobial agent (Chen et al., 2021). The GIT is a valuable source of novel antimicrobial agents. Microcins are bacteriocins produced by bacteria belonging to the *Enterobacterales* group. Microcins have a wide variety of structural compositions and modes of action. This versatility equips them to exert potent antibacterial effects against closely related bacterial species. The antibacterial effect of four different microcins was interrogated against 54 *Enterobacterales* isolates (Telhig et al., 2022). These isolates represented three multidrug-resistant species namely *E. coli*, *K. pneumoniae*, and *Salmonella enterica* (*S. enterica*). The tested microcins were microcin C, microcin J25, microcin B17, and microcin E492. The assessed microcins possessed distinct translational modifications and displayed different modes of action (Telhig et al., 2022). Microcin C achieved the broadest antimicrobial spectrum with 46 inhibited strains, followed by microcin E492 with 38 disrupted strains, while microcin J25 exerted the highest antibacterial activity. Intriguingly, microcin efficacy was reported to be independent of the antibiotic resistance profile and strain genus (Telhig et al., 2022). Of the 42 drug combinations tested, 20 displayed enhanced antibacterial activity, and 18 of 20 were microcin–antibiotic combinations with two acting synergistically. Due to their wide range of structural compositions and modes of action, microcins have been proven to possess impressive antibacterial potency against multidrug-resistant *Enterobacterales*. Furthermore, they are adaptable to combination with antibiotics, especially colistin to robustly reduce the persistence of multidrug-resistant mutant strains belonging to the *Enterobacterales* family and highlight their importance as probiotic agents (Telhig et al., 2022).

Clinical manifestations caused by multidrug-resistant *Enterobacterales* are devastating to patients. The commensal microbiota in the gut offers protection against invading *Enterobacterales*. Unfortunately, antibiotics disrupt the normal flora of the gut. This allows colonisation and infection by members of the *Enterobacterales*. A prospective multiomic survey in mice and hospitalised patients indicated that *Lactobacillus* species restricts the entry and colonisation by *Enterobacterales* (Djukovic et al., 2022). The mechanism of action is that *L. rhamnosus* and *Lactobacillus murinus* (*L. murinus*) enhance the proliferation of *Clostridiales* bacteria in the gut. This successively creates a harsh environment for *Enterobacterales* via enhanced butyrate production and nutrient depletion (Djukovic et al., 2022). It was documented that this strategy was conserved in mice and hospitalised patients and was critical for repelling invasion and colonisation by *Enterobacterales*. The data reveals the necessity of employing *Lactobacillus* bacteria as probiotics to maintain the gut microbiota and provision of a hostile environment for multidrug-resistant *Enterobacterales* in mice and hospitalised individuals (Djukovic et al., 2022).

3.4 CONCLUSION AND PERSPECTIVES

A large number of microorganisms can apply selection pressure against conventional drugs. This allows the infectious microbes to adapt, survive, and persist in the presence of conventional therapies, thus rendering the drugs ineffective. Alarmingly, appreciable increases in antimicrobial drug resistance cause the emergence of diseases that do not respond to recommended drugs. This, in turn, harms infected individuals as it prolongs patient admission and therapeutic application. In some cases, it necessitates the induction of patients into combination therapy. Although effective, combination therapy is frequently accompanied by the manifestation of unintended medical complications. These complications may frustrate the patients from complying with the treatment schedule. Consequently, other innovative strategies are urgently required to overcome the limitations associated with long-term monotherapy and combination therapy. Therefore, probiotics have been applied to repress the replication of wild-type, drug-resistant, and multidrug-resistant microorganisms. Probiotics can be found in certain fruits, vegetables, fermented foods, and dairy products. Notably, some probiotics also form an integral part of the human microbiota. The use of probiotics to regulate the spread of multidrug-resistant pathogens is advantageous due to their beneficial characteristics such as adaptability, scalability, diversity, and versatility. These useful qualities, coupled with the numerous cheaper production methods, bolster the application of probiotics to disrupt the dissemination of multidrug-resistant pathogenic strains. The culture supernatants containing *L. casei* and *L. acidophilus* have been applied to suppress the growth of various multidrug-resistant *P. aeruginosa* mutant strains. This was achieved by targeting the bacterial quorum sensing mechanism, resulting in a significant reduction in the rate of biofilm formation. Recently, *P. inopinatus* extracted from the fermented food Kimchi has also been demonstrated to markedly repress the persistence of the biofilm-forming multidrug-resistant *P. aeruginosa* pathogens. Reassuringly, the use of *Lactobacillus* probiotic isolates obtained from dairy products induced an astronomical decrease in the replication of carbapenem-resistant *P. aeruginosa*.

In other studies, probiotics proved effective against MRSA strains. In MRSA clinical isolates, bioactive peptides extracted from LAB downregulated the expression of *agrA* and *luxS* genes involved in quorum sensing. Additionally, these probiotics also repressed the transcription level of the *hla* gene, which plays an important role in the host's immune response. This culminated in impaired biofilm formation, attenuated bacterial pathogenicity, and virulence. The probiotic bacillimidazole G isolated from *B. licheniformis* caused appreciable attenuation in the markers of bacterial virulence in wild-type, drug-resistant and multidrug-resistant MRSA, *A. baumannii*, and *E. coli* strains. The administration of the cheese-derived probiotic bacteriocin caused a robust repression in the rate of MRSA growth, dissemination, and biofilm formation. In patients infected with the drug-resistant *E. coli* ST131-H30R, co-application of the *E. coli* microcin C7 probiotic strain and bacteriophages triggered an astronomical decline in markers of microbial virulence such as colonisation, microbial load, and the development of subcutaneous sepsis in cell culture and mouse models. The treatment of cultured cells and Wistar rats with *L. paracasei* remarkably suppressed the replication and persistence of drug-resistant *E. coli* and *S. aureus*. Impressively, the cell-free supernatant of *Lactobacillus* species exerted a potent antimicrobial efficacy against planktonic and biofilm cultures of multidrug-resistant *E. coli*, resulting in impaired growth, biofilm formation, surface topography, and surface compatibility. Encouragingly, the lactic acid and mix acid secreted by *Lactobacillus* species displayed potent antimicrobial efficacy against CPE, particularly *K. pneumoniae*. Interestingly, the application of microcin C, microcin J25, microcin B17, and microcin E492 produced by *Enterobacterales* robustly suppressed indicators of microbial virulence in multidrug-resistant *E. coli*, *K. pneumoniae*, and *S. enterica*.

Remarkably, the treatment of mice and hospitalised patients with *Lactobacillus* species restricted entry and colonisation by multidrug-resistant *Enterobacterales*. The *Lactobacillus* species achieved this by creating an adverse environment for *Enterobacterales* after triggering the production of butyrate and depleting essential nutrients. This mechanism was documented to significantly restrict invasion, colonisation, and infection by *Enterobacterales* in mice and hospitalised patients. Further testing and advanced analysis are urgently needed to verify the dosage, efficacy, durability, and the cytotoxicity of probiotics in various cell cultures and animal models. This practice will allow the determination of precise dose regulation, sustainability, and the minimisation of toxic effects. Ultimately, vigorous *in vitro* and *in vivo* testing may potentially lead to clinical trials in patients infected with multidrug-resistant microorganisms. Satisfactory results in cell culture, animal models, and human clinical trials will allow rapid, effective, and timely treatment of patients afflicted with multidrug-resistant microbes. In conclusion, the administration of various probiotic agents triggered a sharp decline in the proliferation and spread of multidrug-resistant microorganisms without triggering appreciable increases in cytotoxicity. Some probiotic agents displayed versatility and demonstrated synergy with conventional drugs currently used for antimicrobial therapy.

ACKNOWLEDGEMENTS

The authors acknowledge research funding provided by the National Research Foundation (NRF) of South Africa Thuthuka grant (reference TTK210331591452, grant number 138302 for Musa Marimani and reference TTK210318590310, grant number 138279 for Sanelisiwe Thinasonke Duze) and the Carnegie DTA research grant (grant code: RKSMM21 for Musa Marimani and grant code: RKSST23 for Sanelisiwe Thinasonke Duze).

AUTHOR CONTRIBUTIONS

All authors contributed to compiling different sections of the book chapter.

COMPETING INTERESTS

The authors have no competing interests to declare.

REFERENCES

Ahmad N, Joji RM, Shahid M (2022). Evolution and implementation of One Health to control the dissemination of antibiotic-resistant bacteria and resistance genes: A review. *Front Cell Infect Microbiol* **12**: 1065796. doi:10.3389/fcimb.2022.1065796

Al-Fakhrany OM, Elekhnawy E (2024). Next-generation probiotics: The upcoming biotherapeutics. *Mol Biol Rep* **51**: 505. doi:10.1007/s11033-024-09398-5

Al-Shawi SG, Dang DS, Yousif AY, Al-Younis ZK, Najm TA, Matarneh SK (2020). The potential use of probiotics to improve animal health, efficiency, and meat quality: A review. *Agriculture (Switzerland)* **10**(10): 10100452. doi:10.3390/agriculture10100452

Alshehri D, Saadah O, Mosli M, Edris S, Alhindi R, Bahieldin A (2021). Dysbiosis of gut microbiota in inflammatory bowel disease: Current therapies and potential for microbiota-modulating therapeutic approaches. *Bosn J Basic Med Sci* **21**: 270–283. doi:10.17305/bjbms.2020.5016

Amandine E, Sébastien M, Lucie G, M. DN, D. CP (2014). *Saccharomyces boulardii* administration changes gut microbiota and reduces hepatic steatosis, low-grade inflammation, and fat mass in obese and type 2 diabetic db/db mice. *MBio* **5** (3). doi:10.1128/mbio.01011-14

Anedda E, Farrell ML, Morris D, Burgess CM (2023). Evaluating the impact of heavy metals on antimicrobial resistance in the primary food production environment: A scoping review. *Environ Pollut* **320**(1): 121035. doi:10.1016/j.envpol.2023.121035

Arasu MV, Al-Dhabi NA, Ilavenil S, Choi KC, Srigopalram S (2016). *In vitro* importance of probiotic *Lactobacillus plantarum* related to medical field. *Saudi J Biol Sci* **23**: S6–S10. doi:10.1016/j.sjbs.2015.09.022

Azeredo J, Azevedo NF, Briandet R, Cerca N, Coenye T, Costa AR, et al (2017). Critical review on biofilm methods. *Crit. Rev. Microbiol* **43**(3): 313–351. doi:10.1080/1040841X.2016.1208146

Basavaprabhu HN, Sonu KS, Prabha R (2020). Mechanistic insights into the action of probiotics against bacterial vaginosis and its mediated preterm birth: An overview. *Microb Pathog* **141**: 104029. doi:10.1016/j.micpath.2020.104029

Bermudez-Brito M, Borghuis T, Daniel C, Pot B, de Haan BJ, Faas MM, et al (2018). *L. plantarum* WCFS1 enhances Treg frequencies by activating DCs even in absence of sampling of bacteria in the Peyer Patches. *Sci Rep* **8**: 1785. doi:10.1038/s41598-018-20243-1

Bezkorovainy A (2001). Probiotics: Determinants of survival and growth in the gut. *Am J Clin Nutr* **73**: 399S–405S. doi:10.1093/ajcn/73.2.399s

Bhakta JN, Ohnishi K, Munekage Y, Iwasaki K, Wei MQ (2012). Characterization of lactic acid bacteria-based probiotics as potential heavy metal sorbents. *J Appl Microbiol* **112**(6): 93–206. doi:10.1111/j.1365-2672.2012.05284.x

Bull MJ, Plummer NT (2014). Part 1: The human gut microbiome in health and disease. *Integr Med (Encinitas)* **13**: 17–22.

Cella MA, Coulson T, MacEachern S, Badr S, Ahmadi A, Tabatabaei MS, et al (2023). Probiotic disruption of quorum sensing reduces virulence and increases cefoxitin sensitivity in methicillin-resistant *Staphylococcus aureus*. *Sci Rep* **13**: 4373. doi:10.1038/s41598-023-31474-2

Chen C-C, Lai C-C, Huang H-L, Su Y-T, Chiu Y-H, Toh H-S, et al (2021). Antimicrobial ability and mechanism analysis of *Lactobacillus* species against carbapenemase-producing *Enterobacteriaceae*. *J Microbiol Immunol Infect* **54**: 447–456. doi:10.1016/j.jmii.2020.01.005.

Cisneros L, Cattelan N, Villalba MI, Rodriguez C, Serra DO, Yantorno O, et al (2021). Lactic acid bacteria biofilms and their ability to mitigate *Escherichia coli* O157:H7 surface colonization. *Lett Appl Microbiol* **73**(2): 47–256. doi:10.1111/lam.13509

Coelho MC, Malcata FX, Silva CCG (2022). Lactic acid bacteria in raw-milk cheeses: From starter cultures to probiotic functions. *Foods (Basel, Switzerland)* **11**. doi:10.3390/foods11152276

Corthier G, Dubos F, Ducluzeau R (1986). Prevention of *Clostridium difficile* induced mortality in gnotobiotic mice by *Saccharomyces boulardii*. *Can J Microbiol* **32**: 894–896. doi:10.1139/m86-164

Deng W, Dittoe DK, Pavilidis HO, Chaney WE, Yang Y, Ricke SC (2020). Current perspectives and potential of probiotics to limit foodborne *Campylobacter* in poultry. *Front Microbiol* **11**: 583429. doi:10.3389/fmicb.2020.583429

Di Vincenzo F, Del Gaudio A, Petito V, Lopetuso LR, Scaldaferri F (2024). Gut microbiota, intestinal permeability, and systemic inflammation: A narrative review. *Intern Emerg Med* **19**: 275–293. doi:10.1007/s11739-023-03374-w

Díaz MA, González SN, Alberto MR, Arena ME (2020). Human probiotic bacteria attenuate *Pseudomonas aeruginosa* biofilm and virulence by quorum-sensing inhibition. *Biofouling* **36**: 597–609. doi:10.1080/08927014.2020.1783253

Djukovic A, Garzón MJ, Canlet C, Cabral V, Lalaoui R, García-Garcerá M, et al (2022). *Lactobacillus* supports *Clostridiales* to restrict gut colonization by multidrug-resistant *Enterobacteriaceae*. *Nat Commun* **13**: 5617. doi:10.1038/s41467-022-33313-w

Dollive S, Peterfreund GL, Sherrill-Mix S, Bittinger K, Sinha R, Hoffmann C, et al (2012). A tool kit for quantifying eukaryotic rRNA gene sequences from human microbiome samples. *Genome Biol* **13**: R60. doi:10.1186/gb-2012-13-7-r60

Doron S, Gorbach SL (2006). Probiotics: Their role in the treatment and prevention of disease. *Expert Rev. Anti. Ther.* **4**(2): 261–75. doi:10.1586/14787210.4.2.261

Duman DG, Kumral ZNÖ, Ercan F, Deniz M, Can G, Çağlayan Yeğen B (2013). *Saccharomyces boulardii* ameliorates clarithromycin- and methotrexate-induced intestinal and hepatic injury in rats. *Br J Nutr* **110**: 493–499. doi:10.1017/S000711451200517X

Duze ST, Marimani M, Patel M (2021). Tolerance of *Listeria monocytogenes* to biocides used in food processing environments. *Food Microbiol* **97**: 103758. doi:10.1016/j.fm.2021.103758

El Jeni R, Villot C, Koyun OY, Osorio-Doblado A, Baloyi JJ, Lourenco JM, et al (2024). Invited review: "Probiotic" approaches to improving dairy production: Reassessing "magic foo-foo dust." *J Dairy Sci* **107**(4): 1832–1856. doi:10.3168/jds.2023-23831

El-Saadony MT, Alagawany M, Patra AK, Kar I, Tiwari R, Dawood MAO, et al (2021). The functionality of probiotics in aquaculture: An overview. *Fish and Shellfish Immunology* **117**: 36–52. doi:10.1016/j.fsi.2021.07.007

Elsaghir H, Reddivari AKR (2024). *Bacteroides Fragilis*. Treasure Island, FL: StatPearls Publishing.

Elshaghabee FMF, Rokana N, Gulhane RD, Sharma C, Panwar H (2017). *Bacillus* as potential probiotics: Status, concerns, and future perspectives. *Front Microbiol* **8**: 1490. doi:10.3389/fmicb.2017.01490

El-Sohaimy SA, Hussain MA (2023). Functional probiotic foods development: Trends, concepts, and products. *Fermentation* **9**(3). doi:10.3390/fermentation9030249

Engels C, Ruscheweyh H-J, Beerenwinkel N, Lacroix C, Schwab C (2016). The common gut microbe *Eubacterium hallii* also contributes to intestinal propionate formation. *Front Microbiol* **7**: 713. doi:10.3389/fmicb.2016.00713

Fernandes MSM, Rios JC, Vasconcelos BM, Lourenço MLMC, Matos MNC, Cavalcante RMB, et al (2023). Effect of *Lactobacillus* spp. cell-free supernatant against planktonic growth and biofilm formation of foodborne *Escherichia coli* isolates. *Lett Appl Microbiol* **76**: ovac006. doi:10.1093/lambio/ovac006

Ferri M, Ranucci E, Romagnoli P, Giaccone V (2017). Antimicrobial resistance: A global emerging threat to public health systems. *Crit Rev Food Sci Nutr* **57**: 2857–2876. doi:10.1080/10408398.2015.1077192

Freeland G, Hettiarachchy N, Atungulu GG, Apple J, Mukherjee S (2023). Strategies to combat antimicrobial resistance from farm to table. *Food Rev Int* **39**: 27–40. doi:10.1080/87559129.2021.1893744

Ganguly S, Sabikhi L, Singh AK (2019). Effect of whey-pearl millet-barley based probiotic beverage on Shigella-induced pathogenicity in murine model. *J Funct Foods* **54**: 498–505. doi:10.1016/j.jff.2019.01.049

Gao J, Li X, Zhang G, Sadiq FA, Simal-Gandara J, Xiao J, et al (2021). Probiotics in the dairy industry—Advances and opportunities. *Compr Rev Food Sci Food Saf* **20**: 3937–3982. doi:10.1111/1541-4337.12755

Ghiaei A, Ghasemi SM, Shokri D (2023). Investigating the antagonistic effect of indigenous probiotics on carbapenem-resistant *Pseudomonas aeruginosa* strains. *Biomed Res Int* **2023**: 6645657. doi:10.1155/2023/6645657

Giri SS, Sukumaran V, Sen SS, Park SC (2018). Use of a potential probiotic, *Lactobacillus casei* L4, in the preparation of fermented coconut water beverage. *Front Microbiol* **9**: 1976. doi:10.3389/fmicb.2018.01976

Gupta V, Garg R (2009). Probiotics. *Indian J Med Microbiol* **27**: 202–209. doi:10.4103/0255-0857.53201

Guslandi M, Mezzi G, Sorghi M, Testoni PA (2000). *Saccharomyces boulardii* in maintenance treatment of Crohn's disease. *Dig Dis Sci* **45**: 1462–1464. doi:10.1023/a:1005588911207

Ham SL, Lee TH, Kim KJ, Kim JH, Hwang SJ, Lee SH, et al (2023). Discovery and biosynthesis of imidazolium antibiotics from the probiotic *Bacillus licheniformis*. *J Nat Prod* **86**: 850–859. doi:10.1021/acs.jnatprod.2c01032

Hantoushzadeh S, Golshahi F, Javadian P, Khazardoost S, Aram S, Hashemi S, et al (2012). Comparative efficacy of probiotic yoghurt and clindamycin in treatment of bacterial vaginosis in pregnant women: A randomized clinical trial. *J Matern Neonatal Med* **25**(7): 1021–1024. doi:10.3109/14767058.2011.614654

Hasan N, Yang H (2019). Factors affecting the composition of the gut microbiota, and its modulation. *PeerJ* **7**: e7502. doi:10.7717/peerj.7502

He X, Zhao S, Li Y (2021). *Faecalibacterium prausnitzii*: A next-generation probiotic in gut disease improvement. *Can J Infect Dis Med Microbiol* **2021**: 6666114. doi:10.1155/2021/6666114

Hemarajata P, Versalovic J (2013). Effects of probiotics on gut microbiota: Mechanisms of intestinal immunomodulation and neuromodulation. *Therap Adv Gastroenterol* **6**: 39–51. doi:10.1177/1756283X12459294

Huang R, Xing HY, Liu HJ, Chen ZF, Tang BB (2021). Efficacy of probiotics in the treatment of acute diarrhea in children: A systematic review and meta-analysis of clinical trials. *Transl Pediatr* **10**(12): 3248–3260. doi:10.21037/tp-21-511

Hudson LE, McDermott CD, Stewart TP, Hudson WH, Rios D, Fasken MB, et al (2016). Characterization of the probiotic yeast *Saccharomyces boulardii* in the healthy mucosal immune system. *PLoS ONE* **11**: e0153351. doi:10.1371/journal.pone.0153351

Ibraheim HK, Madhi KS, Baqer GK, Gharban HAJ (2023). Effectiveness of raw bacteriocin produced from lactic acid bacteria on biofilm of methicillin-resistant *Staphylococcus aureus*. *Vet world* **16**: 491–499. doi:10.14202/vetworld.2023.491-499

Imperial ICVJ, Ibana JA (2016). Addressing the antibiotic resistance problem with probiotics: Reducing the risk of its double-edged sword effect. *Front Microbiol* **7**: 1983. doi:10.3389/fmicb.2016.01983

Ivec M, Botić T, Koren S, Jakobsen M, Weingartl H, Cencič A (2007). Interactions of macrophages with probiotic bacteria lead to increased antiviral response against vesicular stomatitis virus. *Antiviral Res* **75**: 266–274. doi:10.1016/j.antiviral.2007.03.013

Joseph RJ, Ser HL, Kuai YH, Tan LTH, Arasoo VJT, Letchumanan V, et al (2021). Finding a balance in the vaginal microbiome: How do we treat and prevent the occurrence of bacterial vaginosis? *Antibiotics* **10**(6): 719. doi:10.3390/antibiotics10060719

Kelesidis T, Pothoulakis C (2012). Efficacy and safety of the probiotic *Saccharomyces boulardii* for the prevention and therapy of gastrointestinal disorders. *Therap Adv Gastroenterol* **5**: 111–125. doi:10.1177/1756283X11428502

Khan I, Ullah N, Zha L, Bai Y, Khan A, Zhao T, et al (2019). Alteration of gut microbiota in Inflammatory Bowel Disease (IBD): Cause or consequence? IBD treatment targeting the gut microbiome. *Pathog (Basel, Switzerland)* **8**(3): 126. doi:10.3390/pathogens8030126

Khomeiri M, Taheri S, Nasrollahzadeh A (2023). Non-LAB bacterial probiotics BT. In: Jafari SM, Rashidinejad A, Simal-Gandara J, editors. *Handbook of Food Bioactive Ingredients: Properties and Applications*. Cham: Springer International Publishing. pp. 1505–1532. doi:10.1007/978-3-031-28109-9_42

Kumar M, Kumar A, Nagpal R, Mohania D, Behare P, Verma V, et al (2010). Cancer-preventing attributes of probiotics: An update. *Int J Food Sci Nutr* **61**: 473–496. doi:10.3109/09637480903455971

Kwun S-Y, Yoon J-A, Kim G-Y, Bae Y-W, Park E-H, Kim M-D (2024). Isolation of a potential probiotic *Levilactobacillus brevis* and evaluation of its exopolysaccharide for antioxidant and α-glucosidase inhibitory activities. *J Microbiol Biotechnol* **34**: 167–175. doi:10.4014/jmb.2304.04043

Li C, Peng K, Xiao S, Long Y, Yu Q (2023). The role of *Lactobacillus* in inflammatory bowel disease: From actualities to prospects. *Cell Death Discov* **9**: 361. doi:10.1038/s41420-023-01666-w

Liu X-F, Shao J-H, Liao Y-T, Wang L-N, Jia Y, Dong P-J, et al (2023). Regulation of short-chain fatty acids in the immune system. *Front Immunol* **14**: 1186892. doi:10.3389/fimmu.2023.1186892

Lukasik J, Dierikx T, Besseling-van der Vaart I, de Meij T, Szajewska H (2022). Multispecies probiotic for the prevention of antibiotic-associated diarrhea in children: A randomized clinical trial. *JAMA Pediatr*. **176**: 860–866. doi:10.1001/jamapediatrics.2022.1973

Ma X (2024). Heavy metals remediation through lactic acid bacteria: Current status and future prospects. *Sci Total Environ* **946**: 174455. doi:10.1016/j.scitotenv.2024.174455

Maftei N-M, Raileanu CR, Balta AA, Ambrose L, Boev M, Marin DB, et al (2024). The potential impact of probiotics on human health: An update on their health-promoting properties. *Microorganisms* **12**(2): 234. doi:10.3390/microorganisms12020234

Mantzourani I, Kazakos S, Terpou A, Alexopoulos A, Bezirtzoglou E, Bekatorou A, et al (2019). Potential of the probiotic *Lactobacillus plantarum* ATCC 14917 strain to produce functional fermented pomegranate juice. *Foods* **8**(1): 4. doi:10.3390/foods8010004

Manzanares W, Lemieux M, Langlois PL, Wischmeyer PE (2016). Probiotic and synbiotic therapy in critical illness: A systematic review and meta-analysis. *Crit Care* **20**: 262 doi:10.1186/s13054-016-1434-y

Markowiak P, Śliżewska K (2018). The role of probiotics, prebiotics and synbiotics in animal nutrition. *Gut Pathog* **10**(21). doi:10.1186/s13099-018-0250-0

McFarland L V (2007). Meta-analysis of probiotics for the prevention of traveler's diarrhea. *Travel Med Infect Dis* **5**: 97–105. doi:10.1016/j.tmaid.2005.10.003

McFarland L V (2010). Systematic review and meta-analysis of *Saccharomyces boulardii* in adult patients. *World J Gastroenterol* **16**(18): 2202–2222. doi:10.3748/wjg.v16.i18.2202

McMillan A, Dell M, Zellar MP, Cribby S, Martz S, Hong E, et al (2011). Disruption of urogenital biofilms by *Lactobacilli*. *Colloids Surf B Biointerfaces* **86**(1): 58–64. doi:10.1016/j.colsurfb.2011.03.016

Million M, Raoult D (2013). Species and strain specificity of *Lactobacillus* probiotics effect on weight regulation. *Microb Pathog* **55**: 52–54. doi:10.1016/j.micpath.2012.09.013

Min M, Bunt CR, Mason SL, Hussain MA (2019). Non-dairy probiotic food products: An emerging group of functional foods. *Crit Rev Food Sci Nutr* **59**(16): 2626–2641. doi:10.1080/10408398.2018.1462760

Mirzaei A, Razavi SA, Babazadeh D, Laven R, Saeed M (2022). Roles of probiotics in farm animals: A review. *Farm Anim Heal Nutr* **17**(3): 276–280. doi:10.58803/fahn.v1i1.8

Mitra AK, Asala AF, Malone S, Mridha MK (2023). Effects of probiotics in adults with gastroenteritis: A systematic review and meta-analysis of clinical trials. *Diseases* **11**(4): 138. doi:10.3390/diseases11040138

Morrow LE, Gogineni V, Malesker MA (2012). Probiotics in the intensive care unit. *Nutr Clin Pract* **11**(2): 217. doi:10.1177/0884533612440290

Naidu AS, Bidlack WR, Clemens RA (1999). Probiotic spectra of lactic acid bacteria (LAB). *Crit Rev Food Sci Nutr* **39**: 13–126. doi:10.1080/10408699991279187

Nalla K, Manda NK, Dhillon HS, Kanade SR, Rokana N, Hess M, et al (2022). Impact of probiotics on dairy production efficiency. *Front Microbiol* **13**: 805963. doi:10.3389/fmicb.2022.805963

Nash AK, Auchtung TA, Wong MC, Smith DP, Gesell JR, Ross MC, et al (2017). The gut mycobiome of the Human Microbiome Project healthy cohort. *Microbiome* **5**: 153. doi:10.1186/s40168-017-0373-4

Ney L-M, Wipplinger M, Grossmann M, Engert N, Wegner VD, Mosig AS (2023). Short chain fatty acids: Key regulators of the local and systemic immune response in inflammatory diseases and infections. *Open Biol* **13**: 230014. doi:10.1098/rsob.230014

Olanbiwoninu AA, Popoola BM (2023). Biofilms and their impact on the food industry. *Saudi J Biol Sci* **30**(2): 103523. doi:10.1016/j.sjbs.2022.103523

Papizadeh M, Rohani M, Nahrevanian H, Javadi A, Pourshafie MR (2017). Probiotic characters of *Bifidobacterium* and *Lactobacillus* are a result of the ongoing gene acquisition and genome minimization evolutionary trends. *Microb Pathog* **111**: 118–131. doi:10.1016/j.micpath.2017.08.021

Penumetcha SS, Ahluwalia S, Irfan R, Khan SA, Rohit Reddy S, Vasquez Lopez ME, et al (2021). The efficacy of probiotics in the management of *Helicobacter pylori*: A systematic review. *Cureus* **13**(12): e20483. doi:10.7759/cureus.20483

Porter SB, Johnston BD, Kisiela D, Clabots C, Sokurenko E V, Johnson JR (2022). Bacteriophage cocktail and microcin-producing probiotic *Escherichia coli* protect mice against gut colonization with multidrug-resistant *Escherichia coli* sequence type 131. *Front Microbiol* **13**: 887799. doi:10.3389/fmicb.2022.887799

Prezzi LE, Lee SHI, Nunes VMR, Corassin CH, Pimentel TC, Rocha RS, et al (2020). Effect of *Lactobacillus rhamnosus* on growth of *Listeria monocytogenes* and *Staphylococcus aureus* in a probiotic Minas Frescal cheese. *Food Microbiol* **92**. doi:10.1016/j.fm.2020.103557

Rahman MRT, Fliss I, Biron E (2022). Insights in the development and uses of alternatives to antibiotic growth promoters in poultry and swine production. *Antibiotics* **11**(6): 766. doi:10.3390/antibiotics11060766

Reid G, Charbonneau D, Erb J, Kochanowski B, Beuerman D, Poehner R, et al (2023). Oral use of *Lactobacillus rhamnosus* GR-1 and *L. fermentum* RC-14 significantly alters vaginal flora: Randomized, placebo-controlled trial in 64 healthy women. *FEMS Immunol Med Microbiol* **35**(2): 131–134. doi:10.1016/S0928-8244(02)00465-0.

Report of a Joint FAO WHO Expert Consultation on Evaluation of Health and Nutritional Properties of Probiotics in Food Including Powder Milk (2001). Córdoba RP 1 on Evaluation of Health EC, of Probiotics in Food including Powder Milk with Live Lactic Acid Bacteria (Health and Nutritional Properties of Probiotics in Food Including Powder Milk with Live Lactic Acid Bacteria. Available: https://books.google.co.za/books?id=qJqoygAACAAJ

Rodrigues VF, Elias-Oliveira J, Pereira ÍS, Pereira JA, Barbosa SC, Machado MSG, et al (2022). *Akkermansia muciniphila* and gut immune system: A good friendship that attenuates inflammatory bowel disease, obesity, and diabetes. *Front Immunol* **13**: 934695. doi:10.3389/fimmu.2022.934695

Saha S, Fukuyama K, Debnath M, Namai F, Nishiyama K, Kitazawa H (2023). Recent advances in the use of probiotics to improve meat quality of small ruminants: A review. *Microorganisms* **11**(7): 1652. doi:10.3390/microorganisms11071652

Schmerold I, van Geijlswijk I, Gehring R (2023). European regulations on the use of antibiotics in veterinary medicine. *Eur J Pharm Sci* **189**: 106473. doi:10.1016/j.ejps.2023.106473

Schultz M, Veltkamp C, Dieleman LA, Grenther WB, Wyrick PB, Tonkonogy SL, et al (2002). *Lactobacillus plantarum* 299V in the treatment and prevention of spontaneous colitis in interleukin-10-deficient mice. *Inflamm Bowel Dis* **8**: 71–80. doi:10.1097/00054725-200203000-00001

Shah NP (2007). Functional cultures and health benefits. *Int Dairy J* **17**: 1262–1277. doi:10.1016/j.idairyj.2007.01.014

Shahverdi S, Barzegari AA, Vaseghi Bakhshayesh R, Nami Y (2023). *In-vitro* and *in-vivo* antibacterial activity of potential probiotic *Lactobacillus paracasei* against *Staphylococcus aureus* and *Escherichia coli*. *Heliyon* **9**: e14641. doi:10.1016/j.heliyon.2023.e14641

Sharif S, Greer A, Skorupski C, Hao Q, Johnstone J, Dionne JC, et al (2022). Probiotics in critical illness: A systematic review and meta-analysis of randomized controlled trials. *Crit Care Med.* **50**(8): 1175–1186. doi:10.1097/CCM.0000000000005580

Sharma S, Kumar S, Sajjad S, Sharma S (2023). Probiotics in irritable bowel syndrome: A review article. *Cureus* **15**: e36565. doi:10.7759/cureus.36565

Sousa LG V, Pereira SA, Cerca N (2023). Fighting polymicrobial biofilms in bacterial vaginosis. *Microb Biotechnol* **16**(7): 1423–1437. doi:10.1111/1751-7915.14261

Srey S, Jahid IK, Ha S Do (2013). Biofilm formation in food industries: A food safety concern. *Food Control* **31**(2): 572–585. doi:10.1016/j.foodcont.2012.12.001

Stecchini ML, Del Torre M, Munari M (2001). Determination of peroxy radical-scavenging of lactic acid bacteria. *Int J Food Microbiol* **64**: 183–188. doi:10.1016/S0168-1605(00)00456-6

Sun Y, Wang C, Wang H, Yuan Y, Lu J, Zhong L (2022). Probiotic in the prevention of ventilator-associated pneumonia in critically ill patients: Evidence from meta-analysis and trial sequential analysis of randomized clinical trials. *BMC Pulm Med* **22**(1): 168. doi:10.1186/s12890-022-01965-5

Szajewska H, Berni Canani R, Domellöf M, Guarino A, Hojsak I, Indrio F, et al (2023). Probiotics for the management of pediatric gastrointestinal disorders: Position paper of the ESPGHAN special interest group on gut microbiota and modifications. *J Pediatr Gastroenterol Nutr* **76**(2): 232–247. doi:10.1097/MPG.0000000000003633

Szajewska H, Kołodziej M, Zalewski BM (2020). Systematic review with meta-analysis: *Saccharomyces boulardii* for treating acute gastroenteritis in children – A 2020 update. *Aliment Pharmacol Ther* **51**: 678–688. doi:10.1111/apt.15659

Tamanai-Shacoori Z, Smida I, Bousarghin L, Loreal O, Meuric V, Fong SB, et al (2017). *Roseburia* spp.: A marker of health? *Future Microbiol* **12**: 157–170. doi:10.2217/fmb-2016-0130

Tazehabadi MH, Algburi A, Popov I V, Ermakov AM, Chistyakov VA, Prazdnova E V, et al (2021). Probiotic *Bacilli* Inhibit *Salmonella* biofilm formation without killing planktonic cells. *Front Microbiol* **12**: 615328. doi:10.3389/fmicb.2021.615328

Telhig S, Ben Said L, Torres C, Rebuffat S, Zirah S, Fliss I (2022). Evaluating the potential and synergetic effects of microcins against multidrug-resistant *Enterobacteriaceae*. *Microbiol Spectr* **10**: e0275221. doi:10.1128/spectrum.02752-21

Thokchom S, Joshi SR (2012). Probiotic and bacteriocin efficacy of lactic acid bacteria from traditionally fermented foods: A review. *Assam Univ J Sci Technol* **10**: 142–155.

Tomé AR, Carvalho FM, Teixeira-Santos R, Burmølle M, Mergulhão FJM, Gomes LC (2023). Use of probiotics to control biofilm formation in food industries. *Antibiotics* **12**(4): 754. doi:10.3390/antibiotics12040754

Vanhoutte T, De Preter V, De Brandt E, Verbeke K, Swings J, Huys G (2006). Molecular monitoring of the fecal microbiota of healthy human subjects during administration of lactulose and *Saccharomyces boulardii*. *Appl Environ Microbiol* **72**: 5990–5997. doi:10.1128/AEM.00233-06

Vats P, Kaur UJ, Rishi P (2022). Heavy metal-induced selection and proliferation of antibiotic resistance: A review. *J. Appl. Microbiol.* **132**(6): 4058–4076. doi:10.1111/jam.15492

Vlasova AN, Kandasamy S, Chattha KS, Rajashekara G, Saif LJ (2016). Comparison of probiotic *lactobacilli* and *bifidobacteria* effects, immune responses and rotavirus vaccines and infection in different host species. *Vet Immunol Immunopathol* **172**: 72–84. doi:10.1016/j.vetimm.2016.01.003

Wang Q, Yang Q, Liu X (2023). The microbiota-gut-brain axis and neurodevelopmental disorders. *Protein Cell* **14**: 762–775. doi:10.1093/procel/pwad026

Wei C-X, Wu J-H, Huang Y-H, Wang X-Z, Li J-Y (2022). *Lactobacillus plantarum* improves LPS-induced Caco2 cell line intestinal barrier damage via cyclic AMP-PKA signaling. *PLoS ONE* **17**: e0267831. doi:10.1371/journal.pone.0267831

Wells JM, Mercenier A (2008). Mucosal delivery of therapeutic and prophylactic molecules using lactic acid bacteria. *Nat Rev Microbiol* **6**: 349–362. doi:10.1038/nrmicro1840

Wilkins T, Sequoia J (2017). Probiotics for gastrointestinal conditions: A summary of the evidence. *Am Fam Physician* **96**: 170–178.

Xiao Y, Feng Y, Zhao J, Chen W, Lu W (2024). Achieving healthy aging through gut microbiota-directed dietary intervention: Focusing on microbial biomarkers and host mechanisms. *J Adv Res*. doi:10.1016/j.jare.2024.03.005

Xu C, Kong L, Gao H, Cheng X, Wang X (2022). A review of current bacterial resistance to antibiotics in food animals. *Front Microbiol* **13**: 822689. doi:10.3389/fmicb.2022.822689

Yang Q, Hu Z, Lei Y, Li X, Xu C, Zhang J, et al (2023). Overview of systematic reviews of probiotics in the prevention and treatment of antibiotic-associated diarrhea in children. *Front. Pharmacol* **14**: 1153070. doi:10.3389/fphar.2023.1153070

Yang S, Qiao J, Zhang M, Kwok L-Y, Matijašić BB, Zhang H, et al (2024). Prevention and treatment of antibiotics-associated adverse effects through the use of probiotics: A review. *J Adv Res*. doi:10.1016/j.jare.2024.06.006

Yi E-J, Kim A-J (2023). Antimicrobial and antibiofilm effect of bacteriocin-producing *Pediococcus inopinatus* K35 isolated from kimchi against multidrug-resistant *Pseudomonas aeruginosa*. *Antibiotic* **12**(4): 676. doi:10.3390/antibiotics12040676

Yoo JY, Groer M, Dutra SVO, Sarkar A, McSkimming DI (2020). Gut microbiota and immune system interactions. *Microorganisms* **8**. doi:10.3390/microorganisms8101587

Yuniarti M, Bidura IGNG, Siti NW (2023). Effect of probiotic supplementation in the diet on the production and physical quality of eggs in laying hens. *World J Biol Pharm Heal Sci* **1**(3): 180–192. doi:10.30574/wjbphs.2023.15.1.0313

Zhang C, Jiang J, Tian F, Zhao J, Zhang H, Zhai Q, et al (2020). Meta-analysis of randomized controlled trials of the effects of probiotics on functional constipation in adults. *Clin Nutr* **39**(10): 2960–2969. doi:10.1016/j.clnu.2020.01.005

Zhang H, Duan Y, Cai F, Cao D, Wang L, Qiao Z, et al (2022). Next-generation probiotics: Microflora intervention to human diseases. *Biomed Res Int* **2022**: 5633403. doi:10.1155/2022/5633403

4 Beyond Antibiotics
The Mechanisms behind Probiotic Action

Priyanka Sharma[1], *Jasmine Kaur*[2], *Hae-Jeung Lee*[1,3,4], *and Anshul Sharma*[1,3]

[1]Department of Food and Nutrition, College of Bionanotechnology, Gachon University, Seongnam-si, Republic of Korea

[2]Clem Jones Centre for Neurobiology and Stem Cell Research, School of Pharmacy and Medical Science, Menzies Health Institute Queensland, Griffith University, Gold Coast, Australia

[3]Institute for Aging and Clinical Nutrition Research, Gachon University, Seongnam-si, Republic of Korea

[4]Department of Health Sciences and Technology, Gachon Advanced Institute for Health Science and Technology (GAIHST), Gachon University, Incheon, Republic of Korea

ABBREVIATIONS

APCs	antigen-presenting cells
C.	*Clostridium*
CD	cluster of differentiation
DSS	dextran sulfate sodium
EHEC	enterohemorrhagic *E. coli*
EPS	exopolysaccharide
FAO	Food and Agriculture Organization
GI	gastro intestine
H.	*Helicobacter*
H_2O_2	hydrogen peroxide
IBS	irritable bowel syndrome
IFN-γ	interferon-gamma
Ig	immunoglobulin
IL	interleukin
L.	*Lactobacillus*
LAB	lactic acid bacteria
LBPs	live biotherapeutic products
MAPK	mitogen-activated protein kinase
MDR	multidrug resistance
MHC	major histocompatibility complex
MIC	minimum inhibitory concentration
NK	natural killer
NO	nitric oxide
RCTs	randomized controlled trials

DOI: 10.1201/9781032676951-4

S.	*Streptococcus*
TGF-β	Transforming growth factor-beta
TLR	toll-like receptor
TNF-α	tumor necrosis factor-alpha
UTIs	urinary tract infections
ZOI	zone of inhibition

4.1 INTRODUCTION

Probiotics refer to a group of microorganisms that, when ingested in sufficient quantities, provide health benefits to the host (Plaza-Diaz et al. 2019). Probiotics can be consumed in the form of supplements or food products; they have gained popularity as functional foods that offer additional health benefits. In 1953, Werner Kollath first described the term "probiotics," a combination of the Latin word "*pro*" and the Greek word "*bio*" which signifies "for life," implying that the species in this category would likely contribute to a healthy life (Gasbarrini, Bonvicini, and Gramenzi 2016). Later in 2001, the Food and Drug Association described them as "live microbes that, when administered in the right quantities, confer health benefits on the host organisms." Thus, bacterial species mostly found in food products, belonging to the genera *Lactobacillus*, *Pediococcus*, *Lactococcus*, *Enterococcus*, *Bifidobacterium*, *Streptococcus*, *Propionibacterium*, and *Bacillus*, and fungal species of the genus *Saccharomyces* rightly fit into the category of probiotics and have earned a place as a friend to *Homo sapiens* (Latif et al. 2023).

Researchers have reported various health-promoting activities of different species of probiotics. They help in the prevention and management of inflammatory bowel disease, allergic diseases, irritable bowel syndrome (IBS), diarrhea, hypercholesterolemia, lactose intolerance, and cancer (Latif et al. 2023). Species such as *Bifidobacterium* and *Lactobacillus*, which are the most commonly used strains, assist in the detoxification of xenobiotic and environmental pollutants by metabolizing these chemical compounds (Srednicka et al. 2021). *Lactobacillus* and *Bifidobacterium* have been reported to be prevalent in the small intestine and large intestine, respectively. Additionally, healthy microorganisms can also synthesize some important vitamins, including folate, riboflavin, and vitamin K, which are critical for the body's systemic and cognitive functions (Latif et al. 2023). *Weissella* is another important genus, and several of its species have been reported to have antimicrobial and other health-promoting properties such as antiobesity, anticancer, anti-inflammatory, and skin care (Ahmed et al. 2022). Probiotics are well-known for their fermentative properties, which help in maintaining the gut microbiome and balancing the gut dysbiosis *via* the metabolization and fermentation of undigested fiber (Warman, Jia, and Kato 2022). Furthermore, probiotics are widely used in the edible film, dairy, beverage, and baking industries.

Antimicrobial or antibacterial agents are substances that can kill or inhibit the growth of pathogenic microorganisms (Aghamohammad and Rohani 2023). The antimicrobial properties of probiotic strains are attributed to the prophylactic agents produced by them that work against enteric infections (van Zyl, Deane, and Dicks 2020). These antagonistic microorganisms actively participate in immune modulation and stimulation of the host immune response, as well as producing antimicrobial agents. The ability of probiotics to survive the acidic environment in the gut created by the gastric juices and withstand the bactericidal properties of bile salts is by contributing prophylactic agents, organic compounds, and lactic acid, which assist them in controlling or inhibiting the growth of pathogenic bacteria (Monika et al. 2021). Several studies have shown the immune regulatory potential of probiotic strains. Probiotic strains produce cytokines that influence the host's immune system by protecting against pathogenic infections as well as reducing allergic symptoms and even cancer (Fijan 2023). This demonstrates that probiotic strains have clinical uses with emerging antimicrobial, preventive, and protective properties against different diseases (Fijan 2023).

4.2 ROLE OF ANTIBIOTICS IN HUMAN HEALTH

Antimicrobials are synthetic or semisynthetic that substances can inhibit the proliferation of bacteria, fungi, viruses, and parasites (Di Martino 2022). On the other hand, antibiotics are defined as low-molecular-weight organic substances produced by live microorganisms or plants that can hinder the growth of other organisms at low concentrations (Clardy, Fischbach, and Currie 2009). Generally, we classify antibacterial substances as either bactericidal agents, which effectively eliminate bacteria, or bacteriostatic agents, which inhibit bacterial growth. Antibiotics have become an important part of human health. A general physician prescribes these drugs to treat infections ranging from strep throat or urinary tract infections (UTIs) to serious conditions like meningitis or pneumonia. Several of other uses include prevention of infectious complications, prophylactic application in surgical procedures, use in cancer treatments, use as immunosuppressive agents, and contribution to increased life expectancy by treating deadly infectious diseases (Davies and Davies 2010).

Antibiotics and probiotics are linked through their impact on the human body's microbiome. Probiotics are beneficial bacteria that maintain or restore a healthy balance of microorganisms in the body, whereas antibiotics aim to kill or inhibit harmful bacteria that cause infections. Disruption of this harmony can have adverse effects on human health, including impaired immunity, increased resistance, and increased risk of disease and illness (Lathakumari et al. 2024). Recent research has established a link between the gut microbiota and the host's neurological function through the gut–brain axis. Any type of antibiotic-mediated alteration of the gut microbiome could affect the functioning of the brain and may lead to psychiatric disorders (Ahmed et al. 2024). Although antibiotics can help against infections, frequent use and the misuse of these drugs lead to a burden of antibiotic resistance in humans (Llor and Bjerrum 2014).

4.3 SIGNIFICANCE OF COMBATING ANTIBIOTIC RESISTANCE

Microorganisms, specifically bacteria, can evolve to resist antibiotics. It was reported in the early 20th century, by Sir Alexander Fleming, who first discovered penicillin, that the misuse or overuse of such an agent would eventually lead to antibiotic resistance. An increasing number of bacterial strains are developing resistance against several antibiotics, resulting in multidrug-resistant (MDR) bacteria (Huemer et al. 2020). Antimicrobial resistance is a significant worldwide health issue (Chinemerem Nwobodo et al. 2022). Indeed, animal farms can serve as a repository for genes conferring antibiotic resistance and microorganisms resistant to antibiotics (Bai et al. 2022). Therefore, the consumption of antibiotic-treated products poses a risk of transmitting bacteria that are resistant to antibiotics. Furthermore, bacteria related to people are more likely to exchange genetic material than bacteria from other sources (Smillie et al. 2011). Resistant pathogens acquire resistance through several mechanisms, including the presence of resistance genes on plasmids, reduced antibiotic uptake due to biofilm formation, antibiotics efflux from cells, and disruption or mutation at the target site of antibiotics (Darby et al. 2023). The consequences of antimicrobial resistance can range from current treatment failures, cost of treatments, increased hospital stays, and overall higher healthcare costs.

Among alternative treatment options for antibiotic resistance, probiotics have garnered significant attention as wonderful microorganisms to be used as novel antimicrobial agents (Wiese and Imhoff 2019). Researchers have documented that a few strains of *Lactobacillus*, an important genus, produce lactic acid and bacteriocins, which are responsible for their antimicrobial properties. The antimicrobial ability may stem from their capacity to impede intracellular processes or disintegrate cell membranes (Kiousi et al. 2023).

4.4 PROBIOTICS AND THEIR ANTIMICROBIAL ACTIVITY

The food industry has utilized several natural substitutes for antibiotics, such as essential oils, to prevent the spread of pathogenic bacteria (Cherrat et al. 2014). As stated above, probiotics interact with the human gut and alter the composition of the host microbiota, affecting various mechanisms, including immune responses, and modifying the intestinal epithelium of the individual. Several studies have shown that the ingestion of certain probiotic strains, specifically from the *Lactobacillus* and *Bifidobacterium* genera, reduces the chances and/or severity of certain foodborne infections (Wan, Forsythe, and El-Nezami 2019). There are many underlying mechanisms by which these probiotics can target and kill pathogenic bacteria. The following sections will discuss briefly the different mechanisms of probiotic action.

4.4.1 Competition for Nutrients or Competitive Exclusion for Adhesion Site

Probiotic bacteria can fight harmful bacteria for nutrients and attachment sites within the gut. By outcompeting harmful bacteria, probiotics can prevent pathogenic colonization and growth, thereby reducing the risk of infections. It is well-known that probiotics produce compounds that are antimicrobial in nature, such as acetic acid and lactic acid. These mechanisms enable them to modify the conditions of the GIT, therefore creating an unfavorable environment for the pathogenic microorganisms (Servin 2004). Their ability to colonize the intestinal lining also plays a significant role. Another factor is the production of bacteriocins by probiotics that significantly influences their functionality. Bacteriocins function as colonizing peptides that enhance the microbial diversity of the bacteriocin-producing strain in the gut. In addition, these bacteriocins possess the ability to eradicate pathogenic bacteria by blocking signaling pathways or suppressing their proliferation, therefore preventing any modification of the host environment and the immune system (Arques et al. 2015). Certain lactic acid bacteria (LAB) have carbohydrate-binding domains, which allow them to compete with enteropathogens for specific receptors, such as toll-like receptors (TLRs), within the host. In this competitive process, the binding of probiotics efficiently blocks the attachment and colonization of enteropathogens to the host's receptors, known as competitive exclusion (Kawai and Akira 2009, Valeriano et al. 2016). This level of competition is governed by the type of probiotic strains and the affinity for target receptors.

Among probiotics, different strains of *Lactiplantibacillus plantarum* have been documented to demonstrate a competitive exclusion strategy by displacing the attachment of enteropathogens to the host, including *Escherichia coli* and *Salmonella* (*S.*) *enterica* (Dhanani and Bagchi 2013). Recently, the entire genome sequence of the *Bacillus coagulans* HS243 strain was functionally annotated. The strain has several probiotic properties, adhesion, and aggregation proteins that aid in preventing the entry of potentially harmful pathogens. The genome annotation showed the presence of 11 adhesion and aggregation-related proteins. In addition, two bacteriocins were also mined from the HS243 genome. One is a gallidermin, a cationic antimicrobial peptide from class I lantipeptide, and the other is a hypothetical protein from class sactipeptide. The HS243 strain also exhibited antibacterial activity against *Streptococcus* and *Shigella*, supporting the outcomes of genome annotation (Kapse et al. 2019). These studies illustrate how probiotics can competitively exclude potentially harmful pathogens from the host body.

4.4.2 Production of Antibacterial Substances

Probiotics release a variety of antimicrobial agents, such as bacteriocins, organic acids, and hydrogen peroxide (H_2O_2). These agents help probiotics combat pathogens, thus guarding their host (Gaspar et al. 2018).

4.4.2.1 Bacteriocins

Bacteriocins are known to function as colonizing peptides, where they populate the epithelial lining of the GI tract and facilitate the dominance of a bacterial colony within the niche. Another function of bacteriocins is their antibacterial activity, in which they directly kill or inhibit the strains of other origins. Furthermore, these may also function as signaling peptides, thereby communicating with host cells or microbial populations through a quorum signaling mechanism (Dobson et al. 2012). Mainly, these compounds act by forming pores in the target bacteria's cell membrane and disrupting the membrane integrity. Furthermore, they interrupt the basic cellular mechanisms, such as DNA replication or protein synthesis, by penetrating the cell, resulting in cell death. LAB strains produce bacteriocins such as nisin and pediocin. When secreted, these substances penetrate the cell membrane, form potassium-selective channels, and collapse the cell membrane. Once these substances enter the cytoplasm of the cell, they destroy the bacterial cell by inhibiting DNA replication and transcription (Mousavi Khaneghah et al. 2020). Certain probiotic strains, such as *Lactobacillus* (now known as *Lacticaseibacillus*) *rhamnosus*, secrete a broad range of active agents that not only target various bacteria but also interfere with the adhesion of pathogenic bacteria to host intestinal cells (Dutra et al. 2016). A separate strain of *B. coagulans* produces coagulin, among other substances that work to inhibit a variety of enteric microbes (Abhari et al. 2015). Bacteriocins actively serve as food-preserving compounds and effectively inhibit a variety of pathogenic microorganisms due to their unique properties.

4.4.2.2 Organic Acids

Organic acids such as lactic acid, acetic acid, and propionic acid are common metabolites produced by LAB. These acids penetrate the cell membrane of the target bacteria and dissociate within the cytoplasm, leading to a decreased cytoplasmic pH. This acidic environment destroys cell membranes, disrupts nutrient transport, and inhibits metabolic activities, ultimately resulting in cell death (Cizeikiene and Jagelaviciute 2021). Numerous studies have investigated the role of lactic acid as a by product. As stated above, the antimicrobial property of LAB is due to the production of a high amount of lactic acid. Mechanistically, lactic acid creates an unfavorable environment for pathogens by lowering pH levels in the host. For example, the largest pathogenic inhibition zones by *Lactobacillus* (now known as *Lactiplantibacillus*) *plantarum* and *Bifidobacterium longum* reflect their ability to produce higher amounts of lactic acid. *Lactobacillus* (now known as *Limosilactobacillus*) *fermentum* and *Bifidobacterium infantis*, on the other hand, have small zones of inhibition because of their limited production capacity (Tejero-Sarinena et al. 2012). Several probiotic strains have been documented to avoid food spoilage by secreting antimicrobial substances, demonstrating their potential in the food industry.

4.4.2.3 Hydrogen Peroxide

It has been documented that probiotic bacterial species produce reactive oxygen species that can target the pathogenic bacteria and disrupt their cellular processes. Previously, it was reported that in an aerobic environment, *Lactobacillus johnsonii* NCC 533 can produce H_2O_2, which led to the killing of *S. enterica* serovar Typhimurium (Pridmore et al. 2008). H_2O_2, being an oxidizing agent, leads to the oxidation of available macromolecules such as proteins, lipids, and carbohydrates that are the building blocks of bacterial cells. Such a mechanism can lead to the destruction of the bacterial cells, resulting in death. H_2O_2 production is considered a key metabolite of LAB that protects against urogenital infections, particularly bacterial vaginosis (Monika et al. 2021).

The production of antimicrobial substances by probiotics plays a crucial role in maintaining gut health and protecting against infections. These substances not only inhibit the growth of harmful bacteria but also help in maintaining a balanced microbial ecosystem. Understanding specific mechanisms and types of antimicrobial substances produced by probiotics can aid in the development of targeted probiotic therapies, as well as the improvement of food safety and preservation techniques.

4.4.3 Immune Modulation

Probiotics exert their immunomodulatory effects in host through several ways, including the activation of the innate and adaptive immune response that synergistically work together, increasing the growth of bioactive components, enhancement of cytokine productions, competitive exclusion

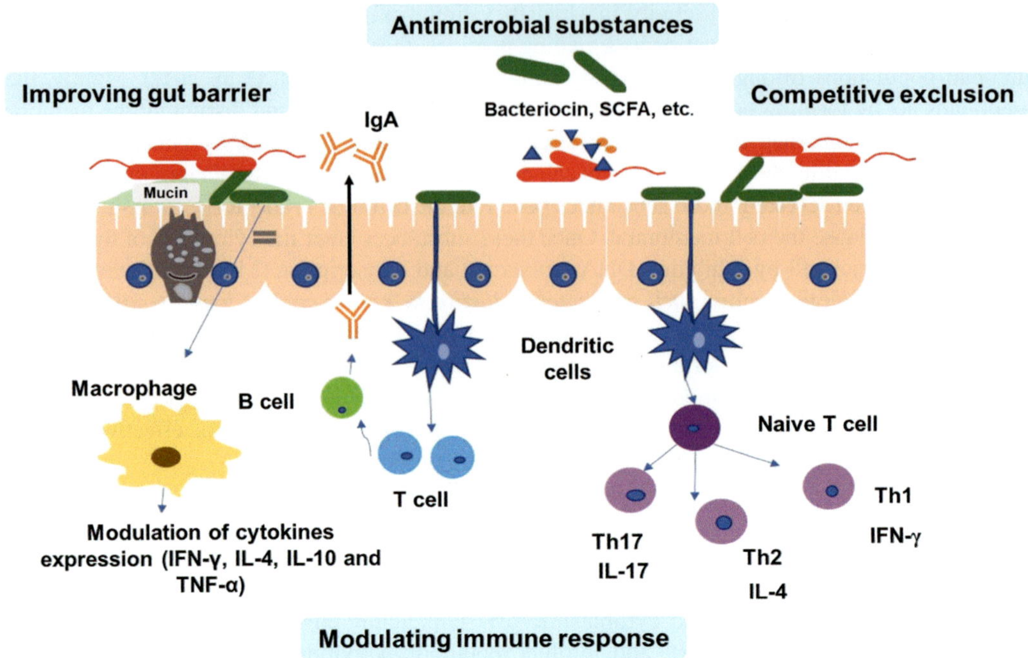

FIGURE 4.1 The schematic representation showing the antimicrobial mechanisms of probiotics against pathogens. (Adapted from Rabetafika et al. 2023.)

and inhibition of pathogenic microorganisms, and disruption of the pathogenic biofilms, as well as maintenance of the gut barrier to protect organisms against pathogens (Mazziotta et al. 2023, Rabetafika et al. 2023) (Figure 4.1).

4.4.3.1 Enhancement of Innate Immunity

It is a first line of defense, also known as a nonspecific response, comprised physical or mechanical barriers (such as epithelial and mucous membranes, and skin), chemical barriers (such as enzymes, mucus, saliva, antimicrobial molecules, and dedicated effector cells stimulating killing of foreign invaders), immune cells, and immunomodulatory cytokines, as well as biological barriers (such as microbiota). Immune cells include monocytes, macrophages, dendritic cells, neutrophils, and natural killer (NK) cells. Probiotics improve innate immunity by increasing the cytotoxicity of NK cells and the phagocytosis of macrophages. They modulate adaptive immunity by interacting and enhancing the functional capacity of immune cells such as dendritic cells, enterocytes, and regulatory T cells. A complex interplay exists between probiotics, their components, and various immune system constituents. However, depending on the probiotic species, probiotics' effects on immunity can vary (Aziz and Bonavida 2016).

4.4.3.1.1 Stimulation of Macrophages and NK Cells

Activated macrophages promote phagocytosis by stimulating the production of cytotoxic molecules such as nitric oxide (NO) and secrete cytokines such as interferon-γ (IFN-γ), interleukin (IL)-1β, IL-6, IL-10, and tumor necrosis factor-alpha (TNF-α). These cytokines initiate pathogen destruction (Noh et al. 2022). NKs constitute 10% of the peripheral lymphocytes and produce cytotoxic molecules, such as granzymes, perforin, and TNF-α, which cause cell-mediated lysis (Aziz and Bonavida 2016). Rocha-Ramírez et al. reported that four probiotic strains, *Lactobacillus* (now known as *Lacticaseibacillus*) *rhamnosus* GG, *Lacticaseibacillus casei* IMAU60214, *L. rhamnosus* KLDS, and *Lactobacillus helveticus* IMAU70129, activated macrophages derived from human monocytes and stimulated inflammatory responses. Pretreatment with *Lactobacillus* increased phagocytosis and the antibacterial activity of macrophages against *S.* Typhimurium, *Staphylococcus aureus*, and *E. coli* (Rocha-Ramírez et al. 2017).

4.4.3.1.2 Production of Cytokines

Cytokines are small protein mediators secreted by immune cells, including NK cells, macrophages, B cells, and T cells to control and modulate the immune response. These proteins contribute significantly to the immunomodulation of the immune system as they regulate cell activation, proliferation, differentiation, and inflammation (Wautier and Wautier 2023).

4.4.3.2 Adaptive Immune Response

Acquired or adaptive immunity is a highly precise and specialized secondary defense mechanism that is associated with antigen-specific reactions and can confer lifelong immunity to the host. This specific immune response is persuaded toward targets by B and T lymphocytes, antibody production and responses, and the cell-mediated immune response (Mazziotta et al. 2023). Probiotics can enhance host defense by increasing the production of specific antibodies against infections and total immunoglobulin (Ig) A. Probiotics help in the maturation of B cells into plasma cells that produce IgA. It has been observed that LAB promotes IL-6 and IL-10 release by dendritic cells, which can upregulate secretory IgA concentration at mucosal locations in humans (Kawashima et al. 2018). In adaptive immune response, major histocompatibility complex (MHC) proteins, which are present on the surface of antigen-presenting cells (APCs), interact with T cell receptors expressed in T helper (cluster of differentiation (CD) 4) cells to regulate, activate, and maintain the homeostasis of the CD4+ lineage. In a cell-mediated response, CD8+ cells (express CD8 on their surface) and CD4+ cells (express CD4 on their surfaces) recognize antigens bound to MHC types I and II complexes, respectively. Moreover, DCs, a part of the innate immune system, connect both innate and adaptive immune responses (Palucka and Banchereau 1999). Naïve CD4+ T lymphocytes can differentiate into subsets of T helper (Th) cells in the presence of specific cytokines secreted through processes that are triggered by antigenic stimulation controlled by APCs such as DCs (Mazziotta et al. 2023).

A research study showed that *L. rhamnosus* S1K3 protects mice against *S. enterica* serovar Typhimurium (*S.* Typhimurium) NCDC infection by regulating both immune and nonimmune responses. During the latter phase of infection, the administration of S1K3 reduces transforming growth factor-beta and maintains IL-4 and IL-12 protein levels. Additionally, it enhances the presence of IgA-secreting cells in the lamina propria, IgA in serum, and secretory IgA levels in intestinal fluids during infection in mice. In addition to lowering *S.* Typhimurium NCDC count in feces, this probiotic inhibits its further dissemination in the spleen, liver, and intestine of mice, thereby promoting overall health. Moreover, in human colorectal adenocarcinoma (Caco-2) cells, *L. rhamnosus* S1K3 improves the tight junction proteins, namely occludin and claudin-1(Kemgang et al. 2016).

4.4.4 Disruption of Pathogenic Biofilms

The biofilm is an aggregate of microorganisms (a single or several species) that are sessile and embedded in a self-produced polymeric matrix. When bacteria get embedded in the biofilm, they become less sensitive to antimicrobial treatments. As a result, they are difficult to eradicate. A recent study reveals that probiotics and their derived compounds have emerged as promising bio tools for combating pathogenic biofilms (Barzegari et al. 2020). Some probiotic strains that are able to inhibit pathogenic biofilm include *B. longum*, *Bifidobacterium lactis*, *L. plantarum*, *L. rhamnosus*, *L. fermentum*, *L. casei*, *Lactobacillus* (now known as *Levilactobacillus*) *brevis*, *Lactobacillus sporogenes*, *Lactobacillus* (now known as *Ligilactobacillus*) *salivarius*, *Lactobacillus delbrueckii*, *Lactobacillus acidophilus*, *Lactobacillus pentosus*, *Streptococcus salivarius*, and *Streptococcus oralis*, are examples of probiotic strains reported to have anti-biofilm activity (Barzegari et al. 2020). Probiotics have anti-biofilm activity, either by interfering with biofilm development or by enzymatically degrading the biofilm matrix.

In a study, *L. fermentum* TCUESC01 and *Lactobacillus* (now known as *Lactiplantibacillus*) *plantarum* TCUESC02 isolated from cocoa seeds were shown to have bactericidal activity against *S. aureus*. A stronger effect was observed for TCUESC02 (minimum inhibitory concentration (MIC), 2.5 mg/mL) compared to TCUESC01 (MIC, 20 mg/mL). However, the supernatant of *L. fermentum* TCUESC01 effectively inhibited *S. aureus* biofilm formation in a dose-dependent manner by modulating the expressions of *icaA* and *icaR* genes that play an important role in biofilm synthesis (Melo et al. 2016). In another study, cultures and cell-free supernatants of *L. casei* DSM 20011 and *Lactobacillus* (now named *Limosilactobacillus*) *reuteri* DSM 20016 showed strong antimicrobial and anti-biofilm activities against MDR *Proteus mirabilis*. Furthermore, the treatments of untreated and pH-adjusted supernatants of both strains reduced biofilm production and adherence of *P. mirabilis*. The study showed the importance of *L.* casei and *L. reuteri* as potential probiotics against UTIs caused by *Proteus* (Shaaban et al. 2020).

Researchers have reported that probiotics interfere mechanistically with the formation of biofilms in pathogenic microorganisms. In a study, *Bifidobacterium* species were reported to inhibit the release of quorum-sensing signals and biofilm formation by enterohemorrhagic *E. coli* (EHEC) O157:H7. Among 13 strains of *Bifidobacterium* species, *B. longum* ATCC 15707 resulted in a 36% reduction in biofilm formation. The survival rate of *Caenorhabditis elegans* (an *in vivo* nematode model) treated with EHEC O157:H7 was significantly enhanced in the presence of cell extracts of *B. longum* ATCC 15707, compared with nematodes not treated with 15707 (Kim et al. 2012).

4.4.5 Enhancement of Intestinal Barrier Function

The gut barrier establishes an essential line of defense for maintaining intestinal homeostasis through surface mucus, the epithelial layer, and immune defense (Camilleri 2019). The compromised integrity of the barrier leads to malfunctions in absorptive capacity, as well as chemical and microbial challenges, and various immune and inflammatory diseases (Bron et al. 2017). Evidence indicates that probiotics contribute to intestinal function by maintaining cell permeability, enhancing the physical mucosal barrier, activating the immune system, and modulating gut microbiota composition.

Recently, Tao et al. reported the screening and isolation of *Lactobacillus* species to evaluate their role in enhancing the intestinal barrier function *in vitro* and *in vivo*. Out of nine strains screened, *L. plantarum* T10 showed enhanced barrier function and exhibited anti-inflammatory and anti-apoptotic activities in Intestinal porcine epithelial cells-jejunum 2 (IPEC-J2) challenged with enterotoxigenic *E. coli*. *L. plantarum* T10 supplantation restored the decrease in tight junction proteins (occludin and claudin-1) in IPEC-J2 challenged with *E. coli*. The bioactive properties of T10 were due to the exopolysaccharide (EPS) from T10 through the modulation of TLR-mediated p38 MAPK pathway as observed by reduced mRNA expressions of TLR2 and TLR4 and reduced levels of p38 MAPK and cleaved-caspase-8/caspase-8 ratio in IPEC-J2 challenged with *E. coli*. Furthermore, the oral supplementation of T10 EPS alleviated dextran sulfate sodium (DSS)–induced intestinal injury and improved intestinal barrier function in male C57BL/6 mice. The study indicates the importance of probiotics and postbiotics in maintaining intestinal barrier functions in *in vitro* and *in vivo* models (Tao et al. 2024).

In another study, following screening of >20 *Lactobacillus* spp. or strains, *L. acidophilus* treatment was documented to enhance the intestinal epithelial tight junction barrier function mediated by the TLR-2 heterodimeric complexes TLR-2/TLR-1 and TLR-2/TLR-6 in Caco-2 cells. The strain also enhanced barrier function in DSS-induced colitis-induced animals, protecting against intestinal inflammation. The study showed strain-specified barrier protection of *L. acidophilus* (Al-Sadi et al. 2021). A meta-analysis including 26 randomized controlled trials (RCTs) indicated that the administration of single- or multistrain probiotics significantly improved intestinal barrier function measured according to specific indicators such as stimulated trans epithelial electrical resistance and reduced serum zonulin (a signature marker for assessing intestinal permeability), endotoxin, and lipopolysaccharide levels. Furthermore, compared to control groups, probiotic-supplemented groups showed better efficacy in reducing C-reactive proteins, TNF-α, and IL-6. Probiotics can also modulate microbial dysbiosis by enriching the abundance of *Bifidobacterium* and *Lactobacillus* (Zheng et al. 2023). The study suggests that probiotics enhance intestinal barrier functions; however, additional RCTs are necessary for a more definitive conclusion.

4.5 CLINICAL APPLICATIONS AND BENEFITS

4.5.1 Prevention of Gastrointestinal Infections

Many gastrointestinal (GI) diseases are associated with alterations in gut microbiota composition, including IBS, *Helicobacter* (*H.*) *pylori* infection, inflammatory bowel diseases, antibiotic-associated diarrhea, and acute infectious diarrhea. Probiotics have shown their potential as both preventive and therapeutic agents against these diseases; however, their mode of action is multifaceted and species-dependent (García-Santos et al. 2023). Different strains of *Lactobacillus*, *Bifidobacterium*, and *Saccharomyces* have been documented to treat GI infections (Wilkins and Sequoia 2017).

A meta-analysis of 9 RCTs involving 1,163 individuals revealed that the intake of probiotics-supplemented *Lactobacillus* along with antibiotics significantly increased the eradication of *H. pylori* infection compared to the control group (Zheng, Lyu, and Mei 2013). Another meta-analysis of 34 RCTs, comprising 9,004 patients, showed that the inclusion of probiotics can promote the removal of *H. pylori* with fewer side effects compared to traditional first-line triple therapy treatment. The triple therapy with probiotics (*Bifidobacterium–Lactobacillus–Saccharomyces*) showed 88.2% of *H. pylori* eradication, while with *Bifidobacterium–Lactobacillus*, 78.3% eradication rate was observed. Further research should focus on clinical studies involving triple therapy procedures with different species of probiotics and treatment regimens of longer durations (Wang et al. 2023).

Reports have shown the preventive and therapeutic effects of probiotics against antibiotic-associated diarrhea in both children and adults. A systematic review and meta-analysis showed the efficacy of probiotics against antibiotic-associated diarrhea. The analysis involved 63 RCTs comprising 11,811 individuals and showed a significant association of probiotics administration with the reduction of antibiotic-associated diarrhea compared to placebo (Hempel et al. 2012). The effects of probiotics on 1,639 IBS patients were investigated in another meta-analysis study that included 21 RCTs. The analysis specifically aimed to evaluate the efficacy of various probiotics on IBS patients. The study also assessed the effects of probiotic supplementation doses and treatment times in IBS patients. Overall, the study implied that, compared to placebo, the intake of a single probiotic supplement at a low dose for a short treatment period effectively modulated overall symptom response and quality of life. Nevertheless, further investigation is essential to obtain more accurate insights about the effects of probiotics on IBS patients (Zhang et al. 2016).

Furthermore, probiotic products such as bacteriocins demonstrated significant efficacy in treating GI disorders by protecting the intestinal barrier, stimulating the immune system, and inhibiting the growth of intestinal tumors. However, using bacteriocins as biological agents is not devoid of challenges. Chemical-based production is not cost-effective and unsuitable for large-scale production. The use of recombinant strains takes a long time and has low yield. Another concern is the stability of bacteriocins under different pH conditions. Also, their stability in GI conditions has not been well evaluated (Mathur et al. 2018).

4.5.2 Urogenital Health

Clinical evidence supports the use of probiotics as a safe and effective strategy for the prevention and treatment of recurrent UTIs. However, choosing suitable probiotic strains for UTIs can be tricky, and the safety and effectiveness of probiotics rely on the specific strain, optimal dose, duration of probiotic use, and timing of delivery. Acute UTIs impact individuals of all age groups globally, particularly young women, children, and the elderly. The most common uropathogens include *Enterococcus faecalis* and *E. coli*; these escape the host immune response and antimicrobial treatment due to their ability to form biofilms. They also affect the normal microflora of the urinary tract. Furthermore, the development of MDR in uropathogens is also an important challenge that requires effective treatment options. *L. fermentum* TIU19, a potential probiotic strain isolated from rice-based fermented beverage Haria, was evaluated as an important biotool against MDR *E. coli* and *E. faecalis*. The strain demonstrated antagonistic and anti-biofilming activity against uropathogens (Das et al. 2022).

A systematic review and meta-analysis of 11 studies involving children showed that probiotic use could be an alternative therapy for children at risk of developing UTIs. However, the study recommended conducting large-scale RCTs to fully explore the potential of probiotics (Emami, Sherwin, and Heidari-Soureshjani 2024). In an RCT that included premenopausal women with a history of recurrent UTIs, 174 individuals were randomly assigned to various groups. Supplementation with oral probiotics or in conjunction with vaginal probiotics demonstrated efficacy in preventing recurrent UTI symptoms when compared to placebo (Gupta, Mastromarino, and Garg 2024).

4.5.3 Oral Health

In addition to gut health, certain studies have also demonstrated the antipathogenic effects of probiotics on oral health. According to an RCT conducted by Babina et al., which included 57 participants, the administration of oral probiotics harboring the *S. salivarius* M18 strain led to a significant improvement in the condition of the gums and the degree of oral hygiene in young adults diagnosed with gingivitis (Babina et al. 2023).

A recent systemic review of 24 clinical trials, including 1,612 individuals, found that the use of probiotic products effectively reduced the number of *Streptococcus* (*S.*) *mutants* in the mouths of participants by 65%. The study further revealed that probiotic supplements were either more effective or equally effective as chlorhexidine in reducing oral infections. Coadministration of probiotics, chlorhexidine, and fluoride appears to be an ideal combination (Navidifar et al. 2023). These studies suggest the role of probiotics in clinical applications.

4.6 CHALLENGES AND CONSIDERATIONS

4.6.1 Strain-Specific Effects

While the general concept of probiotics is well-established, the efficacy of these microorganisms can be highly strain-specific. This specificity introduces several challenges and considerations when selecting and using the probiotic strains. Different health impacts have been observed from different strains of a single probiotic species. For instance, *L. rhamnosus* GG can improve gut health and effectively prevent diarrhea, while other strains of *L. rhamnosus* have not shown the same properties. The differences in activity among the strains of a particular probiotic species are attributed to their differences in genomic, metabolic pathways, and their interactions with host microbial communities (Segers and Lebeer 2014). These differences complicate predicted health outcomes and thus require extensive and in-depth strain-specific research. Moreover, the lack of standard manufacturing processes leads to inconsistencies in the viability and potency of the strain, which ultimately impacts the potential benefits of probiotics.

Broader research based on species-level limits the identification of potential probiotic strains. The interaction between host and probiotic also complicates the activity of probiotics, as the individual responses change based on health status and microbiome composition. The difference in regulations across different regions also adds complexity, affecting consumer trust and product quality.

To overcome these hurdles, precise product labeling, specific probiotic selection, and unified regulations are essential (Fusco, Fanelli, and Chieffi 2022). Continuous monitoring of long-term safety and economic feasibility is important to improve the application and benefits of probiotics as well as their effectiveness and accessibility.

4.6.2 Dosage and Administration

The proper dosage and administration of probiotics poses challenges to the effectiveness and safety. One of the major issues is the dissimilarities among different probiotic strains, which leads to complications in establishing the universal dosage guidelines and requires strain-specific research. Currently, there is no consensus on the appropriate dosages for most of the probiotics, which varies significantly across studies and their products. Determining the correct amount is also made difficult due to lack of standardization, while the individual factors complicate dosage determination even further. Probiotic dose effectiveness is influenced by a variety of parameters, including host factors, age, gender, the specific condition being treated, and health status. (Nielsen and El-Sohemy 2012, Kothari, Patel, and Kim 2019). For example, the elderly and children may require different dosages compared to adults, which adds another layer of complexity. Personalized dosing might potentially address such differences; however, it is challenging to apply on a broad scale.

The probiotic delivery methods also face significant challenges. Probiotic microorganisms need to remain viable and be active upon reaching the intestine, which requires the manufacturing and storage processes to survive the acidic environment of the gut. Probiotics are also sensitive to other factors, such as heat and moisture, which affect their stability and effectiveness (Fenster et al. 2019). Delivery methods, such as capsules, powders, and fermented foods, have different impacts on probiotic survival. Enteric coating or microencapsulation is a method used to ensure the viability and effectiveness of probiotics by the time they reach the gut, despite their added complexity and cost. The pH and bacteria-sensitive coating layers are the most common encapsulation approaches, which facilitate the release of probiotics in the intestine in response to certain bacterial colonies and pH conditions (Dodoo et al. 2017).

4.6.3 Safety and Side Effects

Experimental and clinical studies have mentioned several risks associated with the safe usage of probiotics. The documented adverse effects of probiotics include GI side effects (bloating and flatulence, diarrhea or constipation, nausea, dyspepsia, and abdominal pain), systemic infections, skin problems (mild acne or skin rashes), inflammation of the endocardium, respiratory infections, gene transfer from probiotics to the normal microbial flora, deleterious metabolic impacts of probiotics, and stimulation of the immune system. Primary populations at high risk include the elderly, newborns, hospitalized patients, and immunocompromised persons. Some suggestions for reducing the negative effects of probiotics include lowering the dose, staying hydrated, and using heat or UV radiation has been suggested to inactivate probiotics (Zielińska, Sionek, and Kołożyn-Krajewska 2018, Sotoudegan et al. 2019).

Despite the theoretical potential for gene transfer between probiotics and other microbes, both in their natural environment and inside living organisms, there is no clinical evidence of the transmission of antimicrobial resistance genes. This becomes particularly crucial when using probiotics in conjunction with antibiotics. Therefore, we must evaluate probiotics for resistance to various antibiotics to prevent the transfer of antimicrobial resistance genes. To assess the safety of probiotics, joint Food and Agriculture Organization/World Health Organization (2002) experts developed an arsenal of tests, including antibiotic resistance patterns, potential adverse effects, specific metabolic activities, and potential virulence factors (Zielińska, Sionek, and Kołożyn-Krajewska 2018).

4.7 FUTURE DIRECTIONS AND RESEARCH

4.7.1 Emerging Probiotics

Recently, researchers have shifted their focus on isolating and identifying new or next-generation probiotic strains with health-promoting properties, including antimicrobial activity. The use of advanced sequencing technologies and computational analyses has expedited the discovery and appraisal of these strains. These novel strains can play a fundamental role in reshaping the diversity and richness of the human gut microbiota (Fijan 2023), and could be defined as "living microbes identified on the basis of comparative microbiome investigations that confer health advantages to their host when taken to suitable extents" (Al-Fakhrany and Elekhnawy 2024). Next-generation probiotics are microbial species from unique genera that have not before been proven to be useful as probiotics and are likely to be announced into drug regulatory systems (Cunningham et al. 2021). Next-generation probiotics could include *Eubacterium hallii*, *Parabacteroides goldsteinii*, *Akkermansia muciniphila*, *Christensenella minuta*, *Prevotella copri*, *Bacteroides thetaiotaomicron*, *Bacteroides fragilis*, *Faecalibacterium prausnitzii*, *Roseburia* spp., and others (Kaźmierczak-Siedlecka et al. 2022, Chang et al. 2019).

4.7.2 Novel Strains with Antimicrobial Properties

Next-generation probiotics and conventional probiotics are reported to have some basic differences. Table 4.1 shows the differences between conventional and next-generation probiotic strains adapted with little modification from Abouelela and Helmy (2024). The health modulatory effects of next-generation probiotics could be due to the modulation of the immune response, secretion of bile acids, maintenance of gut health and stability, improved insulin sensitivity, antioxidant effect, anti-inflammatory effects, and secretion of antimicrobials (Abouelela and Helmy 2024).

Spinler et al. (2017) reported the application of *L. reuteri* as next-generation probiotics to combat *Clostridium difficile* infection by reuterin-mediated antimicrobial therapy. Under antibiotic pressure,

TABLE 4.1
Comparison between Traditional Probiotics and Next-Generation Probiotics

	Traditional Probiotics	Next-Generation Probiotics
Origin	Long history of use and are derived from a limited number of species, such as *Lactobacillus* and *Bifidobacterium*	Derived from next-generation microorganisms that have been recently isolated using advanced tools and techniques
Development	Developed based on comparative analysis of microbiota compositions between healthy and diseased individuals	Developed through a top-down screening strategy, which involves screening microbes enriched in healthy individuals compared to those in diseased individuals
Safety	A long history of safe use in humans	Their safety is not yet proven as they are relatively new and have not been used for as long as traditional probiotics
Applications	Mainly used as food ingredients or supplements	Primarily used to treat or cure disease conditions
Regulation	They are not subject to the same level of regulation	Considered to be live biotherapeutic products (LBPs) or drugs, which are subject to pharmaceutical clinical trials and research on their pharmacokinetics and pharmacodynamics
Strain specificity	They are species-specific	Their health-promoting features are more closely tied to specific strains rather than entire species

Source: Adapted from Abouelela and Helmy (2024).

the authors screened a *L. reuteri* strain that was the least susceptible to antibiotics. Furthermore, the authors utilized strain 17938 to produce reuterin, a secondary metabolite, through the active fermentation of glycerol. Overall, *in vitro*, co-treatment of *L. reuteri* 17938 with glycerol obstructs *C. difficile* growth within an antibiotic-treated human fecal microbial community. The co-delivery did not significantly impact the overall composition of the microbial community (Spinler et al. 2017).

4.7.3 Synergy with Antibiotics

While traditional antibiotics prove essential in treating a variety of diseases, they also have related side effects, including allergic responses, nausea, vomiting, disruption of the gut barrier, and gut microbiota dysbiosis (Kesavelu and Jog 2023). People have long regarded probiotics as a supplementary treatment to mitigate the harmful effects of antibiotics. A study evaluated the antibacterial activity of five commercially available probiotics alone as well as in synergism with ten antibiotics against different foodborne pathogens, including *E. coli*, *Klebsiella* spp., *Pseudomonas* spp., *Shigella* spp., *Staphylococcus* spp., and *Vibrio* species. Probiotics demonstrated antibacterial effects against all selected species, except for *Salmonella* spp. and *Bacillus* spp. The observed zone of inhibition (ZOI) ranged from 10 to 30mm. In combination with antibiotics, the ZOI diameter was increased significantly. The study suggests the use of combinational therapy (antibiotics+probiotics) for treating MDR bacteria (Acharjee et al. 2022).

4.7.4 Personalized Probiotics

The beneficial effects of probiotics can vary among individuals due to factors such as heredity, food and daily routine, physiology, and environmental conditions (Singh and Natraj 2021, Rinninella et al. 2019). Furthermore, these differences in the individual's microbial composition, diversity, and metabolic physiognomies significantly affect the efficacy of probiotic supplementations (Zmora et al. 2018). This is the rationale behind the concept of personalized probiotics that have been designed to meet the requirements of individuals' distinct gut microflora (Veiga et al. 2020). The use of artificial intelligence and machine learning tools expedited this novel concept by exploiting information pertaining to people's physiology and gut microbiota, thus facilitating the selection of personalized probiotics (Das et al. 2023).

4.8 CONCLUSIONS

Antibiotic resistance has triggered an alarming situation for human health and the environment. Probiotics have emerged as viable substitutes owing to their beneficial characteristics and properties compared to antibiotics. These beneficial microbes exhibit different modes of action, such as the competitive exclusion against potential pathogens, the modulation of immune responses, and the improvement of gut barrier function. These multimodal actions reduce the likelihood of a pathogen's antimicrobial resistance and enhance the possibility of using probiotics as alternatives to antibiotics. Numerous cell, animal, and clinical studies demonstrate the efficacy of these beneficial microbes in combating pathogenic infections and confirm their role as effective antimicrobial agents. Furthermore, combining these beneficial microbes with antibiotics significantly improves their ability to inhibit pathogens, thereby advocating their use in a synergistic mode. However, additional clinical studies are necessary to ascertain the antimicrobial efficacy of these probiotics, given that their beneficial effects can fluctuate depending on the host's diet, lifestyle, genetics, environment, and gut physiology. Of note, the safety profile and conclusion about a probiotic cannot be generalized. The specific strain and target population should always be taken into account. Further research is needed to explore the antimicrobial properties of next-generation probiotic strains.

REFERENCES

Abhari, K. H, S. S. Shekarforoush, J. Sajedianfard, S. Hosseinzadeh, and S. Nazifi. 2015. The effects of probiotic, prebiotic and synbiotic diets containing *Bacillus coagulans* and inulin on rat intestinal microbiota. *Iranian Journal of Veterinary Research* 16 (3):267–73.

Abouelela, M. E, and Y. A. Helmy. 2024. Next-generation probiotics as novel therapeutics for improving human health: Current trends and future perspectives. *Microorganisms* 12 (3):430. https://doi.org/10.3390/microorganisms12030430

Acharjee, M., F. Hasan, T. Islam, I. Tun Nur, N. Begum, C. Mazumder, M. A. Lubna, N. Zerin, A. Shahriar, and M. R. Mahmud. 2022. In-vitro antibacterial activity of commercially available probiotics on food-borne pathogens along with their synergistic effects with synthetic drugs. *Metabolism Open* 14:100187. https://doi.org/10.1016/j.metop.2022.100187

Aghamohammad, S., and M. Rohani. 2023. Antibiotic resistance and the alternatives to conventional antibiotics: The role of probiotics and microbiota in combating antimicrobial resistance. *Microbiological Research* 267:127275. https://doi.org/10.1016/j.micres.2022.127275

Ahmed, G. K., H. K.-A. Ramadan, K. Elbeh, and N. A. Haridy. 2024. Bridging the gap: Associations between gut microbiota and psychiatric disorders. *Middle East Current Psychiatry* 31 (1). https://doi.org/10.1186/s43045-024-00395-9

Ahmed, S., S. Singh, V. Singh, K. D. Roberts, A. Zaidi, and A. Rodriguez-Palacios. 2022. The weissella genus: Clinically treatable bacteria with antimicrobial/probiotic effects on inflammation and cancer. *Microorganisms* 10 (12). https://doi.org/10.3390/microorganisms10122427

Al-Fakhrany, O. M., and E. Elekhnawy. 2024. Next-generation probiotics: The upcoming biotherapeutics. *Molecular Biology Reports* 51 (1):505. https://doi.org/10.1007/s11033-024-09398-5.

Al-Sadi, R., P. Nighot, M. Nighot, M. Haque, M. Rawat, and T. Y. Ma. 2021. *Lactobacillus acidophilus* induces a strain-specific and toll-like receptor 2–dependent enhancement of intestinal epithelial tight junction barrier and protection against intestinal inflammation. *The American Journal of Pathology* 191 (5):872–884. https://doi.org/10.1016/j.ajpath.2021.02.003

Arques, J. L., E. Rodriguez, S. Langa, J. M. Landete, and M. Medina. 2015. Antimicrobial activity of lactic acid bacteria in dairy products and gut: Effect on pathogens. *Biomedical Research International* 2015:584183. https://doi.org/10.1155/2015/584183

Aziz, N., and B. Bonavida. 2016. Activation of natural killer cells by probiotics. *Forum for Immunopathology and Disease Therapy* 7 (1–2):41–55. https://doi.org/10.1615/ForumImmunDisTher.2016017095

Babina, K., D. Salikhova, V. Doroshina, I. Makeeva, A. Zaytsev, M. Uvarichev, M. Polyakova, and N. Novozhilova. 2023. Antigingivitis and antiplaque effects of oral probiotic containing the *Streptococcus salivarius* M18 strain: A randomized clinical trial. *Nutrients* 15 (18):3882. https://doi.org/10.3390/nu15183882

Bai, H., L.-Y. He, D.-L. Wu, F.-Z. Gao, M. Zhang, H.-Y. Zou, M.-S. Yao, and G.-G. Ying. 2022. Spread of airborne antibiotic resistance from animal farms to the environment: Dispersal pattern and exposure risk. *Environment International* 158:106927. https://doi.org/10.1016/j.envint.2021.106927.

Barzegari, A., K. Kheyrolahzadeh, S. M. Hosseiniyan Khatibi, S. Sharifi, M. Y. Memar, and S. Z. Vahed. 2020. The battle of probiotics and their derivatives against biofilms. *Infection and Drug Resistance* 2020:659–672. https://doi.org/10.2147/IDR.S232982

Bron, P. A., M. Kleerebezem, R.-J. Brummer, P. D. Cani, A. Mercenier, T. T. MacDonald, C. L. Garcia-Ródenas, and J. M. Wells. 2017. Can probiotics modulate human disease by impacting intestinal barrier function? *British Journal of Nutrition* 117 (1):93–107. https://doi.org/10.1017/S0007114516004037

Camilleri, M. 2019. Leaky gut: Mechanisms, measurement and clinical implications in humans. *Gut* 68 (8):1516–1526. https://doi.org/10.1136/gutjnl-2019-318427

Chang, C.-J., T.-L. Lin, Y.-L. Tsai, T.-R. Wu, W.-F. Lai, C.-C. Lu, and H.-C. Lai. 2019. Next generation probiotics in disease amelioration. *Journal of Food and Drug Analysis* 27 (3):615–622. https://doi.org/10.1016/j.jfda.2018.12.011

Cherrat, L., L. Espina, M. Bakkali, D. Garcia-Gonzalo, R. Pagan, and A. Laglaoui. 2014. Chemical composition and antioxidant properties of *Laurus nobilis* L. and *Myrtus communis* L. essential oils from Morocco and evaluation of their antimicrobial activity acting alone or in combined processes for food preservation. *Journal of the Science of Food and Agriculture* 94 (6):1197–204. https://doi.org/10.1002/jsfa.6397

Chinemerem Nwobodo, D., M. C. Ugwu, C. Oliseloke Anie, M. T. S. Al-Ouqaili, J. Chinedu Ikem, U. Victor Chigozie, and M. Saki. 2022. Antibiotic resistance: The challenges and some emerging strategies for tackling a global menace. *Journal of Clinical Laboratory Analysis* 36 (9):e24655. https://doi.org/10.1002/jcla.24655

Cizeikiene, D., and J. Jagelaviciute. 2021. Investigation of antibacterial activity and probiotic properties of strains belonging to Lactobacillus and Bifidobacterium genera for their potential application in functional food and feed products. *Probiotics Antimicrob Proteins* 13 (5):1387–1403. https://doi.org/10.1007/s12602-021-09777-5

Clardy, J., M. A Fischbach, and C. R Currie. 2009. The natural history of antibiotics. *Current Biology* 19 (11):R437–R441. https://doi.org/10.1016/j.cub.2009.04.001

Cunningham, M., M. A. Azcarate-Peril, A. Barnard, V. Benoit, R. Grimaldi, D. Guyonnet, H. D Holscher, K. Hunter, S. Manurung, and D. Obis. 2021. Shaping the future of probiotics and prebiotics. *Trends in Microbiology* 29 (8):667–685. https://doi.org/10.1016/j.tim.2021.01.003

Darby, E. M, E. Trampari, P. Siasat, M. S. Gaya, I. Alav, M. A Webber, and J. M. A. Blair. 2023. Molecular mechanisms of antibiotic resistance revisited. *Nature Reviews Microbiology* 21 (5):280–295. https://doi.org/10.1038/s41579-022-00820-y

Das, A., R. N Behera, A. Kapoor, and K. Ambatipudi. 2023. The potential of meta-proteomics and artificial intelligence to establish the next generation of probiotics for personalized healthcare. *Journal of Agricultural and Food Chemistry* 71 (46):17528–17542. https://doi.org/10.1021/acs.jafc.3c03834

Das, S., K. Vishakha, S. Banerjee, T. Bera, S. Mondal, and A. Ganguli. 2022. A novel probiotic strain of *Lactobacillus fermentum* TIU19 isolated from Haria beer showing both in vitro antibacterial and antibiofilm properties upon two multi resistant uro-pathogen strains. *Current Research in Microbial Sciences* 3:100150. https://doi.org/10.1016/j.crmicr.2022.100150.

Davies, J., and D. Davies. 2010. Origins and evolution of antibiotic resistance. *Microbiology and Molecular Biology Reviews* 74 (3):417–33. https://doi.org/10.1128/MMBR.00016-10

Dhanani, A. S., and T. Bagchi. 2013. The expression of adhesin EF-Tu in response to mucin and its role in Lactobacillus adhesion and competitive inhibition of enteropathogens to mucin. *Journal of Applied Microbiology* 115 (2):546–554. https://doi.org/10.1111/jam.12249

Di Martino, P. 2022. Antimicrobial agents and microbial ecology. *AIMS Microbiology* 8 (1):1–4. https://doi.org/10.3934/microbiol.2022001

Dobson, A., P. D. Cotter, R. P. Ross, and C. Hill. 2012. Bacteriocin production: A probiotic trait? *Applied Environmental Microbiology* 78 (1):1–6. https://doi.org/10.1128/AEM.05576-11

Dodoo, C. C, J. Wang, A. W Basit, P. Stapleton, and S. Gaisford. 2017. Targeted delivery of probiotics to enhance gastrointestinal stability and intestinal colonisation. *International Journal of Pharmaceutics* 530 (1–2):224–229. https://doi.org/10.1016/j.ijpharm.2017.07.068

Dutra, V., A. C. Silva, P. Cabrita, C. Peres, X. Malcata, and L. Brito. 2016. *Lactobacillus plantarum* LB95 impairs the virulence potential of Gram-positive and Gram-negative food-borne pathogens in HT-29 and Vero cell cultures. *Journal of Medical Microbiology* 65 (1):28–35. https://doi.org/10.1099/jmm.0.000196

Emami, E., C. M. T. Sherwin, and S. Heidari-Soureshjani. 2024. Effect of probiotics on urinary tract infections in children: A systematic review and meta-analysis. *Current Reviews in Clinical and Experimental Pharmacology* 19 (1):111–121. https://doi.org/10.2174/2772432817666220501114505

Fenster, K., B. Freeburg, C. Hollard, C. Wong, R. R. Laursen, and A. C Ouwehand. 2019. The production and delivery of probiotics: A review of a practical approach. *Microorganisms* 7 (3):83. https://doi.org/10.3390/microorganisms7030083

Fijan, S. 2023. Probiotics and their antimicrobial effect. *Microorganisms* 11 (2). https://doi.org/10.3390/microorganisms11020528

Fusco, V., F. Fanelli, and D. Chieffi. 2022. Authenticity of probiotic foods and dietary supplements: A pivotal issue to address. *Critical Reviews in Food Science and Nutrition* 62 (25):6854–6871. https://doi.org/10.1080/10408398.2021.1907300

García-Santos, J. A., A. Nieto-Ruiz, M. García-Ricobaraza, T. Cerdó, and C. Campoy. 2023. Impact of probiotics on the prevention and treatment of gastrointestinal diseases in the pediatric population. *International Journal of Molecular Sciences* 24 (11):9427. https://doi.org/10.3390/ijms24119427

Gasbarrini, G., F. Bonvicini, and A. Gramenzi. 2016. Probiotics history. *Journal of Clinical Gastroenterology* 50:S116–S119. https://doi.org/10.1097/MCG.0000000000000697

Gaspar, C., G. G. Donders, R. Palmeira-de-Oliveira, J. A. Queiroz, C. Tomaz, J. Martinez-de-Oliveira, and A. Palmeira-de-Oliveira. 2018. Bacteriocin production of the probiotic *Lactobacillus acidophilus* KS400. *AMB Express* 8 (1):153. https://doi.org/10.1186/s13568-018-0679-z

Gupta, V., P. Mastromarino, and R. Garg. 2024. Effectiveness of prophylactic oral and/or vaginal probiotic supplementation in the prevention of recurrent urinary tract infections: A randomized, double-blind, placebo-controlled trial. *Clinical Infectious Diseases* 78 (5):1154–1161. https://doi.org/10.1093/cid/ciad766

Hempel, S., S. J. Newberry, A. R. Maher, Z. Wang, J. N. Miles, R. Shanman, B. Johnsen, and P. G. Shekelle. 2012. Probiotics for the prevention and treatment of antibiotic-associated diarrhea: A systematic review and meta-analysis. *JAMA* 307 (18):1959–1969. https://doi.org/10.1001/jama.2012.3507

Huemer, M., S. M. Shambat, S. D. Brugger, and A. S. Zinkernagel. 2020. Antibiotic resistance and persistence-Implications for human health and treatment perspectives. *EMBO Reports* 21 (12):e51034. https://doi.org/10.15252/embr.202051034

Kapse, N. G., A. S. Engineer, V. Gowdaman, S. Wagh, and P. K. Dhakephalkar. 2019. Functional annotation of the genome unravels probiotic potential of *Bacillus coagulans* HS243. *Genomics* 111 (4):921–929. https://doi.org/10.1016/j.ygeno.2018.05.022

Kawai, T., and S. Akira. 2009. The roles of TLRs, RLRs and NLRs in pathogen recognition. *International Immunology* 21 (4):317–337. https://doi.org/10.1093/intimm/dxp017

Kawashima, T., N. Ikari, T. Kouchi, Y. Kowatari, Y. Kubota, N. Shimojo, and N. M Tsuji. 2018. The molecular mechanism for activating IgA production by *Pediococcus acidilactici* K15 and the clinical impact in a randomized trial. *Scientific Reports* 8 (1):5065. https://doi.org/10.1038/s41598-018-23404-4

Kaźmierczak-Siedlecka, K., K. Skonieczna-Żydecka, T. Hupp, R. Duchnowska, N. Marek-Trzonkowska, and K. Połom. 2022. Next-generation probiotics – Do they open new therapeutic strategies for cancer patients? *Gut Microbes* 14 (1):2035659. https://doi.org/10.1080/19490976.2022.2035659.

Kemgang, T. S., S. Kapila, V. P. Shanmugam, S. Reddi, and R. Kapila. 2016. Fermented milk with probiotic *Lactobacillus rhamnosus* S1K3 (MTCC5957) protects mice from Salmonella by enhancing immune and nonimmune protection mechanisms at intestinal mucosal level. *The Journal of Nutritional Biochemistry* 30:62–73. https://doi.org/10.1016/j.jnutbio.2015.11.018

Kesavelu, D., and P. Jog. 2023. Current understanding of antibiotic-associated dysbiosis and approaches for its management. *Therapeutic Advances in Infectious Disease* 10:20499361231154443. https://doi.org/10.1177/20499361231154443

Kim, Y., J. W. Lee, S.-G. Kang, S. Oh, and M. W. Griffiths. 2012. Bifidobacterium spp. influences the production of autoinducer-2 and biofilm formation by *Escherichia coli* O157:H7. *Anaerobe* 18 (5):539–545. https://doi.org/10.1016/j.anaerobe.2012.08.006

Kiousi, D. E., C. Efstathiou, V. Tzampazlis, S. Plessas, M. Panopoulou, M. Koffa, and A. Galanis. 2023. Genetic and phenotypic assessment of the antimicrobial activity of three potential probiotic lactobacilli against human enteropathogenic bacteria. *Frontiers in Cellular and Infection Microbiology* 13:1127256. https://doi.org/10.3389/fcimb.2023.1127256

Kothari, D., S. Patel, and S.-K. Kim. 2019. Probiotic supplements might not be universally-effective and safe: A review. *Biomedicine & Pharmacotherapy* 111:537–547. https://doi.org/10.1016/j.biopha.2018.12.104

Lathakumari, R. H., L. K. Vajravelu, A. Satheesan, S. Ravi, and J. Thulukanam. 2024. Antibiotics and the gut microbiome: Understanding the impact on human health. *Medicine in Microecology* 20. https://doi.org/10.1016/j.medmic.2024.100106.

Latif, A., A. Shehzad, S. Niazi, A. Zahid, W. Ashraf, M. W. Iqbal, A. Rehman, T. Riaz, R. M. Aadil, I. M. Khan, et al. 2023. Probiotics: Mechanism of action, health benefits and their application in food industries. *Frontiers in Microbiology* 14:1216674. https://doi.org/10.3389/fmicb.2023.1216674

Llor, C., and L. Bjerrum. 2014. Antimicrobial resistance: Risk associated with antibiotic overuse and initiatives to reduce the problem. *Therapeutic Advances in Drug Safety* 5 (6):229–41. https://doi.org/10.1177/2042098614554919

Mathur, H., D. Field, M. C Rea, P. D Cotter, C. Hill, and R. P. Ross. 2018. Fighting biofilms with lantibiotics and other groups of bacteriocins. *NPJ Biofilms and Microbiomes* 4 (1):9. https://doi.org/10.1038/s41522-018-0053-6

Mazziotta, C., M. Tognon, F. Martini, E. Torreggiani, and J. C. Rotondo. 2023. Probiotics mechanism of action on immune cells and beneficial effects on human health. *Cells* 12 (1):184. https://doi.org/10.3390/cells12010184

Melo, T. A., T. F. Dos Santos, M. E. de Almeida, L. A. Junior, E. F. Andrade, R. P. Rezende, L. M. Marques, and C. C. Romano. 2016. Inhibition of *Staphylococcus aureus* biofilm by Lactobacillus isolated from fine cocoa. *BMC Microbiology* 16 (1):250. https://doi.org/10.1186/s12866-016-0871-8

Monika K., T. Malik, R. Gehlot, R. K, A. Kumari, R. Sindhu, and P. Rohilla. 2021. Antimicrobial Property of Probiotics. *Environment Conservation Journal* 22 (SE):33–48. https://doi.org/10.36953/ecj.2021.Se.2204

Mousavi Khaneghah, A., K. Abhari, I. Eş, M. B. Soares, R. B. A. Oliveira, H. Hosseini, M. Rezaei, C. F. Balthazar, R. Silva, A. G. Cruz, et al. 2020. Interactions between probiotics and pathogenic microorganisms in hosts and foods: A review. *Trends in Food Science & Technology* 95:205–218. https://doi.org/10.1016/j.tifs.2019.11.022

Navidifar, T., M. M. Ari, A. Alipourkermani, R. Afifirad, P. Asadollahi, A. Veisi, R. Ghanavati, and A. Darbandi. 2023. Clinical efficacy of probiotics on oral health: A systematic review of clinical trials. *Current Pharmaceutical Biotechnology* 24 (15):1916–1927. https://doi.org/10.2174/1389201024666230405135457

Nielsen, D. E, and A. El-Sohemy. 2012. Applying genomics to nutrition and lifestyle modification. *Personalized Medicine* 9 (7):739–749. https://doi.org/10.2217/pme.12.79

Noh, H.-J., J. M. Park, Y. J. Kwon, K. Kim, S. Y. Park, I. Kim, J. H. Lim, B. K. Kim, and B.-Y. Kim. 2022. Immunostimulatory effect of heat-killed probiotics on RAW264. 7 macrophages. *Journal of Microbiology and Biotechnology* 32 (5):638. https://doi.org/10.4014/jmb.2201.01015

Palucka, K., and J. Banchereau. 1999. Dendritic cells: A link between innate and adaptive immunity. *Journal of Clinical Immunology* 19 (1):12–25. https://doi.org/10.1023/A:1020558317162

Plaza-Diaz, J., F. J. Ruiz-Ojeda, M. Gil-Campos, and A. Gil. 2019. Mechanisms of action of probiotics. *Advances in Nutrition* 10 (suppl_1):S49–S66. https://doi.org/10.1093/advances/nmy063

Pridmore, R. D., A. C. Pittet, F. Praplan, and C. Cavadini. 2008. Hydrogen peroxide production by *Lactobacillus johnsonii* NCC 533 and its role in anti-Salmonella activity. *FEMS Microbiology Letters* 283 (2):210–5. https://doi.org/10.1111/j.1574-6968.2008.01176.x

Rabetafika, H. N, A. Razafindralambo, B. Ebenso, and H. L Razafindralambo. 2023. Probiotics as antibiotic alternatives for human and animal applications. *Encyclopedia* 3 (2):561–581. https://doi.org/10.3390/encyclopedia3020040

Rinninella, E., P. Raoul, M. Cintoni, F. Franceschi, G. A. D. Miggiano, A. Gasbarrini, and M. C. Mele. 2019. What is the healthy gut microbiota composition? A changing ecosystem across age, environment, diet, and diseases. *Microorganisms* 7 (1). https://doi.org/10.3390/microorganisms7010014

Rocha-Ramírez, L. M., R. A. Pérez-Solano, S. L. Castañón-Alonso, S. S. Moreno Guerrero, A. Ramírez Pacheco, M. García Garibay, and C. Eslava. 2017. Probiotic *Lactobacillus* strains stimulate the inflammatory response and activate human macrophages. *Journal of Immunological Research* 2017:4607491. https://doi.org/10.1155/2017/4607491

Segers, M. E, and S. Lebeer. 2014. Towards a better understanding of *Lactobacillus rhamnosus* GG-host interactions. *Microbial Cell Factories* 13 (Suppl 1):S7. https://doi.org/10.1186/1475-2859-13-S1-S7

Servin, A. L. 2004. Antagonistic activities of lactobacilli and bifidobacteria against microbial pathogens. *FEMS Microbiology Reviews* 28 (4):405–440. https://doi.org/10.1016/j.femsre.2004.01.003

Shaaban, M., O. A. Abd El-Rahman, B. Al-Qaidi, and H. M Ashour. 2020. Antimicrobial and antibiofilm activities of probiotic Lactobacilli on antibiotic-resistant *Proteus mirabilis*. *Microorganisms* 8 (6):960. https://doi.org/10.3390/microorganisms8060960

Singh, T. P., and B. H. Natraj. 2021. Next-generation probiotics: A promising approach towards designing personalized medicine. *Critical Reviews in Microbiology* 47 (4):479–498. https://doi.org/10.1080/1040841x.2021.1902940

Smillie, C. S, M. B Smith, J. Friedman, O. X Cordero, L. A David, and E. J Alm. 2011. Ecology drives a global network of gene exchange connecting the human microbiome. *Nature* 480 (7376):241–244. https://doi.org/10.1038/nature10571

Sotoudegan, F., M. Daniali, S. Hassani, S. Nikfar, and M. Abdollahi. 2019. Reappraisal of probiotics' safety in human. *Food and Chemical Toxicology* 129:22–29. https://doi.org/10.1016/j.fct.2019.04.032

Spinler, J. K, J. Auchtung, A. Brown, P. Boonma, N. Oezguen, C. L Ross, R. A. Luna, J. Runge, J. Versalovic, and A. Peniche. 2017. Next-generation probiotics targeting *Clostridium difficile* through precursor-directed antimicrobial biosynthesis. *Infection and Immunity* 85 (10):e00303–17. https://doi.org/10.1128/IAI.00303-17.

Srednicka, P., E. Juszczuk-Kubiak, M. Wojcicki, M. Akimowicz, and M. L. Roszko. 2021. Probiotics as a biological detoxification tool of food chemical contamination: A review. *Food and Chemical Toxicology* 153:112306. https://doi.org/10.1016/j.fct.2021.112306

Tao, T., L. Zhang, T. Yu, J. Ma, S. Lu, J. Ren, X. Li, and X. Guo. 2024. Exopolysaccharide production by *Lactobacillus plantarum* T10 is responsible for the probiotic activity in enhancing intestinal barrier function *in vitro* and *in vivo*. *Food & Function* 15 (7):3583–3599. https://doi.org/10.1039/D4FO00526K

Tejero-Sarinena, S., J. Barlow, A. Costabile, G. R. Gibson, and I. Rowland. 2012. In vitro evaluation of the antimicrobial activity of a range of probiotics against pathogens: Evidence for the effects of organic acids. *Anaerobe* 18 (5):530–538. https://doi.org/10.1016/j.anaerobe.2012.08.004

Valeriano, V. D., B. B. Bagon, M. P. Balolong, and D. K. Kang. 2016. Carbohydrate-binding specificities of potential probiotic Lactobacillus strains in porcine jejunal (IPEC-J2) cells and porcine mucin. *Journal of Microbiology* 54 (7):510–519. https://doi.org/10.1007/s12275-016-6168-7

van Zyl, W. F., S. M. Deane, and L. M. T. Dicks. 2020. Molecular insights into probiotic mechanisms of action employed against intestinal pathogenic bacteria. *Gut Microbes* 12 (1):1831339. https://doi.org/10.1080/19490976.2020.1831339

Veiga, P., J. Suez, M. Derrien, and E. Elinav. 2020. Moving from probiotics to precision probiotics. *Nature Microbiology* 5 (7):878–880. https://doi.org/10.1038/s41564-020-0721-1.

Wan, M. L. Y., S. J. Forsythe, and H. El-Nezami. 2019. Probiotics interaction with foodborne pathogens: A potential alternative to antibiotics and future challenges. *Critical Reviews in Food Science and Nutrition* 59 (20):3320–3333. https://doi.org/10.1080/10408398.2018.1490885

Wang, Y., X. Wang, X.-Y. Cao, H.-L. Zhu, and L. Miao. 2023. Comparative effectiveness of different probiotics supplements for triple *Helicobacter pylori* eradication: A network meta-analysis. *Frontiers in Cellular and Infection Microbiology* 13:1120789. https://doi.org/10.3389/fcimb.2023.1120789

Warman, D. J., H. Jia, and H. Kato. 2022. The potential roles of probiotics, resistant starch, and resistant proteins in ameliorating inflammation during aging (inflammaging). *Nutrients* 14 (4). https://doi.org/10.3390/nu14040747.

Wautier, J.-L., and M.-P. Wautier. 2023. Pro- and anti-inflammatory prostaglandins and cytokines in humans: A mini review. *International Journal of Molecular Sciences* 24 (11):9647. https://doi.org/10.3390/ijms24119647

Wiese, J., and J. F. Imhoff. 2019. Marine bacteria and fungi as promising source for new antibiotics. *Drug Development Research* 80 (1):24–27. https://doi.org/10.1002/ddr.21482

Wilkins, T., and J. Sequoia. 2017. Probiotics for gastrointestinal conditions: A summary of the evidence. *American Family Physician* 96 (3):170–178.

Zhang, Y., L. Li, C. Guo, D. Mu, B. Feng, X. Zuo, and Y. Li. 2016. Effects of probiotic type, dose and treatment duration on irritable bowel syndrome diagnosed by Rome III criteria: A meta-analysis. *BMC Gastroenterology* 16:1–11. https://doi.org/10.1186/s12876-016-0470-z

Zheng, X., L. Lyu, and Z. Mei. 2013. *Lactobacillus*-containing probiotic supplementation increases *Helicobacter pylori* eradication rate: Evidence from a meta-analysis. *Revista Espanola de Enfermedades Digestivas* 105 (8):445–53. https://doi.org/10.4321/s1130-01082013000800002

Zheng, Y., Z. Zhang, P. Tang, Y. Wu, A. Zhang, D. Li, C.-Z. Wang, J.-Y. Wan, H. Yao, and C.-S. Yuan. 2023. Probiotics fortify intestinal barrier function: A systematic review and meta-analysis of randomized trials. *Frontiers in Immunology* 14:1143548. https://doi.org/10.3389/fimmu.2023.1143548

Zielińska, D., B. Sionek, and D. Kołożyn-Krajewska. 2018. "Chapter 6 – Safety of Probiotics." In *Diet, Microbiome and Health*, edited by Alina Maria Holban and Alexandru Mihai Grumezescu, 131–161. Academic Press.

Zmora, N., G. Zilberman-Schapira, J. Suez, U. Mor, M. Dori-Bachash, S. Bashiardes, E. Kotler, M. Zur, D. Regev-Lehavi, and R. Ben-Zeev Brik. 2018. Personalized gut mucosal colonization resistance to empiric probiotics is associated with unique host and microbiome features. *Cell* 174 (6):1388–1405. e21. https://doi.org/10.1016/j.cell.2018.08.041

5 Animal Health and Antimicrobial Resistance

A One-Health Perspective on Probiotics

Koushlesh Ranjan, R. A. Siddique, and M. K. Bharti
Department of Veterinary Physiology and Biochemistry, College of Veterinary and Animal Sciences, Sardar Vallabhbhai Patel University of Agriculture & Technology, Meerut, Uttar Pradesh, India

5.1 INTRODUCTION

Antibiotics have revolutionized modern medicine, transforming the treatment and management of bacterial infections. However, their extensive and often improper use has led to the emergence of antibiotic resistance, posing a serious threat to public health and veterinary practices (Sabtu et al., 2015). To combat this growing challenge, it is essential to minimize antibiotic misuse and explore alternative strategies. Among these alternatives, probiotics have emerged as a promising solution. Probiotics have been shown to reduce the risk of infectious diseases in both humans and animals (King et al., 2014), thereby potentially decreasing the dependence on antibiotics.

Probiotics are living microorganisms that, when administered in adequate amounts, confer health benefits to the host. They are most commonly found in the intestines, where they contribute for maintaining a healthy microbiome. The intestinal microbiome plays a critical role in resisting infections, supporting immune system development, and synthesizing essential nutrients (Ubeda & Pamer, 2012). Probiotics have also been effective in reducing the incidence of antibiotic-associated diarrhea (AAD), a common side effect of antibiotic use, through the stabilization of intestinal microbiota. This stabilization may also help mitigate the spread of antibiotic resistance (Hempel et al., 2012). Recent studies suggest that probiotics are effective in both preventing and treating infectious diseases (Yang et al., 2019). Despite the reliance on antibiotics as the primary treatment for infections, irrational use has resulted in adverse effects, such as patient-specific side effects and the widespread emergence of multidrug-resistant bacteria, which pose significant public health risks (Yang et al., 2019). These challenges underscore the urgent need for alternative antimicrobial therapies, particularly those based on natural products (Silva et al., 2019).

Probiotics are widely used to promote gut health, enhance immune function, and improve overall well-being in humans and animals. However, selecting the appropriate strains and dosages tailored to specific needs is crucial. Common probiotic strains in humans include *Bifidobacterium bifidum*, *Lactobacillus acidophilus*, and *Saccharomyces boulardii*, while in animals, strains such as *Lactobacillus reuteri*, *Enterococcus faecium*, and *Bacillus subtilis* are commonly utilized. This chapter provides a comprehensive overview of the diverse roles of probiotics in human and animal health and productivity. The information presented is based on an in-depth analysis of data gathered from reputable sources, including Google Scholar, PubMed, and PubMed Central.

DOI: 10.1201/9781032676951-5

5.2 ANTIBIOTICS AND ANTIBIOTIC RESISTANCE

Antibiotics are among the most critical drugs in modern medicine, enabling the treatment of life-threatening infectious diseases. They are the drugs that kill the microorganisms, specifically targeting bacterial infections. However, due to their antimicrobial properties, antibiotics can impact not only the pathogenic microbes they are meant to treat but also other useful microbes, particularly within the gastrointestinal tract, where most of the body's natural microbes are found. Moreover, several pathogenic microbes are acquiring resistance to antibiotics. Therefore, to ensure their continued effectiveness, it is essential to reduce the emergence and spread of antibiotic resistance. This effort begins with protecting the intestinal microbiota, which is often an unintended casualty of indiscriminate antibiotic use.

Scientifically, antibiotics target specific molecular components within the microbial cells. In some bacteria, these targets may be absent or altered, leading to intrinsic resistance, which is common across most or all strains of a species (Reveles et al., 2014). However, resistance can also be acquired through various mechanisms (Table 5.1). For instance, a mutation in an existing transport system may enable the bacterium to expel intracellular antibiotics, reducing sensitivity or causing

TABLE 5.1
Mechanisms of Action and Resistance Strategies of Various Antibiotic Classes

Antibiotic Class	Mechanisms of Action	Mechanisms of Antibiotic Resistance	Reference
Sulfonamides	Inhibit bacterial growth by competitive inhibition of p-aminobenzoic acid in folic acid synthesis	• Mutations in sul1, sul2, folP, sulfonamide monooxygenase gene	Kim et al. (2019)
Lactams	Interfere with peptidoglycan synthesis followed by inhibition in cell wall synthesis	• Change in the permeability of antibiotics to microbial cell • Production of β-Lactamases by microbes	Fernandes et al. (2013)
Chloramphenicols	Interfere with 50S ribosomal subunit and inhibit the protein synthesis	• Mutations in 23S rRNA of the 50S ribosomal subunit • Enzymatic inactivation of acetyltransferases	Schwarz et al. (2016)
Tetracyclines	Interfere with 30S ribosomal subunit and inhibit the protein synthesis	• Mutation in the binding site of the antibiotic • Enzymatic inactivation of antibiotic	Grossman (2016)
Aminoglycosides	Interfere with 30S ribosomal subunit and inhibit the protein synthesis	• 16S rRNA methylation • Inhibition of aminoglycosides by adenylation, phosphorylation, and acetylation • By action of aminoglycoside-modifying enzymes	Krause et al. (2016)
Glycopeptides	Interfere with peptidoglycan synthesis followed by inhibition in cell wall synthesis	• Membrane impermeability of Gram-negative bacteria • Enzymatic hydrolysis and modification of peptidoglycan precursors	Zeng et al. (2016)
Fluoroquinolones	Inhibition of nucleic acid synthesis	• Mutations in topoisomerase IV or DNA gyrase	Redgrave et al. (2014)

resistance. Similarly, a mutation in a gene coding for an antibiotic target can render a previously susceptible microbe resistant to known antibiotics. These acquired resistances typically occur in only one or a few strains within a species and are generally not transferable, except to daughter cells. Subsequently, the population of such microbes may increase, and they may become the cause of new pandemics.

Resistance can also emerge from specific genes that facilitate mechanisms to inactivate antibiotics, with β-lactamases, a well-known example, or otherwise neutralize the antibiotics. For antibiotics that work intracellularly, specialized export systems may evolve to expel the drug. When these resistance genes are located on or near transferable elements, such as plasmids or transposons, the resistance can spread to other nonresistant bacteria. This form of transferable antibiotic resistance is particularly concerning, as it can spread to pathogenic bacteria, making infections potentially untreatable.

5.3 PROBIOTICS

Probiotics are living microorganisms that, when given in adequate quantities, offer health benefits to the host organisms (Hill et al., 2014). While *Lactobacilli* and *Bifidobacteria* are the most prevalent types of probiotics, other genera like *Saccharomyces cerevisiae*, *Escherichia coli*, and *Bacillus coagulans* are also available commercially. Additionally, new species of beneficial microbes, such as *Faecalibacterium prausnitzii*, *Eubacterium halii*, and *Akkermansia muciniphila*, are identified and explored for their probiotic potential. Safety is a crucial factor for any probiotic strain commercially available in the market. Many probiotics strains have regulatory safety approval due to their long history of consumption (Bourdichon et al., 2012).

Most probiotic bacteria are Gram-positive and primarily function in maintaining and improving intestinal health (e.g., *Lactobacillus* and *Bifidobacterium*) (Marco et al., 2006). Lactic acid bacteria (LAB), including species like *Lactobacillus*, *Lactococcus*, *Enterococcus*, and *Streptococcus*, as well as *Bifidobacterium*, are among the commonly available probiotics (Prado et al., 2015; Soccol et al., 2015). These microorganisms can endure harsh conditions within the host, such as acidity and enzymatic action, enabling them to colonize the host. They positively influence the health of the host by regulating the microbiome and performing essential biological functions (de Melo Pereira et al., 2018).

Probiotics exert health benefits to the host in various ways, including the improvement in digestive, immune, metabolic, and even mental health. However, the study shows that these benefits are strain-specific and cannot be generalized to other strains, even within the same species (Fijan, 2014). However, some properties related to probiotics' action may be shared across different strains due to similarities in their metabolic and ecological functions (Bermudez-Brito et al., 2012).

5.3.1 Mechanisms of Antimicrobial Action of Probiotics against Pathogens

The probiotics elicits its antimicrobial action through several known mechanisms, including competitive exclusion, the enhancement of intestinal barrier function by increasing mucin and tight junction (TJ) protein expression, the secretion of antimicrobial molecules, and the modulation of the host immune system (Figure 5.1).

5.3.1.1 Competitive Exclusion of Pathogens

Probiotic bacteria establish themselves in the gastrointestinal tract, competing for nutrients and adhesion sites, thereby preventing the overgrowth of pathogenic microbes. The competitive exclusion mechanism works via two separate mechanisms, viz., exploitation competition and interference competition (Knipe et al., 2021).

5.3.1.2 Exploitation Competition

This indirect mechanism involves rapid nutrient consumption through the secretion of extracellular molecules like proteases and iron-chelating siderophores, which break down complex

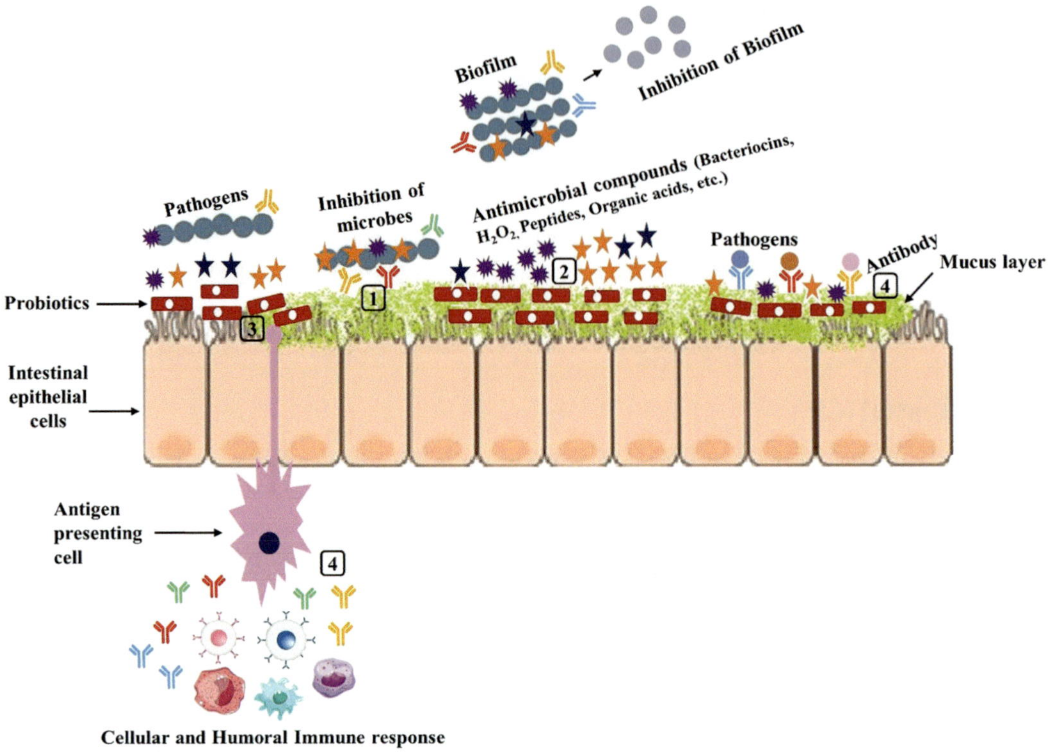

FIGURE 5.1 Mechanisms of antimicrobial action of probiotics. 1. Competitive exclusion of pathogens, 2. Secretion of antimicrobial compounds, 3. Enhancement of the intestinal barrier, 4. Modulation of the host immune system.

macromolecules, limiting resources for pathogenic microbes. Probiotics can also colonize uninhabited niches quickly and compete with pathogenic microbes by producing adhesins and receptors that bind to specific surface features (Schluter et al., 2015). Several studies have shown that LAB exhibit antagonistic effects on pathogen adhesion (Siedler et al., 2020; Zuo et al., 2019).

5.3.1.3 Interference Competition

This direct mechanism involves the production of antimicrobial biomolecules, such as bacteriocins, that inhibit the growth of pathogenic microbes. It also helps reduce antibiotic-induced superinfections and restores the desired microbial balance in the body. For example, *Lactiplantibacillus plantarum* strains effectively compete with and displace or exclude the pathogenic strains of *Salmonella enterica* and *E. coli* (Dhanani & Bagchi, 2013).

5.3.1.4 Enhancement of the Intestinal Barrier

The intestinal barrier plays a crucial role in maintaining health and preventing disease. It serves as a vital defense system against microbial infections. However, its functions are compromised when the mucosa experiences structural damage and dysregulation (Vancamelbeke & Vermeire, 2017). Probiotics offer a promising strategy to support mucosal barrier function and outcompete pathogenic organisms. The mucosal barrier is maintained by intestinal epithelial cells (IECs) and intercellular junction complexes. TJs at the apical side of IECs regulate the passage of small and ionic molecules, preserving the intestinal barrier against harmful bacteria and substances (Gou et al., 2022). Probiotics help restore gut barrier integrity by enhancing the expression of genes and proteins related to TJ signaling, regulating IEC apoptosis, and promoting IEC proliferation. A study shows that *L. acidophilus* enhances the TJ barrier function in IECs, which is mediated by Toll-like receptor-2 (TLR-2) heterodimeric complexes TLR-2/TLR-1 and TLR-2/TLR-6. This system protects against intestinal inflammation (Al-Sadi et al., 2021).

Additionally, goblet cells of intestinal epithelium also secrete a mucus layer composed primarily of glycoproteins known as mucins, which have a known role in nutrient absorption and in the prevention of pathogenic bacteria penetration. They also provide the adhesion sites for beneficial bacteria in the intestine (Vancamelbeke & Vermeire, 2017). Probiotics can stimulate mucin expression and mucus secretion from goblet cells. In one of the studies, mucus-secreting colon epithelial cells (HT29-MTX) were treated with supernatants probiotic yogurt mixes consisting of *Lactobacillus bulgaricus*, *B. bifidum*, *Streptococcus thermophilus*, *L. acidophilus*, and *Lactobacillus gasseri* increased the expression of CDX2 and MUC2 along with mucin protein production. MUC2 is an established mucin protein in the mucus layer, whereas CDX2 regulates the expression of MUC2 (Chang et al., 2021).

5.3.1.5 Secretion of Antimicrobial Molecules

Probiotic bacteria release antimicrobial molecules, including diacetyl, organic acids, peptides, and hydrogen peroxide, which selectively target several microbial strains in the gut (Rabetafika et al., 2023). The bacteriocins or antimicrobial peptides (AMPs) are a diverse group of ribosomally synthesized peptides that can inhibit the growth of pathogens in the gut lumen (Hassan et al., 2012).

The bacteriocins can be further classified into three categories, Class I to Class III. Class I compounds consist of heat-stable peptides known as lantibiotics, Class II consists of heat-stable peptides that do not contain lanthionine, and Class III consists of heat-labile, high-molecular-weight bacteriocins (Negash & Tsehai, 2020). The antimicrobial actions of bacteriocins depend on amino acid sequence and net charge that act through various mechanisms like enzyme activity modulation, pore formation, and quorum sensing (Bharti et al., 2015).

Class I bacteriocins can disrupt cell integrity by penetrating the cell membrane and inhibiting cell wall synthesis. Class II bacteriocins depolarize cell membranes by binding to membrane pore receptors, such as the mannose phosphotransferase system, while Class III bacteriocins directly lyse cells (Darvishi et al., 2021).

Class I bacteriocins, for example, Nisin, is produced by *Lactococcus lactis* and belong to the positively charged A-lantibiotics. The AMP nisin interacts with membrane-bound lipid II proteins, causing pore formation in the cell membrane and leading to bacterial lysis (Panina et al., 2021). Similarly, *Bacillus* species produces a neutral or negatively charged Class I B-lantibiotic peptide (mersacidin) that interferes with cell wall biosynthesis (Wang et al., 2020).

Class II bacteriocin pediocin, produced by *Pediococcus pentosaceus* GS4, exhibits antibacterial and antagonistic effects against *Pseudomonas aeruginosa*, *E.coli*, *Staphylococcus aureus* and, *Listeria monocytogenes* (Ghosh et al., 2019).

Class III bacteriocins include several compounds such as enterolysin, megacin, klebicin, colicin, helveticin I produced by bacterial species *Enterococcus faecalis*, *Klebsiella pneumoniae*, *Lactobacillus helveticus*, and *Bacillus megaterium*, respectively. They catalyze the cell wall hydrolysis (Bharti et al., 2015).

5.3.1.6 Modulation of the Host Immune System

Probiotic bacteria can influence the immune system by promoting the growth of beneficial components in the gut microbiome. The nonspecific or innate immunity serves as the body's first line of defense. It consists of physical and chemical barriers in the body, immune system cells, and immunomodulatory agents like cytokines. In contrast, specific or adaptive immunity targets specific threats through lymphocytes and includes immunoglobulin generation, antibody production, and cell-mediated immune responses (Mazziotta et al., 2023).

Probiotics enhance innate immunity by boosting the cytotoxic activity of natural killer (NK) cells and improving macrophage-dependent phagocytosis. They also modulate adaptive immunity by interacting with intestinal immune cells, including dendritic cells, enterocytes, and regulatory T cells (Azad et al., 2018).

Overall, probiotics improve the host defense mechanisms by the stimulation of phagocytic activity, balancing pro-inflammatory and anti-inflammatory cytokines and increasing the production of immunoglobulin IgA.

5.3.2 Techniques for Assessing the Antimicrobial Activity of Probiotics

Various *in vitro* methods can be used for the assessment of the antimicrobial activity of probiotics. One approach is to evaluate the direct antagonism between a probiotic culture and a pathogenic strain (Fijan, 2016). The antagonism test method involves detecting the growth inhibition of an indicator microbial strain by the probiotics extract (Balouiri et al., 2016). The common *in vitro* methods currently used for evaluating the antimicrobial activity of probiotics are the agar spot test, microdilution test, etc.

5.3.2.1 The Agar Spot Test

This test has been described by several researchers (Choi & Chang, 2015; Macaluso et al., 2016), with various modifications over time. The commonly used modifications are direct and deferred antagonism tests. In the direct assay, both the test and indicator microbial cultures are grown simultaneously, and antagonism is demonstrated through the release of a diffusible inhibitor early in the test culture's growth (Tagg et al., 1976). However, in the deferred antagonism method, the probiotic culture is allowed to grow on an agar plate and then inactivated. An overlay of the indicator strain is subsequently applied on top of the agar. This method is considered more sensitive and allows independent variation of the incubation time and conditions for the test and indicator cultures (Tagg et al., 1976). After incubation, the antimicrobial activity is quantified either as a zone of inhibition (in mm) or in arbitrary units (AU/mL).

5.3.2.2 The Agar Well Diffusion Assay

This technique is used to assess the antagonistic effects of cell-free supernatants. In this assay, agar plates are inoculated with the indicator microorganism, and wells of 6 or 7 mm in diameter are created in each plate. Subsequently, the probiotic supernatant is centrifuged, diluted to various concentrations, and poured into the wells. Following the incubation period, the antimicrobial activity is expressed either as an inhibition zone or in arbitrary units (AU/mL) (Parente et al., 1995).

5.3.2.3 Microdilution Test

This is one of the simplest and most reproducible methods for screening antimicrobial susceptibility. This method involves preparing 1:2 dilutions of the cell-free antimicrobial agent in growth medium, which is then dispensed into a 96-well microplate. Each well is inoculated with a standard inoculum of the pathogenic strain and incubated for possible microbial growth. The broth microdilution method determines the minimum inhibitory concentration (MIC), which is the lowest concentration of the antimicrobial agent that completely inhibits microbial growth (Balouiri et al., 2016).

However, the above-mentioned methods do not necessarily confirm the antimicrobial activity of probiotic microorganisms, especially against the biofilm-forming microbes. Bacterial biofilms are composed of communities embedded in a self-produced matrix of extracellular polymeric substances. Biofilms act as physical barriers and possess the genotypes that increase the virulence, making them up to 1,000 times more resistant to antimicrobial compounds compared to planktonic bacteria (Flemming et al., 2016). Currently, there is no standardized assay to evaluate the activity of antibiofilm agents. However, some researchers have assessed the antibiofilm activity of probiotics using different methods, such as the Calgary biofilm device, modified Robbins device, disk reactor, perfused biofilm fermenter, bladder model, etc. (Hager et al., 2019).

5.3.3 Targeted Delivery of Probiotics to the Site of Action

The health benefits of probiotics are well-established. However, for these beneficial bacteria to effectively perform their biological functions, their concentration must be at least 10^7 CFU/mL or gram of the product being used (Serna-Cock & Vallejo-Castillo, 2013). The survival of probiotics is influenced by various factors, such as temperature, pH, and peroxide production, making it challenging to deliver probiotics cells to their intended site of action. The survival and delivery of probiotics to the site of action can be greatly enhanced by the encapsulation technologies.

Encapsulation technologies have been studied for drug delivery and have shown significant advantages, including improved therapeutic efficacy and reduced dose-dependent toxicity (Singh et al., 2010). Apart from traditional drugs, probiotics can also be encapsulated to deliver at targeted site. However, probiotics need a biocompatible matrix for this purpose. This matrix encapsulates and immobilizes viable cells, protecting them from harsh environments, such as chemical and physical stress and the host's immune response (Orive et al., 2013). Moreover, the matrix acts as a semipermeable membrane, facilitating the bidirectional transport of nutrients (Gurruchaga et al., 2015). These matrixes are usually made of various types of biocompatible materials, such as fibrin, agarose, hyaluronic acid, and collagen (Vrana et al., 2009). Alginate is the most suitable biomaterial due to its ability to withstand stomach acids, biocompatibility, and ease of use (Vrana et al., 2009). Encapsulating probiotics with alginate has been associated with the formation of beneficial biofilm by these bacteria (Li et al., 2018).

Encapsulating probiotic microorganisms to treat infections beyond intestinal conditions could lead to new alternative antimicrobial therapies and adjuvant or synergistic approaches with conventional antibiotic treatments.

5.3.4 Synergism between Probiotics and Conventional Drugs in Antimicrobial Therapy

While antimicrobial drugs currently used in clinical practice are effective, they possess certain drawbacks such as side effects, high costs, and the risk of antimicrobial resistance (Vitor & Vale, 2011). As a result, the combination of probiotic microorganisms with conventional drugs has gained attention. A study shows that probiotics may elicit a synergistic effect with traditional antimicrobials, which is evidenced by the reduced dose requirement of the conventional drug, faster healing, lesser side effects from standard therapy and increased eradication rates of certain microbial infections (Kosgey et al., 2019).

In one of the studies, probiotics were combined with standard antimicrobials (two antibiotics and a proton pump inhibitor) for treatment of *H. pylori* infection. The study revealed the better eradication rate of 81% compared with 71% with conventional therapy alone, along with reduced therapy-associated side effects (Lesbros-Pantoflickova et al., 2007).

In another study, oral formulation containing probiotics and lactoferrin as an adjuvant were evaluated with topical application of clotrimazole for vulvovaginal candidiasis (VVC) (Russo et al., 2018). The findings revealed that the combined application of probiotics (*Lactobacillus rhamnosus* HN001 and *L. acidophilus*) and lactoferrin along with clotrimazole, reduced the symptoms and recurrence of VVC. Further study also suggests that probiotics offer an effective alternative treatment for VVC, as they can resist *Candida* spp. and help maintain or restore normal vaginal microbiota (Matsubara et al., 2016). The co-aggregation of lactobacilli with *Candida* spp. inhibits the yeast cells from binding to vaginal epithelial receptors.

However, other studies did not observe changes in the MIC values of antibiotics like clavulanic acid, ampicillin, and cotrimoxazole against *E. coli* isolates before and after treatment with lactobacilli supernatants (Naderi et al., 2014). This suggests that despite previous findings, lactobacilli supernatants may not alter the antibiotic resistance patterns of *E. coli* strains. In summary, probiotics are being investigated as innovative adjuvants to antimicrobial therapies, but these novel strategies require further testing and validation.

5.4 APPLICATIONS OF PROBIOTICS IN HUMANS AND ANIMALS

Probiotics are important for both human and animal health. In humans, they are primarily used to treat digestive issues like diarrhea, constipation, and irritable bowel syndrome (IBS), as they help maintain a balanced gut microbiome and enhance digestive health. Additionally, probiotics contribute to boosting immunity, supporting mental health, and managing cholesterol levels. In animals, probiotics are mainly used to improve gut health, strengthen immunity, manage stress, and enhance reproductive health, among other benefits.

5.4.1 Applications of Probiotics in Humans

In recent years, studies have demonstrated various benefits of probiotics, from inhibiting pathogenic microorganisms to enhancing the host's immunity (Moraes et al., 2019). Currently, the most commonly used probiotics in human applications are *L. plantarum*, *Lactobacillus johnsonii*, *L. acidophilus*, *Lactobacillus paracasei*, *Lactobacillus casei*, *L. reuteri*, *L. rhamnosus*, *Bifidobacterium infantis*, *B. bifidum*, *S. boulardii*, *Propionibacterium freudenreichii* (Markowiak & Śliżewska, 2017; Silva et al., 2020). Most of these probiotics play their role in the maintenance of intestinal health (Underwood, 2019). Several probiotics strains have been studied both *in vitro* as well as *in vivo* for control of microbial infections in humans (Table 5.2).

Probiotics exhibit their beneficial role by the production of inhibitory substances like hydrogen peroxide and bacteriocins, which inhibit both Gram-negative and Gram-positive pathogenic microbes. They also block adhesion sites and compete for the nutrients with pathogenic organisms (Markowiak & Śliżewska, 2017; Moraes et al., 2019). Moreover, probiotics also modulate immune responses by increasing nonspecific phagocytic activity through macrophage activation and influencing the release of pro- and anti-inflammatory cytokines (Plaza-Díaz et al., 2017). Thus, probiotics are considered a promising strategy for preventing and controlling various infectious diseases (Moraes et al., 2019).

Multiple studies have demonstrated that probiotics offer a wide range of health benefits, including antidiarrheal effects, the stimulation of the immune system's antimutagenic properties, anticarcinogenic effects, the ability to lower blood cholesterol levels, and the prevention of atopic dermatitis (Lise et al., 2018; Silva et al., 2020).

Probiotic formulations appear to improve human health by enhancing the immune system and encouraging competition between beneficial and harmful bacteria (Piewngam et al., 2019). As a result, probiotics are increasingly being seen as a promising strategy for preventing and managing various infectious diseases.

5.4.2 Applications of Probiotics in Animal

Antibiotics are commonly used in animal farming to promote growth, boost production, and control diseases (Roth et al., 2019). There is increasing interest in using probiotics as an alternative, especially for promoting animal growth by enhancing overall health rather than targeting specific infections (Zhang et al., 2021; Lokapirnasari et al., 2019). Numerous probiotic strains have shown effectiveness in inhibiting various animal pathogens, positioning them as a promising alternative to antibiotics in industries like poultry, swine, and cattle farming, where they can

TABLE 5.2
Studies on Antimicrobial Activity of Probiotics for Human Application

Probiotic Strain	Experiment	Outcome	Reference
L. casei	Identification of a probiotics against *C. albicans* infection	Fungicidal effect of *L. casei* was observed in VVC in murine model	Liao et al. (2019)
L. salivarius	Development of a mouthwash for the treatment of periodontitis condition	*L. salivarius*-based mouthwash showed beneficial effects on maintenance of dental health.	Sajedinejad et al. (2018)
L. reuteri DSM 17648	Identification of probiotics strain for controlling gastritis caused by *H. pylori* infection	Probiotics reduced the *H. pylori* load and improved the gastrointestinal condition in children and adults	Liang et al. (2022)
L. casei	Application of probiotics in control of *C. difficile*-associated diarrhea (CDAD)	Probiotics significantly reduced the incidence rates of CDAD	Ma et al. (2020)
L. crispatus CTV-05 (Lactin-V)	Treatment of female urogenital tract infection by probiotics	Probiotics application prevented the recurrence of bacterial vaginosis	Cohen et al. (2020)
L. johnsonii No.1088 (HK-LJ88)	Investigation of probiotics activity against difficult-to-treat *H. pylori* infection	Probiotics showed favorable effects against *H. pylori* by improvement in mucosal barrier and immunomodulation	Aiba et al. (2017)
L. paracasei, *L. plantarum*, and *L. salivarius*	Identification of probiotics strain for control of Intestines Inflammatory bowel syndrome (IBS)	Improvement in IBS symptoms and reduced abdominal pain without significant adverse effect	Oh et al. (2019)
L. casei	Study of synergistic effect of *L. casei* strain with antiprotozoal medicine against giardia infection in murine model.	Oral administration of albendazole along with probiotic *L. casei* reduced Giardia infection.	Shukla et al. (2013)
L. paracasei	Probiotics *L. paracasei* was tested against *C. albicans* infection in larvae of *Galleria mellonella*	*L. paracasei* showed a protective effect against fungal infection in *G. mellonella* by immune modulation.	Rossoni et al. (2017)
L. plantarum CIRM653	Evaluation of antibiofilm activity against *K. pneumoniae*	Probiotics significantly reduced the biofilm formation, and improve the host tolerability	Lagrafeuille et al. (2018)
L. reuteri DSM 17938	Study of antimicrobial activity of *L. reuteri* against skin-related microbes	*L. reuteri* showed an antimicrobial effect against *Cutibacterium acnes*, *S. aureus*, *Streptococcus pyogenes*, and *P. aeruginosa* and decreased the inflammatory condition	Khmaladze et al. (2019)
L. paracasei	Evaluation of antibiofilm activity of probiotics against *Candida* spp.	*L. paracaseis* showed antimicrobial activity against Gram-negative and Gram-positive bacteria along with yeast in *in vitro* microdilution tests	Nyanzi et al. (2014); Wannun et al. (2016)

(Continued)

TABLE 5.2 (*Continued*)
Studies on Antimicrobial Activity of Probiotics for Human Application

Probiotic Strain	Experiment	Outcome	Reference
L. plantarum PBS067 *L. reuteri* PBS072 *L. rhamnosus* LRH020	Treatment of atopic dermatitis in skin	Probiotics application improves the condition of skin smoothness, moisturization, and decrease in levels of inflammatory markers and scoring atopic dermatitis index	Michelotti et al. (2021)
L. acidophilus W55, *L. casei* W56, *B. lactis* W51, *B. lactis* W52	Study of the role of probiotics in the treatment of acne vulgaris	Oral probiotic elevated the serum IL-10 levels and acted as adjuvant in acne vulgaris therapy	Rahmayani et al. (2019)
B. lactis Probio-M8	Study of probiotics action in acute respiratory tract infection	Reduced dose of antibiotic use, decreased level of TNF-α and increased IL-10 level were observed over 4 weeks postdischarge in patient	Mageswary et al. (2022)
Bifidobacterium strains	Investigation of the antibacterial activity against pathogens	The probiotics strains inhibited the growth of *S. aureus* and showed the best antistaphylococcal activity in comparison to other probiotics strains of different genera	Lahtinen et al. (2007)
B. longum 5	Evaluation of antibacterial activity of *Bifidobacterium* against *K. pnemoniae* infection	The probiotic reduced the lethality rate in mice by 50% from *K. pnemoniae* lung infection, by enhanced secretion of pro-inflammatory cytokines and neutrophil recruitment	Vieira et al. (2016)
Bifidobacterium, *Lactobacillus* species	Evaluation of the probiotics against rotavirus diarrhea	*Lactobacillus* and *Bifidobacterium* showed an improved antirotavirus response which is evidenced by increased level of rotavirus-specific anti-IgA	Gonzalez-Ochoa et al. (2017)
Bifidobacterium animalis subsp. *Lactis*	Trial of *B. animalis* on 51 patients supplemented in yogurt for 4 weeks for periodontal benefit	*B. animalis* showed beneficial effect against the bacterial plaque formation and inflammatory conditions	Kuru et al. (2017)
B. animalis subsp. Lactis BB-12, *L. acidophilus* LA-5, *Saccharomyces boulardii*	Identification of probiotics for control of ventilator-associated pneumonia (VAP) in respiratory infection	Decrease in incidence of VAP caused by *P. aeruginosa* and *Acinobacter baumannii* in patients due to severe multiple trauma such as brain injury	Tsilika et al. (2022)
S. boulardii and *Bifidobacterium*	Treatment of acute diarrhea in human	Probiotics application shortened the duration of diarrhea, and enhanced cellular immunity	Wang and Feng. (2019)
B. subtilis MB40	Study of probiotics to control *S. aureus* infection	Significant reduction of *S. aureus* colonization in humans without modification of host body microbiome	Piewngam et al. (2023)
Pediococcus acidilactici HW01	*In vitro* study of antibiofilm activity of probiotics against *P. aeruginosa*	Probiotics inhibited the biofilm formation and decreases the production of virulence factors viz., protease, pyocyanin, and rhamnolipid	Lee et al. (2020)
P. acidilactici HW01	Study of antibiofilm activity of probiotics against *L. monocytogenes*	Probiotics inhibited the bacterial biofilm formation, adhesion, and invasion in human–intestinal–epithelial cell line (HT-29 cell line)	Bin Lee et al. (2022)

improve immune function and help prevent diseases (Table 5.3). In addition to inhibiting animal pathogens, probiotics play a significant role in boosting the immune system, reducing stress, improving reproductive health, increasing milk production, and reducing methane emissions in ruminants (Ncho et al., 2024).

TABLE 5.3
Animal and Poultry Applications of Probiotics

Species	Probiotics	Mode of Administration	Effects	Reference
		Ruminant		
Dairy cows	*B. subtilis*, *B. lichenformis*, *S. cerevisiae*, *E. faecium*, *L. acidophilus*, *L. plantarum*, *B. tedium*	Oral feeding	Improvement in reproductive performance. Increase in milk yield, milk fat, and protein	Merati and Towhidi. (2022)
Dairy cows	*L. reuteri*, *L. rhamnosus*, *P. acidilactici*	Bovine endometrial explants	Reduction in acute inflammation by *E. coli* infection	Genís et al. (2017)
Cattle	*S. cerevisiae*, *S. salivarius*, *Saccharomyces equinus*, *Lactobacillus gallinarum.*	Oral feeding	Improvement in hemato-biochemical parameters of growing cattle	Yasmin et al. (2021)
Buffalo calf	*L. acidophilus*	Oral feeding	Increased dry matter intake and feed conversion efficiency	Sharma et al. (2018)
Sheep	*B. subtilis* and *Bacillus licheniformis*	Oral feeding	Improvement in intestinal microbiota and body weight gain	Devyatkin et al. (2021)
Goat kid	*S. cerevisiae*, *Lactobacillus Sporogenes*	Oral feeding	Improvement in body weight gain and feed conversion ratio due to better feed utilization and suppression of pathogenic bacteria by bacteriocins and enhanced immunity	Mamta and Sharma (2008)
Camel	*Bacillus subtlis*, *L. acidophilus*	Liquid feeding	Enhancing dry matter digestibility and intake	Arafa et al. (2023)
		Equine		
Horse	*S. cerevisiae*, *L. acidophilus*, *E. faecium*	Oral feeding	Increased sand output after 4 weeks of application in comparison to baseline treatment	Landes et al. (2008)
Horse	*L. casei*, *L. acidophilus*, *E. faecium*	Oral feeding	Reduced incidence of *Salmonella* shedding	Ward et al. (2004)
Mule	*L. bulgaricus*, *L. acidophilus*, *L. helveticus*, *L. lactis*, *S. thermophilus and E. faecium*	Oral feeding	Inhibition of *Salmonella* and coli-forms, increased body weight, and lower mortality rate in comparison to control group	Djouvinov et al. (2005)
		Swine		
Weaning piglets	*B. subtilis*, *E. faecium*	Liquid feed	Improve growth performance	Zhang et al. (2020)
Piglets	*L. plantarum*	Oral feeding	Protection of weaned pigs from severe infection of enterotoxigenic *E. coli*	Pupa et al. (2022)

(Continued)

TABLE 5.3 (*Continued*)
Animal and Poultry Applications of Probiotics

Species	Probiotics	Mode of Administration	Effects	Reference
		Poultry		
Hatching chicks	*L. plantarum* LTC-113	Oral vaccination	Protection from *Salmonella* infection by enhanced immunity	Wang et al. (2018)
Chickens	*L. paracasei, L. rhamnosus*	Oral feeding	Improvement in growth parameter	Fesseha et al. (2021)
Chickens	*B. licheniformis*	Oral feeding	Improvement in intestinal mucosal immunity and modulation in intestinal microflora	Kan et al. (2021)
Chickens	Clostridium *butyricum*	Oral feeding	Improvement in the condition of chickens infected with *Clostridium perfringens* by inhibition of pro-inflammatory genes	Huang et al. (2019)
Laying hens	*Bifidobacterium* spp., *L. casei*	Oral feeding	Improvement in egg weight, feed efficiency, and growth performance	Lokapirnasari et al. (2019)
Broilers	*L. acidophilus, L. casei,* and *Bifidobacterium*	Water supplementation	Improvement of carcass traits, immunity, gut microbiota, growth performance, and antioxidant activity	Zhang et al. (2021)

5.5 ROLE OF PROBIOTICS IN DECREASING ANTIBIOTIC RESISTANCE

The production of bacteriocins by probiotics offers the potential to reduce pathogenic bacterial populations at mucosal sites and disrupt biofilms (Lin et al., 2015), enhancing the effectiveness of antibiotics (Vuotto et al., 2014). Additionally, probiotics can strengthen mucosal immunity, helping to eliminate pathogens from these areas.

However, the role of probiotics in preventing drug-resistant infections in humans remains uncertain. The Centers for Disease Control and Prevention (CDC) is investigating this area, and although some studies indicate potential benefits, the evidence is not yet robust enough to support specific guidelines (Ouwehand et al., 2016). As a result, probiotics are currently seen as a possible supplement or partial alternative to antibiotic treatment.

Most research on antibiotic resistance has centered on pathogenic bacteria in healthcare settings, such as hospitals and nursing homes. However, antibiotic resistance is a broader concern, also impacting food, animal production, and aquaculture. Since antibiotic resistance genes (ARGs) often reside on mobile genetic elements, understanding how these elements spread through the food ecosystem is critical. Despite growing awareness of antibiotic resistance leading to the ban of antibiotic growth promoters (AGPs) in animals in several countries, their use persists in other regions. Antibiotics are continued to be widely and indiscriminately used in veterinary medicine around the world to treat diseases in livestock. To address antibiotic resistance effectively, more targeted interventions are needed to reduce the selection pressure that drives the emergence, evolution, and spread of resistance. Incorporating probiotics into animal feed and improving animal health could help mitigate the spread of antibiotic resistance.

5.5.1 Spread of Antibiotic Resistance and the Role of Probiotics

The dissemination of ARGs via probiotics is an increasing public health concern (Table 5.4). Probiotics, commonly beneficial bacteria like *Lactobacillus* and *Bifidobacterium*, are often used to enhance gut health. Nevertheless, recent research indicates that these microorganisms may harbor ARGs, which can be transferred to pathogenic bacteria in the human gut, thus contributing to the spread of antibiotic resistance.

This spread occurs through horizontal gene transfer (HGT), a process in which bacteria exchange genetic material, including ARGs, through methods such as conjugation, transformation, and transduction. When probiotics contain ARGs, they can pass these genes on to other bacteria in the gut, including harmful pathogens. This can diminish the effectiveness of antibiotics and complicate the treatment of bacterial infections.

TABLE 5.4
Antibiotic Resistance Genes in Common Probiotics Strain

Source	Probiotic Strain	Gene	Antibiotic Resistance Against	Location	References
Fermented dry sausage	*Lb. plantarum DG507*	erm(B)	Erythromycin	Plasmid	Gevers et al. (2003)
Probiotic bacteria	*Lactobacillus, Staphylococcus, Bifidobacterium, Pediococcus, Lactococci, Streptococci*	tet(K), tet(W)	Tetracycline	Plasmid	Kastner et al. (2006)
Fermented food and commercial probiotic	*Lb. salivarius BFE 7441*	erm(B)	Ciprofloxacin, erythromycin, gentamicin, streptomycin	Chromosome	Hummel et al. (2007)
Italian sola cheese	*Lb. sakei Rits 9*	tet(M), tet(L)	Tetracycline, erythromycin	Transposon, plasmid	Ammor et al. (2008)
Commercial probiotic strain	*Lb. reuteri ATCC 55730*	lnu(A)	Lincosamide	Plasmid	Rosander et al. (2008)
Raw milk, starter-free cheese	*Lc. lactis*	tet(M)	Tetracycline	Plasmid	Florez et al. (2008)
Chinese fermented foods	*Lb. plantarum, Lb. fermentum, Lb. helveticus, E. faecium*	tet(M), erm(B), aph A3, mef A	Tetracycline, erythromycin, chloramphenicol, kanamycin	Plasmid, chromosome	Pan et al. (2011)
Dairy product	*Lb. vaginalis NWL35*	erm(B)	Erythromycin	Plasmid	Nawaz et al. (2011)
Fermented dry sausage	*Lb. plantarum*	erm(B), erm(C)	Erythromycin	Plasmid	Zonenschain et al. (2009)
Probiotic cultures	*Lactococcus, Lactobacillus, Pediococcus*	erm(B), tet(W), tet(M)	Streptomycin, erythromycin, clindamycin, oxytetracycline	Plasmid	Klare et al. (2007)
LAB strains	*LAB* spp.	tet(M), tet(W), tet (O)	Tetracycline	Plasmid	Ammor et al. (2008)

(Continued)

TABLE 5.4 (*Continued*)
Antibiotic Resistance Genes in Common Probiotics Strain

Source	Probiotic Strain	Gene	Antibiotic Resistance Against	Location	References
Commercial dairy and pharmaceutical products	*Lactobacilli*, *Lactococci*, *Enterococci*, *S. thermophilus* and *Bifidobacterium* spp.	msrC, vanX, dfrA	Vancomycin, rifampicin, streptomycin, bacitracin, erythromycin	Plasmid	Liu et al. (2009)
Italian fermented products	*Lb. paracasei*	tet(M), erm(B)	Tetracycline, erythromycin	Plasmid	Comunian et al. (2010)
Chinese yogurts	*S. thermophilus and Lb. delbruekii ssp. bulgaricus*	tet (M), ant 6, aph 3′-IIIa	Ampicillin, kanamycin, chloramphenicol, chlortetracycline, tetracyclines, neomycin, gentamicin	Plasmid	Zhou et al. (2012)
Yakult	*Bifidobacterium breve*	rpsL	Streptomycin	Plasmid	Kiwaki and Sato (2009)
Human, animal, and probiotics	*Bifidobacterium* spp.	tet(W)	Tetracycline	Plasmid	Masco et al. (2006)
Human, animal, and probiotics	*Bifidobacterium* spp.	tet(W)	Tetracycline	Plasmid	Masco et al. (2006)
Humans, animals, and probiotics	*Bifidobacterium longum B36*	tet(W)	Tetracycline	Plasmid	Moubareck et al. (2005)
Fermented foods and curd	*E. faecium*, *E. durans*, *P. pentosaceus*	erm(B), msr(C)	Erythromycin	Plasmid	Thumu and Halami (2012)
Probiotic product	*E. faecium*	erm(B), cat(pC194), cat(piP501), aad(E), aad(E), aph(A)	Chloramphenicol, erthromycin, gentamicin, streptomycin	Plasmid	Vankerckhoven et al. (2008)

Colonic bacteria can act as reservoirs for resistance genes acquired from ingested bacteria. While it is plausible that gene transfer among bacteria happens in the gastrointestinal tract, intrinsic resistance generally has a lower potential for horizontal spread compared to acquired resistance, which has a higher likelihood of lateral transfer (Devirgiliis et al., 2011). Bacteria from the genera *Lactococcus* and *Lactobacillus* are generally recognized as safe (GRAS) or hold Qualified Presumption of Safety (QPS) status. However, other genera of LAB, such as *Streptococcus* and *Enterococcus*, include some opportunistic pathogens.

Probiotics containing resistant genes are not inherently harmful. However, there is concern about the possibility of resistance genes being transferred from probiotics to other gut microorganisms and eventually to opportunistic pathogens in the same intestinal environment, which could have significant clinical implications (Broaders et al., 2013; Imperial and Ibana, 2016). *Enterococcal* species have been studied as probiotics in various studies due to their beneficial properties, including resistance to gastric juice and bile salts, stability during storage and handling, and the production of bacteriocins like enterocin (Pieniz et al., 2014). Despite their low pathogenicity, enterococci are increasingly associated with hospital-acquired infections, particularly in immunocompromised patients. This may be due in part to their intrinsic tolerance to several antimicrobial agents and harsh conditions, as well as their capacity to acquire resistance genes from other bacteria. Consequently, enterococci are more prone to acquiring ARGs compared to other bacteria in the same environment (Hollenbeck and Rice, 2012).

Moreover, enterococcal species may carry undesirable traits for probiotics, such as the production of β-hemolysin, aggregation substances and, gelatinase (Belgacem et al., 2010). Some resistance genes, such as those for cephalosporins, sulfonamides, and aminoglycosides, are intrinsic and located in chromosomes. However, enterococci may also carry extrinsic resistance to antibiotics like chloramphenicol, erythromycin, tetracycline, and vancomycin. These extrinsic resistance genes are found on plasmids or near transposons, which facilitates their horizontal or vertical transfer to other bacterial groups, including *Streptococcus* spp., *S. aureus*, and *B. subtilis*, complicating antimicrobial treatment (Donabedian et al., 2003). Virulence genes could also be transferred to human endogenous strains in the gastrointestinal tract, potentially increasing the virulence of this genus and endogenous strains. Therefore, ensuring the safety of enterococcal strains used as probiotics is crucial, and their use should be carefully monitored through a thorough risk/benefit analysis (Franz et al., 2011).

Commercial probiotic products may also carry the risk of harboring ARGs. A comprehensive study of various commercial products from the USA, Australia, and Malaysia, which contained probiotic strains such as *Bifidobacterium longum*, *B. bifidum*, *B. infantis*, *L. bulgaricus*, *L. rhamnosus*, *L. casei*, *L. acidophilus*, *L. gasseri*, *Lactobacillus salivarius*, and *S. thermophilus*, revealed resistance to several antibiotics, including vancomycin, streptomycin, aztreonam, gentamicin, and ciprofloxacin (Wong et al., 2015).

Lactobacillus fermentum is frequently used in the production of fermented products and as a probiotics. However, this bacterium may act as a vector for ARGs against antibiotics such as tetracycline and erythromycin, which can be transferred to pathogenic bacteria (Anisimova and Yarullina, 2018). Therefore, it is essential to monitor and control the presence of transferable antibiotic resistance determinants in probiotics to ensure their safety.

To reduce this risk, probiotic strains should be screened for ARGs before used in supplements and food products. Regulatory authorities should also establish guidelines to ensure that probiotics do not contribute to the spread of antibiotic resistance.

5.6 CONCLUSIONS AND FUTURE PERSPECTIVE

Human and animal bodies are vulnerable to various pathogenic microorganisms. This chapter presents clear evidence of the antimicrobial effects of probiotics, their role in enhancing host immunity, and their use as growth promoters in animals. Probiotics offer multiple benefits, particularly in preventing AAD, maintaining the balance of intestinal and other microbiomes, and improving the overall health of the host. One of the most well-established advantages of probiotics is their positive impact when used alongside antibiotics. Additionally, probiotics can help reduce the development of antibiotic resistance.

Probiotic microorganisms produce certain biomolecules, such as lactic acid, biosurfactants, acetic acid, hydrogen peroxide, and bacteriocins, which are believed to contribute to their antimicrobial properties. However, further research is needed to identify new molecules and clarify their inhibitory mechanisms against pathogenic bacteria and fungi.

Probiotic research should be completed through the entire process until it reaches clinical application. Currently, there is no consensus or standardization for using probiotics as antimicrobial therapy in human and animal medicine. Key aspects such as dosage, mechanism of action, and clinical efficacy remain undetermined. The targeted delivery and nanotechnology-based encapsulation of probiotic-derived molecules should be promoted to develop more effective therapeutic approaches.

Probiotic studies related to antibiotic use should prioritize the detection of relevant resistance genes in probiotic strains to better assess potential benefits. More attention should be given to this area, as reducing the global spread of resistance through probiotics could lead to significant advantages.

ACKNOWLEDGMENTS

The authors express their gratitude to SVP University of Agriculture and Technology, Meerut, Uttar Pradesh, for offering the necessary facilities to prepare this manuscript.

REFERENCES

Aiba, Y., Ishikawa, H., Tokunaga, M., & Komatsu, Y. (2017). Anti-*Helicobacter pylori* activity of non-living, heat-killed form of lactobacilli including *Lactobacillus johnsonii* No. 1088. *FEMS Microbiology Letters*, 364(11). https://doi.org/10.1093/femsle/fnx102

Al-Sadi, R., Nighot, P., Nighot, M., Haque, M., Rawat, M., & Ma, T. Y. (2021). *Lactobacillus acidophilus* induces a strain-specific and Toll-like receptor 2-dependent enhancement of intestinal epithelial tight junction barrier and protection against intestinal inflammation. *American Journal of Pathology*, 191(5), 872–884.

Ammor, M. S., Gueimonde, M., Danielsen, M., Zagorec, M., van Hoek, A. H., de Los Reyes-Gavilán, C. G., Mayo, B., & Margolles, A. (2008). Two different tetracycline resistance mechanisms, plasmid-carried tet(L) and chromosomally located transposon-associated tet(M), coexist in *Lactobacillus sakei* rits 9. *Applied and Environmental Microbiology*, 74, 1394–1401.

Anisimova, E., & Yarullina, D. (2018). Characterization of erythromycin and tetracycline resistance in *Lactobacillus fermentum* strains. *International Journal of Microbiology*, 2018, 3912326. https://doi.org/10.1155/2018/3912326

Arafa, A. M., Gado, H. M., Ghandour, M. M., Metwally, H. M., & Askar, A. R. (2023). Effect of using probiotics on Arabian camels (*Camelus dromedarius*) feeding: An in vitro study. *Egyptian Journal of Nutrition and Feeds*, 26(2), 159–166.

Azad, M., Kalam, A., Sarker, M., & Wan, D. (2018). Immunomodulatory effects of probiotics on cytokine profiles. *BioMed Research International*, 2018, 8063647.

Balouiri, M., Sadiki, M., & Ibnsouda, S. K. (2016). Methods for in vitro evaluating antimicrobial activity: A review. *Journal of Pharmaceutical Analysis*, 6(2), 71–79.

Belgacem, Z. B., Abriouel, H., Omar, N. B., Lucas, R., Martínez-Canamero, M., Gálvez, A., & Manai, M. (2010). Antimicrobial activity, safety aspects, and some technological properties of bacteriocinogenic *Enterococcus faecium* from artisanal Tunisian fermented meat. *Food Control*, 21(4), 462–470. https://doi.org/10.1016/j.foodcont.2009.07.007

Bermudez-Brito, M., Plaza-Diaz, J., Munoz-Quezada, S., Gomez-Llorente, C., & Gil, A. (2012). Probiotic mechanisms of action. *Annals of Nutrition & Metabolism*, 61(2), 160–174.

Bharti, V., Mehta, A., Singh, S., Jain, N., Ahirwal, L., & Mehta, S. (2015). Bacteriocin: A novel approach for preservation of food. *International Journal of Pharmacy and Pharmaceutical Sciences*, 7(2), 20–29.

Bin Lee, H., Kim, K. H., Kang, G. A., Lee, K.G., & Kang, S.-S. (2022). Antibiofilm, antiadhesive, and anti-invasive activities of bacterial lysates extracted from *Pediococcus acidilactici* against *Listeria monocytogenes*. *Foods*, 11(19), 2948.

Bourdichon, F., Casaregola, S., Farrokh, C., Frisvad, J. C., Gerds, M. L., Hammes, W. P., Harnett, J., Huys, G., Laulund, S., Ouwehand, A., Powell, I. B., Prajapati, J. B., Seto, Y., Ter Schure, E., Van Boven, A., Vankerckhoven, V., Zgoda, A., Tuijtelaars, S., & Hansen, E. B. (2012). Food fermentations: Microorganisms with technological beneficial use. *International Journal of Food Microbiology*, 154(3), 87–97. https://doi.org/10.1016/j.ijfoodmicro.2011.12.030

Broaders, E., Gahan, C. G., & Marchesi, J. R. (2013). Mobile genetic elements of the human gastrointestinal tract: Potential for spread of antibiotic resistance genes. *Gut Microbes*, 4(4), 271–280. https://doi.org/10.4161/gmic.24627

Chang, Y. H., Jeong, C. H., Cheng, W. N., Choi, Y., Shin, D. M., Lee, S., & Han, S. G. (2021). Quality characteristics of yogurts fermented with short-chain fatty acid-producing probiotics and their effects on mucin production and probiotic adhesion onto human colon epithelial cells. *Journal of Dairy Science*, 104(7), 7415–7425.

Choi, E. A., & Chang, H. C. (2015). Cholesterol-lowering effects of a putative probiotic strain *Lactobacillus plantarum* EM isolated from kimchi. *LWT – Food Science and Technology*, 62(1), 210–217.

Cohen, C. R., Wierzbicki, M. R., French, A. L., Morris, S., Newmann, S., Reno, H., Green, L., Miller, S., Powell, J., Parks, T., & Hemmerling, A. (2020). Randomized trial of lactin-V to prevent recurrence of bacterial vaginosis. *The New England Journal of Medicine*, 382(20), 1906–1915. https://doi.org/10.1056/NEJMoa1915254

Comunian, R., Daga, E., Dupré, I., Paba, A., Devirgiliis, C., Piccioni, V., Perozzi, G., Zonenschain, D., Rebecchi, A., Morelli, L., De Lorentiis, A., & Giraffa, G. (2010). Susceptibility to tetracycline and erythromycin of *Lactobacillus paracasei* strains isolated from traditional Italian fermented foods. *International Journal of Food Microbiology*, 138, 151–156.

Darvishi, N., Fard, N. A., & Sadrnia, M. (2021). Genomic and proteomic comparisons of bacteriocins in probiotic species *Lactobacillus* and *Bifidobacterium* and inhibitory ability of *Escherichia coli* MG 1655. *Biotechnology Reports*, 31, e00654.

de Melo Pereira, G. V., de Oliveira Coelho, B., Magalhães Junior, A. I., Thomaz-Soccol, V., & Soccol, C. R. (2018). How to select a probiotic? A review and update of methods and criteria. *Biotechnology Advances*, 36(8), 2060–2076.

Devirgiliis, C., Barile, S., & Perozzi, G. (2011). Antibiotic resistance determinants in the interplay between food and gut microbiota. *Genes & Nutrition*, 6(3), 275–284.

Devyatkin, V., Mishurov, A., & Kolodina, E. (2021). Probiotic effect of *Bacillus subtilis* B-2998D, B-3057D, and *Bacillus licheniformis* B-2999D complex on sheep and lambs. *Journal of Advanced Veterinary and Animal Research*, 8, 146–157.

Dhanani, A., & Bagchi, T. (2013). The expression of adhesin EF-Tu in response to mucin and its role in *Lactobacillus* adhesion and competitive inhibition of enteropathogens to mucin. *Journal of Applied Microbiology*, 115(2), 546–554.

Djouvinov, D., Boicheva, S., Simeonova, T., & Vlaikova, T. (2005). Effect of feeding Lactina® probiotic on performance, some blood parameters, and caecal microflora of mule ducklings. *Trakia Journal of Sciences*, 3(2), 22–28.

Donabedian, S. M., Thal, L. A., Hershberger, E., Perri, M. B., Chow, J. W., Bartlett, P., Jones, R., Joyce, K., Rossiter, S., Gay, K., Johnson, J., Mackinson, C., Debess, E., Madden, J., Angulo, F., & Zervos, M. J. (2003). Molecular characterization of gentamicin-resistant enterococci in the United States: Evidence of spread from animals to humans through food. *Journal of Clinical Microbiology*, 41(3), 1109–1113. https://doi.org/10.1128/JCM.41.3.1109-1113.2003

Fernandes, R., Amador, P., & Prudêncio, C. (2013). β-Lactams: Chemical structure, mode of action, and mechanisms of resistance. *Reviews in Medical Microbiology*, 24, 7–17.

Fesseha, H., Demlie, T., Mathewos, M., & Eshetu, E. (2021). Effect of *Lactobacillus* species probiotics on growth performance of dual-purpose chicken. *Veterinary Medicine: Research and Reports*, 12, 75–83.

Fijan, S. (2014). Microorganisms with claimed probiotic properties: An overview of recent literature. *International Journal of Environmental Research and Public Health*, 11(5), 4745–4767.

Fijan, S. (2016). Antimicrobial effect of probiotics against common pathogens. In V. Rao & L. G. Rao (Eds.), *Probiotics and Prebiotics in Human Nutrition and Health* (pp. 191–221). InTech.

Flemming, H. C., Wingender, J., Szewzyk, U., Steinberg, P., Rice, S. A., & Kjelleberg, S. (2016). Biofilms: An emergent form of bacterial life. *Nature Reviews Microbiology*, 14(9), 563–575.

Florez, A., Tosi, L., Danielsen, M., von Wright, A., Bardowski, L., & Mayo, B. (2008). Resistance-susceptibility profiles of *Lactococcus lactis* and *Streptococcus thermophilus* strains to eight antibiotics and proposition of new cut-offs. *International Journal of Probiotics and Prebiotics*, 3, 249–256.

Franz, C. M., Huch, M., Abriouel, H., Holzapfel, W., & Gálvez, A. (2011). Enterococci as probiotics and their implications in food safety. *International Journal of Food Microbiology*, 151(2), 125–140.

Genís, S., Sánchez-Chardi, A., Bach, À., Fàbregas, F., & Arís, A. (2017). A combination of lactic acid bacteria regulates *Escherichia coli* infection and inflammation of the bovine endometrium. *Journal of Dairy Science*, 100, 479–492.

Gevers, D., Danielson, M., Huys, G., & Swings, J. (2003). Molecular characterization of tet(M) genes in *Lactobacillus* isolates from different types of fermented dry sausage. *Applied and Environmental Microbiology*, 69, 1270–1275.

Ghosh, B., Sukumar, G., & Ghosh, A. R. (2019). Purification and characterization of pediocin from probiotic *Pediococcus pentosaceus* GS4, MTCC 12683. *Folia Microbiologica*, 64(6), 765–778.

Gonzalez-Ochoa, G., Flores-Mendoza, L. K., Icedo-Garcia, R., Gomez-Flores, R., & Tamez-Guerra, P. (2017). Modulation of rotavirus severe gastroenteritis by the combination of probiotics and prebiotics. *Archives of Microbiology*, 199(7), 953–961. https://doi.org/10.1007/s00203-017-1400-3

Gou, H.-Z., Zhang, Y.-L., Ren, L.-F., Li, Z.-J., & Zhang, L. (2022). How do intestinal probiotics restore the intestinal barrier? *Frontiers in Microbiology*, 13, 929346.

Grossman, T. H. (2016). Tetracycline antibiotics and resistance. *Cold Spring Harbor Perspectives in Medicine*, 6, a025387.

Gurruchaga, H., Saenz del Burgo, L., Ciriza, J., Orive, G., Hernández, R. M., & Pedraz, J. L. (2015). Advances in cell encapsulation technology and its application in drug delivery. *Expert Opinion on Drug Delivery*, 12(8), 1251–1267. https://doi.org/10.1517/17425247.2015.1001362

Hager, C. L., Isham, N., Schrom, K. P., Chandra, J., McCormick, T., Miyagi, M., & Ghannoum, M. A. (2019). Effects of a novel probiotic combination on pathogenic bacterial-fungal polymicrobial biofilms. *mBio*, 10(2), e00338–19.

Hassan, M., Kjos, M., Nes, I. F., Diep, D. B., & Lotfipour, F. (2012). Natural antimicrobial peptides from bacteria: Characteristics and potential applications to fight against antibiotic resistance. *Journal of Applied Microbiology*, 113(4), 723–736.

Hempel, S., Newberry, S. J., Maher, A. R., Wang, Z., Miles, J. N., Shanman, R., Johnsen, B., & Shekelle, P. G. (2012). Probiotics for the prevention and treatment of antibiotic-associated diarrhea: A systematic review and meta-analysis. *JAMA*, 307(18), 1959–1969. https://doi.org/10.1001/jama.2012.3507

Hill, C., Guarner, F., Reid, G., Gibson, G. R., Merenstein, D. J., Pot, B., Morelli, L., Canani, R. B., Flint, H. J., Salminen, S., Calder, P. C., & Sanders, M. E. (2014). Expert consensus document. The International Scientific Association for Probiotics and Prebiotics consensus statement on the scope and appropriate use of the term probiotic. *Nature Reviews Gastroenterology & Hepatology*, 11(8), 506–514. https://doi.org/10.1038/nrgastro.2014.66

Hollenbeck, B. L., & Rice, L. B. (2012). Intrinsic and acquired resistance mechanisms in Enterococcus. *Virulence*, 3(5), 421–569.

Huang, T., Peng, X.-Y., Gao, B., Wei, Q.-L., Xiang, R., Yuan, M.-G., & Xu, Z.-H. (2019). The effect of *Clostridium butyricum* on gut microbiota, immune response, and intestinal barrier function during the development of necrotic enteritis in chickens. *Frontiers in Microbiology*, 10, 2309.

Hummel, A. S., Hertel, C., Holzapfel, W. H., & Franz, C. M. (2007). Antibiotic resistances of starter and probiotic strains of lactic acid bacteria. *Applied and Environmental Microbiology*, 73, 730–739.

Imperial, I. C., & Ibana, J. A. (2016). Addressing the antibiotic resistance problem with probiotics: Reducing the risk of its double-edged sword effect. *Frontiers in Microbiology*, 7, 01983. https://doi.org/10.3389/fmicb.2016.01983

Kan, L., Guo, F., Liu, Y., Pham, V. H., Guo, Y., & Wang, Z. (2021). Probiotics *Bacillus licheniformis* improves intestinal health of subclinical necrotic enteritis-challenged broilers. *Frontiers in Microbiology*, 12, 623739.

Kastner, S., Perreten, V., Bleuler, H., Hugenschmidt, G., Lacroix, C., & Meile, L. (2006). Antibiotic susceptibility patterns and resistance genes of starter cultures and probiotic bacteria used in food. *Systematic and Applied Microbiology*, 29, 145–155.

Khmaladze, I., Butler, E., Fabre, S., & Gillbro, J. M. (2019). *Lactobacillus reuteri* DSM 17938-A comparative study on the effect of probiotics and lysates on human skin. *Experimental Dermatology*, 28, 822–828. https://doi.org/10.1111/exd.13950

Kim, D.W., Thawng, C. N., Lee, K., Wellington, E. M., & Cha, C.J. (2019). A novel sulfonamide resistance mechanism by two-component flavin-dependent monooxygenase system in sulfonamide-degrading actinobacteria. *Environment International*, 127, 206–215.

King, S., Glanville, J., Sanders, M. E., Fitzgerald, A., & Varley, D. (2014). Effectiveness of probiotics on the duration of illness in healthy children and adults who develop common acute respiratory infectious conditions: A systematic review and meta-analysis. *British Journal of Nutrition*, 112(1), 41–54.

Kiwaki, M., & Sato, T. (2009). Antimicrobial susceptibility of *Bifidobacterium* breve strains and genetic analysis of streptomycin resistance of probiotic B. breve strain Yakult. *International Journal of Food Microbiology*, 134, 211–215.

Klare, I., Konstabel, C., Werner, G., Huys, G., Vankerckhoven, V., & Kahlmeter, G. (2007). Antimicrobial susceptibilities of *Lactobacillus*, *Pediococcus* and *Lactococcus* human isolates and cultures intended for probiotic or nutritional use. *Journal of Antimicrobial Chemotherapy*, 59, 900–912.

Knipe, H., Temperton, B., Lange, A., Bass, D., & Tyler, C. R. (2021). Probiotics and competitive exclusion of pathogens in shrimp aquaculture. *Reviews in Aquaculture*, 13(1), 324–352.

Kosgey, J. C., Jia, L., Fang, Y., Yang, J., Gao, L., Wang, J., Nyamao, R., Cheteu, M., Tong, D., Wekesa, V., Vasilyeva, N., & Zhang, F. (2019). Probiotics as antifungal agents: Experimental confirmation and future prospects. *Journal of Microbiological Methods*, 162, 28–37. https://doi.org/10.1016/j.mimet.2019.05.001

Krause, K. M., Serio, A. W., Kane, T. R., & Connolly, L. E. (2016). Aminoglycosides: An overview. *Cold Spring Harbor Perspectives in Medicine*, 6, a027029.

Kuru, B. E., Laleman, I., Yalnızoğlu, T., Kuru, L., & Teughels, W. (2017). The influence of a *Bifidobacterium animalis* probiotic on gingival health: A randomized controlled clinical trial. *Journal of Periodontology*, 88(11), 1115–1123. https://doi.org/10.1902/jop.2017.170213

Lagrafeuille, R., Miquel, S., Balestrino, D., Vareille-Delarbre, M., Chain, F., Langella, P., & Forestier, C. (2018). Opposing effect of *Lactobacillus* on in vitro *Klebsiella pneumoniae* in biofilm and in an in vivo intestinal colonisation model. *Beneficial Microbes*, 9, 87–100.

Lahtinen, S. J., Jalonen, L., Ouwehand, A. C., & Salminen, S. J. (2007). Specific Bifidobacterium strains isolated from elderly subjects inhibit growth of *Staphylococcus aureus*. *International Journal of Food Microbiology*, 117(1), 125–128. https://doi.org/10.1016/j.ijfoodmicro.2007.02.023

Landes, A. D., Hassel, D. M., Funk, J. D., & Hill, A. (2008). Fecal sand clearance is enhanced with a product combining probiotics, prebiotics, and psyllium in clinically normal horses. *Journal of Equine Veterinary Science*, 28, 79–84.

Lee, D.H., Kim, B. S., & Kang, S.-S. (2020). Bacteriocin of *Pediococcus acidilactici* HW01 inhibits biofilm formation and virulence factor production by *Pseudomonas aeruginosa*. *Probiotics and Antimicrobial Proteins*, 12, 73–81.

Lesbros-Pantoflickova, D., Corthesy-Theulaz, I., & Blum, A. L. (2007). *Helicobacter pylori* and probiotics. *Journal of Nutrition*, 137(3 Suppl.), 812S–818S.

Li, Z., Behrens, A. M., Ginat, N., Tzeng, S. Y., Lu, X., Sivan, S., Langer, R., & Jaklenec, A. (2018). Biofilm-inspired encapsulation of probiotics for the treatment of complex infections. *Advanced Materials*, 30(51), e1803925. https://doi.org/10.1002/adma.201803925

Liang, B., Yuan, Y., Peng, X.J., Liu, X.L., Hu, X.K., & Xing, D.M. (2022). Current and future perspectives for *Helicobacter pylori* treatment and management: From antibiotics to probiotics. *Frontiers in Cellular and Infection Microbiology*, 12, 1740.

Liao, H., Liu, S., Wang, H., Su, H., & Liu, Z. (2019). Enhanced antifungal activity of bovine lactoferrin-producing probiotic *Lactobacillus casei* in the murine model of vulvovaginal candidiasis. *BMC Microbiology*, 19, 7. https://doi.org/10.1186/s12866-018-1370-x

Lin, X., Chen, X., Chen, Y., Jiang, W., & Chen, H. (2015). The effect of five probiotic Lactobacilli strains on the growth and biofilm formation of Streptococcus mutans. *Oral Diseases*, 21(4), e128–e134.

Lise, M., Mayer, I., & Silveira, M. (2018). Use of probiotics in atopic dermatitis. *Revista Da Associação Médica Brasileira*, 64(11), 997–1001. https://doi.org/10.1590/1806-9282.64.11.997

Liu, C., Zhang, Z. Y., Dong, K., Yuan, J. P., & Guop, X. K. (2009). Antibiotic resistance of probiotic strains of lactic acid bacteria isolated from marketed foods and drugs. *Biomedical and Environmental Sciences*, 22, 401–412.

Lokapirnasari, W. P., Pribadi, T. B., Arif, A. A., Soeharsono, S., Hidanah, S., Harijani, N., Najwan, R., Huda, K., Wardhani, H. C. P., Rahman, N. F. N., & Yulianto, A. B. (2019). Potency of probiotics *Bifidobacterium* spp. and *Lactobacillus casei* to improve growth performance and business analysis in organic laying hens. *Veterinary World*, 12(6), 860–867. https://doi.org/10.14202/vetworld.2019.860-867

Ma, Y., Yang, J. Y., Peng, X., Xiao, K. Y., Xu, Q., & Wang, C. (2020). Which probiotic has the best effect on preventing *Clostridium difficile*-associated diarrhea? A systematic review and network meta-analysis. *Journal of Digestive Diseases*, 21, 69–80.

Macaluso, G., Fiorenza, G., Gaglio, R., Mancuso, I., & Scatassa, M. L. (2016). In vitro evaluation of bacteriocin-like inhibitory substances produced by lactic acid bacteria isolated during traditional Sicilian cheese making. *Italian Journal of Food Safety*, 5(2), 5503.

Mageswary, M. U., Ang, X. Y., Lee, B. K., Chung, Y. F., Azhar, S. N. A., Hamid, I. J. A., Bakar, H. A., Roslan, N. S., Liu, X., Kang, X., Dai, L., Sreenivasan, S., Taib, F., Zhang, H., & Liong, M. T. (2022). Probiotic *Bifidobacterium lactis* Probio-M8 treated and prevented acute RTI, reduced antibiotic use and hospital stay in hospitalized young children: A randomized, double-blind, placebo-controlled study. *European Journal of Nutrition*, 61(3), 1679–1691. https://doi.org/10.1007/s00394-021-02689-8.

Mamta, M., & Sharma P. (2008). Effect of probiotics on body weight gain and feed conversion ratio in goat kids. *Haryana Veterinarian*, 47, 39–40.

Marco, M. L., Pavan, S., & Kleerebezem, M. (2006). Towards understanding molecular modes of probiotic action. *Current Opinion in Biotechnology*, 17(2), 204–210.

Markowiak, P., & Śliżewska, K. (2017). Effects of probiotics, prebiotics, and synbiotics on human health. *Nutrients*, 9(9), 1021. https://doi.org/10.3390/nu9091021

Masco, L., Van Hoorde, K., De Brandt, E., Swings, J., & Huys, G. (2006). Antimicrobial susceptibility of *Bifidobacterium* strains from humans, animals, and probiotic products. *Journal of Antimicrobial Chemotherapy*, 58, 85–94.

Matsubara, V. H., Bandara, H. M., Ishikawa, K. H., Mayer, M. P., & Samaranayake, L. P. (2016). The role of probiotic bacteria in managing periodontal disease: A systematic review. *Expert Review of Anti-infective Therapy*, 14(7), 643–655.

Mazziotta, C., Tognon, M., Martini, F., Torreggiani, E., & Rotondo, J. C. (2023). Probiotics mechanism of action on immune cells and beneficial effects on human health. *Cells*, 12(2), 184.

Merati, Z., & Towhidi, A. (2022). Effect of a multispecies probiotic on productive and reproductive performance of Holstein cows. *Iranian Journal of Applied Animal Science*, 12, 237–247.

Michelotti, A., Cestone, E., De Ponti, I., Giardina, S., Pisati, M., Spartà, E., & Tursi, F. (2021). Efficacy of a probiotic supplement in patients with atopic dermatitis: A randomized, double-blind, placebo-controlled clinical trial. *European Journal of Dermatology*, 31, 225–232.

Moraes, M. C. C., Costa, P. J. C., Segundo, A. S. G., & Peruzzo, D. C. (2019). Assessment of probiotic bacterial strains effect on *S. aureus* biofilm on titanium discs with treated surfaces. *Revista de Odontologia da UNESP*, 48, e20190096.

Moubareck, C., Gavini, F., Vaugien, L., Butel, M. J., & Doucet-Populaire, F. (2005). Antimicrobial susceptibility of bifidobacteria. *Journal of Antimicrobial Chemotherapy*, 55, 38–44.

Naderi, A., Kasra-Kermanshahi, R., Gharavi, S., Imani Fooladi, A. A., Abdollahpour Alitappeh, M., & Saffarian, P. (2014). Study of antagonistic effects of *Lactobacillus* strains as probiotics on multi-drug resistant (MDR) bacteria isolated from urinary tract infections (UTIs). *Iranian Journal of Basic Medical Sciences*, 17(3), 201–208.

Nawaz, M., Wang, J., Zhou, A., Ma, C., Wu, X., Moore, J. E., Millar, B. C., & Xu, J. (2011). Characterization and transfer of antibiotic resistance in lactic acid bacteria from fermented food products. *Current Microbiology*, 62(3), 1081–1089. https://doi.org/10.1007/s00284-010-9856-2.

Ncho, C. M., Kim, S. H., Rang, S. A., & Lee, S. S. (2024). A meta-analysis of probiotic interventions to mitigate ruminal methane emissions in cattle: Implications for sustainable livestock farming. *Animal: An International Journal of Animal Bioscience*, 18(6), 101180. https://doi.org/10.1016/j.animal.2024.101180

Negash, A. W., & Tsehai, B. A. (2020). Current applications of bacteriocin. *International Journal of Microbiology*, 2020, 4374891.

Nyanzi, R., Awouafack, M. D., Steenkamp, P., Jooste, P. J., & Eloff, J. N. (2014). Anticandidal activity of cell extracts from 13 probiotic *Lactobacillus* strains and characterization of lactic acid and a novel fatty acid derivative from one strain. *Food Chemistry*, 164, 470–475.

Oh, J. H., Jang, Y. S., Kang, D., Chang, D. K., & Min, Y. W. (2019). Efficacy and safety of new *Lactobacilli* probiotics for unconstipated irritable bowel syndrome: A randomized, double-blind, placebo-controlled trial. *Nutrients*, 11(12), 2887.

Orive, G., Santos, E., Pedraz, J. L., & Hernández, R. M. (2013). Application of cell encapsulation for controlled delivery of biological therapeutics. *Advanced Drug Delivery Reviews*, 67–68, 3–14.

Ouwehand, A. C., Forssten, S., Hibberd, A. A., Lyra, A., & Stahl, B. (2016). Probiotic approach to prevent antibiotic resistance. *Annals of Medicine*, 48(4), 246–255. https://doi.org/10.3109/07853890.2016.1161232

Pan, L., Hu, X., & Wang, X. (2011). Assessment of antibiotic resistance of lactic acid bacteria in Chinese fermented foods. *Food Control*, 22, 1316–1321.

Panina, I., Taldaev, A., Efremov, R., & Chugunov, A. (2021). Molecular dynamics insight into the lipid II recognition by type A lantibiotics: Nisin, epidermin, and gallidermin. *Micromachines*, 12(9), 1169.

Parente, E., Brienza, C., Moles, M., & Ricciardi, A. (1995). A comparison of methods for the measurement of bacteriocin activity. *Journal of Microbiological Methods*, 22(1), 95–108.

Pieniz, S., Andreazza, R., Anghinoni, T., Camargo, F., & Brandelli, A. (2014). Probiotic potential, antimicrobial and antioxidant activities of *Enterococcus durans* strain LAB18s. *Food Control*, 37(1), 251–256.

Piewngam, P., Khongthong, S., Roekngam, N., Theapparat, Y., Sunpaweravong, S., Faroongsarng, D., & Otto, M. (2023). Probiotic for pathogen-specific *Staphylococcus aureus* decolonization in Thailand: A phase 2, double-blind, randomized, placebo-controlled trial. *Lancet Microbe*, 4, e75–e83.

Piewngam, P., Quiñones, M., Thirakittiwatthana, W., Yungyuen, T., Otto, M., & Kiratisin, P. (2019). Composition of the intestinal microbiota in extended-spectrum β-lactamase-producing Enterobacteriaceae carriers and non-carriers in Thailand. *International Journal of Antimicrobial Agents*, 53(4), 435–441. https://doi.org/10.1016/j.ijantimicag.2018.12.006

Plaza-Díaz, J., Ruiz-Ojeda, F. J., Vilchez-Padial, L. M., & Gil, A. (2017). Evidence of the anti-inflammatory effects of probiotics and synbiotics in intestinal chronic diseases. *Nutrients*, 9(6), 555. https://doi.org/10.3390/nu9060555

Prado, F. C., Lindner, J. de D., Inaba, J., Thomaz-Soccol, V., Brar, S. K., & Soccol, C. R. (2015). Development and evaluation of a fermented coconut water beverage with potential health benefits. *Journal of Functional Foods*, 12, 489–497.

Pupa, P., Apiwatsiri, P., Sirichokchatchawan, W., Pirarat, N., Nedumpun, T., Hampson, D. J., Muangsin, N., & Prapasarakul, N. (2022). Microencapsulated probiotic *Lactiplantibacillus plantarum* and/or *Pediococcus acidilactici* strains ameliorate diarrhea in piglets challenged with enterotoxigenic *Escherichia coli*. *Scientific Reports*, 12, 7210.

Rabetafika, H. N., Razafindralambo, A., Ebenso, B., & Razafindralambo, H. L. (2023). Probiotics as antibiotic alternatives for human and animal applications. *Encyclopedia*, 3(2), 561–581. https://doi.org/10.3390/encyclopedia3020040

Rahmayani, T., Putra, I. B., & Jusuf, N. K. (2019). The effect of oral probiotic on the interleukin-10 serum levels of acne vulgaris. Open *Access Macedonian Journal of Medical Sciences*, 7, 3249–3252.

Redgrave, L. S., Sutton, S. B., Webber, M. A., & Piddock, L. J. V. (2014). Fluoroquinolone resistance: Mechanisms, impact on bacteria, and role in evolutionary success. *Trends in Microbiology*, 22, 438–445.

Reveles, K. R., Lee, G. C., Boyd, N. K., & Frei, C. R. (2014). The rise in Clostridium difficile infection incidence among hospitalized adults in the United States: 2001–2010. *American Journal of Infection Control*, 42(10), 1028–1032.

Rosander, A., Connolly, E., & Roos, S. (2008). Removal of antibiotic resistance gene-carrying plasmids from *Lactobacillus reuteri* ATCC 55730 and characterization of the resulting daughter strain, *L. reuteri* DSM 17938. *Applied and Environmental Microbiology*, 74, 6032–6040.

Rossoni, R. D., Fuchs, B. B., de Barros, P. P., Velloso, M. D., Jorge, A. O., Junqueira, J. C., & Mylonakis, E. (2017). *Lactobacillus paracasei* modulates the immune system of *Galleria mellonella* and protects against *Candida albicans* infection. *PLoS ONE*, 12, e0173332.

Roth, N., Käsbohrer, A., Mayrhofer, S., Zitz, U., Hofacre, C., & Domig, K. J. (2019). The application of antibiotics in broiler production and the resulting antibiotic resistance in *Escherichia coli*: A global overview. *Poultry Science*, 98(4), 1791–1804.

Russo, R., Superti, F., Karadja, E., & De Seta, F. (2018). Randomized clinical trial in women with recurrent vulvovaginal candidiasis: Efficacy of probiotics and lactoferrin as maintenance treatment. *Mycoses*. https://doi.org/10.1111/myc.12883

Sabtu, N., Enoch, D. A., & Brown, N. M. (2015). Antibiotic resistance: What, why, where, when and how? *British Medical Bulletin*, 116(1), 105–113.

Sajedinejad, N., Paknejad, M., Houshmand, B., Sharafi, H., Jelodar, R., Shahbani Zahiri, H., & Noghabi, K. A. (2018). *Lactobacillus salivarius* NK02: A potent probiotic for clinical application in mouthwash. *Probiotics and Antimicrobial Proteins*, 10(3), 485–495. https://doi.org/10.1007/s12602-017-9296-4

Schluter, J., Nadell, C. D., Bassler, B. L., & Foster, K. R. (2015). Adhesion as a weapon in microbial competition. *ISME Journal*, 9(1), 139–149.

Schwarz, S., Shen, J., Kadlec, K., Wang, Y., Michael, G. B., Feßler, A. T., & Vester, B. (2016). Lincosamides, streptogramins, phenicols, and pleuromutilins: Mode of action and mechanisms of resistance. *Cold Spring Harbor Perspectives in Medicine*, 6, a027037.

Serna-Cock, L., & Vallejo-Castillo, V. (2013). Probiotic encapsulation. *African Journal of Microbiological Research*, 7(43), 4743–4753. https://doi.org/10.5897/AJMR2013.5718

Sharma, A. N., Kumar, S., & Tyagi, A. K. (2018). Effects of mannan-oligosaccharides and *Lactobacillus acidophilus* supplementation on growth performance, nutrient utilization, and fecal characteristics in Murrah buffalo calves. *Journal of Animal Physiology and Animal Nutrition*, 102, 679–689. https://doi.org/10.1111/jpn.12878

Shukla, G., Kaur, H., & Sharma, L. (2013). Comparative therapeutic effect of probiotic *Lactobacillus casei* alone and in conjunction with antiprotozoal drugs in murine giardiasis. *Parasitology Research*, 112, 2143–2149. https://doi.org/10.1007/s00436-013-3394-3.

Siedler, S., Rau, M. H., Bidstrup, S., Vento, J. M., Aunsbjerg, S. D., Bosma, E. F., McNair, L. M., Beisel, C. L., & Neves, A. R. (2020). Competitive exclusion is a major bioprotective mechanism of lactobacilli against fungal spoilage in fermented milk products. *Applied and Environmental Microbiology*, 86(7), e02312–19.

Silva, D. R., Rosalen, P. L., Freires, I. A., Sardi, J. C. O., Lima, R. F., Lazarini, J. G., Lopes da Costa, T. K. V., Pereira, J. V., Godoy, G. P., & de Brito Costa, E. M. (2019). Anadenanthera Colubrina vell Brenan: Anti-Candida and antibiofilm activities, toxicity and therapeutical action. *Brazilian Oral Research*, 33(23), 1–11. https://doi.org/10.1590/1807-3107bor-2019.vol33.0023

Silva, D. R., Sardi, J. C. O., Pitangui, N. S., Roque, S. M., da Silva, A. C. B., & Rosalen, P. L. (2020). Probiotics as an alternative antimicrobial therapy: Current reality and future directions. *Journal of Functional Foods*, 73, 104080. https://doi.org/10.1016/j.jff.2020.104080

Singh, M. N., Hemant, K. S., Ram, M., & Shivakumar, H. G. (2010). Microencapsulation: A promising technique for controlled drug delivery. *Research in Pharmaceutical Sciences*, 5(2), 65–77.

Soccol, C. R., Prado, M. R. M., Garcia, L. M. B., Rodrigues, C., Medeiros, A. B. P., & Thomaz-Soccol, V. (2015). Current developments in probiotics. *Journal of Microbial & Biochemical Technology*, 7(1), 11–20.

Tagg, J. R., Dajani, A. S., & Wannamaker, L. W. (1976). Bacteriocins of gram-positive bacteria. *Bacteriological Reviews*, 40(3), 722–756.

Thumu, S. C., & Halami, P. M. (2012). Presence of erythromycin and tetracycline resistance genes in lactic acid bacteria from fermented foods of Indian origin. *Antonie Van Leeuwenhoek*, 102, 541–551.

Tsilika, M., Thoma, G., Aidoni, Z., Tsaousi, G., Fotiadis, K., Stavrou, G., Malliou, P., Chorti, A., Massa, H., Antypa, E., Vasiliadou, G., Pagdatoglou, K., Voudouris, A., Vasiliagou, S., Mitos, G., Kontopoulou, N., Paraforou, N., Antoniadou, E., Mouloudi, H., Gkeka, E., Grosomanidis, V., Giamarellos-Bourboulis, E. J., & Kotzampassi, K. (2022). A four-probiotic preparation for ventilator-associated pneumonia in multi-trauma patients: Results of a randomized clinical trial. *International Journal of Antimicrobial Agents*, 59(1), 106471. https://doi.org/10.1016/j.ijantimicag.2021.106471

Ubeda, C., & Pamer, E. G. (2012). Antibiotics, microbiota, and immune defense. *Trends in Immunology*, 33(9), 459–466.

Underwood, M. A. (2019). Probiotics and the prevention of necrotizing enterocolitis. *Journal of Pediatric Surgery*, 54(3), 405–412. https://doi.org/10.1016/j.jpedsurg.2018.08.055

Vancamelbeke, M., & Vermeire, S. (2017). The intestinal barrier: A fundamental role in health and disease. *Expert Review of Gastroenterology & Hepatology*, 11(9), 821–834.

Vankerckhoven, V., Huys, G., Vancanneyt, M., Vael, C., Klare, I., Romond, M. B., Entenza, J. M., Moreillon, P., Wind, R. D., Knol, J., Wiertz, E., Pot, B., Vaughan, E. E., Kahlmeter, G., & Goossens, H. (2008). Biosafety assessment of probiotics used for human consumption: Recommendations from the EU-PROSAFE project. *Trends in Food Science and Technology*, 19(2), 102–114.

Vieira, A. T., Rocha, V. M., Tavares, L., Garcia, C. C., Teixeira, M. M., Oliveira, S. C., Cassali, G. D., Gamba, C., Martins, F. S., & Nicoli, J. R. (2016). Control of *Klebsiella pneumoniae* pulmonary infection and immunomodulation by oral treatment with the commensal probiotic *Bifidobacterium longum*. *Microbes and Infection*, 18(3), 180–189. https://doi.org/10.1016/j.micinf.2015.10.008

Vítor, J. M. B., & Vale, F. F. (2011). Alternative therapies for *Helicobacter pylori*: Probiotics and phytomedicine. *FEMS Immunology & Medical Microbiology*, 63(2), 153–164.

Vrana, N. E., O'Grady, A., Kay, E., Cahill, P. A., & McGuinness, G. B. (2009). Cell encapsulation within PVA-based hydrogels via freeze-thawing: A one-step scaffold formation and cell storage technique. *Journal of Tissue Engineering and Regenerative Medicine*, 3(8), 567–572.

Vuotto, C., Longo, F., & Donelli, G. (2014). Probiotics to counteract biofilm-associated infections: Promising and conflicting data. *International Journal of Oral Science*, 6(4), 189–194.

Wang, G., & Feng, D. (2019). Therapeutic effect of *Saccharomyces boulardii* combined with Bifidobacterium on cellular immune function in children with acute diarrhea. *Experimental and Therapeutic Medicine*, 18, 2653–2659.

Wang, L., Li, L., Lv, Y., Chen, Q., Feng, J., & Zhao, X. (2018). *Lactobacillus plantarum* restores intestinal permeability disrupted by Salmonella infection in newly-hatched chicks. *Scientific Reports*, 8, 2229.

Wang, X., Gu, Q., & Breukink, E. (2020). Non-lipid II targeting lantibiotics. *Biochimica et Biophysica Acta (BBA) – Biomembranes*, 1862(3), 183244.

Wannun, P., Piwat, S., & Teanpaisan, R. (2016). Purification, characterization, and optimum conditions of fermenc in SD11, a bacteriocin produced by human orally *Lactobacillus fermentum* SD11. *Applied Biochemistry and Biotechnology*, 179, 572–582.

Ward, M. P., Alinovi, C. A., Couëtil, L. L., Glickman, L. T., & Wu, C. C. (2004). A randomized clinical trial using probiotics to prevent Salmonella fecal shedding in hospitalized horses. *Journal of Equine Veterinary Science*, 24(6), 242–247.

Wong, A., Ngu, D. Y., Dan, L. A., Ooi, A., & Lim, R. L. (2015). Detection of antibiotic resistance in probiotics of dietary supplements. *Nutrition Journal*, 14, 95. https://doi.org/10.1186/s12937-015-0084-2

Yang, H., Sun, Y., Cai, R., Chen, Y., & Gu, B. (2019). The impact of dietary fiber and probiotics in infectious diseases. *Microbial Pathogenesis*, 137, 103931.

Yasmin, F., Alam, M. J., Kabir, M. E., Al Maruf, A., Islam, M. A., & Hossain, M. M. (2021). Influence of probiotics supplementation on growth and haemato-biochemical parameters in growing cattle. *International Journal of Livestock Research*, 11, 36–42.

Zeng, D., Debabov, D., Hartsell, T. L., Cano, R. J., Adams, S., Schuyler, J. A., McMillan, R., & Pace, J. L. (2016). Approved glycopeptide antibacterial drugs: Mechanism of action and resistance. *Cold Spring Harbor Perspectives in Medicine*, 6, a026989.

Zhang, L., Zhang, R., Jia, H., Zhu, Z., Li, H., & Ma, Y. (2021). Supplementation of probiotics in water beneficial growth performance, carcass traits, immune function, and antioxidant capacity in broiler chickens. *Open Life Sciences*, 16(1), 311–322.

Zhang, S., Yoo, D. H., Ao, X., & Kim, I. H. (2020). Effects of dietary probiotic, liquid feed, and nutritional concentration on the growth performance, nutrient digestibility, and fecal score of weaning piglets. *Asian-Australasian Journal of Animal Sciences*, 33, 1617–1623.

Zhou, N., Zhang, J. X., Fan, M. T., Wang, J., Guo, G., & Wei, X. Y. (2012). Antibiotic resistance of lactic acid bacteria isolated from Chinese yogurts. *Journal of Dairy Science*, 95, 4775–4783.

Zonenschain, D., Rebecchi, A., & Morelli, L. (2009). Erythromycin and tetracycline-resistant lactobacilli in Italian fermented dry sausage. *Journal of Applied Microbiology*, 107, 1559–1568.

Zuo, F., Appaswamy, A., Gebremariam, H. G., & Jonsson, A.-B. (2019). Role of Sortase A in *Lactobacillus gasseri* Kx110A1 adhesion to gastric epithelial cells and competitive exclusion of *Helicobacter pylori*. *Frontiers in Microbiology*, 10, 2770.

6 The Probiotic Revolution

Ensuring Safety and Compliance in the Age of Antimicrobial Resistance

Maryam Shakeel[1] *and Eijaz Ahmed Bhat*[2*]
[1]Microbiology/Molecular Physiology of Prokaryotes, Institute of Biology, University of Freiburg, Freiburg, Germany
[2]Faculty of Biology, Department of Microbiology, University of Freiburg, Freiburg, Germany
*Corresponding author: Eijaz Ahmed Bhat

6.1 INTRODUCTION

Scientific efforts to uncover the mysteries of nature are becoming increasingly evident. With advancements in scientific methods, our understanding of probiotics has greatly improved. Commonly referred to as "good microbes," probiotics have gained significant recognition in the food and pharmaceutical industries. However, despite their widespread use, recent studies have raised concerns about probiotics being a major reservoir of antimicrobial resistance genes (Varzakas et al., 2018b). This chapter explores guidelines for regulating their effects on human health. While the term "probiotics" may have been introduced in the mid-20th century by Lilley and Stillwell, their use dates back to ancient times (Al Matrood, 2016). Molecular evidence from ancient Chinese pottery suggests that microbes were used as early as 7,000 BC in the fermentation of beverages (Alvarez-Olmos & Oberhelman, 2001). Though microbes were first discovered in the 17th century, historical evidence shows they were utilized in fermentation and food preservation long before (Figure 6.1) (Alvarez-Olmos & Oberhelman, 2001). Elie Metchnikoff, often regarded as the father of probiotics, made significant contributions to the field by highlighting the health benefits of fermented foods in the early 20th century, particularly among Bulgarian populations (Santacroce et al., 2019). The term "probiotic" combines Latin and Greek roots, and in 1965, Lilly and Stillwell officially defined it as a growth-promoting substance (Bernardeau et al., 2006).

In probiotics, "Pro" comes from Latin, which means "for," whereas "biotic" originates from the Greek word bios, which means "life" (Varzakas et al., 2018a). Since its birth, many attempts have been made to design the most suitable definition for this term. Still, the most widely accepted latest definition is "probiotics, living microbes that confer health benefits to its host when administered adequately (Sánchez et al., 2017)." Over the last few decades, scientists have been actively conducting their studies to familiarize themselves with probiotics. Numerous studies have highlighted its effect on the digestive and immune system (Figure 6.2) (La Fata et al., 2018). Probiotics help to maintain microflora equilibrium in GIT (gastrointestinal tract) by suppressing pathogens (Vitetta et al., 2014). Its popularity also increased with modernization, and probiotics established its foundation in the pharmaceutical industry. In the National Health Interview survey conducted in 2012, probiotics were reported as the third most popular food supplement (Wu et al., 2014). In 2007, a significant turn was witnessed in the probiotic story when the first case of sepsis was reported in immunocompromised patients (Alvarez-Olmos & Oberhelman, 2001). Since ancient times, probiotics like lactic acid bacteria have been utilized mainly for food storage; therefore, the effects of overuse and

 DOI: 10.1201/9781032676951-6

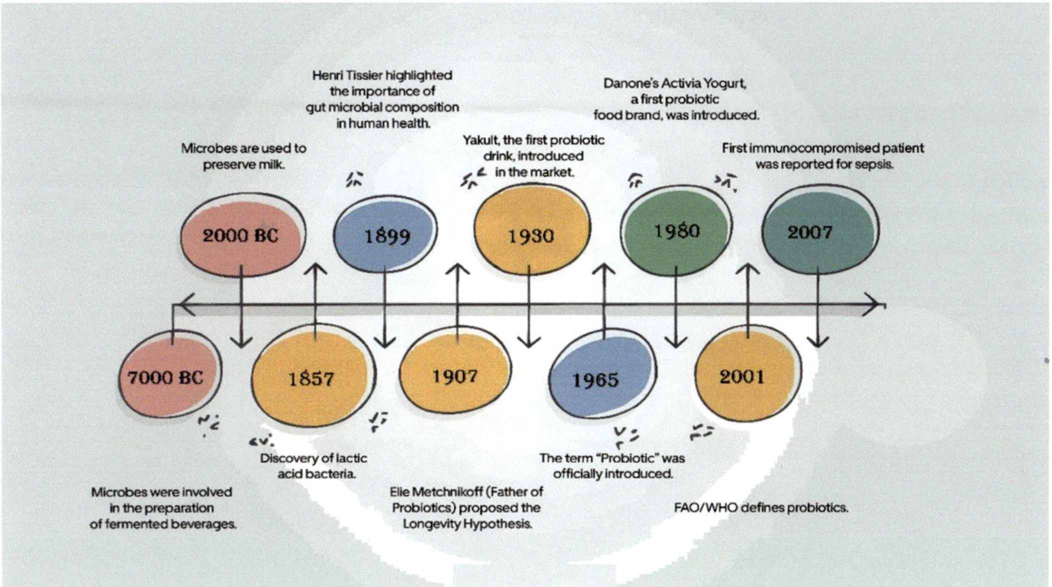

FIGURE 6.1 Historical events that play an essential role in the birth and understanding of probiotics (Alvarez-Olmos & Oberhelman, 2001).

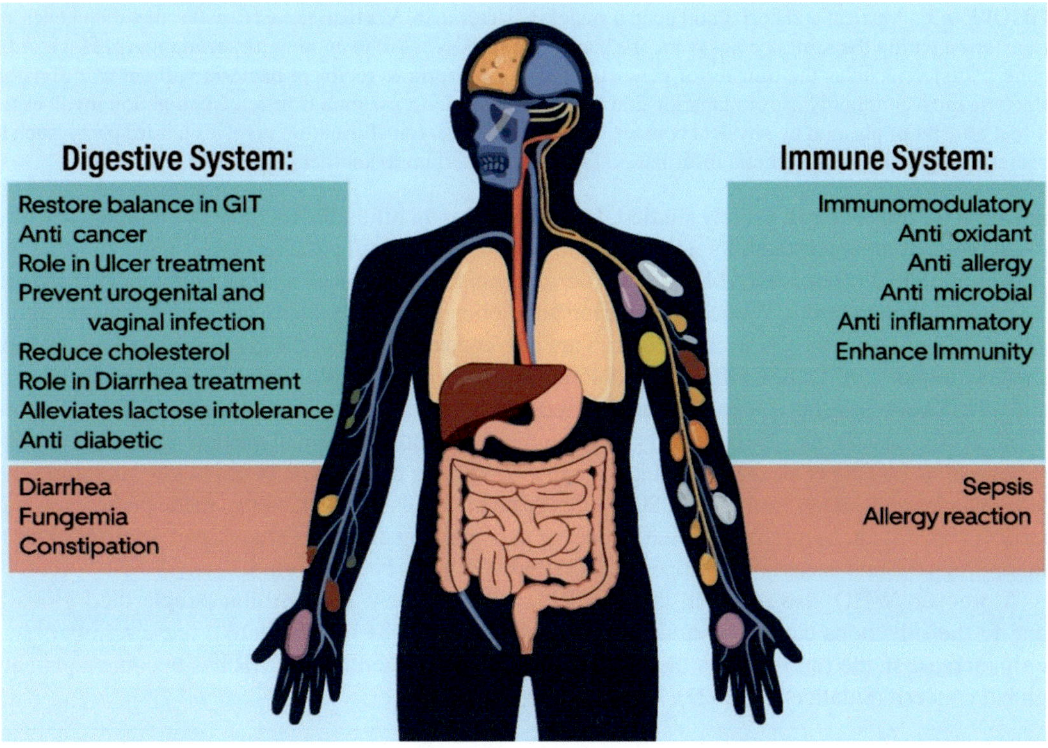

FIGURE 6.2 Effects of probiotics on human health. The role of probiotics in immune and digestive system has been highlighted. Adverse effects due to misuse of probiotics have been enlisted in the red area, while the green area shows positive impacts on health (La Fata et al., 2018).

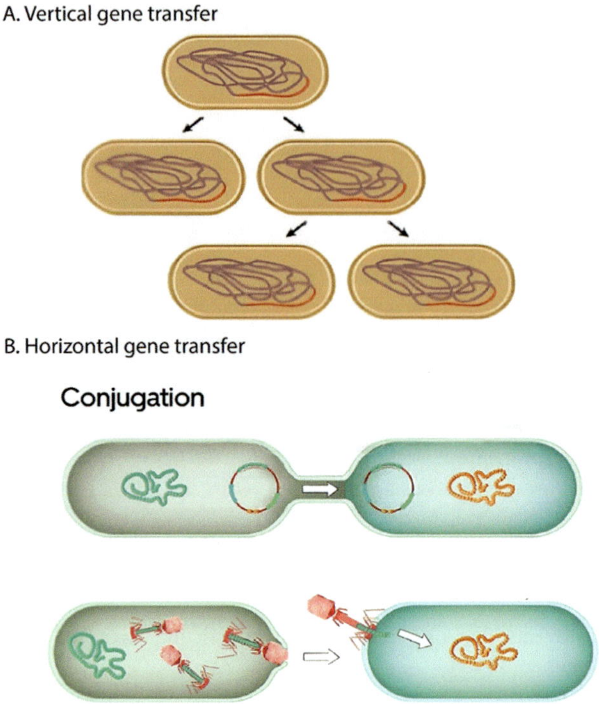

FIGURE 6.3 Vertical and horizontal gene transfer in bacteria. (a) **Vertical gene transfer**, in which genes are transferred within the same species from one bacterium (predecessor) to another (descendant). (b) **Horizontal gene transfer** includes the transfer of genes from donor bacteria to recipient bacteria without cell division. Bacteria most commonly use conjugation and transformation for horizontal transfer. Conjugation involves the direct transfer of plasmid by physical contact. At the same time, transformation involves a third party, such as bacteriophage, to transfer genetic information from one bacterium to another (La Fata et al., 2018).

misuse have not been extensively studied. Last year, after an infant death, the FDA issued a warning to probiotic companies about its use on newborns (Embleton et al., 2023).

Antimicrobial resistance (AMR), a defense mechanism of microbes against antibiotics, is a major concern for human health. While encountering antibiotics, microbes undergo gene alteration to survive against this unfavorable condition. Probiotics are also considered one of the great reservoirs of antibiotics (Gueimonde et al., 2013). Different probiotic strains such as Lactobacillus sp., Bifidobacterium sp., and Streptococcus sp. have already been reported for containing antimicrobial-resistant genes (Sharma et al., 2016). Due to their extraordinary intrinsic ability to horizontal and vertical gene transmission (Figure 6.3) (La Fata et al., 2018), probiotics can also transfer these AMR genes to gut microflora that also includes pathogens (Paul & Das, 2022). Overusing and misusing antibiotics made pathogenic bacteria more resistant, becoming a primary concern in combating diseases (Mancuso et al., 2021). Due to these drug-resistant bacteria in Europe, 33,000 deaths/year has been reported.

Moreover, WHO also reported that in 2019, approximately 1.27 million people died globally due to the infections caused by antibiotic-resistant bacteria. If not regulated, these numbers will only increase in the future due to the misuse of antibiotics. Therefore, AMR has become a primary global concern (Salam et al., 2023).

6.2 PROBIOTICS AND ANTIBIOTIC RESISTANCE

Probiotics are widely recognized for their numerous health benefits, many of which are backed by scientific research. Consequently, they are commonly found in foods such as yogurt, cheese, milk, and dietary supplements (Kechagia et al., 2013). Additionally, probiotics are added to animal feed, which means they can end up in our bodies through the food chain. However, despite their benefits, there are growing concerns about the long-term safety of consuming probiotics or probiotic-enriched foods, creating a "double-edged sword" scenario (Bernardeau et al., 2006). These concerns include the risk of serious infections like bacteremia or endocarditis, potential adverse effects on the digestive system, and the risk of transferring antibiotic-resistant bacteria to humans. This article focuses specifically on the issue of antibiotic resistance.

Antibiotics are frequently viewed as a one-stop shop for infections in modern medicine, and they are predicted to continue to play a major role in the fight against microbial illnesses for some time to come (Olshaker & Osterholm, 2017). However, pathogenic bacteria are under a great deal of selective pressure due to the increasing dependence on antibiotics and their abuse and overuse. This is mainly because the increased transfer of resistance genes through mobile genetic elements has accelerated their adaptation to antibiotics. Probiotics are becoming more prevalent in our diets, and as their demand grows worldwide, it's critical to take advantage of their benefits without adding to the rise in antibiotic resistance (Sharma et al., 2014). The GRAS (Generally Recognized as Safe) designation for probiotics, particularly lactic acid bacteria, indicates that they are generally considered safe (Negm, 2018). Probiotics containing resistant genes are not intrinsically hazardous. Still, there is a chance that these genes will spread to other gut flora and possibly even to opportunistic pathogens within the same intestinal environment, which could have detrimental clinical effects.

It has become evident that many antibiotic resistance traits in clinical bacteria are caused by gene transfer rather than mutations since the first known case of a resistant gene being transferred from *Shigella* to *E. coli* in 1959. Mutations are not the only reason for the rapid emergence of resistance; the sequences of resistance genes, including those for β-lactamases and aminoglycoside inactivating enzymes, differ significantly (Negm, 2018). On the other hand, molecular studies suggest that a more plausible explanation for the presence of these resistance traits in pathogenic bacteria is gene exchange between bacteria and bacteria and their animal or human hosts in environments like soil and the human gut. Numerous characteristics of these organisms resemble those of microbes that naturally produce antibiotics. This demonstrates how human activities can aid in the spread of resistance genes from a variety of sources. These activities include traditional agricultural and veterinary practices, more recent food preparation, and commercial probiotics. Through processes like conjugation, natural transformation, or transduction—discussed in other reviews—bacteria can transfer genes horizontally (Thomas & Nielsen, 2005). The ability of conjugative transfer, which involves transposons and integrons, to transfer genes freely across species barriers within the bacterial genome and between the genome and plasmids makes it stand out among these (Gillings, 2017).

Numerous settings, including hospitals, food, animal feed, water treatment facilities, and human samples, have been linked to antibiotic-resistant bacteria (Boonyasiri et al., 2014). Notably, antibiotic-resistant characteristics have also been found in probiotic strains from human and animal sources, dairy products, starter cultures, and probiotic foods (Ashraf & Shah, 2011). The profiles of antibiotic resistance of commercial probiotic bacteria, especially those found in well-known dietary supplements, are poorly understood. Owing to their extensive application, a number of Lactobacillus resistance genes have been thoroughly studied. For instance, the genes that confer resistance to vancomycin, ciprofloxacin, and tetracycline have been identified recently: van(X), van(E), gyr(A), and tet(M) (Das et al., 2020). According to another study, the erm(B) and tet(W), tet(M) genes are linked to resistance to oxytetracycline, clindamycin, and/or erythromycin in Lactobacillus. In contrast, the aac(6′)-aph(2′)-Ia gene is linked to resistance to gentamicin (Muñoz-Atienza et al., 2013). Furthermore, resistance to tetracycline is conferred by genes such as tet(L), tet(M), and tet(S), whereas resistance to streptomycin and spectinomycin is provided by the aadA gene (Stasiak et al., 2021).

Determining how probiotic bacteria create defense mechanisms against antibiotics requires understanding the antibiotic resistance profiles of these bacteria from various sources, including dietary supplements. Using efflux pumps, bacteria can, for instance, alter the sites at which antibiotics are targeted or lower the amount of free antibiotics present in the cell. These pumps aid in efficiently excluding antibiotics in conjunction with the outer membrane barrier (Stasiak et al., 2021).

Some bacteria produce enzymes that break down antibiotics, such as β-lactamase, to resist them. Other bacteria change the target sites of antibiotics, like penicillin-binding proteins, to render the medications ineffective (Eiamphungporn et al., 2018). Comprehensive analyses of various mechanisms of antibiotic resistance are accessible.

6.3 SAFETY REGULATION

Various regulatory committees have been formed in different regions of the world to overcome possible drawbacks associated with the misuse of probiotics containing AMR genes (Freeland et al., 2023). In 1997, Japan introduced the regulatory committee FOSHU to discuss basic rules for probiotic marketing strategy (Fukushima & Iino, 2006). Similarly, the European Food Safety Authority, Food and Drug Administration, and natural health products discussed regulatory frameworks for their region in Europe, the United States, and Canada. Still, no single unit body has been introduced to discuss this matter globally. European Food Safety Authority has rejected numerous health claims related to probiotics due to insufficient evidence submitted by the companies. Therefore, it banned labeling food with probiotics in Europe to prevent any misleading. In the United States, approval from a local body is required before certain claims, particularly health-related ones, can be used on product labels. To draw a framework for regulation, probiotics have been divided into three categories according to their intended use, which are probiotics food (PF), live biotherapeutic product (LBP), and probiotic dietary supplements (PDS) (Spacova et al., 2023). The United States of America has issued a list of microbes used as probiotics GRAS systems that are safe to use under specific conditions. For PF products in the United States, no minimum viable count of each strain at the end of the product's shelf life is required to be labeled (Williams, 2010).

6.4 SAFETY CONSIDERATIONS FOR MANUFACTURING AND PROCESSING

Dietary supplements, such as probiotics, must be produced in facilities that follow strict Good Manufacturing Practices (GMP), an international standard that is acknowledged globally (De Simone, 2022). This ensures that the supplements are safe for human use. A probiotic may pose dangers beyond these parameters even if it passes release testing. Adhering to GMP requirements guarantees traceability of operations and systems, appropriate design of facilities and equipment, appropriate selection of materials for the product, and risk-based monitoring and control of production processes. This ensures the items' potency, safety, identification, purity, and quality. A risk management system must be put in place throughout the production process to keep an eye out for any possible chemical, allergic, physical, and biological/microbiological contamination. A facility can more successfully identify essential operational steps and crucial control points in specific phases when hazards need to be addressed to assure product safety and quality by integrating hazard analysis with GMPs. When necessary, corrective and preventive actions should be taken in addition to monitoring and adjusting these crucial control points. Comprehensive recordkeeping is necessary for effective management.

Another critical factor in the production of probiotics is purity. Raw materials can allow chemical hazards such as pesticides, leftover solvents, heavy metals, polycyclic aromatic hydrocarbons, allergies, and microbiological pathogens to enter the supply chain (Kochhar et al., 2022). A surveillance plan that is both effective and practical should be put into action considering the risks involved with raw materials, production, and finished goods. This strategy is necessary to monitor, manage, and deal with pollutants to guarantee consumer safety on items. Qualifiability and specification of raw materials are prerequisites for GMP. A comprehensive study of new raw materials is required, which includes identification, physical property, assay, and potential contamination analyses. This assessment is a component of the vendor qualification program, which guarantees the reliability and security of the raw materials.

Various ingredients are used in the creation of probiotic products, including elements of culture media, such as proteins, carbohydrates, nitrogen sources, vitamins, and minerals—all of which are critical for the best possible growth of probiotics (Heenan et al., 2002). Additives are added to the mixes to standardize them to the necessary concentration, and cryoprotectants are employed to protect the strains during the freeze-drying process. To guarantee that the finished product satisfies safety, quality, identity, purity, and potency (SQIPP) requirements and is fit for human consumption, all raw materials must be assessed and put through quality control analysis before use (Heenan et al., 2002). Additional procedures and in-process inspections are performed to guarantee product safety in addition to these assessments. For instance, to reduce microbial burden, media culture is usually pasteurized or sterilized before inoculation, and testing is carried out for microbiological pollutants and hygiene. It is rare for media components to persist following centrifugation of the probiotic cell mass because most probiotic manufacturers optimize growing media to ensure it is fully utilized during fermentation. The safety evaluation and probiotic product labeling must consider any residual media components if centrifugation is not used during manufacturing (Roe et al., 2022).

To ensure that a product maintains its promised concentration throughout its shelf-life, GMP guidelines state that the required overage of constituents in a product is determined based on its stability, recommended storage temperature, and final packaging (Sirois et al., 2023). To ensure that the overages are sufficient for the entire shelf-life, real-time stability studies must be carried out on the completed product in its final packaging and under its planned storage conditions (Touw et al., 2023). It is important to consider any overage in addition to the daily serving amount labeled when assessing probiotics' safety. Although probiotics at high concentrations usually don't cause problems in terms of safety, it's important to record actual exposure levels (Mandal & Mandal, 2023). Major food allergies must be identified on product labels clearly and concisely to guarantee consumer safety. A supplier may be subject to fines and product recalls if they conceal the existence of a priority food allergy. Due to their frequent use in production, milk, soy, and gluten are common allergies detected in probiotics. A firm allergen management policy and proof that the product is free of specific priority allergens as listed by local laws are requirements for a product to be labeled as allergen-free. The same rules apply whether a product is labeled gluten-free, dairy-free, or soy-free.

6.5 ANTIBIOTIC-DEPENDENT BACTERIAL GROWTH

Not long after antibiotics were introduced into medicine, bacterial mutants requiring an antibiotic to develop were discovered (Martinez et al., 2008). However, this reliance does not mean that bacteria utilize antibiotics as a source of nutrition. Instead, this reliance is frequently associated with antibiotic resistance, where the resistance mechanism has changed the metabolism of the bacteria, permitting development only when the antibiotic is present (Blake et al., 2021). The causes of antibiotic-dependent growth in vancomycin-resistant enterococci have been investigated in studies (Martinez et al., 2008). For instance, producing a D-Ala-D-Lac depsipeptide is a sign of resistance to glycopeptides. Instead of using the host's D-Ala: D-Ala ligase to link D-Lac to D-Ala, the bacteria use an alternative ligase (VanA or VanB) in the presence of significant selective pressure and the vancomycin-resistance genes (Woodford et al., 1995). Because the antibiotic causes the development of the alternative ligase linked to resistance, the host ligase's ddl gene accumulates mutations, preventing the bacteria from growing outside of the presence of the antibiotic (Reynolds, 2001).

6.6 GLOBAL REGULATORY NETWORKS

A distinguishing characteristic of some bacterial species is their intrinsic resistance to antibiotics, which is probably incorporated into more extensive global regulatory networks. The protein IgaA serves as a prominent illustration of the significance of researching antibiotic resistance from a worldwide viewpoint (Abokor et al., 2021). Studies conducted on *Salmonella enterica* mucoid mutants resistant to mecillinam showed that mucoidity and antibiotic resistance were caused by mutations in the mucM gene, which overexpress the RcsCDB system (Martinez et al., 2008). IgaA is a protein that controls intracellular development and is essential for *S. enterica* pathogenicity, according to a different study (Cano et al., 2002). Later research revealed that mucM and igaA are the same genes, acting as both virulence and resistance (Mariscotti & Garcia-del Portillo, 2009).

Subsequent investigations revealed that IgaA unstable forms activate the RcsCDB system, which results in excess colanic acid production (Mariscotti & Garcia-del Portillo, 2009). In IgaA-deficient strains, this overactivation lowers the acute virulence of *S. enterica* (Mariscotti & Garcia-del Portillo, 2009). Additionally, it was discovered that RcsCDB, which comprises a response regulator, phosphotransferase, and hybrid sensor kinase, operates as a phospho-relay system. Many bacterial genes, including those involved in capsule synthesis, which are essential for intrinsic resistance to β-lactams like mecillinam, are regulated by RcsCDB (Laubacher & Ades, 2008). This example emphasizes the need for better-integrated research on these areas by showing how two independent studies, one focusing on virulence and the other on resistance, converged on the same gene and regulatory components. Two networks that simultaneously affect different bacterial properties, such as resistance, will be discussed in the following.

Antibiotic resistance has been shown to integrate into more extensive regulatory networks and cause particular alterations in bacterial metabolism. This implies that the metabolic status of the bacteria may have an impact on resistance. Antibiotic indifference occurring when antibiotics are ineffective against latent, non-dividing cells, is a prime example of this form of non-inherited phenotypic resistance (Al Matrood, 2016). This phenomenon was observed shortly after the initial use of antibiotics, highlighting that illnesses caused by actively growing bacteria are generally less affected by antibiotic indifference. However, other diseases might persist and return because of dormant cells. Bacterial populations susceptible to antibiotics can have long-lasting subpopulations, and these dormant cells explain phenotypic resistance in bacterial biofilms. Non-inherited resistance is also significantly influenced by the global regulatory responses of bacteria to various lifestyles (Jayaraman, 2008).

6.7 REGULATORY SYSTEMS

Effectiveness and safety regulation are essential to "consumer protection law." There is currently no comprehensive regulatory or legislative framework addressing the safety of probiotics, despite the growing desire for transparency in the labeling, marketing, and advertising of products containing microorganisms known as probiotics. However, creating a legal framework is a top priority for people working on probiotic supplement research and development (Britton et al., 2021). While there are notable distinctions in the regulatory strategies of the United States, Canada, Japan, and the European Union, certain nations have put in place their legal frameworks to guarantee the safety of these products in compliance with consumer protection regulations (Vogel, 2009). The most extensive guidelines on this subject have been established by the European Food Safety Authority (EFSA), the Food and Drug Administration (FDA) of the United States, and the Natural Health Product Directorate of Canada. A list of microorganisms declared safe for use is part of Europe's Qualified Presumption of Safety (QPS) concept (Hazards et al., 2023). To provide a consistent risk assessment technique for assessing the safety of microorganisms introduced into the food chain, the Scientific Committee of EFSA suggested in 2007 that the QPS approach be implemented by all EFSA panels (Hazards et al., 2023). The QPS list is periodically reviewed and updated by the EFSA Panel on Biological Hazards. The environment, target animals, and consumers of animal products treated with these chemicals are all guaranteed safety if a microorganism is added to the QPS list (de Benito et al., 2017). The "8651 Probiotics" Working Group of the Superior Health Council in Belgium produced a study that outlined a three-step procedure for evaluating the quality of microorganisms (Huys et al., 2013). This entails verifying the identity of the species, assessing the probiotic strains' safety, and ensuring the finished product's microbial composition, concentration, stability, and labeling are accurate. As an illustration, consider the European probiotic product Bactisubtil, which was discovered to include *Bacillus cereus* spores despite its initial claim to contain *Bacillus subtilis* (Sharma & Devi, 2014). The animal feed product Paciflor, which was withdrawn in 2002 because of the strain's propensity to create toxic substances, was also made with this same strain (Nunes, 2018). 2009, the Scientific Committee for Food Safety in Norway released a comparable national regulation.

The Norwegian Scientific Committee for Food Safety (2009) states that critically unwell individuals with antibiotic-associated diarrhea (AAD) or *Clostridium difficile* infections are not eligible to receive probiotics. Regulations pertaining to probiotics differ depending on the product category in the United States, which includes food, food additives, medications, nutritional supplements, and cosmetics. Foods are often exempt from FDA approval requirements before marketing, which means the FDA has less authority in this area. Because live probiotics are meant to be used for illness diagnosis, treatment, or prevention, the FDA classed them as medicines in 2010. These products cannot be incorporated into food items and must be categorized as Investigational New Drugs (INDs). For example, a probiotic food product would be considered a medicine if advertised as helping with gastrointestinal issues. Probiotics may also be classified as medications if clinical data demonstrates their therapeutic advantages when taken as food or supplements. Food companies can escape the strict IND restrictions and the time-consuming and costly approval process by staying in non-drug categories. Compared to pharmaceuticals, foods and dietary supplements typically need less proof to be approved, which could deter clinical research. The FDA, which started collecting GRAS notices in 1998, has historically recognized microorganisms used in food fermentation as safe and frequently classifies them as "food grade" or generally recognized as safe (GRAS) (Russo et al., 2017). If trained experts determine that GRAS substances are secure, they are not subject to premarket evaluation.

Only when supplements include novel dietary components, they are required to report to the FDA. Health Canada, which is in charge of natural health products under the Canadian Food and Drugs Act, regulates probiotics in Canada. The Joint FAO/WHO Working Group and EFSA guidelines align with Health Canada's laws. In Canada, probiotics are classified as natural health products, and product license applications and a monograph guide labeling. Long-term safe use of probiotics needs to fulfill particular requirements, such as evaluations of detrimental metabolic activities and antibiotic resistance specific to a strain. Randomized clinical trials are not required for probiotics because they are usually licensed as food supplements rather than pharmaceuticals in Canada. The product's effectiveness and safety must be emphasized in both labeling and advertising. Probiotics prepared as nutritional supplements are considered "foods," whereas probiotics prepared as ointments, nasal sprays, or cosmetics might be subject to stricter safety guidelines (Liang et al., 2024).

6.8 CONCLUSION

Probiotics are well known for their positive health effects, especially in supporting immune function and preserving gut microbiota equilibrium. Their capacity to support intestinal health has led to a notable increase in their appeal in the food and pharmaceutical sectors. However, recent worries about probiotics, particularly their possible contribution to the spread of genes that cause antibiotic resistance (AMR), have raised serious safety concerns. Probiotics are generally regarded as safe; however, there is concern that they may act as genetic reservoirs for antibiotic-resistant genes. These genes could then transfer to other gut bacteria, including harmful pathogens, and potentially leading to the emergence of resistant infections. Probiotics' dual nature emphasizes the need for cautious control, especially regarding their usage in susceptible groups like babies and immunocompromised people. It has been discovered that certain probiotic strains contain AMR genes, which can increase *via* processes such as horizontal gene transfer. Probiotic abuse and overuse combined with antibiotic overuse may make the worldwide problem of antibiotic resistance worse by increasing the risk of drug-resistant bacterial infections. Around the world, attempts have been made to control the manufacturing, labeling, and application of probiotics. Guidelines have been created by regulatory authorities such as the EFSA, the FDA in the United States, and Health Canada to guarantee the safety and quality of probiotic products. Still, there is cause for concern regarding the absence of a single global regulatory framework. These guidelines emphasize the importance of stability testing, adherence to GMP, and contamination monitoring to ensure the efficacy and safety of probiotics. Given the possible hazards, there is an increasing need for more research on the safety of probiotics, especially concerning their function in AMR. A more coordinated international effort, along with stronger regulations and comprehensive safety assessments, will be required to maximize the health benefits of probiotics while minimizing potential risks.

REFERENCES

Abokor, A. A., McDaniel, G. H., Golonka, R. M., Campbell, C., Brahmandam, S., Yeoh, B. S., Joe, B., Vijay-Kumar, M., & Saha, P. (2021). Immunoglobulin A, an active liaison for host-microbiota homeostasis. *Microorganisms*, *9*(10), 2117.

Al Matrood, W. (2016). *Factors that Enhance the Ability of Pseudomonas aeruginosa to Resist the Action of Antibiotics*. Liverpool John Moores University (United Kingdom).

Alvarez-Olmos, M. I., & Oberhelman, R. A. (2001). Probiotic agents and infectious diseases: A modern perspective on a traditional therapy. *Clinical Infectious Diseases*, *32*(11), 1567–1576.

Ashraf, R., & Shah, N. P. (2011). Antibiotic resistance of probiotic organisms and safety of probiotic dairy products. *International Food Research Journal*, *18*(3), 837–853.

Bernardeau, M., Guguen, M., & Vernoux, J. P. (2006). Beneficial lactobacilli in food and feed: Long-term use, biodiversity and proposals for specific and realistic safety assessments. *FEMS Microbiology Reviews*, *30*(4), 487–513.

Blake, K. S., Choi, J., & Dantas, G. (2021). Approaches for characterizing and tracking hospital-associated multidrug-resistant bacteria. *Cellular and Molecular Life Sciences*, *78*(6), 2585–2606.

Boonyasiri, A., Tangkoskul, T., Seenama, C., Saiyarin, J., Tiengrim, S., & Thamlikitkul, V. (2014). Prevalence of antibiotic resistant bacteria in healthy adults, foods, food animals, and the environment in selected areas in Thailand. *Pathogens and Global Health*, *108*(5), 235–245.

Britton, R. A., Hoffmann, D. E., & Khoruts, A. (2021). Probiotics and the microbiome—How can we help patients make sense of probiotics? *Gastroenterology*, *160*(2), 614–623.

Cano, D. A., Domínguez-Bernal, G., Tierrez, A., Portillo, F. G.-d., & Casadesús, J. (2002). Regulation of capsule synthesis and cell motility in Salmonella enterica by the essential gene igaA. *Genetics*, *162*(4), 1513–1523.

Das, D. J., Shankar, A., Johnson, J. B., & Thomas, S. (2020). Critical insights into antibiotic resistance transferability in probiotic Lactobacillus. *Nutrition*, *69*, 110567.

de Benito, A., Ibáñez, C., Moncho, W., Martínez, D., Vettorazzi, A., & de Cerain, A. L. (2017). Database on the taxonomical characterisation and potential toxigenic capacities of microorganisms used for the industrial production of food enzymes and feed additives, which do not have a recommendation for Qualified Presumption of Safety. *EFSA Supporting Publications*, *14*(8), 1274E.

De Simone, C. (2022). The authenticity of probiotic foods and dietary supplements: Facts and reflections from a court case. *CyTA-Journal of Food*, *20*(1), 366–373.

Eiamphungporn, W., Schaduangrat, N., Malik, A. A., & Nantasenamat, C. (2018). Tackling the antibiotic resistance caused by class A β-lactamases through the use of β-lactamase inhibitory protein. *International Journal of Molecular Sciences*, *19*(8), 2222.

Embleton, N. D., Berrington, J., Clarke, P., Deierl, A., Luyt, K., Spruce, M., & Oddie, S. J. (2023). Probiotics for preterm infants and the recent FDA alert in the USA. *Archives of Disease in Childhood-Fetal and Neonatal Edition*, *109*(6), e1.

Freeland, G., Hettiarachchy, N., Atungulu, G. G., Apple, J., & Mukherjee, S. (2023). Strategies to combat antimicrobial resistance from farm to table. *Food Reviews International*, *39*(1), 27–40.

Fukushima, Y., & Iino, H. (2006). Probiotics in food safety and human health: Current status of regulations on the use of probiotics in foods in Japan. In I. Goktepe, V. K. Juneja, & M. Ahmedna (Eds.), *Probiotics in Food Safety and Human Health* (pp. 431–459). CRC Press, ISBN 9780367391997.

Gillings, M. R. (2017). Lateral gene transfer, bacterial genome evolution, and the Anthropocene. *Annals of the New York Academy of Sciences*, *1389*(1), 20–36.

Gueimonde, M., Sánchez, B., G. de los Reyes-Gavilán, C., & Margolles, A. (2013). Antibiotic resistance in probiotic bacteria. *Frontiers in Microbiology*, *4*, 202.

Hazards, E. P. o. B., Koutsoumanis, K., Allende, A., Alvarez-Ordóñez, A., Bolton, D., Bover-Cid, S., Chemaly, M., De Cesare, A., Hilbert, F., & Lindqvist, R. (2023). Update of the list of qualified presumption of safety (QPS) recommended microbiological agents intentionally added to food or feed as notified to EFSA 18: Suitability of taxonomic units notified to EFSA until March 2023. *EFSA Journal*, *21*(7), e08092.

Heenan, C., Adams, M., Hosken, R., & Fleet, G. (2002). Growth medium for culturing probiotic bacteria for applications in vegetarian food products. *LWT-Food Science and Technology*, *35*(2), 171–176.

Huys, G., Botteldoorn, N., Delvigne, F., De Vuyst, L., Heyndrickx, M., Pot, B., Dubois, J. J., & Daube, G. (2013). Microbial characterization of probiotics–Advisory report of the W orking G roup "8651 Probiotics" of the B elgian S uperior H ealth C ouncil (SHC). *Molecular Nutrition & Food Research*, *57*(8), 1479–1504.

Jayaraman, R. (2008). Bacterial persistence: Some new insights into an old phenomenon. *Journal of Biosciences*, *33*(5), 795–805.

Kechagia, M., Basoulis, D., Konstantopoulou, S., Dimitriadi, D., Gyftopoulou, K., Skarmoutsou, N., & Fakiri, E. M. (2013). Health benefits of probiotics: A review. *International Scholarly Research Notices*, *2013*(1), 481651.

Kochhar, N., Shrivastava, S., Ghosh, A., Rawat, V. S., Sodhi, K. K., & Kumar, M. (2022). Perspectives on the microorganism of extreme environments and their applications. *Current Research in Microbial Sciences*, *3*, 100134.

La Fata, G., Weber, P., & Mohajeri, M. H. (2018). Probiotics and the gut immune system: Indirect regulation. *Probiotics and Antimicrobial Proteins*, *10*, 11–21.

Laubacher, M. E., & Ades, S. E. (2008). The Rcs phosphorelay is a cell envelope stress response activated by peptidoglycan stress and contributes to intrinsic antibiotic resistance. *Journal of Bacteriology*, *190*(6), 2065–2074.

Liang, D., Wu, F., Zhou, D., Tan, B., & Chen, T. (2024). Commercial probiotic products in public health: Current status and potential limitations. *Critical Reviews in Food Science and Nutrition*, *64*(19), 6455–6476.

Mancuso, G., Midiri, A., Gerace, E., & Biondo, C. (2021). Bacterial antibiotic resistance: The most critical pathogens. *Pathogens*, *10*(10), 1310.

Mandal, S., & Mandal, N. C. (2023). Current trends and future perspectives of probiotics on human health: An overview. In S. De Mandal, X. Xu, F. Jin, A. K. Panda, & K. S. Ibrahim (Eds.), *Recent Advances and Future Perspectives of Microbial Metabolites: Applications in Biomedicine* (pp. 81–122). Academic Press. https://doi.org/10.1016/C2020-0-02635-6. ISBN 978-0-323-90113-0.

Mariscotti, J. F., & Garcia-del Portillo, F. (2009). Genome expression analyses revealing the modulation of the Salmonella Rcs regulon by the attenuator IgaA. *Journal of Bacteriology*, *191*(6), 1855–1867.

Martinez, J. L., Fajardo, A., Garmendia, L., Hernandez, A., Linares, J. F., Martínez-Solano, L., & Sánchez, M. B. (2008). A global view of antibiotic resistance. *FEMS Microbiology Reviews*, *33*(1), 44–65.

Muñoz-Atienza, E., Gómez-Sala, B., Araújo, C., Campanero, C., Del Campo, R., Hernández, P. E., Herranz, C., & Cintas, L. M. (2013). Antimicrobial activity, antibiotic susceptibility and virulence factors of lactic acid bacteria of aquatic origin intended for use as probiotics in aquaculture. *BMC Microbiology*, *13*, 1–22.

Negm, G. E. S. (2018). Classification, antimicrobial potential, industrial applications and probiotic capability of lactic acid bacteria: A review article. *Research Journal of Applied Sciences*, *13*(12), 742–757.

Nunes, C. S. (2018). Probiotics and enzymes in the gastrointestinal tract. In C. S. Nunes & V. Kumar (Eds.), *Enzymes in Human and Animal Nutrition: Principles and Perspectives* (pp. 413–427). Academic Press. ISBN: 9780128054192 Elsevier.

Olshaker, M., & Osterholm, M. T. (2017). *Deadliest Enemy: Our War against Killer Germs*. Hachette UK.

Paul, D., & Das, B. (2022). Gut microbiome in the emergence of antibiotic-resistant bacterial pathogens. *Progress in Molecular Biology and Translational Science*, *192*(1), 1–31.

Reynolds, P. (2001). The genetics and biochemistry of resistance to glycopeptide antibiotics: Glycopeptide resistance mechanism. In *Novel Frontiers in the Production of Compounds for Biomedical Use* (pp. 99–115). Springer.

Roe, A. L., Boyte, M.-E., Elkins, C. A., Goldman, V. S., Heimbach, J., Madden, E., Oketch-Rabah, H., Sanders, M. E., Sirois, J., & Smith, A. (2022). Considerations for determining safety of probiotics: A USP perspective. *Regulatory Toxicology and Pharmacology*, *136*, 105266.

Russo, P., Spano, G., & Capozzi, V. (2017). Safety evaluation of starter cultures. In B. Speranza, A. Bevilacqua, M. R. Corbo, & M. Sinigaglia (Eds.), *Starter Cultures in Food Production*. John Wiley & Sons, Ltd (pp. 101–128).

Salam, M. A., Al-Amin, M. Y., Salam, M. T., Pawar, J. S., Akhter, N., Rabaan, A. A., & Alqumber, M. A. (2023). Antimicrobial resistance: A growing serious threat for global public health. Healthcare, *11*(13), 1946.

Sánchez, B., Delgado, S., Blanco-Míguez, A., Lourenço, A., Gueimonde, M., & Margolles, A. (2017). Probiotics, gut microbiota, and their influence on host health and disease. *Molecular Nutrition & Food Research*, *61*(1), 1600240.

Santacroce, L., Charitos, I. A., & Bottalico, L. (2019). A successful history: Probiotics and their potential as antimicrobials. *Expert Review of Anti-infective Therapy*, *17*(8), 635–645.

Sharma, M., & Devi, M. (2014). Probiotics: A comprehensive approach toward health foods. *Critical Reviews in Food Science and Nutrition*, *54*(4), 537–552.

Sharma, P., Tomar, S. K., Goswami, P., Sangwan, V., & Singh, R. (2014). Antibiotic resistance among commercially available probiotics. *Food Research International*, *57*, 176–195.

Sharma, P., Tomar, S. K., Sangwan, V., Goswami, P., & Singh, R. (2016). Antibiotic resistance of Lactobacillus sp. isolated from commercial probiotic preparations. *Journal of Food Safety*, *36*(1), 38–51.

Sirois, J., Reddy, S., Nguyen, T., Walker, H., Rendall, J., Bergen, G., Reimers, M., Cermak, E., Tiwary, A., & Helmes, E. (2023). Safety considerations for dietary supplement manufacturers in the United States. *Regulatory Toxicology and Pharmacology*, *147*, 105544. https://doi.org/10.1016/j.yrtph.2023.105544.

Spacova, I., Binda, S., Ter Haar J.A., Henoud, S., Legrain-Raspaud, S., Dekker, J., Espadaler-Mazo, J., Langella, P., Martín, R., Pane, M., & Ouwehand, A.C. (2023) Comparing technology and regulatory landscape of probiotics as food, dietary supplements and live biotherapeutics. *Frontiers in Microbiology, 14*, 1272754. https://doi.org/10.3389/fmicb.2023.1272754.

Stasiak, M., Maćkiw, E., Kowalska, J., Kucharek, K., & Postupolski, J. (2021). Silent genes: Antimicrobial resistance and antibiotic production. *Polish Journal of Microbiology*, *70*(4), 421–429.

Thomas, C. M., & Nielsen, K. M. (2005). Mechanisms of, and barriers to, horizontal gene transfer between bacteria. *Nature Reviews Microbiology*, *3*(9), 711–721.

Touw, D., Thiesen, J., & Vigneron, J. (2023). Stability. In M. E. Ali & N. N. A. Nizar (Eds.), *Practical Pharmaceutics: An International Guideline for the Preparation, Care and Use of Medicinal Products* (pp. 809–837). Springer.

Varzakas, T., Kandylis, P., Dimitrellou, D., Salamoura, C., Zakynthinos, G., & Proestos, C. (2018). Innovative and fortified food: Probiotics, prebiotics, GMOs, and superfood. In *Preparation and Processing of Religious and Cultural Foods* (pp. 67–129). Woodhead Publishing.

Vitetta, L., Briskey, D., Alford, H., Hall, S., & Coulson, S. (2014). Probiotics, prebiotics and the gastrointestinal tract in health and disease. *Inflammopharmacology*, *22*, 135–154.

Vogel, D. (2009). *Trading Up: Consumer and Environmental Regulation in a Global Economy*. Harvard University Press.

Williams, N. T. (2010). Probiotics. *American Journal of Health-System Pharmacy*, *67*(6), 449–458.

Woodford, N., Johnson, A. P., Morrison, D., & Speller, D. (1995). Current perspectives on glycopeptide resistance. *Clinical Microbiology Reviews*, *8*(4), 585–615.

Wu, C.-H., Wang, C.-C., Tsai, M.-T., Huang, W.-T., & Kennedy, J. (2014). Trend and pattern of herb and supplement use in the United States: Results from the 2002, 2007, and 2012 national health interview surveys. *Evidence-based Complementary and Alternative Medicine*, *2014*(1), 872320.

7 Probiotics and Antibiotic Resistance

Navigating the Potential Challenges and Risks

Kuldeep Lahry[1] *and Eijaz Ahmed Bhat*[2*]

[1]Institut de Recherche en Cancérologie de Montpellier (IRCM), Inserm, Institut du Cancer Montpellier (ICM), Montpellier, France

[2]Faculty of Biology, Department of Microbiology, University of Freiburg, Freiburg, Germany

*Corresponding author: Eijaz Ahmed Bhat

7.1 INTRODUCTION

The term "probiotics" is derived from the Greek words "pro bios," which translate to "for life" (Hamilton-Miller, Gibson, & Bruck, 2003). This term refers to live microorganisms that, when administered in sufficient quantities, offer health benefits to the host. This definition is officially recognized by both the World Health Organization (WHO) and the U.S. Food and Drug Administration (FDA) (FAO/WHO, 2002, 2006). In 2014, an amendment to this definition further specified that probiotics should have a well-defined composition, maintain optimal viable cell counts throughout their shelf life, and provide demonstrable health benefits while ensuring safety for their intended use (Hill et al., 2014) (Figure 7.1). Typically derived from human sources, probiotics are safe, free from pathogenic or toxic vectors, and devoid of antibiotic resistance (AR) genes. They are capable of enduring the harsh conditions of the intestines, including acidic pH, digestive enzymes, and bile salts, while demonstrating antagonistic effects against pathogens, boosting the immune system, and enhancing overall health (Plaza-Diaz, Ruiz-Ojeda, Gil-Campos, & Gil, 2019; Vandenplas, Huys, & Daube, 2015). Also, during technological processing, probiotics must retain their activity, viability, and efficacy (Tripathi & Giri, 2014). In 2014, the International Scientific Association for Probiotics and Prebiotics (ISAPP) clarified that although non-viable microbial products, dead bacteria, or metabolic byproducts might offer health benefits, they are not considered probiotics (Binda et al., 2020; Hill et al., 2014; Salminen et al., 2021). These components, referred to as "postbiotics," are soluble components with biological activity and are considered safer alternatives to live bacteria (Vinderola, Sanders, & Salminen, 2022). The concept of probiotics originates from Elie Metchnikoff's pioneering research, which emphasized the health benefits of lactic acid-producing bacteria (LAB) (Metchnikoff E, 1907). LAB, a diverse group of bacteria prevalent in the human gut, are commonly consumed through fermented milk products, cheese, and fermented meats (De Filippis, Pasolli, & Ercolini, 2020; De Vuyst et al., 2014; Derrien & van Hylckama Vlieg, 2015). Certain LAB strains, particularly *Lactobacillus* and *Streptococcus* species, are naturally part of the human and animal microbiota, are considered "Generally Recognized As Safe" (GRAS), and have been utilized as probiotics (De Filippis et al., 2020). Most commercialized probiotic strains include *Lactobacillus, Bifidobacterium, Enterococcus, Saccharomyces boulardii, and Escherichia* (Bermudez-Brito, Plaza-Diaz, Muñoz-Quezada, Gómez-Llorente, & Gil, 2012; Fijan, 2014; Manzoor, Wani, Ahmad Mir, & Rizwan, 2022; Plaza-Diaz et al., 2019), along with *Lactococcus, Leuconostoc,*

DOI: 10.1201/9781032676951-7

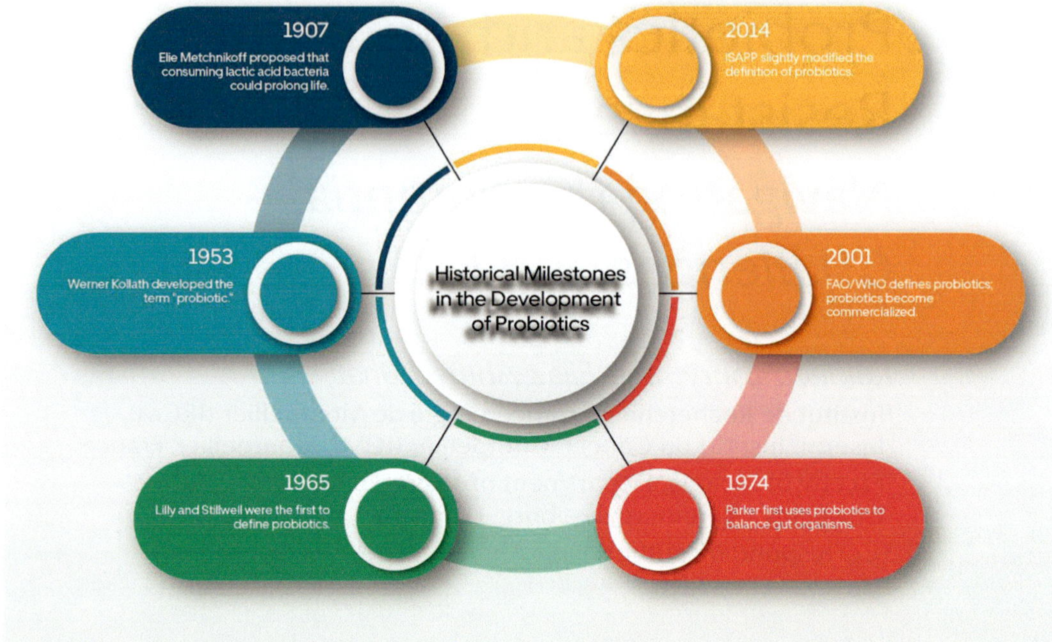

FIGURE 7.1 History of the development of probiotics: FAO: Food and Agriculture Organization of the United Nations; WHO: World Health Organization; ISAPP, International Scientific Association for Probiotics and Prebiotics (Ji, Jin, Liu, Jiao, & Li, 2023).

Pediococcus, and *Streptococcus* (Khalighi et al., 2016). Sources for these microorganisms include human feces, dairy products, fruits, grains, and waste. These microorganisms must maintain their original characteristics during manufacturing and transit through the gastrointestinal tract (Lima et al., 2009; Ranadheera et al., 2019; Shori, 2017; Tripathi & Giri, 2014; Varga, Süle, & Nagy, 2014). The recommended daily dose for effective probiotic action ranges from 10^8 to 10^9 viable microorganisms (Karimi, Mortazavian, & Da Cruz, 2011; Knorr, 1998). Strains such as *Lactobacillus*, *Bifidobacterium*, and *Propionibacterium* are employed in fermented dairy products and pharmaceuticals, whereas *Escherichia coli* and *Enterococcus* are used in non-food applications (Plaza-Diaz et al., 2019). *Saccharomyces boulardii*, the sole fungal probiotic frequently used in the dairy industry, was originally isolated from fruits (Pais, Almeida, Yılmaz, & Teixeira, 2020).

The FDA regulates probiotic products based on their intended use and effect, categorizing them as dietary supplements, food ingredients, or drugs. Probiotics intended to treat or prevent diseases must meet stringent drug regulatory standards, with health claims substantiated through appropriate trials meeting drug regulatory standards (Degnan King, Washington, & Degnan, 2008; Szajewska & Vinderola, 2024). Probiotics are being more frequently utilized to address dysbiosis, restore microbial diversity, and alter disrupted intestinal microbiota through mechanisms that remain not fully understood (Mendes et al., 2018; Vieira, Fukumori, & Ferreira, 2016). Their growing popularity is due to potential health benefits, such as improved nutrient absorption (Bielik & Kolisek, 2021) and boosted immunity (Varvara & Vodnar, 2024; Yang & Yang, 2023). Probiotics generate functional molecules like antioxidants, enzymes, short-chain fatty acids (SCFAs), and peptides, which help maintain nutrient bioavailability (Chugh & Kamal-Eldin, 2020). Their role in regulating gut health, especially following broad-spectrum antibiotic treatments (Grazul, Kanda, & Gondek, 2016), has made probiotics valuable for addressing various pathological conditions, for example, antibiotic-associated diarrhea (AAD) (Goodman, Keating, Georgousopoulou, Hespe, & Levett, 2021), irritable bowel syndrome (IBS) (Sharma, Kumar, Sajjad, & Sharma, 2023), inflammatory bowel disease (IBD) (Li, Peng, Xiao, Long, & Yu, 2023), Clostridium difficile-associated diarrhea

(Johnston et al., 2012), and vaginal and pulmonary infections (Du, Lei, Zhang, & Zhu, 2022; Ma, Wang, Su, Sun, & Ma, 2023). Additionally, probiotics demonstrate antihypertensive effects, reduce cholesterol levels and obesity, and improve symptoms of diabetes and metabolic disorders (Kocsis et al., 2020; Lye, Kuan, Ewe, Fung, & Liong, 2009; Zikou, Dovrolis, Dimosthenopoulos, Gazouli, & Makrilakis, 2023). They also show protective effects against cancers, particularly colon, bladder, and mammary cancer (Górska, Przystupski, Niemczura, & Kulbacka, 2019; Thu et al., 2023; Wollowski, Rechkemmer, & Pool-Zobel, 2001), and help reduce dental caries (Abikshyeet et al., 2022), alleviate allergy symptoms in infants (Di Costanzo et al., 2024), treat atopic dermatitis and allergic rhinitis (Iftikhar et al., 2022; Luo, Peng, Li, Ao, & Liu, 2022; Rather et al., 2016), and mitigate viral-associated pulmonary damage (Shahbazi, Yasavoli-Sharahi, Alsadi, Ismail, & Matar, 2020). Furthermore, probiotics address mental health conditions by reversing symptoms of anxiety, depression, insomnia, and obsessive-compulsive disorder and improve lactose intolerance symptoms (Merkouris et al., 2024; Wallace & Milev, 2017). Certain strains, such as *L. rhamnosus* SHA113, have shown potential in treating multidrug-resistant *S. aureus* infections (Liu et al., 2020), while *Bifidobacterium* strains are employed in managing *C. difficile* infections and alleviating IBD inflammation (Wei et al., 2018). Probiotic supplements are employed for conditions beyond the gastrointestinal tract, including depression, Parkinson's disease, and rheumatoid arthritis (Oroojzadeh, Bostanabad, & Lotfi, 2022; Sabit et al., 2023; H. Yang, Liu, Cai, Li, & Gu, 2021). For instance, *L. rhamnosus* increases glutamine and glutathione to alleviate depression symptoms (Chudzik et al., 2022; Kochalska et al., 2020), and *L. acidophilus* and *L. salivarius* offer neuroprotective effects for Parkinson's patients (Lorente-picón & Laguna, 2021; Zhu et al., 2022). Certain *Bifidobacterium* strains help regulate luteinizing hormone in polycystic ovary syndrome (Sun, Gao, Ye, & Zhao, 2023). Furthermore, probiotic bacteria exhibit several advantageous properties that help control pathogenic bacteria. These include improving intestinal barrier function, competitive exclusion by reducing pathogen adherence to cells, co-aggregation, and producing organic acids that inhibit pathogenic bacteria (Lau & Quek, 2024; Ouwehand, Forssten, Hibberd, Lyra, & Stahl, 2016). Additionally, many probiotics produce antimicrobial compounds such as SCFAs, hydrogen peroxide, nitric oxide, and bacteriocins, which enhance their competitive advantage against other gastrointestinal microorganisms and potentially inhibit pathogens (Dobson, Cotter, Paul Ross, & Hill, 2012; Ouwehand et al., 2016). The production of these antimicrobials is crucial for both bacterial fitness and the efficacy of probiotics, with bacteriocin production being a key factor in selecting probiotic strains. In the last two decades, there has been substantial progress in the number and quality of clinical trials assessing the benefits of probiotics, fueled by advancements in efficacy measurement (Merenstein et al., 2023). Early probiotics faced less scrutiny regarding adverse events (AEs) due to their traditional use in fermented foods. Current research indicates minimal acute safety issues for the general population and non-immunocompromised individuals. However, the long-term safety effects on microbiota, immunity, cardiometabolic health, and other physiological parameters are not well-documented. The health benefits of probiotics are strain-specific, although marketing often emphasizes broad effects (Lau & Quek, 2024; Merenstein et al., 2023). Therefore, it is essential for probiotics to undergo thorough testing and comply with safety standards established by regulatory authorities. The FDA regulates probiotics as dietary supplements, food ingredients, or drugs depending on their intended use and claims. If a probiotic is meant to cure, alleviate, treat, or prevent disease, it is classified as a drug and must comply with stricter requirements (Venugopalan, Shriner, & Wong-Beringer, 2010). While regulations vary by country, any claim must meet the standards for drug evidence. Probiotics categorized as food or supplements without specific health claims must align with the Qualified Presumption of Safety (QPS) in the EU or the GRAS category of FDA, supported by evidence of a general benefit to human health (Astashkina, Khudyakova, & Kolbysheva, 2014). For probiotics with specific health claims, they must be of a defined strain and provide proof of delivering viable doses at the product's shelf life end (Binda et al., 2020). In most EU countries, the term "probiotic" cannot be used for advertising or on food supplement labels due to its classification as an unauthorized health claim without adequate scientific backing (European

Commission (EC) Regulation (EC) No. 1924/2006, n.d.). Probiotics can be categorized under medicines, dietary supplements, food for special medical purposes, functional foods, and food ingredients, each subject to different regulatory frameworks.

The global market for probiotics has seen significant growth, with probiotic supplements driving the functional foods industry (Wiegers, van Beek, & Larsen, 2023). The interest of the public in probiotics is mirrored by a substantial increase in scholarly research, indicating a need for continued exploration of probiotic benefits and risks. Given the widespread use and potential health benefits of probiotics, it is essential to assess and analyze the associated risks, particularly concerning AR. This chapter highlights the risk of AR spread through probiotics, providing a comprehensive overview of current knowledge and regulations.

7.2 SAFETY ISSUES AND RISKS ASSOCIATED WITH PROBIOTICS

Probiotics are often advertised as natural treatments with few side effects. Nonetheless, as live microorganisms, they contain enzymes and chemicals that can interact with both the host and its microbial community. These interactions involve regulating gene expression in the host or other microbial species through miRNA modulation, producing bioactive compounds such as SCFAs, bile acid derivatives, lipopolysaccharides, and antimicrobial peptides, and transferring resistance genes to other microbes (Oh et al., 2021; Ruíz-Ramírez, Guadarrama-Mendoza, Escalante, Giles-Gómez, & Valadez-Blanco, 2022; Sehrawat et al., 2021; Tóth et al., 2021; Wang et al., 2021; Zhao et al., 2021).

The widespread use of probiotics has raised safety concerns among clinicians, researchers, and policymakers that is primarily related to the strains used, product quality, and administration protocols. The 2002 FAO/WHO report suggested that probiotics could potentially cause four types of adverse effects: systemic infections, harmful metabolic activities, excessive immune stimulation in susceptible individuals, and gene transfer (FAO/WHO, 2002). The identification of risks associated with probiotic strains can be achieved through whole genome sequencing, which helps detect genes related to toxigenicity, pathogenicity, and AR. Additionally, formulation aspects such as purity, potency, and composition are critical, requiring rigorous testing to identify microbial contaminants, allergens, and other impurities. Safe administration of probiotics is similarly essential, with specific concerns about cross-contamination in hospital settings and risks associated with certain methods. Studies, such as PROPATRIA, highlight these risks (Sanders et al., 2010).

AEs of probiotics include a higher risk for immunocompromised individuals, those under stress, the elderly, and newborns (Kothari, Patel, & Kim, 2019; Sanders et al., 2010), as well as D-lactic acidosis and brain fogginess (Rao, Rehman, Yu, & De Andino, 2018; Rao, Yu, Tetangco, & Yan, 2018), bacteremia (Kullar, Goldstein, Johnson, & McFarland, 2023), and AR gene transfer (Li et al., 2020; Shahali, Soltani, & Akbari, 2023). Probiotic consumption can lead to infectious diseases such as endocarditis (Ioannou et al., 2024). For instance, Lactobacillus has frequently been associated with endocarditis and bacteremia, with certain studies reporting a mortality rate of nearly 30% (Cannon, Lee, Bolanos, & Danziger, 2005). Likewise, Enterococcus and Leuconostoc species also pose risks of bacterial endocarditis and bloodstream infections (Hernández & Juliá, 2023). In patients with acute pancreatitis, probiotics have not reduced the incidence of infectious complications and may even increase bowel ischemia (Van Baal et al., 2012). In developed countries, Lactobacilli and Bifidobacteria account for 0.05%–0.4% of cases of infective bacteremia and endocarditis, indicating that probiotics are not always harmless additions to enteral nutrition (Borriello et al., 2003). The risks of probiotics can be categorized into acute and long-term effects, with most research focusing on acute risks while long-term risks remain less well-studied.

7.2.1 Acute Risks

The acute risks linked to probiotics include alterations in microbiota function or composition, invasive infections, effects on drug effectiveness, and the horizontal transfer of AR. To thoroughly assess the impact of probiotics on the host microbiome, their health benefits, safety issues, and the differentiation between responders and non-responders, it is essential to analyze baseline microbiota composition, as well as transcriptional and metabolic outputs (microbiome profiling), both before and after probiotic use (Suez, Zmora, & Elinav, 2020; Suez, Zmora, Segal, & Elinav, 2019). Microbiome profiling can reveal individual responses to probiotics and underlying mechanisms, yet its infrequent practice and limited research leave the association between microbiome characteristics and AEs underexplored (Bafeta, Koh, Riveros, & Ravaud, 2018; De Vos, Tilg, Van Hul, & Cani, 2022; Merenstein et al., 2023; Suez et al., 2020).

The acute risks linked to probiotics include alterations in microbiota function or composition, invasive infections, effects on drug effectiveness, and the horizontal transfer of AR. To thoroughly assess the impact of probiotics on the host microbiome, their health benefits, safety issues, and the differentiation between responders and non-responders, it is essential to analyze baseline microbiota composition, as well as transcriptional and metabolic outputs (microbiome profiling), both before and after probiotic use.

In rare instances, probiotics may translocate from the gastrointestinal tract, causing invasive infections. A systematic review of probiotics in children identified sepsis as the most frequent invasive infection among 49 cases, primarily affecting those under 2 years old with conditions like prematurity or the presence of intravenous catheters; 94% of these cases were successfully treated with antimicrobial therapy (D'agostin et al., 2021). Accurately determining the frequency of probiotic-related invasive infections is difficult due to underreporting outside of scientific studies. Furthermore, reporting of AEs in probiotic trials is often inadequate, and clinical laboratories do not consistently culture and identify probiotic organisms from clinical samples (Bafeta et al., 2018). Meta-analyses suggest that probiotics do not increase sepsis rates in preterm neonates (Morgan et al., 2020), although the exact frequency of probiotic-associated sepsis remains unclear. Similarly, no increase in AEs was observed in critically ill and high-risk individuals using probiotics compared to control groups (Hempel et al., 2011). Therefore, thorough documentation and monitoring of invasive infections and sepsis in clinical settings and probiotic trials are essential.

Probiotics may influence drug safety and effectiveness through interactions (Dashnyam et al., 2018; Wilson & Nicholson, 2017). For instance, azo drugs like Prontosil or neoprontosil can be converted to sulfanilamide by microbial azoreductase (Carmody & Turnbaugh, 2014; Gingell & Bridges, 1971). The microbiota can also affect drug metabolism processes such as acetylation, sulfation, dehydroxylation, demethylation (Klaassen & Cui, 2015), dehalogenation (Atashgahi, Shetty, Smidt, & de Vos, 2018), and decarboxylation (Rekdal, Bess, Bisanz, Turnbaugh, & Balskus, 2019). Additional research is necessary to identify and understand drug-modifying enzymes in vivo. The presence and activity of these enzymes in probiotics and their impact on the host can vary. It is crucial to screen for enzymes that influence drug efficacy and consult genomic databases to advance this field and detect possible probiotic-drug interactions. Toximicrobiomics, also known as pharmacomicrobiomics, is an emerging field that examines the interactions between microbiota and xenobiotics (Abdelsalam, Ramadan, ElRakaiby, & Aziz, 2020; Dikeocha, Al-Kabsi, Miftahussurur, & Alshawsh, 2022). Understanding the effects of food, drug metabolism, response, and microbiome diversity is essential for personalized treatment (Dikeocha et al., 2022).

The horizontal transfer of AR genes is another major acute risk linked to probiotics, which is explored in depth in Section 7.3 of this chapter.

7.2.2 Long-Term Colonization Risks

Most probiotics do not colonize the gut for a long term due to competition with resident microbiota and challenging conditions in the gastrointestinal tract (Khalesi et al., 2019; Sanders et al., 2011). However, external microbes can colonize if an open niche is available. Their tolerance to these conditions is strain-specific and often only tested in vitro (Frese et al., 2017; Janina Krumbeck, Hutkins, & Walter Lincoln, 2016; Maldonado-Gómez et al., 2016; O'Brien et al., 2022; Panigrahi et al., 2017; Perraudeau et al., 2020; Samara et al., 2022). Moreover, different strategies have been optimized to enhance the persistence of probiotics in host including manipulating host diet, administering probiotics, precise strain targeting, and dose optimization (Aggarwala et al., 2021; Rattanaprasert, Roos, Hutkins, & Walter, 2014; Tachon, Lee, & Marco, 2014; Taverniti et al., 2019). The long-term colonization of probiotics raises safety concerns, such as displacing essential endemic microbes, negatively affecting adjacent microbiota structure and function, and translocation to systemic circulation, leading to invasive infections (Merenstein et al., 2023; Yelin et al., 2019).

Despite potential adverse effects, long-term colonization of probiotics can offer significant health benefits, such as providing a missing metabolic function like metabolizing human milk oligosaccharides in infants. Long-term colonizing probiotics could compensate for metabolic disorders, offering superior benefits. Developing such probiotics should have clear objectives, such as balancing risks against benefits and considering relevant long-term safety data. Acute exposure testing and biomarkers could be used in research to evaluate the safety of long-term colonization probiotics (Merenstein et al., 2023).

7.3 PROBIOTICS AND ANTIBIOTIC RESISTANCE

Probiotics can reduce pathogenic bacterial populations at mucosal sites by producing bacteriocins, disrupting biofilms, and enhancing antibiotic efficacy (Ouwehand et al., 2016). However, most bacteriocin activity has been observed in vitro, with limited in vivo evidence from human double-blind randomized controlled trials (Hegarty, Guinane, Ross, Hill, & Cotter, 2016). The role of probiotics in preventing antibiotic-resistant infections in humans is still under investigation. The CDC is actively researching this topic, and while some studies indicate benefits, the evidence is not yet strong enough to make specific recommendations. Currently, probiotics may serve as a partial replacement or adjunct to antibiotic treatment, potentially aiding in the treatment of multidrug-resistant UTIs (Ouwehand et al., 2016). Research on AR has primarily focused on pathogenic bacteria in hospitals and nursing homes, but it also impacts food, animal production, and aquaculture.

Contrary to the idea that probiotics can help reduce the spread of AR, there are also significant concerns about the horizontal gene transfer (HGT) of AR from probiotics to potential gut pathogens. Regulators, industry experts, and researchers have raised these concerns, supported by evidence of in vivo transfer of AR genes through conjugation from viable probiotic bacteria acting as donors. This phenomenon was first documented in 1998 and has been confirmed by subsequent studies (Crits-Christoph, Hallowell, Koutouvalis, & Suez, 2022; Daniali, Nikfar, & Abdollahi, 2020; Morelli, Sarra, & Bottazzi, 1988; Shahali et al., 2023; Tóth et al., 2021).

The extensive use of antibiotics and the possibility that probiotics could transfer resistance genes to other microorganisms underscore the urgency of this issue (Figure 7.2). Most of the LABs are resistant to aminoglycosides, such as streptomycin, kanamycin, and gentamicin (Danielsen & Wind, 2003). Additionally, almost all *Lactobacillus* strains show resistance to vancomycin, ampicillin, and cephalosporins (Anisimova & Yarullina, 2019). Because of their transferable AR genes, species of *Bacillus, Enterococcus*, and *Streptococcus* also exhibit resistance to most antibiotics (Alves-Barroco, Rivas-García, Fernandes, & Baptista, 2020; Miller, Munita, & Arias, 2014; Wash et al., 2022). In contrast, *Bifidobacterium* and *Lactococcus lactis* exhibit low levels of antibiotic resistance, likely because their AR genes are non-transferable (Rozman et al., 2022, 2023). The determinants and characteristics of AR in *Lactobacilli, Bifidobacteria,* and *Bacillus* strains are detailed in Table 7.1. Therefore, different probiotics display unique and complex drug resistance traits, influenced by their AR gene composition. The potential for HGT of AR genes from probiotics to gut pathogens (Figure 7.2) necessitates thorough evaluation and regulation to mitigate associated risks. This includes screening probiotic strains for AR genes and understanding the mechanisms behind their resistance profiles to ensure they do not exacerbate the broader issue of antibiotic resistance.

Probiotic strains can be evaluated for AR using two main approaches: phenotypic and genotypic. The phenotypic approach measures AR gene expression using techniques like minimum inhibitory concentration (MIC), antibiotic susceptibility tests such as the Kirby-Bauer disk-diffusion method, or the Epsilometer test (E-test). This analysis helps determine the susceptibility patterns of various

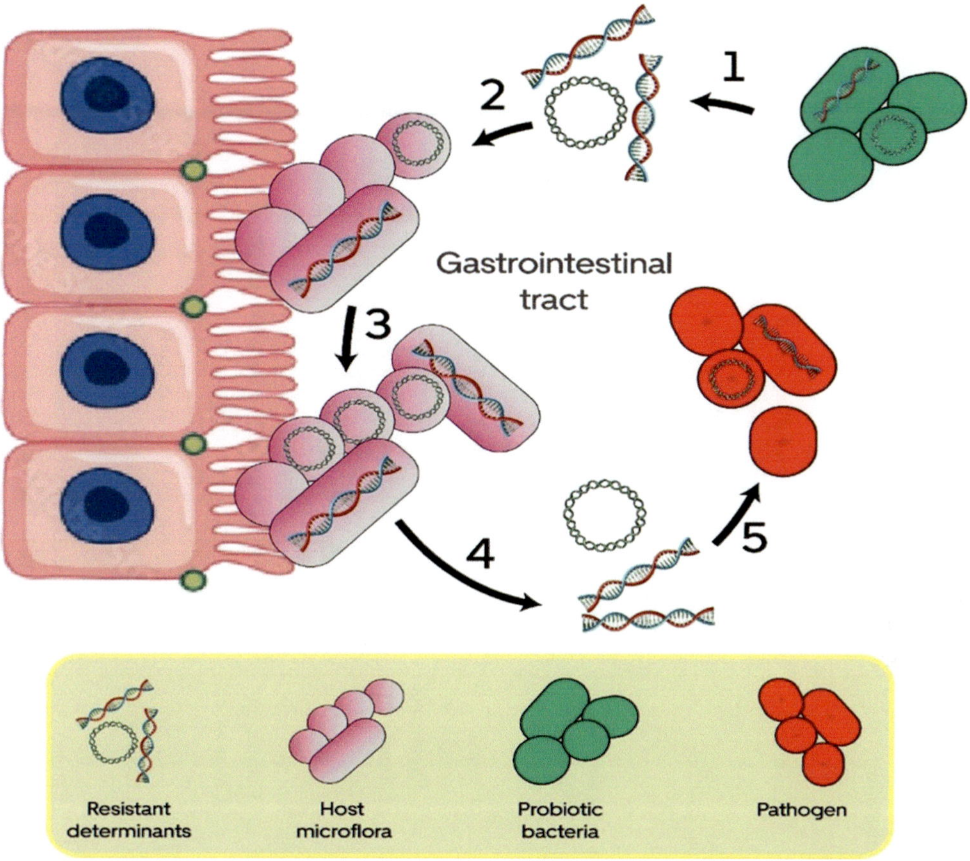

FIGURE 7.2 A conceptual model for the dissemination of AR genes from probiotics to pathogenic bacteria within the human gastrointestinal tract. (1) Probiotic bacteria present in dietary supplements that carry resistance genes (2) have the potential to transfer these genetic elements to the host microflora (3) Over time, this can lead to the accumulation of a reservoir of antibiotic resistance genes within the gut flora (4, 5) These resistance genes may subsequently be transferred to opportunistic pathogens within the gut, thereby compromising the efficacy of antibiotic treatments (Zheng et al., 2017).

species to different antibiotics (Andriani et al., 2021; Salam, Al-Amin, Pawar, Akhter, & Lucy, 2023). In some cases, such as intrinsic resistance of *Lactobacilli* to vancomycin, it does not represent a health risk (Stogios & Savchenko, 2020). However, any phenotypic resistance that is atypical for a species requires additional examination.

The genotypic approach involves identifying AR gene(s) in the genome of a probiotic strain using methods such as PCR, whole genome sequencing, microarray analysis, metagenomics, and bioinformatics tools. This type of analysis aids in detecting specific AR genes, correlating them with phenotypic traits, and assessing the potential for HGT (Andriani et al., 2021; Shahali et al., 2023). Generally, strains that have plasmid-encoded or mobile element flanked AR genes should not be commercialized unless a risk/benefit analysis determines it to be safe (Pariza, Gillies, Kraak-Ripple, Leyer, & Smith, 2015). Resistance to non-clinically relevant antibiotics poses a lower risk. For example, the FDA maintained the GRAS status of *Bifidobacterium animalis subsp. lactis* BB-12 for use in infant formula, despite the presence of the tet(W) gene, as it was not considered clinically or environmentally significant (FDA, 2005). Similarly, the EFSA found no new information to change the QPS status of this subspecies (Koutsoumanis et al., 2022). However, non-intrinsic resistance to clinically important antibiotics, such as vancomycin, suggests that the strain should not be commercialized unless it is deemed safe (WHO, 2018a, 2022).

TABLE 7.1
Antibiotic Resistance Determinants Identified and Characterized in *Lactobacilli, Bifidobacteria,* and *Bacillus* Strains

Gene(s)	Resistance	Resistance Mechanism	Location	Reference
		Lactobacillus		
blaZ	β-Lactams	Antibiotic hydrolysis	–	Aquilanti et al. (2007)
***vat*(E)**	Quinupristin–dalfopristin	Antibiotic acetylation	–	Mayrhofer et al. (2010)
Cat	Chloramphenicol	Antibiotic acetylation	Plasmid	Ahn, Collins-Thompson, Duncan, and Stiles (1992); Hummel et al. (2007); Lin, Fung, Wu, and Chung (1996); Mayrhofer et al. (2010)
msrC	MLS	Efflux	–	Rao Thumu and Halami (2012)
***mef*(A)**	Macrolide	Efflux	–	Cauwerts et al. (2006)
***aac*(6′)-*aph*(2″), *ant*(6), *aph*(3′)-IIIa**	Aminoglycoside	Enzymatic modification	–	Rojo-Bezares et al. (2006)
***erm*(B), *erm*(C), *erm*(T), *erm*(LF), *erm*(GT), *erm*(A)**	MLS	Ribosomal methylation	Plasmid, transposon, chromosome	Ammor, Flórez, Van Hoek, et al. (2008); Aquilanti et al. (2007); Cauwerts et al. (2006); Egervärn, Roos, and Lindmark (2009); Gfeller, Roth, Meile, and Teuber (2003); Hummel et al. (2007); Klare et al. (2007); Mayrhofer et al., 2010; Rao Thumu and Halami (2012); Tannock et al. (1994)
***tet*(W), *tet*(M), *tet* (S), *tet* (O), *tet* (Q), *tet* (36), *tet* (Z), *tet*(W/O), *tet*(O/W/32/O/W/O)**	Tetracycline	Ribosomal protection	Plasmid, transposon, chromosome	Ammor, Flórez, Van Hoek, et al. (2008); Aquilanti et al. (2007); Klare et al. (2007)
***tet* (K)/*tet* (L)**	Tetracycline	Efflux	Plasmid	Ammor, Flórez, Álvarez-Martín, Margolles, and Mayo (2008); Aquilanti et al. (2007); Devirgiliis, Coppola, Barile, Colonna, and Perozzi (2009); Rao Thumu, and Halami (2012)
		Bifidobacterium		
***erm*(X)**	MLS	Ribosomal methylation	Transposon	van Hoek, Mayrhofer, Domig, and Aarts (2008)
***tet*(W), *tet*(M), *tet* (O), *tet*(W/32/O), *tet*(O/W)**	Tetracycline	Ribosomal protection	Chromosome	Ammor, Flórez, Álvarez-Martín, et al. (2008); Flórez, Ammor, Álvarez-Martín, Margolles, and Mayo (2006); Kazimierczak, Flint, and Scott (2006); Van Hoek et al. (2008)
***tet* (L)**	Tetracycline	Efflux	Chromosome	Van Hoek et al. (2008)
		Bacillus		
aaD2	Aminoglycoside	Antibiotic adenylation	Chromosome	Bozdogan, Galopin, Gerbaud, Courvalin, and Leclercq (2003)
***erm*(34)**	MLS	Ribosomal protection	Chromosome	Bozdogan, Galopin, and Leclercq (2004)
BCL-1	β-lactams	Antibiotic hydrolysis	Chromosome	Girlich, Leclercq, Naas, and Nordmann (2007)
***cat*(Bcl)**	Chloramphenicol	Antibiotic acetylation	Chromosome	Galopin, Cattoir, and Leclercq (2009)

Source: Adopted from Gueimonde et al. (2013).

In the absence of non-intrinsic AR genes transfer through conjugation, there is a possibility of transfer via transformation of naked DNA or transduction through phage (McInnes, McCallum, Lamberte, & van Schaik, 2020; Von Wintersdorff et al., 2016; Winter, Buckling, Harms, Johnsen, & Vos, 2021). The transformation of DNA from non-viable cells that perished during production or storage is possible. This concern is crucial in the regulatory approach to postbiotic safety, which contains dead but useful microorganisms (Salminen et al., 2021). The EFSA found no AR genes of concern in the genome of *Akkermansia muciniphila* strain for assessment as a novel food, even after following a defensive approach (Turck et al., 2021). This analysis was based on screening of AR genes in the genome of the novel strain, using data from the Comprehensive Antibiotic Resistance Database (Alcock et al., 2020) and National Database of Antibiotic Resistant Organisms ("National Database of Antibiotic Resistant Organisms (NDARO) – Pathogen Detection – NCBI," n.d.).

7.3.1 Antibiotic Resistance Mechanism in Probiotics

Probiotic AR can be classified like that in pathogenic bacteria: intrinsic and acquired. Bacteria develop intrinsic resistance naturally and can be advantageous by aiding probiotics in restoring gut microbiota during and after antibiotic use. Conversely, acquired resistance develops from DNA mutations or HGT.

7.3.1.1 Intrinsic Resistance

In probiotics, intrinsic resistance is primarily due to chromosome-encoded genes and is rarely transferable to other bacterial species. For instance, mutations in chromosomal genes in LAB strains and *Bifidobacterium sp.* are common mechanisms of resistance. These mutations aid in restoring natural microbiota after antibiotic treatment and in reducing AAD (Rozman et al., 2023). Similarly, genes *msrC* in *Enterococcus faecium* and *aadD2* in *Bacillus clausii* provide resistance against macrolides and aminoglycosides, respectively (Gueimonde, Sánchez, de los Reyes-Gavilán, & Margolles, 2013; Tóth et al., 2021). *Bifidobacterium*, *Lactobacillus*, and *Enterococcus* possess resistance against vancomycin which typically inhibits peptidoglycan synthesis crucial for Gram-positive bacteria (Rubinstein & Keynan, 2014). *Lactobacillus sp.* has been shown to possess natural resistance to vancomycin, bacitracin, ciprofloxacin, cefoxitin, kanamycin, fusidic acid, metronidazole, gentamicin, sulfadiazine, streptomycin, and nalidixic acid (Blandino, Milazzo, & Fazio, 2008; Kastner et al., 2006; Temmerman, Pot, Huys, & Swings, 2003). Specific resistance profiles are also well-documented: *Bifidobacterium* resists norfloxacin, metronidazole, nalidixic acid, and polymyxin B; *Lactobacillus* resists nalidixic acid, and metronidazole; *Lactococcus lactis* resists polymyxin B, and norfloxacin; *Streptococcus thermophilus* resists neomycin (P. Sharma, Tomar, Goswami, Sangwan, & Singh, 2014). Additionally, over 70% of 45 LAB strains in one study were resistant to ciprofloxacin and aminoglycosides (gentamicin and streptomycin), suggesting these resistances are intrinsic (Hummel, Hertel, Holzapfel, & Franz, 2007).

Intrinsic resistance in bacteria can arise from two main factors: either they have resistance genes encoded in their DNA, or they lack the specific target that an antibiotic is designed to affect. This natural resistance can often be anticipated based on the species and genetic relatedness of probiotic strains, with these enduring resistance genes being inherited by future generations.

7.3.1.2 Acquired Resistance

Acquired resistance in probiotics is often due to genes found on conjugative plasmids or transposons, a pattern observed in *Lactobacilli* and *Enterococci* (Colautti, Arnoldi, Comi, & Iacumin, 2022; Ojha, Shah, & Mishra, 2021; Rozman et al., 2023; Shahali et al., 2023). This type of resistance arises when initially susceptible bacteria become resistant through HGT or gene mutation (Tóth, Judge, Nagy, Papp, & Solymosi, 2023). It frequently results from the selective pressure of prolonged or frequent antibiotic use, leading to genetic mutations or the acquisition of exogenous resistance genes. These changes alter bacterial structure and physiological or biochemical processes, reducing sensitivity to similar antibiotics (Gueimonde et al., 2013; Li et al., 2020; Montassier et al.,

2021). Acquired resistance is more concerning than intrinsic resistance because it can be transferred between bacteria, posing a broader threat (Flórez, Ammor, & Mayo, 2008; Gueimonde et al., 2013; Mathur & Singh, 2005; Van Reenen, Dicks, & Chikindas, 1998).

7.3.1.2.1 Genetic Mutation

AR in certain probiotics results from both long-term antibiotic selection pressure and spontaneous mutations rather than direct antibiotic exposure (Li et al., 2020). In *Lactobacilli*, mutation rates for nitrofurazone, streptomycin, and kanamycin range from 10^{-4} to 10^{-5}, correlating with increased AR and highlighting the variability in resistance profiles (Curragh & Collins, 1992). Additionally, antibiotics like amoxicillin and gentamicin can increase the mutation frequency by four-fold in *L. casei*, reaching 3.5×10^{-9}/base pair/generation (Wang et al., 2017). Mutation frequencies vary across different species and strains. While spontaneous chromosomal mutations confer low-level resistance, they can pave the way for high-level resistance through HGT.

7.3.1.2.2 Horizontal Gene Transfer

Resistance genes that are acquired, can spread between commensals and probiotics both in laboratory settings and in living organisms, crossing various species and genera (Imperial & Ibana, 2016; Kent, Vill, Shi, Satlin, & Brito, 2020; Montassier et al., 2021; Shahali et al., 2023; Zavišić et al., 2023). The three primary mechanisms of HGT are transformation (uptake and integration of free DNA from donor to recipient cells), transduction (DNA transfer via bacteriophages), and conjugation (direct DNA transfer through cell contact), with conjugation being the most prevalent (Gueimonde et al., 2013; Huddleston, 2014; Li et al., 2020; Von Wintersdorff et al., 2016) (Table 7.2). The ycdB gene, for instance, transfers throughout distinct LAB or between LAB and enteric bacteria (Bolotin, Quinquis, Sorokin, & Ehrlich, 2004). *E. faecalis* 181 has received the tet(M) genes from *L. brevis* NWL59 and *L. plantarum* NWL22, as well as the erm(B) genes from *L. salivarius* NWL33 and *L. fermentum* NWL24 (Ramos, Thorsen, Schwan, & Jespersen, 2013; Wegmann et al., 2007). Even in the absence of selection pressure, the vancomycin resistance gene cluster, vanA, is transferred in vivo between *Enterococci* and *L. acidophilus* (Arredondo-Alonso, Top, Corander, Willems, & Schürch, 2021; Ligozzi, Lo Cascio, & Fontana, 1998). Table 7.2 provides details on HGT in LAB strains, resistance genes, and their transmission modes. Additionally, the vanA gene can spread more easily from animals to humans via *E. faecium* than *E. faecalis*, having been transferred from porcine to human *Enterococci* in the gastrointestinal tract of mice (Li et al., 2019; Mater, Langella, Corthier, & Flores, 2005).

7.3.2 Methods to Detect AR Genes in Probiotics

The common methods to identify the presence and expression of resistance genes are PCR, Southern hybridization, and metagenomics; PCR being the most used (Galhano, Ferrari, Panzenhagen, de Jesus, & Conte-Junior, 2021). For example, conjugation, Southern hybridization, and PCR have all been used to identify the vancomycin resistance gene (vanB) in *L. rhamnosus* GG, *E. faecium*, and *E. faecalis* (Tynkkynen, Singh, & Varmanen, 1998). Tet(M), Tet(L), and Tet(S) genes which confer tetracycline resistance were also found in *Enterococcus* food isolates, as verified by PCR (Choi & Woo, 2015). PCR identified the cfxA gene in Bacteroides, which confers resistance to β-lactams and cefoxitin (Avelar et al., 2003). PCR was also used to confirm AR in isolates of *Pediococcus*, *Lactobacillus*, and *Lactococcus* from humans and cultures. Owing to the presence of the erm(B), tet(M), and tet(W) genes, some of these isolates are resistant to erythromycin, oxytetracycline, and clindamycin, respectively (Duche et al., 2023). In antibiotic-susceptible *Lactobacillus* isolates from chickens and calves, multiplex PCR was utilized to find resistance genes for tetracycline, kanamycin, gentamicin, spectinomycin, and streptomycin (Bujnakova, Strakova, & Kmet, 2014). Moreover, filter mating studies were used to evaluate the transferability of AR genes in *Lactobacillus* strains from fermented milk using PCR. According to the study, all strains of *Lactobacillus* exhibited resistance to vancomycin (Guo et al., 2017).

TABLE 7.2
Horizontal Gene Transfer of Antibiotic Resistance from Probiotic LAB to Other Species (Sharma et al., 2014)

Donor Strain	Recipient Strain	Conjugal Mating Method	Mode of Transmission	Resistance Gene	References
***Lb. plantarum* DG507**	*E. faecalis*	In vitro	Plasmid	erm(B)	Gevers, Danielsen, Huys, and Swings (2003)
***E. faecalis* LMG20790**	*E. faecalis* JH2-2	In vitro	Tn916–Tn1545	erm(B)	Huys, D'Haene, Collard, and Swings (2004)
***E. faecalis* LMG20927**	*E. faecalis* JH2-2	In vitro	Plasmid/ Tn916–Tn1545	erm(B)	Huys et al. (2004)
***Lb. plantarum* DG 522, *Lb. plantarum* DG 507**	*E. faecalis*	In vivo	Plasmid	tet(M), erm(B)	Jacobsen et al. (2007)
***Lb. reuteri* L4:12002**	*E. faecalis* JH2-2	In vitro	Plasmid	erm(B)	Ouoba, Lei, and Jensen (2008)
***Lb. plantarum* pLFE1**	*E. faecalis*	In vitro	–	erm (B)	Feld et al. (2008)
***Lc. lactis* SH4174**	*Listeria monocytogenes* (H7)	In vitro	pAMβ-1 Plasmid	erm(B)	Toomey, Monaghan, Fanning, and Bolton (2009)
S. thermophilus	*Listeria monocytogenes* (H7)	In vitro	Plasmid	erm(B)	Toomey et al. (2009)
***Lc. lactis* BU-2–60**	*E. faecalis* JH2-2	In vitro	–	tet(M)	Toomey, Bolton, and Fanning (2010)
***Lb. fermentum* NWL24, *Lb. salivarius* NWL33**	*E. faecalis* 181	In vitro	–	erm(B), tet(M)	Nawaz et al. (2011)
***E. faecalis* CM5V, CM6V**	*E. faecalis* OG1RF	In vitro	Plasmid	erm(B)	Vignaroli, Zandri, Aquilanti, Pasquaroli, and Biavasco (2011)
***E. durans* PF1V**	*E. faecalis* 64/3	In vitro	Plasmid	erm(B)	Vignaroli et al. (2011)
***E. durans* PF3V**	*E. faecalis* OG1RF, *E. faecalis* 64/3	In vitro	Plasmid	erm(B)	Vignaroli et al. (2011)

Probiotic resistance mechanisms have been better understood thanks to whole genome sequencing. For instance, it has made it possible to investigate genetic mutations in *L. plantarum* under gentamicin selection pressure and streptomycin resistance in *L. plantarum* ATCC14917 (Zhang et al., 2018). One aminoglycoside resistance gene and 12 putative resistance genes against beta-lactams, lincosamide, trimethoprim, tetracycline, aminocoumarin, rifampin, and glycopeptides are present in the genome of *E. durans* KLDS6.0930 (Li et al., 2018). *E. faecium* 17OM39 may be a viable probiotic candidate since, according to comparative genomics, it lacks genes that confer tetracycline or vancomycin resistance (Ghattargi et al., 2018).

Even while PCR and whole genome sequencing are useful techniques, they are not sufficient to verify whether probiotics containing resistance genes are resistant to antibiotics. Probiotic AR is becoming better to understand and research, owing to developments in bioinformatics, which have made it possible to identify probiotic resistance genes using molecular identification tools.

7.4 PROBIOTICS IN CLINICAL PRACTICE: NAVIGATING GUIDELINES AND SAFETY CONCERNS

Probiotics are becoming increasingly beneficial, but using them in medicine needs to be done with caution. In 2002, the WHO released "Guidelines for the Evaluation of Probiotics in Food," which provided guidelines for evaluating the safety and effectiveness of probiotics (Table 7.3). According to

TABLE 7.3
Probiotic Safety Considerations at the Levels of Strain(s), Product Quality, and Administration to the Patients (Roe et al., 2022)

Category	Subcategory	Details
Probiotic strain	Traditional species with a history of safe use	Strain-specific characterization
	Next-gen species	Full safety assessment needed
	Whole genome sequence	• Taxonomy • Strain-level identity • Toxin or virulence genes present?
	Antibiotic resistance assessment	**Phenotype** • MIC within range for species? **Genotype** If AR genes are present: • Does the gene encode resistance to a critically important antibiotic? • Is gene functional? • Is gene likely to be transferable?
	Monitoring to the extent possible	**Acute** • Invasive infection • Drug metabolism • Microbiome alteration • Transfer of AR genes **Long-term** • Long-term colonization • Drug metabolism • Microbiome alteration • Physiological impact on host
Probiotic product quality and safety		• GMP-compliant production • Safe and efficacious dose • Microbiological testing fit for end-user • Potency guaranteed through the end of the shelf life
Probiotic product administration	Risks with method of administration	• Contamination of hospital environment • Contamination of catheters • Exposure of nearby patients • How dispensed to patient (e.g., nasojejunal feeding)
	Suitability for target population	**Vulnerable**: • Preterm infant • Obese and pregnant • Short bowel syndrome

these recommendations, probiotics should be assessed for their antimicrobial activity, safety, ability to adhere to epithelial tissues, and resilience in dealing with hardships within the human body. A comprehensive strain characterization, safety data from clinical trials, and assuring the bacteria survive the whole shelf life are crucial requirements for probiotic products. Probiotics have to be devoid of pathogenic or opportunistic pathogens and have to be minimally contaminated with molds, yeasts, and aerobic bacteria (FAO/WHO, 2002). The absence of genes for AR is a major aspect of safety. To identify living microorganisms used in foods with a history of safe usage, the EFSA employed the QPS approach ("Qualified Presumption of Safety (QPS) | EFSA," n.d.). To assess the safety of products without such a background, Pariza et al. (2015) devised a 15-step decision tree. One of their major studies involved *Enterococcus faecium* SF68. Best practices for evaluating the safety and quality of probiotics were evaluated by Roe et al. (2022), who also highlighted the regional variations in regulatory frameworks. During its 2022 annual conference, the ISAPP also discussed new safety concerns. These included acute and long-term risks, the necessity of more studies, concerns for vulnerable groups, product quality, and the significance of thorough AEs reporting (Merenstein et al., 2023).

FDA and EFSA do not endorse probiotics for disease prevention or treatment, even though many probiotic strains are known to be GRAS or QPS category (FDA, 2016; Louis Bresson et al., 2016). The EFSA has not approved any health claims related to probiotics, despite the fact that more than 300 applications have been submitted for more than 200 probiotic strains or combinations that offer more than 60 different benefits. These rejections are mainly due to insufficient strain characterization, vague or unproven claims, a lack of pertinent human studies, and subpar study quality (Louis Bresson et al., 2016). Similarly, the FDA has not approved any probiotics for the prevention or treatment of disease, even though it recognizes probiotic strains as dietary supplements, ingredients in food, or drugs (Degnan King et al., 2008). However, as a natural health product, Health Canada has licensed a multistrain probiotic that includes different strains of *Lactobacillus* and *Bifidobacterium*, as well as a single strain of *B. animalis spp. lactis* LAFTI B94, to assist in relieving symptoms of IBS, including bloating, gas, and discomfort in the abdomen ("Bifidobacterium lactis Lafti® B94 – MicrobiomeHUB," n.d.).

The current guidelines encourage evaluating the phenotypes and genotypes of AR in all probiotic strains. Probiotics containing genes for AR classified as critically important by the WHO should undergo a thorough risk-benefit analysis, considering public health and consumer safety (WHO, 2018b, 2022). While the risk of DNA transfer through transformation is considered low, further research is required to better understand this process. Historical probiotic strains that have not been evaluated for AR risks should be reassessed to ensure compliance with current safety standards. The majority of naturally occurring strains that are frequently consumed and utilized in food fermentation are not subject to the same level of assessment, despite these precautions (Campbell-Platt, 1994). Similarly, there are no restrictions on fecal microbial transplants, which may contain transferable AR genes. Studies reveal that LABs employed in food fermentation may have AR genes, possibly through HGT from surrounding microorganisms (Maeusli et al., 2020; Stefańska et al., 2021). However, since the impact of AR genes in these microbes is not fully understood, there is no scientific or regulatory reason for prohibiting the consumption of fermented foods. To confirm that the AR genes in probiotics are safe, a thorough assessment must consider several key factors:

1. **Horizontal Gene Transfer**: Assess the association of AR genes with elements that facilitate horizontal transfer.
2. **Functional Expression**: Evaluate whether these genes can be functionally expressed if transferred to a naïve host.
3. **Intrinsic Resistance**: Assess to determine if the resistance characteristic is specific or inherent to the probiotic species.
4. **Environmental Contribution**: Assess the potential impact of the probiotic on the AR gene reservoir among environmental microbes.
5. **Clinical Relevance**: Consider whether the resistance is to a clinically significant antibiotic.

A thorough assessment is crucial to weigh the advantages of probiotics against possible risks to patients and the safety of consumers, public health issues, and their intended application – whether as restricted or widely distributed drugs, foods, or supplements. Since probiotics can both add to and help reduce the human AR gene pool, evaluating their safety on a case-by-case basis is essential (Sanders et al., 2010).

Clear product specifications that consider the intended use and acceptable risk levels must be established by manufacturers. Reputable producers meet the unique requirements of the intended usage and intended consumers by keeping a close eye on identification, potency, and purity. These standards call for extensive procedures for quality assurance and control. Multiple studies have shown problems with the quality of probiotics available on the market, frequently because of strains that are not reported, bacterial numbers that are lower than promised, or inaccurate strain identification. Moreover, several strains of probiotics have genes linked to AR. With the laws already in place to guarantee the safety and quality of probiotic goods, reports of non-compliance have led to demands for improved product quality (Merenstein et al., 2023).

Supplements are not meant to treat or prevent illnesses; rather, they are used to improve the diet of people who are generally healthy. Dietary supplement safety standards assure the safety of products when used as prescribed; however, testing for especially susceptible individuals is not necessary (Sanders et al., 2010). Different jurisdictions have varying rules and regulations for good manufacturing practices, making intended application critical in identifying fundamental efficacy and safety requirements (Arora & Baldi, 2015). The FDA distinguishes between food and dietary supplements in terms of current good manufacturing practices (cGMPs), ensuring drug identity, strength, quality, and purity, as well as food safety (Bailey, 2020).

Although quality standards specific to vulnerable populations can be developed, they will require more testing than those for healthy individuals (Shane & Preidis, 2023). These standards are verifiable by unbiased third parties possessing pertinent competence (Jackson et al., 2019). Certificates of analysis ought to correspond with the results of testing, even if product labels do not mention adhering to stricter guidelines. Hospital pharmacists and researchers who work with vulnerable populations can benefit from supplier cooperation in order to ensure more product testing and results that are transparent (Merenstein et al., 2023).

In randomized controlled trials (RCTs), it is essential to report AEs to comprehend how therapies balance risks and benefits. Inadequate or absent harm-related data affects not only probiotics but other therapies as well (Cornelius et al., 2013; Ioannidis & Lau, 2001; Phillips, Hazell, Sauzet, & Cornelius, 2019; Smith et al., 2013). Clinical researchers must follow established protocols when reporting harms, particularly the CONSORT Extension for Harms, which encourages thorough and specific reporting (Ioannidis et al., 2004). Data on AEs in probiotic clinical studies should be carefully collected, listed, and described using standard guidelines. It is necessary to record the absolute risk of each adverse event for each trial arm using appropriate criteria for recurring occurrences, as well as the number of participants who withdrew owing to harm. Researchers must present a fair analysis of the intervention's advantages and disadvantages to ensure a more precise and transparent evaluation of the safety and effectiveness of probiotic therapies (Merenstein et al., 2023).

7.5 DISCUSSION AND CONCLUSION

The interplay between probiotics and AR presents a complex landscape with significant benefits and notable risks. Probiotics are widely recognized for supporting gut health and mitigating the side effects of antibiotics. However, their potential to harbor and transfer AR genes necessitates stringent regulation and careful consideration. The prevalence of AR in probiotic supplements is documented and poses potential health risks (Tables 7.1 and 7.2). The primary severe risk of probiotics is their potential to contribute to the AR gene pool in the gut microbiota, which can then spread to pathogenic bacteria and cause more difficult-to-treat infections (Figure 7.2). Probiotic risk evaluations for AR genotypes and phenotypes are evolving with scientific advances, particularly distinguishing between gene presence versus gene transferability. Genetic modification techniques offer the possibility of modifying or eliminating resistance factors of concern, making next-generation probiotics suitable for use in both pharmaceutical and food applications. For instance, plasmid curing can effectively address resistance issues, enhancing the safety profile of these probiotics.

The rapid expansion of the probiotic industry highlights the need for updated regulatory guidelines to ensure safety (Table 7.3). Probiotics must undergo thorough inspection for mobile AR genes contributing to acquired resistance before approval for human use. This evaluation should occur at the genome level, complemented by molecular identification of probiotic strains. This requires comprehensive screening protocols, including both genotypic and phenotypic assessments. Collaborative research projects such as the Joint International Organization for Standardization-International Dairy Federation Action Team on Probiotics (ISO-IDF), Assessment and Critical Evaluation of Antibiotic Resistance Transferability in the Food Chain (ACE-ART) (Sanders et al., 2010), and Biosafety Assessment of Probiotics used for Human Consumption (PROSAFE) (Moretti et al., 2024), provide useful insights and standardized screening methodologies. Adhering to safety guidelines, such as the QPS by EFSA and the GRAS status by the FDA, is crucial. The EFSA emphasizes the importance of testing bacterial strains intended for ingestion for antibiotic susceptibility. Additional analysis is required for strains exhibiting acquired resistance to determine if it arises from chromosomal mutations or the acquisition of genetic elements. These data demonstrate that precautions against the use of probiotic strains having antibiotic-resistant genes within mobile genetic elements for medicinal and nutritional purposes must be taken.

The sensitivity and resistance profiles of common antibiotics for both natural gut flora and pathogens have been the subject of extensive research. Gut commensals benefit from intrinsic drug resistance, enhancing their resilience to antimicrobial stresses. However, acquired resistance presents a "double-edged sword." While it aids beneficial bacteria in surviving hostile environments, it also poses a major health risk by potentially spreading to pathogens within the gut microbiome (Lerner, Shoenfeld, & Matthias, 2019).

Probiotics, alone or in combination with antibiotics, show promise for treating infectious diseases and restoring gut microbiota post-antibiotic treatment. However, questions persist regarding the survival of probiotic bacteria in such combinations. Regulatory bodies should ensure that treatment plans include comprehensive information on the susceptibility patterns of probiotics. This information should be clearly indicated on product labels to help clinicians in prescribing appropriate probiotic and antibiotic combinations.

REFERENCES

Abdelsalam, N. A., Ramadan, A. T., ElRakaiby, M. T., & Aziz, R. K. (2020). Toxicomicrobiomics: The human microbiome vs. pharmaceutical, dietary, and environmental xenobiotics. *Frontiers in Pharmacology*, *11*. https://doi.org/10.3389/FPHAR.2020.00390

Abikshyeet, P., Mishra, P., Bhuyan, L., Kumar, V., Mahapatra, N., & Adhikary, T. (2022). Probiotics: Dawn of a new era in dental caries management. *Journal of Pharmacy & Bioallied Sciences*, *14*(Suppl 1), S34. https://doi.org/10.4103/JPBS.JPBS_801_21

Aggarwala, V., Mogno, I., Li, Z., Yang, C., Britton, G. J., Chen-Liaw, A., … Faith, J. (2021). Precise quantification of bacterial strains after fecal microbiota transplantation delineates long-term engraftment and explains outcomes. *Nature Microbiology*, *6*(10), 1309–1318. https://doi.org/10.1038/S41564-021-00966-0

Ahn, C., Collins-Thompson, D., Duncan, C., & Stiles, M. E. (1992). Mobilization and location of the genetic determinant of chloramphenicol resistance from Lactobacillus plantarum caTC2R. *Plasmid*, *27*(3), 169–176. https://doi.org/10.1016/0147-619X(92)90018-6

Alcock, B. P., Raphenya, A. R., Lau, T. T. Y., Tsang, K. K., Bouchard, M., Edalatmand, A., … McArthur, A. G. (2020). CARD 2020: Antibiotic resistome surveillance with the comprehensive antibiotic resistance database. *Nucleic Acids Research*, *48*(D1), D517–D525. https://doi.org/10.1093/NAR/GKZ935

Alves-Barroco, C., Rivas-García, L., Fernandes, A. R., & Baptista, P. V. (2020). Tackling multidrug resistance in streptococci – From novel biotherapeutic strategies to nanomedicines. *Frontiers in Microbiology*, *11*. https://doi.org/10.3389/FMICB.2020.579916

Ammor, M. S., Flórez, A. B., Álvarez-Martín, P., Margolles, A., & Mayo, B. (2008). Analysis of tetracycline resistance tet(W) genes and their flanking sequences in intestinal Bifidobacterium species. *The Journal of Antimicrobial Chemotherapy*, *62*(4), 688–693. https://doi.org/10.1093/JAC/DKN280

Ammor, M. S., Flórez, A. B., Van Hoek, A. H. A. M., De Los Reyes-Gavilán, C. G., Aarts, H. J. M., Margolles, A., & Mayo, B. (2008). Molecular characterization of intrinsic and acquired antibiotic resistance in lactic acid bacteria and bifidobacteria. *Journal of Molecular Microbiology and Biotechnology*, *14*(1–3), 6–15. https://doi.org/10.1159/000106077

Andriani, D., Hasan, P. N., Utami, T., Suroto, D. A., Wikandari, R., & Rahayu, E. S. (2021). Genotypic and phenotypic analyses of antibiotic resistance in Indonesian indigenous lactobacillus probiotics. *Applied Food Biotechnology*, *8*(4), 267–274. https://doi.org/10.22037/afb.v8i4.34448

Anisimova, E. A., & Yarullina, D. R. (2019). Antibiotic resistance of Lactobacillus strains. *Current Microbiology*, *76*(12), 1407–1416. https://doi.org/10.1007/s00284-019-01769-7

Aquilanti, L., Garofalo, C., Osimani, A., Silvestri, G., Vignaroli, C., & Clementi, F. (2007). Isolation and molecular characterization of antibiotic-resistant lactic acid bacteria from poultry and swine meat products. *Journal of Food Protection*, *70*(3), 557–565. https://doi.org/10.4315/0362-028X-70.3.557

Arora, M., & Baldi, A. (2015). Regulatory categories of probiotics across the globe: A review representing existing and recommended categorization. *Indian Journal of Medical Microbiology*, *33*, S2–S10. https://doi.org/10.4103/0255-0857.150868

Arredondo-Alonso, S., Top, J., Corander, J., Willems, R. J. L., & Schürch, A. C. (2021). Mode and dynamics of vanA-type vancomycin resistance dissemination in Dutch hospitals. *Genome Medicine*, *13*(1), 1–18. https://doi.org/10.1186/s13073-020-00825-3

Astashkina, A. P., Khudyakova, L. I., & Kolbysheva, Y. V. (2014). Microbiological quality control of probiotic products. *Procedia Chemistry*, *10*, 74–79. https://doi.org/10.1016/J.PROCHE.2014.10.014

Atashgahi, S., Shetty, S. A., Smidt, H., & de Vos, W. M. (2018). Flux, Impact, and Fate of Halogenated Xenobiotic Compounds in the Gut. *Frontiers in Physiology*, *9*(JUL). https://doi.org/10.3389/FPHYS.2018.00888

Avelar, K. E. S., Otsuki, K., Vicente, A. C. P., Vieira, J. M. B. D., De Paula, G. R., Domingues, R. M. C. P., & Ferreira, M. C. D. S. (2003). Presence of the cfxA gene in Bacteroides distasonis. *Research in Microbiology*, *154*(5), 369–374. https://doi.org/10.1016/S0923-2508(03)00093-7

Bafeta, A., Koh, M., Riveros, C., & Ravaud, P. (2018). Harms reporting in randomized controlled trials of interventions aimed at modifying microbiota: A systematic review. *Annals of Internal Medicine*, *169*(4), 240–247. https://doi.org/10.7326/M18-0343

Bailey, R. L. (2020). Current regulatory guidelines and resources to support research of dietary supplements in the United States. *Critical Reviews in Food Science and Nutrition*, *60*(2), 298. https://doi.org/10.1080/10408398.2018.1524364

Bermudez-Brito, M., Plaza-Díaz, J., Muñoz-Quezada, S., Gómez-Llorente, C., & Gil, A. (2012). Probiotic mechanisms of action. *Annals of Nutrition and Metabolism*, *61*(2), 160–174. https://doi.org/10.1159/000342079

Bielik, V., & Kolisek, M. (2021). Bioaccessibility and Bioavailability of Minerals in Relation to a Healthy Gut Microbiome. *International Journal of Molecular Sciences*, *22*(13), 6803. https://doi.org/10.3390/IJMS22136803

Bifidobacterium lactis Lafti® B94 – MicrobiomeHUB. (n.d.). Retrieved 31 July 2024, from https://microbiome-hub.com/probiotic-strains/bifidobacterium-lactis-lafti-b94/

Binda, S., Hill, C., Johansen, E., Obis, D., Pot, B., Sanders, M. E., … Ouwehand, A. C. (2020). Criteria to qualify microorganisms as 'probiotic' in foods and dietary supplements. *Frontiers in Microbiology*, *11*. https://doi.org/10.3389/FMICB.2020.01662

Blandino, G., Milazzo, I., & Fazio, D. (2008). Antibiotic susceptibility of bacterial isolates from probiotic products available in Italy. *Microbial Ecology in Health and Disease*, *20*(4), 199–203. https://doi.org/10.1080/08910600802408111

Bolotin, A., Quinquis, B., Sorokin, A., & Ehrlich, D. S. (2004). Recent genetic transfer between Lactococcus lactis and enterobacteria. *Journal of Bacteriology*, *186*(19), 6671–6677. https://doi.org/10.1128/JB.186.19.6671-6677.2004

Borriello, S. P., Hammes, W. P., Holzapfel, W., Marteau, P., Schrezenmeir, J., Vaara, M., & Valtonen, V. (2003). Safety of probiotics that contain lactobacilli or bifidobacteria. *Clinical Infectious Diseases: An Official Publication of the Infectious Diseases Society of America*, *36*(6), 775–780. https://doi.org/10.1086/368080

Bozdogan, B., Galopin, S., Gerbaud, G., Courvalin, P., & Leclercq, R. (2003). Chromosomal aadD2 encodes an aminoglycoside nucleotidyltransferase in Bacillus clausii. *Antimicrobial Agents and Chemotherapy*, *47*(4), 1343–1346. https://doi.org/10.1128/AAC.47.4.1343-1346.2003

Bozdogan, B., Galopín, S., & Leclercq, R. (2004). Characterization of a new erm-related macrolide resistance gene present in probiotic strains of Bacillus clausii. *Applied and Environmental Microbiology*, *70*(1), 280–284. https://doi.org/10.1128/AEM.70.1.280-284.2004

Bujnakova, D., Strakova, E., & Kmet, V. (2014). In vitro evaluation of the safety and probiotic properties of Lactobacilli isolated from chicken and calves. *Anaerobe*, *29*, 118–127. https://doi.org/10.1016/J.ANAEROBE.2013.10.009

Campbell-Platt, G. (1994). Fermented foods – A world perspective. *Food Research International*, *27*(3), 253–257. https://doi.org/10.1016/0963-9969(94)90093-0

Cannon, J. P., Lee, T. A., Bolanos, J. T., & Danziger, L. H. (2005). Pathogenic relevance of Lactobacillus: A retrospective review of over 200 cases. *European Journal of Clinical Microbiology & Infectious Diseases: Official Publication of the European Society of Clinical Microbiology*, *24*(1), 31–40. https://doi.org/10.1007/S10096-004-1253-Y

Carmody, R. N., & Turnbaugh, P. J. (2014). Host-microbial interactions in the metabolism of therapeutic and diet-derived xenobiotics. *The Journal of Clinical Investigation*, *124*(10), 4173–4181. https://doi.org/10.1172/JCI72335

Cauwerts, K., Pasmans, F., Devriese, L. A., Martel, A., Haesebrouck, F., & Decostere, A. (2006). Cloacal Lactobacillus isolates from broilers show high prevalence of resistance towards macrolide and lincosamide antibiotics. *Avian Pathology: Journal of the World Veterinary Poultry Association*, *35*(2), 160–164. https://doi.org/10.1080/03079450600598137

Choi, J. M., & Woo, G. J. (2015). Transfer of tetracycline resistance genes with aggregation substance in food-borne Enterococcus faecalis. *Current Microbiology*, *70*(4), 476. https://doi.org/10.1007/S00284-014-0742-1

Chudzik, A., Słowik, T., Kochalska, K., Pankowska, A., Łazorczyk, A., Andres-mach, M., … Orzyłowska, A. (2022). Continuous ingestion of Lacticaseibacillus rhamnosus JB-1 during chronic stress ensures neurometabolic and behavioural stability in rats. *International Journal of Molecular Sciences*, *23*(9), 5173. https://doi.org/10.3390/ijms23095173

Chugh, B., & Kamal-Eldin, A. (2020). Bioactive compounds produced by probiotics in food products. *Current Opinion in Food Science*, *32*, 76–82. https://doi.org/10.1016/J.COFS.2020.02.003

Colautti, A., Arnoldi, M., Comi, G., & Iacumin, L. (2022). Antibiotic resistance and virulence factors in lactobacilli: Something to carefully consider. *Food Microbiology*, *103*, 103934. https://doi.org/10.1016/J.FM.2021.103934

Cornelius, V. R., Sauzet, O., Williams, J. E., Ayis, S., Farquhar-Smith, P., Ross, J. R., … Peacock, J. L. (2013). Adverse event reporting in randomised controlled trials of neuropathic pain: Considerations for future practice. *Pain*, *154*(2), 213–220. https://doi.org/10.1016/J.PAIN.2012.08.012

Crits-Christoph, A., Hallowell, H. A., Koutouvalis, K., & Suez, J. (2022). Good microbes, bad genes? The dissemination of antimicrobial resistance in the human microbiome. *Gut Microbes*, *14*(1). https://doi.org/10.1080/19490976.2022.2055944

Curragh, H. J., & Collins, M. A. (1992). High levels of spontaneous drug resistance in Lactobacillus. *Journal of Applied Bacteriology*, *73*(1), 31–36. https://doi.org/10.1111/J.1365-2672.1992.TB04965.X

D'agostin, M., Squillaci, D., Lazzerini, M., Barbi, E., Wijers, L., & Da Lozzo, P. (2021). Invasive infections associated with the use of probiotics in children: A systematic review. *Children*, *8*(10). https://doi.org/10.3390/CHILDREN8100924

Daniali, M., Nikfar, S., & Abdollahi, M. (2020). Antibiotic resistance propagation through probiotics. *Expert Opinion on Drug Metabolism & Toxicology*, *16*(12), 1207–1215. https://doi.org/10.1080/17425255.2020.1825682

Danielsen, M., & Wind, A. (2003). Susceptibility of Lactobacillus spp. to antimicrobial agents. *International Journal of Food Microbiology*, *82*(1), 1–11. https://doi.org/10.1016/S0168-1605(02)00254-4

Dashnyam, P., Mudududdla, R., Hsieh, T. J., Lin, T. C., Lin, H. Y., Chen, P. Y., … Lin, C. H. (2018). β-Glucuronidases of opportunistic bacteria are the major contributors to xenobiotic-induced toxicity in the gut. *Scientific Reports*, *8*(1). https://doi.org/10.1038/S41598-018-34678-Z

De Filippis, F., Pasolli, E., & Ercolini, D. (2020). The food-gut axis: Lactic acid bacteria and their link to food, the gut microbiome and human health. *FEMS Microbiology Reviews*, *44*(4), 454. https://doi.org/10.1093/FEMSRE/FUAA015

De Vos, W. M., Tilg, H., Van Hul, M., & Cani, P. D. (2022). Gut microbiome and health: Mechanistic insights. *Gut*, *71*(5), 1020–1032. https://doi.org/10.1136/GUTJNL-2021-326789

De Vuyst, L., Van Kerrebroeck, S., Harth, H., Huys, G., Daniel, H. M., & Weckx, S. (2014). Microbial ecology of sourdough fermentations: Diverse or uniform? *Food Microbiology*, *37*, 11–29. https://doi.org/10.1016/J.FM.2013.06.002

Degnan King, F. H., Washington, S., & Degnan, F. H. (2008). The US food and drug administration and probiotics: Regulatory categorization. *Clinical Infectious Diseases*, *46*(Supplement_2), S133–S136. https://doi.org/10.1086/523324

Derrien, M., & van Hylckama Vlieg, J. E. T. (2015). Fate, activity, and impact of ingested bacteria within the human gut microbiota. *Trends in Microbiology*, *23*(6), 354–366. https://doi.org/10.1016/J.TIM.2015.03.002

Devirgiliis, C., Coppola, D., Barile, S., Colonna, B., & Perozzi, G. (2009). Characterization of the Tn916 conjugative transposon in a food-borne strain of Lactobacillus paracasei. *Applied and Environmental Microbiology*, *75*(12), 3866–3871. https://doi.org/10.1128/AEM.00589-09

Di Costanzo, M., Vella, A., Infantino, C., Morini, R., Bruni, S., Esposito, S., & Biasucci, G. (2024). Probiotics in infancy and childhood for food allergy prevention and treatment. *Nutrients*, *16*(2), 297. https://doi.org/10.3390/NU16020297

Dikeocha, I. J., Al-Kabsi, A. M., Miftahussurur, M., & Alshawsh, M. A. (2022). Pharmacomicrobiomics: Influence of gut microbiota on drug and xenobiotic metabolism. *FASEB Journal: Official Publication of the Federation of American Societies for Experimental Biology*, *36*(6). https://doi.org/10.1096/FJ.202101986R

Dobson, A., Cotter, P. D., Paul Ross, R., & Hill, C. (2012). Bacteriocin production: A probiotic trait? *Applied and Environmental Microbiology*, *78*(1), 1. https://doi.org/10.1128/AEM.05576-11

Du, T., Lei, A., Zhang, N., & Zhu, C. (2022). The beneficial role of probiotic lactobacillus in respiratory diseases. *Frontiers in Immunology*, *13*, 908010. https://doi.org/10.3389/fimmu.2022.908010

Duche, R. T., Singh, A., Wandhare, A. G., Sangwan, V., Sihag, M. K., Nwagu, T. N. T., … Ezeogu, L. I. (2023). Antibiotic resistance in potential probiotic lactic acid bacteria of fermented foods and human origin from Nigeria. *BMC Microbiology*, *23*(1). https://doi.org/10.1186/S12866-023-02883-0

Egervärn, M., Roos, S., & Lindmark, H. (2009). Identification and characterization of antibiotic resistance genes in Lactobacillus reuteri and Lactobacillus plantarum. *Journal of Applied Microbiology*, *107*(5), 1658–1668. https://doi.org/10.1111/J.1365-2672.2009.04352.X

European Commission (EC) Regulation (EC) No. 1924/2006. (n.d.). European Commission (EC) Regulation (EC) No. 1924/2006 of the European Parliament and of the Council of 20 December 2006 on Nutrition and Health Claims Made on Foods.

FAO/WHO. (2002). *Guidelines for the Evaluation of Probiotics in Food*. Paris: FAO, pp. 1–11.

FAO/WHO. (2006). Health and nutritional properties and guidelines for evaluation – Report of a Joint FAO/WHO Expert Consultation on Evaluation of Health and Nutritional Properties of Probiotics in Food including Powder Milk with Live Lactic Acid Bacteria.

FDA. (2005). U.S. Food & Drug Administration. GRN No. 49. Bifidobacterium lactis strain Bb12 and streptococcus thermophilus strain Th4.

FDA. (2016). FDA Dietary supplements: New dietary ingredient notifications and related issues: Guidance for industry.

Feld, L., Schjørring, S., Hammer, K., Licht, T. R., Danielsen, M., Krogfelt, K., & Wilcks, A. (2008). Selective pressure affects transfer and establishment of a Lactobacillus plantarum resistance plasmid in the gastrointestinal environment. *The Journal of Antimicrobial Chemotherapy*, *61*(4), 845–852. https://doi.org/10.1093/JAC/DKN033

Fijan, S. (2014). Microorganisms with claimed probiotic properties: An overview of recent literature. *International Journal of Environmental Research and Public Health*, *11*(5), 4745. https://doi.org/10.3390/IJERPH110504745

Flórez, A. B., Ammor, M. S., Álvarez-Martín, P., Margolles, A., & Mayo, B. (2006). Molecular analysis of tet(W) gene-mediated tetracycline resistance in dominant intestinal Bifidobacterium species from healthy humans. *Applied and Environmental Microbiology*, *72*(11), 7377–7379. https://doi.org/10.1128/AEM.00486-06

Flórez, A. B., Ammor, M. S., & Mayo, B. (2008). Identification of tet(M) in two Lactococcus lactis strains isolated from a Spanish traditional starter-free cheese made of raw milk and conjugative transfer of tetracycline resistance to lactococci and enterococci. *International Journal of Food Microbiology*, *121*(2), 189–194. https://doi.org/10.1016/J.IJFOODMICRO.2007.11.029

Frese, S. A., Hutton, A. A., Contreras, L. N., Shaw, C. A., Palumbo, M. C., Casaburi, G., … Underwood, M. A. (2017). Persistence of supplemented bifidobacterium longum subsp. infantis EVC001 in breastfed infants. *MSphere*, *2*(6). https://doi.org/10.1128/MSPHERE.00501-17

Galhano, B. S. P., Ferrari, R. G., Panzenhagen, P., de Jesus, A. C. S., & Conte-Junior, C. A. (2021). Antimicrobial resistance gene detection methods for bacteria in animal-based foods: A brief review of highlights and advantages. *Microorganisms*, *9*(5). https://doi.org/10.3390/MICROORGANISMS9050923

Galopin, S., Cattoir, V., & Leclercq, R. (2009). A chromosomal chloramphenicol acetyltransferase determinant from a probiotic strain of Bacillus clausii. *FEMS Microbiology Letters*, *296*(2), 185–189. https://doi.org/10.1111/J.1574-6968.2009.01633.X

Gevers, D., Danielsen, M., Huys, G., & Swings, J. (2003). Molecular characterization of tet(M) genes in Lactobacillus isolates from different types of fermented dry sausage. *Applied and Environmental Microbiology*, *69*(2), 1270–1275. https://doi.org/10.1128/AEM.69.2.1270-1275.2003

Gfeller, K. Y., Roth, M., Meile, L., & Teuber, M. (2003). Sequence and genetic organization of the 19.3-kb erythromycin- and dalfopristin-resistance plasmid pLME300 from Lactobacillus fermentum ROT1. *Plasmid*, *50*(3), 190–201. https://doi.org/10.1016/j.plasmid.2003.08.001

Ghattargi, V. C., Gaikwad, M. A., Meti, B. S., Nimonkar, Y. S., Dixit, K., Prakash, O., … Dhotre, D. P. (2018). Comparative genome analysis reveals key genetic factors associated with probiotic property in Enterococcus faecium strains. *BMC Genomics*, *19*(1). https://doi.org/10.1186/S12864-018-5043-9

Gingell, R., & Bridges, J. W. (1971). Intestinal azo reduction and glucuronide conjugation of prontosil. *The Biochemical Journal*, *125*(2). https://doi.org/10.1042/BJ1250024PA

Girlich, D., Leclercq, R., Naas, T., & Nordmann, P. (2007). Molecular and biochemical characterization of the chromosome-encoded class A beta-lactamase BCL-1 from Bacillus clausii. *Antimicrobial Agents and Chemotherapy*, *51*(11), 4009–4014. https://doi.org/10.1128/AAC.00537-07

Goodman, C., Keating, G., Georgousopoulou, E., Hespe, C., & Levett, K. (2021). Probiotics for the prevention of antibiotic-associated diarrhoea: A systematic review and meta-analysis. *BMJ Open*, *11*(8). https://doi.org/10.1136/BMJOPEN-2020-043054

Górska, A., Przystupski, D., Niemczura, M. J., & Kulbacka, J. (2019). Probiotic Bacteria: A Promising Tool in Cancer Prevention and Therapy. *Current Microbiology*, *76*(8), 939. https://doi.org/10.1007/S00284-019-01679-8

Grazul, H., Kanda, L. L., & Gondek, D. (2016). Impact of probiotic supplements on microbiome diversity following antibiotic treatment of mice. *Gut Microbes*, *7*(2), 101. https://doi.org/10.1080/19490976.2016.1138197

Gueimonde, M., Sánchez, B., de los Reyes-Gavilán, C. G., & Margolles, A. (2013a). Antibiotic resistance in probiotic bacteria. *Frontiers in Microbiology*, *4*(JUL). https://doi.org/10.3389/FMICB.2013.00202

Guo, H., Pan, L., Li, L., Lu, J., Kwok, L., Menghe, B., … Zhang, W. (2017). Characterization of antibiotic resistance genes from lactobacillus isolated from traditional dairy products. *Journal of Food Science*, *82*(3), 724–730. https://doi.org/10.1111/1750-3841.13645

Hamilton-Miller, J. M. T., Gibson, G. R., & Bruck, W. (2003). Some insights into the derivation and early uses of the word 'probiotic'. *The British Journal of Nutrition*, *90*(4), 845–845. https://doi.org/10.1079/BJN2003954

Hegarty, J. W., Guinane, C. M., Ross, R. P., Hill, C., & Cotter, P. D. (2016). Bacteriocin production: A relatively unharnessed probiotic trait? *F1000Research*, *5*. https://doi.org/10.12688/F1000RESEARCH.9615.1

Hempel, S., Newberry, S., Ruelaz, A., Wang, Z., Miles, J. N. V., Suttorp, M. J., … Shekelle, P. G. (2011). Safety of probiotics used to reduce risk and prevent or treat disease. *Evidence Report/Technology Assessment*, (200), 1. Retrieved from /pmc/articles/PMC4780970/

Hernández, R. B., & Juliá, M. B. (2023). Enterococcus spp. and Streptococcus spp. bloodstream infections: Epidemiology and therapeutic approach. *Revista Española de Quimioterapia*, *36*(Suppl 1), 2. https://doi.org/10.37201/REQ/S01.01.2023

Hill, C., Guarner, F., Reid, G., Gibson, G. R., Merenstein, D. J., Pot, B., … Sanders, M. E. (2014). Expert consensus document. The International Scientific Association for Probiotics and Prebiotics consensus statement on the scope and appropriate use of the term probiotic. *Nature Reviews. Gastroenterology & Hepatology*, *11*(8), 506–514. https://doi.org/10.1038/NRGASTRO.2014.66

Huddleston, J. R. (2014). Horizontal gene transfer in the human gastrointestinal tract: Potential spread of antibiotic resistance genes. *Infection and Drug Resistance*, *7*, 167. https://doi.org/10.2147/IDR.S48820

Hummel, A. S., Hertel, C., Holzapfel, W. H., & Franz, C. M. A. P. (2007). Antibiotic resistances of starter and probiotic strains of lactic acid bacteria. *Applied and Environmental Microbiology*, *73*(3), 730–739. https://doi.org/10.1128/AEM.02105-06

Huys, G., D'Haene, K., Collard, J. M., & Swings, J. (2004). Prevalence and molecular characterization of tetracycline resistance in Enterococcus isolates from food. *Applied and Environmental Microbiology*, *70*(3), 1555–1562. https://doi.org/10.1128/AEM.70.3.1555-1562.2004

Iftikhar, H., Awan, M. O., Awan, M. S., Mustafa, K., Das, J. K., & Ahmed, S. K. (2022). Role of probiotics in patients with allergic rhinitis: A systematic review of systematic reviews. *International Archives of Otorhinolaryngology*, *26*(4), e744. https://doi.org/10.1055/S-0042-1749370

Imperial, I. C. V. J., & Ibana, J. A. (2016). Addressing the antibiotic resistance problem with probiotics: Reducing the risk of its double-edged sword effect. *Frontiers in Microbiology*, *7*(DEC). https://doi.org/10.3389/FMICB.2016.01983

Ioannidis, J. P. A., Evans, S. J. W., Gøtzsche, P. C., O'Neill, R. T., Altman, D. G., Schulz, K., & Moher, D. (2004). Better reporting of harms in randomized trials: An extension of the CONSORT statement. *Annals of Internal Medicine*, *141*(10), 781–788. https://doi.org/10.7326/0003-4819-141-10-200411160-00009

Ioannidis, J. P. A., & Lau, J. (2001). Completeness of safety reporting in randomized trials: An evaluation of 7 medical areas. *JAMA*, *285*(4), 437–443. https://doi.org/10.1001/JAMA.285.4.437

Ioannou, P., Ziogou, A., Giannakodimos, I., Giannakodimos, A., Baliou, S., & Samonis, G. (2024). Infective endocarditis by lactobacillus species—A narrative review. *Antibiotics*, *13*(1), 53. https://doi.org/10.3390/antibiotics13010053

Jackson, S. A., Schoeni, J. L., Vegge, C., Pane, M., Stahl, B., Bradley, M., … Sanders, M. E. (2019). Improving end-user trust in the quality of commercial probiotic products. *Frontiers in Microbiology*, *10*(MAR). https://doi.org/10.3389/FMICB.2019.00739

Jacobsen, L., Wilcks, A., Hammer, K., Huys, G., Gevers, D., & Andersen, S. R. (2007). Horizontal transfer of tet(M) and erm(B) resistance plasmids from food strains of Lactobacillus plantarum to Enterococcus faecalis JH2-2 in the gastrointestinal tract of gnotobiotic rats. *FEMS Microbiology Ecology*, *59*(1), 158–166. https://doi.org/10.1111/J.1574-6941.2006.00212.X

Janina Krumbeck, A., Hutkins, R. W., & Walter Lincoln, J. (2016). *Characterization of the Role of Host and Dietary Factors in the Establishment of Bacteria in the Gastrointestinal Tract*. The University of Nebraska-Lincoln

Ji, J., Jin, W., Liu, S. J., Jiao, Z., & Li, X. (2023). Probiotics, prebiotics, and postbiotics in health and disease. *MedComm*, *4*(6). https://doi.org/10.1002/MCO2.420

Johnston, B. C., Ma, S. S. Y., Goldenberg, J. Z., Thorlund, K., Vandvik, P. O., Loeb, M., & Guyatt, G. H. (2012). Probiotics for the prevention of Clostridium difficile-associated diarrhea: A systematic review and meta-analysis. *Annals of Internal Medicine*, *157*(12), 878–888. https://doi.org/10.7326/0003-4819-157-12-201212180-00563

Karimi, R., Mortazavian, A. M., & Da Cruz, A. G. (2011). Viability of probiotic microorganisms in cheese during production and storage: A review. *Dairy Science and Technology*, *91*(3), 283–308. https://doi.org/10.1007/s13594-011-0005-x

Kastner, S., Perreten, V., Bleuler, H., Hugenschmidt, G., Lacroix, C., & Meile, L. (2006). Antibiotic susceptibility patterns and resistance genes of starter cultures and probiotic bacteria used in food. *Systematic and Applied Microbiology*, *29*(2), 145–155. https://doi.org/10.1016/J.SYAPM.2005.07.009

Kazimierczak, K. A., Flint, H. J., & Scott, K. P. (2006). Comparative analysis of sequences flanking tet(W) resistance genes in multiple species of gut bacteria. *Antimicrobial Agents and Chemotherapy*, *50*(8), 2632–2639. https://doi.org/10.1128/AAC.01587-05

Kent, A. G., Vill, A. C., Shi, Q., Satlin, M. J., & Brito, I. L. (2020). Widespread transfer of mobile antibiotic resistance genes within individual gut microbiomes revealed through bacterial Hi-C. *Nature Communications*, *11*(1), 1–9. https://doi.org/10.1038/s41467-020-18164-7

Khalesi, S., Bellissimo, N., Vandelanotte, C., Williams, S., Stanley, D., & Irwin, C. (2019). A review of probiotic supplementation in healthy adults: Helpful or hype? *European Journal of Clinical Nutrition*, *73*(1), 24–37. https://doi.org/10.1038/S41430-018-0135-9

Khalighi, A., Behdani, R., & Kouhestani, S. (2016). Probiotics a Comprehensive Review of Their Classification, Mode of Action and Role in Human Nutrition. In A. Rao & L. Rao (Eds.), *Probiotics and Prebiotics in Human Nutrition and Health*. IntechOpen. https://doi.org/10.5772/61495

Klaassen, C. D., & Cui, J. Y. (2015). Review: Mechanisms of how the intestinal microbiota alters the effects of drugs and bile acids. *Drug Metabolism and Disposition: The Biological Fate of Chemicals*, *43*(10), 1505–1521. https://doi.org/10.1124/DMD.115.065698

Klare, I., Konstabel, C., Werner, G., Huys, G., Vankerckhoven, V., Kahlmeter, G., … Goossens, H. (2007). Antimicrobial susceptibilities of Lactobacillus, Pediococcus and Lactococcus human isolates and cultures intended for probiotic or nutritional use. *The Journal of Antimicrobial Chemotherapy*, *59*(5), 900–912. https://doi.org/10.1093/JAC/DKM035

Knorr, D. (1998). Technology aspects related to microorganisms in functional foods. *Trends in Food Science & Technology*, *9*(8–9), 295–306. https://doi.org/10.1016/S0924-2244(98)00051-X

Kochalska, K., Oakden, W., Słowik, T., Chudzik, A., Pankowska, A., Łazorczyk, A., … Orzylowska, A. (2020). Dietary supplementation with Lactobacillus rhamnosus JB-1 restores brain neurochemical balance and mitigates the progression of mood disorder in a rat model of chronic unpredictable mild stress. *Nutrition Research*, *82*, 44–57. https://doi.org/10.1016/J.NUTRES.2020.06.019

Kocsis, T., Molnár, B., Németh, D., Hegyi, P., Szakács, Z., Bálint, A., … Solymár, M. (2020). Probiotics have beneficial metabolic effects in patients with type 2 diabetes mellitus: A meta-analysis of randomized clinical trials. *Scientific Reports*, *10*(1), 1–14. https://doi.org/10.1038/s41598-020-68440-1

Kothari, D., Patel, S., & Kim, S. K. (2019). Probiotic supplements might not be universally-effective and safe: A review. *Biomedicine & Pharmacotherapy = Biomedecine & Pharmacotherapie*, *111*, 537–547. https://doi.org/10.1016/J.BIOPHA.2018.12.104

Koutsoumanis, K., Allende, A., Alvarez-Ordóñez, A., Bolton, D., Bover-Cid, S., Chemaly, M., … Herman, L. (2022). Update of the list of QPS-recommended microbiological agents intentionally added to food or feed as notified to EFSA 16: Suitability of taxonomic units notified to EFSA until March 2022. *EFSA Journal. European Food Safety Authority*, *20*(7). https://doi.org/10.2903/J.EFSA.2022.7408

Kullar, R., Goldstein, E. J. C., Johnson, S., & McFarland, L. V. (2023). Lactobacillus bacteremia and probiotics: A review. *Microorganisms*, *11*(4). https://doi.org/10.3390/MICROORGANISMS11040896

Lau, L. Y. J., & Quek, S. Y. (2024). Probiotics: Health benefits, food application, and colonization in the human gastrointestinal tract. *Food Bioengineering*, *3*(1), 41–64. https://doi.org/10.1002/FBE2.12078

Lerner, A., Shoenfeld, Y., & Matthias, T. (2019). Probiotics: If it does not help it does not do any harm. Really? *Microorganisms* *7*(4), 104. https://doi.org/10.3390/MICROORGANISMS7040104

Li, B., Zhan, M., Evivie, S. E., Jin, D., Zhao, L., Chowdhury, S., … Liu, F. (2018). Evaluating the safety of potential probiotic Enterococcus durans KLDS6.0930 using whole genome sequencing and oral toxicity study. *Frontiers in Microbiology*, *9*, 1943. https://doi.org/10.3389/fmicb.2018.01943

Li, C., Peng, K., Xiao, S., Long, Y., & Yu, Q. (2023). The role of Lactobacillus in inflammatory bowel disease: From actualities to prospects. *Cell Death Discovery*, *9*(1), 1–12. https://doi.org/10.1038/s41420-023-01666-w

Li, N., Yu, H., Liu, H., Wang, Y., Zhou, J., Ma, X., … Qiao, S. (2019). Horizontal transfer of vanA between probiotic Enterococcus faecium and Enterococcus faecalis in fermented soybean meal and in digestive tract of growing pigs. *Journal of Animal Science and Biotechnology*, *10*(1). https://doi.org/10.1186/S40104-019-0341-X

Li, T., Teng, D., Mao, R., Hao, Y., Wang, X., & Wang, J. (2020). A critical review of antibiotic resistance in probiotic bacteria. *Food Research International*, *136*, 109571. https://doi.org/10.1016/J.FOODRES.2020.109571

Ligozzi, M., Lo Cascio, G., & Fontana, R. (1998). vanA gene cluster in a vancomycin-resistant clinical isolate of Bacillus circulans. *Antimicrobial Agents and Chemotherapy*, *42*(8), 2055. https://doi.org/10.1128/AAC.42.8.2055

Lima, K. G. de C., Kruger, M. F., Behrens, J., Destro, M. T., Landgraf, M., & Gombossy de Melo Franco, B. D. (2009). Evaluation of culture media for enumeration of Lactobacillus acidophilus, Lactobacillus casei and Bifidobacterium animalis in the presence of Lactobacillus delbrueckii subsp bulgaricus and Streptococcus thermophilus. *LWT – Food Science and Technology*, *42*(2), 491–495. https://doi.org/10.1016/J.LWT.2008.08.011

Lin, C. F., Fung, Z. F., Wu, C. L., & Chung, T. C. (1996). Molecular characterization of a plasmid-borne (pTC82) chloramphenicol resistance determinant (cat-TC) from Lactobacillus reuteri G4. *Plasmid*, *36*(2), 116–124. https://doi.org/10.1006/PLAS.1996.0039

Liu, G., Pang, B., Li, N., Jin, H., Li, J., Wu, W., … Shi, J. (2020). Therapeutic effect of Lactobacillus rhamnosus SHA113 on intestinal infection by multi-drug-resistant Staphylococcus aureus and its underlying mechanisms. *Food & Function*, *11*(7), 6226–6239. https://doi.org/10.1039/D0FO00969E

Lorente-picón, M., & Laguna, A. (2021). New avenues for Parkinson's disease therapeutics: Disease-modifying strategies based on the gut microbiota. *Biomolecules*, *11*(3), 1–39. https://doi.org/10.3390/BIOM11030433

Louis Bresson, J., Burlingame, B., Dean, T., Fairweather-Tait, S., Heinonen, M., Ildico Hirsch-Ernst, K., … Strain, S. (2016). Guidance on the scientific requirements for health claims related to the immune system, the gastrointestinal tract and defence against pathogenic microorganisms. *EFSA Journal*, *14*(1), 4369. https://doi.org/10.2903/J.EFSA.2016.4369

Luo, C., Peng, S., Li, M., Ao, X., & Liu, Z. (2022). The efficacy and safety of probiotics for allergic rhinitis: A systematic review and meta-analysis. *Frontiers in Immunology*, *13*, 848279. https://doi.org/10.3389/fimmu.2022.848279

Lye, H. S., Kuan, C. Y., Ewe, J. A., Fung, W. Y., & Liong, M. T. (2009). The improvement of hypertension by probiotics: Effects on cholesterol, diabetes, renin, and phytoestrogens. *International Journal of Molecular Sciences*, *10*(9), 3755. https://doi.org/10.3390/IJMS10093755

Ma, S., Wang, W., Su, Y., Sun, W., & Ma, L. (2023). Antibiotics therapy combined with probiotics administered intravaginally for the treatment of bacterial vaginosis: A systematic review and meta-analysis. *Open Medicine (Warsaw, Poland)*, *18*(1). https://doi.org/10.1515/MED-2023-0644

Maeusli, M., Lee, B., Miller, S., Reyna, Z., Lu, P., Yan, J., … Luna, B. (2020). Horizontal gene transfer of antibiotic resistance from Acinetobacter baylyi to Escherichia coli on lettuce and subsequent antibiotic resistance transmission to the gut microbiome. *MSphere*, *5*(3). https://doi.org/10.1128/MSPHERE.00329-20

Maldonado-Gómez, M. X., Martínez, I., Bottacini, F., O'Callaghan, A., Ventura, M., van Sinderen, D., … Walter, J. (2016). Stable engraftment of bifidobacterium longum AH1206 in the human gut depends on individualized features of the resident microbiome. *Cell Host & Microbe*, *20*(4), 515–526. https://doi.org/10.1016/J.CHOM.2016.09.001

Manzoor, S., Wani, S. M., Ahmad Mir, S., & Rizwan, D. (2022). Role of probiotics and prebiotics in mitigation of different diseases. *Nutrition (Burbank, Los Angeles County, Calif.)*, *96*. https://doi.org/10.1016/J.NUT.2022.111602

Mater, D. D. G., Langella, P., Corthier, G., & Flores, M. J. (2005). Evidence of vancomycin resistance gene transfer between enterococci of human origin in the gut of mice harbouring human microbiota. *The Journal of Antimicrobial Chemotherapy*, *56*(5), 975–978. https://doi.org/10.1093/JAC/DKI336

Mathur, S., & Singh, R. (2005). Antibiotic resistance in food lactic acid bacteria: A review. *International Journal of Food Microbiology*, *105*(3), 281–295. https://doi.org/10.1016/J.IJFOODMICRO.2005.03.008

Mayrhofer, S., van Hoek, A. H. A. M., Mair, C., Huys, G., Aarts, H. J. M., Kneifel, W., & Domig, K. J. (2010). Antibiotic susceptibility of members of the Lactobacillus acidophilus group using broth microdilution and molecular identification of their resistance determinants. *International Journal of Food Microbiology*, *144*(1), 81–87. https://doi.org/10.1016/J.IJFOODMICRO.2010.08.024

McInnes, R. S., McCallum, G. E., Lamberte, L. E., & van Schaik, W. (2020). Horizontal transfer of antibiotic resistance genes in the human gut microbiome. *Current Opinion in Microbiology*, *53*, 35–43. https://doi.org/10.1016/J.MIB.2020.02.002

Mendes, M. C. S., Paulino, D. S. M., Brambilla, S. R., Camargo, J. A., Persinoti, G. F., & Carvalheira, J. B. C. (2018). Microbiota modification by probiotic supplementation reduces colitis associated colon cancer in mice. *World Journal of Gastroenterology*, *24*(18), 1995–2008. https://doi.org/10.3748/WJG.V24.I18.1995

Merenstein, D., Pot, B., Leyer, G., Ouwehand, A. C., Preidis, G. A., Elkins, C. A., … Sanders, M. E. (2023). Emerging issues in probiotic safety: 2023 perspectives. *Gut Microbes*, *15*(1). https://doi.org/10.1080/19490976.2023.2185034

Merkouris, E., Mavroudi, T., Miliotas, D., Tsiptsios, D., Serdari, A., Christidi, F., … Tsamakis, K. (2024). Probiotics' effects in the treatment of anxiety and depression: A comprehensive review of 2014–2023 clinical trials. *Microorganisms*, *12*(2), 411. https://doi.org/10.3390/microorganisms12020411

Metchnikoff E, H. W. (1907). *The Prolongation of Life: Optimistic Studies*. London: Springer Publishing Company.

Miller, W. R., Munita, J. M., & Arias, C. A. (2014). Mechanisms of antibiotic resistance in enterococci. *Expert Review of Anti-Infective Therapy*, *12*(10), 1221. https://doi.org/10.1586/14787210.2014.956092

Montassier, E., Valdés-Mas, R., Batard, E., Zmora, N., Dori-Bachash, M., Suez, J., & Elinav, E. (2021). Probiotics impact the antibiotic resistance gene reservoir along the human GI tract in a person-specific and antibiotic-dependent manner. *Nature Microbiology*, *6*(8), 1043–1054. https://doi.org/10.1038/S41564-021-00920-0

Morelli, L., Sarra, P. G., & Bottazzi, V. (1988). In vivo transfer of pAM beta 1 from Lactobacillus reuteri to Enterococcus faecalis. *The Journal of Applied Bacteriology*, *65*(5), 371–375. https://doi.org/10.1111/J.1365-2672.1988.TB01905.X

Moretti, F., Colpo, S., Montresor, S., Facchinello, D., Raffaella, R. S. P. A., Roche, R., … Onlus -Fedemo, E. (2024). patients' associations as co-creators of knowledge: Tools and insights into stakeholders' engagement process. The experience of the ProSafe project for advancing treatment safety in proximity medicine. https://doi.org/10.21203/RS.3.RS-4007129/V1

Morgan, R. L., Preidis, G. A., Kashyap, P. C., Weizman, A. V., Sadeghirad, B., Chang, Y., … Zeraatkar, D. (2020). Probiotics reduce mortality and morbidity in preterm, low-birth-weight infants: A systematic review and network meta-analysis of randomized trials. *Gastroenterology*, *159*(2), 467–480. https://doi.org/10.1053/J.GASTRO.2020.05.096

National Database of Antibiotic Resistant Organisms (NDARO) – Pathogen Detection – NCBI. (n.d.). Retrieved 30 July 2024, from https://www.ncbi.nlm.nih.gov/pathogens/antimicrobial-resistance/

Nawaz, M., Wang, J., Zhou, A., Ma, C., Wu, X., Moore, J. E., … Xu, J. (2011). Characterization and transfer of antibiotic resistance in lactic acid bacteria from fermented food products. *Current Microbiology*, *62*(3), 1081–1089. https://doi.org/10.1007/S00284-010-9856-2

O'Brien, C. E., Meier, A. K., Cernioglo, K., Mitchell, R. D., Casaburi, G., Frese, S. A., … Smilowitz, J. T. (2022). Early probiotic supplementation with B. infantis in breastfed infants leads to persistent colonization at 1 year. *Pediatric Research*, *91*(3), 627–636. https://doi.org/10.1038/S41390-020-01350-0

Oh, J. K., Vasquez, R., Kim, S. H., Hwang, I. C., Song, J. H., Park, J. H., … Kang, D. K. (2021). Multispecies probiotics alter fecal short-chain fatty acids and lactate levels in weaned pigs by modulating gut microbiota. *Journal of Animal Science and Technology*, *63*(5), 1142–1158. https://doi.org/10.5187/JAST.2021.E94

Ojha, A. K., Shah, N. P., & Mishra, V. (2021). Conjugal transfer of antibiotic resistances in Lactobacillus spp. *Current Microbiology*, *78*(8), 2839–2849. https://doi.org/10.1007/s00284-021-02554-1

Oroojzadeh, P., Bostanabad, S. Y., & Lotfi, H. (2022). Psychobiotics: The influence of gut microbiota on the gut-brain axis in neurological disorders. *Journal of Molecular Neuroscience*, *72*(9), 1952. https://doi.org/10.1007/S12031-022-02053-3

Ouoba, L. I. I., Lei, V., & Jensen, L. B. (2008). Resistance of potential probiotic lactic acid bacteria and bifidobacteria of African and European origin to antimicrobials: Determination and transferability of the resistance genes to other bacteria. *International Journal of Food Microbiology*, *121*(2), 217–224. https://doi.org/10.1016/J.IJFOODMICRO.2007.11.018

Ouwehand, A. C., Forssten, S., Hibberd, A. A., Lyra, A., & Stahl, B. (2016). Probiotic approach to prevent antibiotic resistance. *Annals of Medicine*, *48*(4), 246–255. https://doi.org/10.3109/07853890.2016.1161232

Pais, P., Almeida, V., Yılmaz, M., & Teixeira, M. C. (2020). Saccharomyces boulardii: What makes it tick as successful probiotic? *Journal of Fungi* *6*(2), 78. https://doi.org/10.3390/JOF6020078

Panigrahi, P., Parida, S., Nanda, N. C., Satpathy, R., Pradhan, L., Chandel, Di. S., … Gewolb, I. H. (2017). A randomized synbiotic trial to prevent sepsis among infants in rural India. *Nature*, *548*(7668), 407–412. https://doi.org/10.1038/NATURE23480

Pariza, M. W., Gillies, K. O., Kraak-Ripple, S. F., Leyer, G., & Smith, A. B. (2015). Determining the safety of microbial cultures for consumption by humans and animals. *Regulatory Toxicology and Pharmacology : RTP*, *73*(1), 164–171. https://doi.org/10.1016/J.YRTPH.2015.07.003

Perraudeau, F., McMurdie, P., Bullard, J., Cheng, A., Cutcliffe, C., Deo, A., … Kolterman, O. (2020). Improvements to postprandial glucose control in subjects with type 2 diabetes: A multicenter, double blind, randomized placebo-controlled trial of a novel probiotic formulation. *BMJ Open Diabetes Research & Care*, *8*(1). https://doi.org/10.1136/BMJDRC-2020-001319

Phillips, R., Hazell, L., Sauzet, O., & Cornelius, V. (2019). Analysis and reporting of adverse events in randomised controlled trials: A review. *BMJ Open*, *9*(2). https://doi.org/10.1136/BMJOPEN-2018-024537

Plaza-Diaz, J., Ruiz-Ojeda, F. J., Gil-Campos, M., & Gil, A. (2019). Mechanisms of action of probiotics. *Advances in Nutrition*, *10*(Suppl 1), S49. https://doi.org/10.1093/ADVANCES/NMY063

Qualified Presumption of Safety (QPS) | EFSA. (n.d.). Retrieved 31 July 2024, from https://www.efsa.europa.eu/en/topics/topic/qualified-presumption-safety-qps

Ramos, C. L., Thorsen, L., Schwan, R. F., & Jespersen, L. (2013). Strain-specific probiotics properties of Lactobacillus fermentum, Lactobacillus plantarum and Lactobacillus brevis isolates from Brazilian food products. *Food Microbiology*, *36*(1), 22–29. https://doi.org/10.1016/J.FM.2013.03.010

Ranadheera, C. S., Evans, C. A., Baines, S. K., Balthazar, C. F., Cruz, A. G., Esmerino, E. A., … Vasiljevic, T. (2019). Probiotics in goat milk products: Delivery capacity and ability to improve sensory attributes. *Comprehensive Reviews in Food Science and Food Safety*, *18*(4), 867–882. https://doi.org/10.1111/1541-4337.12447

Rao, S. S. C., Rehman, A., Yu, S., & De Andino, N. M. (2018). Brain fogginess, gas and bloating: A link between SIBO, probiotics and metabolic acidosis. *Clinical and Translational Gastroenterology*, *9*(6). https://doi.org/10.1038/S41424-018-0030-7

Rao, S. S. C., Yu, S., Tetangco, E. P., & Yan, Y. (2018). Probiotics can cause D-lactic acidosis and brain fogginess: Reply to Quigley et al. *Clinical and Translational Gastroenterology*, *9*(11), 207. https://doi.org/10.1038/S41424-018-0077-5

Rao Thumu, S. C., & Halami, P. M. (2012). Presence of erythromycin and tetracycline resistance genes in lactic acid bacteria from fermented foods of Indian origin. *Antonie van Leeuwenhoek*, *102*(4), 541–551. https://doi.org/10.1007/S10482-012-9749-4

Rather, I. A., Bajpai, V. K., Kumar, S., Lim, J., Paek, W. K., & Park, Y. H. (2016). Probiotics and atopic dermatitis: An overview. *Frontiers in Microbiology*, *7*(APR). https://doi.org/10.3389/FMICB.2016.00507

Rattanaprasert, M., Roos, S., Hutkins, R. W., & Walter, J. (2014). Quantitative evaluation of synbiotic strategies to improve persistence and metabolic activity of Lactobacillus reuteri DSM 17938 in the human gastrointestinal tract. *Journal of Functional Foods*, *10*, 85–94. https://doi.org/10.1016/J.JFF.2014.05.017

Rekdal, V. M., Bess, E. N., Bisanz, J. E., Turnbaugh, P. J., & Balskus, E. P. (2019). Discovery and inhibition of an interspecies gut bacterial pathway for Levodopa metabolism. *Science (*New York, N.Y.), *364*(6445). https://doi.org/10.1126/SCIENCE.AAU6323

Roe, A. L., Boyte, M. E., Elkins, C. A., Goldman, V. S., Heimbach, J., Madden, E., … Smith, A. (2022). Considerations for determining safety of probiotics: A USP perspective. *Regulatory Toxicology and Pharmacology*, *136*, 105266. https://doi.org/10.1016/J.YRTPH.2022.105266

Rojo-Bezares, B., Sáenz, Y., Poeta, P., Zarazaga, M., Ruiz-Larrea, F., & Torres, C. (2006). Assessment of antibiotic susceptibility within lactic acid bacteria strains isolated from wine. *International Journal of Food Microbiology*, *111*(3), 234–240. https://doi.org/10.1016/J.IJFOODMICRO.2006.06.007

Rozman, V., Lorbeg, P. M., Treven, P., Accetto, T., Janežič, S., Rupnik, M., & Matijašić, B. B. (2023). Genomic insights into antibiotic resistance and mobilome of lactic acid bacteria and bifidobacteria. *Life Science Alliance*, *6*(4). https://doi.org/10.26508/LSA.202201637

Rozman, V., Mohar Lorbeg, P., Treven, P., Accetto, T., Golob, M., Zdovc, I., & Bogovič Matijašić, B. (2022). Lactic acid bacteria and bifidobacteria deliberately introduced into the agro-food chain do not significantly increase the antimicrobial resistance gene pool. *Gut Microbes*, *14*(1). https://doi.org/10.1080/19490976.2022.2127438

Rubinstein, E., & Keynan, Y. (2014). Vancomycin revisited – 60 years later. *Frontiers in Public Health*, *2*, 217. https://doi.org/10.3389/fpubh.2014.00217

Ruíz-Ramírez, Y., Guadarrama-Mendoza, P. C., Escalante, A., Giles-Gómez, M., & Valadez-Blanco, R. (2022). Probiotic activity traits in vitro and production of antimicrobial peptides by Lactobacillaceae isolates from pulque using Lactobacillus acidophilus NCFM as control. *Brazilian Journal of Microbiology: [Publication of the Brazilian Society for Microbiology]*, *53*(2), 921–933. https://doi.org/10.1007/S42770-022-00684-7

Sabit, H., Kassab, A., Alaa, D., Mohamed, S., Abdel-Ghany, S., Mansy, M., … Abushady, A. M. (2023). The effect of probiotic supplementation on the gut–brain axis in psychiatric patients. *Current Issues in Molecular Biology 45*(5), 4080–4099. https://doi.org/10.3390/CIMB45050260

Salam, M. A., Al-Amin, M. Y., Pawar, J. S., Akhter, N., & Lucy, I. B. (2023). Conventional methods and future trends in antimicrobial susceptibility testing. *Saudi Journal of Biological Sciences*, *30*(3). https://doi.org/10.1016/J.SJBS.2023.103582

Salminen, S., Collado, M. C., Endo, A., Hill, C., Lebeer, S., Quigley, E. M. M., … Vinderola, G. (2021). The International Scientific Association of Probiotics and Prebiotics (ISAPP) consensus statement on the definition and scope of postbiotics. *Nature Reviews. Gastroenterology & Hepatology*, *18*(9), 649–667. https://doi.org/10.1038/S41575-021-00440-6

Samara, J., Moossavi, S., Alshaikh, B., Ortega, V. A., Pettersen, V. K., Ferdous, T., … Arrieta, M. C. (2022). Supplementation with a probiotic mixture accelerates gut microbiome maturation and reduces intestinal inflammation in extremely preterm infants. *Cell Host & Microbe*, *30*(5), 696–711.e5. https://doi.org/10.1016/J.CHOM.2022.04.005

Sanders, M. E., Akkermans, L. M. A., Haller, D., Hammerman, C., Heimbach, J., Hörmannsperger, G., … Vaughan, E. (2010). Safety assessment of probiotics for human use. *Gut Microbes*, *1*(3), 164. https://doi.org/10.4161/GMIC.1.3.12127

Sanders, M. E., Heimbach, J. T., Pot, B., Tancredi, D., Lenoir-Wijnkoop, I., Lähteenmäki-Uutela, A., … Bañares, S. (2011). Health claims substantiation for probiotic and prebiotic products. *Gut Microbes*, *2*(3), 127–133. https://doi.org/10.4161/GMIC.2.3.16174

Sehrawat, N., Yadav, M., Singh, M., Kumar, V., Sharma, V. R., & Sharma, A. K. (2021). Probiotics in microbiome ecological balance providing a therapeutic window against cancer. *Seminars in Cancer Biology*, *70*, 24–36. https://doi.org/10.1016/J.SEMCANCER.2020.06.009

Shahali, A., Soltani, R., & Akbari, V. (2023). Probiotic Lactobacillus and the potential risk of spreading antibiotic resistance: A systematic review. *Research in Pharmaceutical Sciences*, *18*(5), 468. https://doi.org/10.4103/1735-5362.383703

Shahbazi, R., Yasavoli-Sharahi, H., Alsadi, N., Ismail, N., & Matar, C. (2020). Probiotics in treatment of viral respiratory infections and neuroinflammatory disorders. *Molecules*, *25*(21). https://doi.org/10.3390/MOLECULES25214891

Shane, A. L., & Preidis, G. A. (2023). Probiotics in the neonatal intensive care unit—A framework for optimizing product standards. *JAMA Pediatrics*, *177*(9), 879–880. https://doi.org/10.1001/JAMAPEDIATRICS.2023.2165

Sharma, P., Tomar, S. K., Goswami, P., Sangwan, V., & Singh, R. (2014). Antibiotic resistance among commercially available probiotics. *Food Research International*, *57*, 176–195. https://doi.org/10.1016/J.FOODRES.2014.01.025

Sharma, S., Kumar, S., Sajjad, S., & Sharma, S. (2023). Probiotics in irritable bowel syndrome: A review article. *Cureus*, *15*(3). https://doi.org/10.7759/CUREUS.36565

Shori, A. B. (2017). Microencapsulation improved probiotics survival during gastric transit. *HAYATI Journal of Biosciences*, *24*(1), 1–5. https://doi.org/10.1016/J.HJB.2016.12.008

Smith, S. M., Wang, A. T., Katz, N. P., McDermott, M. P., Burke, L. B., Coplan, P., … Dworkin, R. H. (2013). Adverse event assessment, analysis, and reporting in recent published analgesic clinical trials: ACTTION systematic review and recommendations. *Pain*, *154*(7), 997–1008. https://doi.org/10.1016/J.PAIN.2013.03.003

Stefańska, I., Kwiecień, E., Jóźwiak-Piasecka, K., Garbowska, M., Binek, M., & Rzewuska, M. (2021). Antimicrobial susceptibility of lactic acid bacteria strains of potential use as feed additives – The basic safety and usefulness criterion. *Frontiers in Veterinary Science*, *8*. https://doi.org/10.3389/FVETS.2021.687071

Stogios, P. J., & Savchenko, A. (2020). Molecular mechanisms of vancomycin resistance. *Protein Science: A Publication of the Protein Society*, *29*(3), 654–669. https://doi.org/10.1002/PRO.3819

Suez, J., Zmora, N., & Elinav, E. (2020). Probiotics in the next-generation sequencing era. *Gut Microbes*, *11*(1), 77–93. https://doi.org/10.1080/19490976.2019.1586039

Suez, J., Zmora, N., Segal, E., & Elinav, E. (2019). The pros, cons, and many unknowns of probiotics. *Nature Medicine*, *25*(5), 716–729. https://doi.org/10.1038/S41591-019-0439-X

Sun, Y., Gao, S., Ye, C., & Zhao, W. (2023). Gut microbiota dysbiosis in polycystic ovary syndrome: Mechanisms of progression and clinical applications. *Frontiers in Cellular and Infection Microbiology*, *13*. https://doi.org/10.3389/FCIMB.2023.1142041

Szajewska, H., & Vinderola, G. (2024). *Current Regulatory Issues for the Use of Probiotics*. 187–193. https://doi.org/10.1007/978-3-031-58572-2_12

Tachon, S., Lee, B., & Marco, M. L. (2014). Diet alters probiotic Lactobacillus persistence and function in the intestine. *Environmental Microbiology*, *16*(9), 2915–2926. https://doi.org/10.1111/1462-2920.12297

Tannock, G. W., Luchansky, J. B., Miller, L., Connell, H., Thode-Andersen, S., Mercer, A. A., & Klaenhammer, T. R. (1994). Molecular characterization of a plasmid-borne (pGT633) erythromycin resistance determinant (ermGT) from Lactobacillus reuteri 100–63. *Plasmid*, *31*(1), 60–71. https://doi.org/10.1006/PLAS.1994.1007

Taverniti, V., Koirala, R., Dalla Via, A., Gargari, G., Leonardis, E., Arioli, S., & Guglielmetti, S. (2019). Effect of cell concentration on the persistence in the human intestine of four probiotic strains administered through a multispecies formulation. *Nutrients*, *11*(2). https://doi.org/10.3390/NU11020285

Temmerman, R., Pot, B., Huys, G., & Swings, J. (2003). Identification and antibiotic susceptibility of bacterial isolates from probiotic products. *International Journal of Food Microbiology*, *81*(1), 1–10. https://doi.org/10.1016/S0168-1605(02)00162-9

Thu, M. S., Ondee, T., Nopsopon, T., Farzana, I. A. K., Fothergill, J. L., Hirankarn, N., … Pongpirul, K. (2023). Effect of probiotics in breast cancer: A systematic review and meta-analysis. *Biology*, *12*(2), 280. https://doi.org/10.3390/biology12020280

Toomey, N., Bolton, D., & Fanning, S. (2010). Characterisation and transferability of antibiotic resistance genes from lactic acid bacteria isolated from Irish pork and beef abattoirs. *Research in Microbiology*, *161*(2), 127–135. https://doi.org/10.1016/J.RESMIC.2009.12.010

Toomey, N., Monaghan, Á., Fanning, S., & Bolton, D. J. (2009). Assessment of antimicrobial resistance transfer between lactic acid bacteria and potential foodborne pathogens using in vitro methods and mating in a food matrix. *Foodborne Pathogens and Disease*, *6*(8), 925–933. https://doi.org/10.1089/FPD.2009.0278

Tóth, A. G., Csabai, I., Judge, M. F., Maróti, G., Becsei, Á., Spisák, S., & Solymosi, N. (2021). Mobile antimicrobial resistance genes in probiotics. *Antibiotics (*Basel, Switzerland), *10*(11). https://doi.org/10.3390/ANTIBIOTICS10111287

Tóth, A. G., Judge, M. F., Nagy, S. Á., Papp, M., & Solymosi, N. (2023). A survey on antimicrobial resistance genes of frequently used probiotic bacteria, 1901 to 2022. *Eurosurveillance*, *28*(14). https://doi.org/10.2807/1560-7917.ES.2023.28.14.2200272

Tripathi, M. K., & Giri, S. K. (2014). Probiotic functional foods: Survival of probiotics during processing and storage. *Journal of Functional Foods*, *9*(1), 225–241. https://doi.org/10.1016/J.JFF.2014.04.030

Turck, D., Bohn, T., Castenmiller, J., De Henauw, S., Hirsch-Ernst, K. I., Maciuk, A., … Knutsen, H. K. (2021). Safety of pasteurised Akkermansia muciniphila as a novel food pursuant to regulation (EU) 2015/2283. *EFSA Journal*, *19*(9), e06780. https://doi.org/10.2903/J.EFSA.2021.6780

Tynkkynen, S., Singh, K. V., & Varmanen, P. (1998). Vancomycin resistance factor of Lactobacillus rhamnosus GG in relation to enterococcal vancomycin resistance (van) genes. *International Journal of Food Microbiology*, *41*(3), 195–204. https://doi.org/10.1016/S0168-1605(98)00051-8

Van Baal, M. C., Kohout, P., Besselink, M. G., Van Santvoort, H. C., Benes, Z., Zazula, R., … Gooszen, H. G. (2012). Probiotic treatment with Probioflora in patients with predicted severe acute pancreatitis without organ failure. *Pancreatology*, *12*(5), 458–462. https://doi.org/10.1016/J.PAN.2012.08.004

van Hoek, A. H. A. M., Mayrhofer, S., Domig, K. J., & Aarts, H. J. M. (2008). Resistance determinant erm(X) is borne by transposon Tn5432 in Bifidobacterium thermophilum and Bifidobacterium animalis subsp. lactis. *International Journal of Antimicrobial Agents*, *31*(6), 544–548. https://doi.org/10.1016/J.IJANTIMICAG.2008.01.025

Van Hoek, A. H. A. M., Mayrhofer, S., Domig, K. J., Flórez, A. B., Ammor, M. S., Mayo, B., & Aarts, H. J. M. (2008). Mosaic tetracycline resistance genes and their flanking regions in Bifidobacterium thermophilum and Lactobacillus johnsonii. *Antimicrobial Agents and Chemotherapy*, *52*(1), 248–252. https://doi.org/10.1128/AAC.00714-07

Van Reenen, C. A., Dicks, L. M. T., & Chikindas, M. L. (1998). Isolation, purification and partial characterization of plantaricin 423, a bacteriocin produced by Lactobacillus plantarum. *Journal of Applied Microbiology*, *84*(6), 1131–1137. https://doi.org/10.1046/J.1365-2672.1998.00451.X

Vandenplas, Y., Huys, G., & Daube, G. (2015). Probiotics: An update. *Jornal de Pediatria*, *91*(1), 6–21. https://doi.org/10.1016/J.JPED.2014.08.005

Varga, L., Süle, J., & Nagy, P. (2014). Short communication: Survival of the characteristic microbiota in probiotic fermented camel, cow, goat, and sheep milks during refrigerated storage. *Journal of Dairy Science*, *97*(4), 2039–2044. https://doi.org/10.3168/JDS.2013-7339

Varvara, R. A., & Vodnar, D. C. (2024). Probiotic-driven advancement: Exploring the intricacies of mineral absorption in the human body. *Food Chemistry: X*, *21*, 101067. https://doi.org/10.1016/J.FOCHX.2023.101067

Venugopalan, V., Shriner, K. A., & Wong-Beringer, A. (2010). Regulatory oversight and safety of probiotic use. *Emerging Infectious Diseases*, *16*(11), 1661–1665. https://doi.org/10.3201/EID1611.100574

Vieira, A. T., Fukumori, C., & Ferreira, C. M. (2016). New insights into therapeutic strategies for gut microbiota modulation in inflammatory diseases. *Clinical & Translational Immunology*, *5*(6). https://doi.org/10.1038/CTI.2016.38

Vignaroli, C., Zandri, G., Aquilanti, L., Pasquaroli, S., & Biavasco, F. (2011). Multidrug-resistant enterococci in animal meat and faeces and co-transfer of resistance from an Enterococcus durans to a human Enterococcus faecium. *Current Microbiology*, *62*(5), 1438–1447. https://doi.org/10.1007/S00284-011-9880-X

Vinderola, G., Sanders, M. E., & Salminen, S. (2022). The concept of postbiotics. *Foods*, *11*(8). https://doi.org/10.3390/FOODS11081077

Von Wintersdorff, C. J. H., Penders, J., Van Niekerk, J. M., Mills, N. D., Majumder, S., Van Alphen, L. B., … Wolffs, P. F. G. (2016). Dissemination of antimicrobial resistance in microbial ecosystems through horizontal gene transfer. *Frontiers in Microbiology*, *7*(FEB). https://doi.org/10.3389/FMICB.2016.00173

Wallace, C. J. K., & Milev, R. (2017). The effects of probiotics on depressive symptoms in humans: A systematic review. *Annals of General Psychiatry*, *16*(1), 14. https://doi.org/10.1186/S12991-017-0138-2

Wang, J., Dong, X., Shao, Y., Guo, H., Pan, L., Hui, W., … Zhang, W. (2017). Genome adaptive evolution of Lactobacillus casei under long-term antibiotic selection pressures. *BMC Genomics*, *18*(1). https://doi.org/10.1186/S12864-017-3710-X

Wang, Y., Liang, Q., Lu, B., Shen, H., Liu, S., Shi, Y., … Chen, H. (2021). Whole-genome analysis of probiotic product isolates reveals the presence of genes related to antimicrobial resistance, virulence factors, and toxic metabolites, posing potential health risks. *BMC Genomics*, *22*(1). https://doi.org/10.1186/S12864-021-07539-9

Wash, P., Batool, A., Mulk, S., Nazir, S., Yasmin, H., Mumtaz, S., … Hassan, M. N. (2022). Prevalence of antimicrobial resistance and respective genes among Bacillus spp., a versatile bio-fungicide. *International Journal of Environmental Research and Public Health*, *19*(22), 14997. https://doi.org/10.3390/ijerph192214997

Wegmann, U., O'Connell-Motherway, M., Zomer, A., Buist, G., Shearman, C., Canchaya, C., … Kok, J. (2007). Complete genome sequence of the prototype lactic acid bacterium Lactococcus lactis subsp. cremoris MG1363. *Journal of Bacteriology*, *189*(8), 3256. https://doi.org/10.1128/JB.01768-06

Wei, Y., Yang, F., Wu, Q., Gao, J., Liu, W., Liu, C., … Tang, R. (2018). Protective effects of bifidobacterial strains against toxigenic Clostridium difficile. *Frontiers in Microbiology*, *9*, 888. https://doi.org/10.3389/fmicb.2018.00888

WHO. (2018a). WHO Advisory Group on Integrated Surveillance of Antimicrobial Resistance.

WHO. (2018b). WHO Advisory Group on Integrated Surveillance of Antimicrobial Resistance. Critically important antimicrobials for human medicine.

WHO. (2022). The WHO Essential Medicines List Antibiotic Book: Improving Antibiotic Awareness.

Wiegers, C., van Beek, E. H. T., & Larsen, O. F. A. (2023). Clinical research with probiotics as an indicator of global valorization since the year 2000. *Frontiers in Microbiology*, *14*. https://doi.org/10.3389/FMICB.2023.1323920

Wilson, I. D., & Nicholson, J. K. (2017). Gut microbiome interactions with drug metabolism, efficacy, and toxicity. *Translational Research : The Journal of Laboratory and Clinical Medicine*, *179*, 204–222. https://doi.org/10.1016/J.TRSL.2016.08.002

Winter, M., Buckling, A., Harms, K., Johnsen, P. J., & Vos, M. (2021). Antimicrobial resistance acquisition via natural transformation: Context is everything. *Current Opinion in Microbiology*, *64*, 133–138. https://doi.org/10.1016/J.MIB.2021.09.009

Wollowski, I., Rechkemmer, G., & Pool-Zobel, B. L. (2001). Protective role of probiotics and prebiotics in colon cancer. *The American Journal of Clinical Nutrition*, *73*(2), 451s–455s. https://doi.org/10.1093/AJCN/73.2.451S

Yang, H., Liu, Y., Cai, R., Li, Y., & Gu, B. (2021). A narrative review of relationship between gut microbiota and neuropsychiatric disorders: Mechanisms and clinical application of probiotics and prebiotics. *Annals of Palliative Medicine*, *10*(2), 2304313–2302313. https://doi.org/10.21037/APM-20-1365

Yang, J., & Yang, H. (2023). Recent development in Se-enriched yeast, lactic acid bacteria and bifidobacteria. *Critical Reviews in Food Science and Nutrition*, *63*(3), 411–425. https://doi.org/10.1080/10408398.2021.1948818

Yelin, I., Flett, K. B., Merakou, C., Mehrotra, P., Stam, J., Snesrud, E., … Priebe, G. P. (2019). Genomic and epidemiological evidence of bacterial transmission from probiotic capsule to blood in ICU patients. *Nature Medicine*, *25*(11), 1728–1732. https://doi.org/10.1038/S41591-019-0626-9

Zavišić, G., Popović, M., Stojkov, S., Medić, D., Gusman, V., Jovanović Lješković, N., & Jovanović Galović, A. (2023). Antibiotic resistance and probiotics: Knowledge gaps, market overview and preliminary screening. *Antibiotics*, *12*(8), 1281. https://doi.org/10.3390/antibiotics12081281

Zhang, F., Gao, J., Wang, B., Huo, D., Wang, Z., Zhang, J., & Shao, Y. (2018). Whole-genome sequencing reveals the mechanisms for evolution of streptomycin resistance in Lactobacillus plantarum. *Journal of Dairy Science*, *101*(4), 2867–2874. https://doi.org/10.3168/JDS.2017-13323

Zhao, Y., Zeng, Y., Zeng, D., Wang, H., Zhou, M., Sun, N., … Ni, X. (2021). Probiotics and MicroRNA: Their roles in the host-microbe interactions. *Frontiers in Microbiology*, *11*. https://doi.org/10.3389/FMICB.2020.604462

Zheng, M., Zhang, R., Tian, X., Zhou, X., Pan, X., & Wong, A. (2017). Assessing the risk of probiotic dietary supplements in the context of antibiotic resistance. *Frontiers in Microbiology*, *8*(MAY). https://doi.org/10.3389/FMICB.2017.00908

Zhu, M., Liu, X., Ye, Y., Yan, X., Cheng, Y., Zhao, L., … Ling, Z. (2022). Gut microbiota: A novel therapeutic target for Parkinson's disease. *Frontiers in Immunology*, *13*, 937555. https://doi.org/10.3389/FIMMU.2022.937555

Zikou, E., Dovrolis, N., Dimosthenopoulos, C., Gazouli, M., & Makrilakis, K. (2023). The effect of probiotic supplements on metabolic parameters of people with type 2 diabetes in Greece—A randomized, double-blind, placebo-controlled study. *Nutrients*, *15*(21), 4663. https://doi.org/10.3390/nu15214663

8 Synergistic Approaches
Probiotics in Combination with Traditional Antibiotics

Suhail Maqbool, Nasir Nisar, Darakshan Bhat, and Nahida Tabassum
Department of Pharmaceutical Sciences, Pharmacology Division, University of Kashmir, Hazratbal, Srinagar, Jammu & Kashmir, India

8.1 INTRODUCTION

8.1.1 Probiotics and Antibiotics

Probiotics and antibiotics depict two different yet correlative strategies in modern healthcare, each having a vital role in upholding and recovering health. Probiotics are living microorganisms, primarily useful bacteria and yeast, that, when taken in adequate amounts, possess numerous health benefits to the host (Reid 2016). These microorganisms are required for maintaining a balanced gut microbiota, which plays an important role in digestion, immune function, and overall health (Li et al. 2020). Probiotics are widely available in fermented foods like yogurt, dietary supplements kefir, sauerkraut, etc. They enhance gut health by combating pathogens, secrete antimicrobial bioactive chemicals, and elevate the immune system response to infections (Azad, Sarker, and Wan 2018). Moreover, probiotics assist in the formation of short-chain fatty acids and vitamins, also maintaining gut and metabolic health (Markowiak-Kopeć and Śliżewska 2020).

Antibiotics are substances that are either derived from microorganisms or semi-synthetic chemicals developed to counter bacterial infections, either by killing the bacteria (bactericidal) or by inhibiting their growth (bacteriostatic) (Hutchings, Truman, and Wilkinson 2019). From the discovery of penicillin in the 20th century, antibiotics have transformed the field of medicine, turning fatal infectious pathogens into controllable conditions. Although antibiotics do possess their drawbacks, they disturb the natural balance of the GIT microbiome, leading to conditions such as antibiotic-associated diarrhea (AAD), resistance of burn wound bacteria *Pseudomonas aeruginosa* species, *Clostridioides difficile* infection, as well as attenuating the immune system (Stavropoulou and Bezirtzoglou 2020). Also, the overexploitation of traditional antibiotics has influenced the worldwide rise of multiple antibiotic drug-resistant bacteria, presenting a notable challenge to public health (Sagar et al. 2019).

The synergistic approaches of integrating the effectiveness of probiotics with antibiotics are drawing interest to increase the efficacy of probiotic-antibiotic therapy while alleviating adverse effects. Probiotics help in protecting the gut flora from destabilizing due to antibiotics, remarkably prohibiting the colonization of antibiotic-resistant bacteria, preventing AAD; synergism can also augment immune function and improve the restoration of gut microbiota post antibiotic treatment (Kwoji et al. 2021). Given this context, probiotics are progressively appraised not only as an aiding therapy for traditional antibiotics but also as a supportive measure to maintain and balance gut health and decrease the adverse effects associated with the use of traditional antibiotic therapy. The growing knowledge of the probiotics and their importance in biomedical prioritizes its approach to synergism by combining probiotics into antibiotics to improve public health.

 DOI: 10.1201/9781032676951-8

8.1.2 Synergism – Necessary Approach

Traditional antibiotics are crucial for curing bacterial infections, but their drawbacks, such as an increase in resistance, their adverse effects, etc., have evolved as a crucial health challenge worldwide as bacteria have developed mechanisms against traditional treatments. Improper use or overuse of conventional antibiotics in both developed and developing countries has become a serious matter of concern. The rise of carbapenem-resistant Enterobacteriaceae (CRE) and extensively drug-resistant *Mycobacterium tuberculosis*, methicillin-resistant *Staphylococcus aureus* (MRSA), resistance of burn wound bacteria *Pseudomonas aeruginosa* species, similar more examples are there that are worrying for biomedical scientists (Chinemerem Nwobodo et al. 2022, Bhowmick, Oishi, and Aishy 2022). Moreover, traditional antibiotics can destroy the gut microbiota, causing complications like AAD and immune imbalance (Maier et al. 2021). To address such complications, novel approaches are always the need of the hour. The synergistic approach, combining traditional antibiotics with probiotics, has been among the first choices in medical science. It has definitely revolutionized the antibacterial approach for pathogenic infection.

The synergism of probiotics with traditional antibiotics resolves the limitations of conventional antibiotic treatment by elevating overall health benefits and decreasing adverse effects. Traditional antibiotics in combination with probiotics have shown promising results in combating harmful pathogens. *Saccharomyces boulardii*, a probiotic yeast, has been shown to mitigate the condition of AAD in patients taking antibiotic medication (McFarland 2017). Probiotics help to increase and retain the beneficial gut microbiome, reversing the negative effect of antibiotics and restoring GIT health. *Lactobacillus rhamnosus* GG in synergism with antibiotics (such as amoxicillin and clarithromycin) reduced the effect of *Helicobacter pylori* infection (Ji and Yang 2020). Similarly, the zone of inhibition was found to be more against resistant burn wound bacteria *Pseudomonas aeruginosa* species when probiotics were combined with tetracycline and gentamycin (Moghadam et al. 2018). Thus, suggesting increased efficacy of traditional antibiotics when given in combination with probiotics. This admirable method not only reduces the side effects but also provides a better choice for treating antimicrobial-resistant bacteria and improving gut microbiota.

8.2 EFFECT OF ANTIBIOTICS: DYSBIOSIS AND GUT MICROBIOME

Traditional antibiotics, even though effective against pathogens, cause dysbiosis by disturbing the normal GIT flora balance. The effect of traditional antibiotics on the microbiome persists from weeks to several months; some reports have even found an imbalance in normal microbiota persisting for years, influencing both children and adults. Such microbiome disruption leads to prolonged effects of antibiotic resistance and physiologically related changes, for example, immune system or metabolism imbalance, lasting for years even post antibiotic therapy (Kesavelu and Jog 2023, McDonnell et al. 2021).

Traditional antibiotics significantly affect the gut microbiome by destroying microbial flora, causing an imbalance. Broad-spectrum antibiotics, such as ciprofloxacin and amoxicillin, are effective against a wide range of bacteria, including both harmful and beneficial bacteria. Important mechanisms for antimicrobial action involve bacterial cell wall inhibition and disrupting of protein synthesis and DNA replication, resulting in either killing or inhibiting the growth of bacteria. This nonselective action decreases populations of beneficial microorganisms as well, like *Lactobacillus*, *Bifidobacterium*, *Pediococcus*, *Propionibacterium*, *Saccharomyces,* and species of spore-forming Bacillus, which are important for keeping normal gut health and immune function (Maier et al. 2021). The reduction of these useful microbes results in GIT disorders, such as diarrhea, and enhances the chance of infections like Clostridium difficile. The consequent overall dysbiosis leads to immediate health problems like AAD and an unhealthy gut environment. Chronic outcomes include disturbed metabolism, altered immune responses, and an elevated risk of chronic disorders like obesity and inflammatory bowel disease (Patangia et al. 2022).

Traditional antibiotics have drastic effects, even in children and infants as well. Studies have proven the altered microbiome composition in children using different traditional antibiotics like macrolides and other broad-spectrum antibiotics like ampicillin, gentamycin, etc. (Yassour et al. 2016). The richness of beneficial gut microbiome species like *Lactobacillus*, *Bifidobacterium,* and *Candidatus Arthromitus* was shown to be reduced greatly, and overexpression of drug-resistant bacteria was developed among infants given traditional antibiotics, including ticarcillin-clavulanate, cefotaxime, and meropenem (Korpela et al. 2016). These findings highlight the necessity of introducing novel approaches, like a synergistic approach with probiotics, to overcome such complications.

8.3 PROBIOTICS: ALTERNATIVE APPROACH FOR ANTIBACTERIAL COMPLICATIONS

Probiotics are among the primary alternatives in biomedicine to counteract harmful pathogens and also reduce the lethal complications and limitations of traditional antibiotics. Probiotics have been explored to be compatible with antibiotics, showing tremendous health benefits. They have their own antimicrobial activity as well, hence showing a remarkable synergistic effect with traditional antibiotics (Acharjee et al. 2022).

One of the essential mechanisms is the competitive elimination of pathogenic bacteria. Probiotics fight with infectious microbes for binding sites on the GIT epithelium and necessary nutrients, resulting in the prevention of the production and implantation of pathogenic bacteria like *Clostridium difficile* and *Salmonella* (Javanshir et al. 2021). Moreover, probiotics secrete antimicrobial chemicals, like bacteriocins, hydrogen peroxide, and organic acids, which directly inhibit or kill pathogenic bacteria.

Probiotics also play a vital role in strengthening the intestinal barrier function. They enhance the secretion of mucus by goblet cells and intensify tight junctions between epithelial cells, which decreases intestinal permeability and stops the translocation of harmful pathogens and toxins entering the systemic circulation. This barrier strengthening is crucial in blocking conditions like leaky gut syndrome and related inflammatory disorders (Potrykus et al. 2021).

Immunomodulation is yet another important mechanism in probiotics through which they combat pathogens. Probiotics engage with the gut-associated lymphoid tissue (GALT) to regulate the host's immune response. They enhance the secretion of anti-inflammatory cytokines, like IL-10, and diminish the levels of pro-inflammatory cytokines, thus helping to sustain immune homeostasis (Raheem et al. 2021). Probiotics also increase the functions of many immune cells, including macrophages, dendritic cells, and natural killer (NK) cells, amplifying the host's ability against harmful bacteria (Yousefi et al. 2019). Additionally, probiotics help in the metabolism of indigestible dietary substances, which leads to the formation of short-chain fatty acids (SCFAs) like butyrate, which aid as energy sources for colonocytes and have anti-inflammatory properties (Fernández et al. 2016). They also help in the production of essential vitamins, such as vitamin K and B-complex (thiamin, pyridoxine, cobalamin, etc.), which are important for a healthy being (Batta, Thakur, and Meghwal 2024, Kang et al. 2022). These diverse mechanisms point out the significance of probiotics in preserving gut health, improving immune function, and potentially protecting from pathogenic bacteria, hence proving them to be important components in modern biomedicine for treating various infectious conditions in an efficient way.

8.4 SYNERGISTIC MECHANISMS

The synergistic mechanisms of probiotics and antibiotics have been widely evaluated in several studies, exploring a multidimensional aspect that improves therapeutic outcomes. Probiotics, in synergistic compatibility with traditional antibiotics, enhance antimicrobial activity and mitigate the adverse effects. One such example is from a study where the combination of probiotic with antibiotics is used against multidrug-resistant *P. aeruginosa* isolated from burn wounds and it was found that synergistic approach showed better results than antibiotics used alone, possible mechanism could be the production of antimicrobial agents, pathogen elimination, and immunomodulation (Moghadam et al. 2018). Probiotics have been found to enhance the number of immunoglobulin-secreting cells in the gut mucosa by stimulating interferon release and, in turn, protect epithelium from bacterial attack (Galdeano et al. 2007). Another important mechanism has been found to be the revival of the GIT microbiome environment. Traditional antibiotics disturb the normal balance of gut microbiota, resulting in GIT issues and other gut-related complications. Probiotics restore the gut microbiota by nullifying the effect on beneficial bacteria and counteract the harmful effect on gut flora, such as *Lactobacillus* and *Bifidobacterium*, which can outplay the effect of pathogenic microbes and thereby restore microbial equilibrium.

Studies have proven that probiotics do augment the efficacy of traditional antibiotics via other mechanisms, too. They secrete antibacterial peptides and organic acids that suppress the growth of drug-resistant pathogens, clinically potentiating the antimicrobial effect of traditional antibiotics at lower doses (Elshaghabee and Rokana 2022). Probiotics enhance the gut barrier, which in turn prevents the transit of harmful pathogens and toxins into the blood circulation. Based on studies, eradication of pathogenic infection by *Pseudomonas aeruginosa* and *Klebsiella pneumoniae* was seen in children treated with probiotic+traditional antibiotic prophylaxis (Madden-Fuentes et al. 2015). Similarly, other human clinical trials showed treatment with *L. rhamnosus* or *L. fermentum* via vaginal route in women decreases the reoccurrence rate of urinary tract infections (UTIs). Possible mechanisms could be the secretion of bacteriocin, biosurfactant barrier and hydrogen peroxide, and adhesive property at the sites of uropathogenic bacteria (Falagas et al. 2006, Akgül and Karakan 2018).

Moreover, probiotics modify the immune system of the patient, thereby elevating the immune defense mechanism and pathogenic response. This immunomodulatory property acts via a synergistic effect with traditional antibiotics, therefore increasing the antagonistic effect against infectious bacteria and decreasing the drug-resistant bacterial potential (Eslami et al. 2020).

Probiotic given in combination with traditional antibiotics has been shown to overcome adverse effects due to antibiotics, like AAD, and restoration of traditional antibiotic-associated dysbiosis. Probiotics synergistic with traditional antibiotics reduce the mouth and vaginal fungal infection, side effect associated with antibiotic. They do so by maintaining the pH and natural flora of mouth and vagina therefore combating the adverse effects of traditional antibiotics (Pendharkar et al. 2023). GIT side effects like bloating, flatulence, and nausea associated with antibiotics are also found to be relieved when given in combination with probiotics (Shafaghi et al. 2016).

8.5 CLINICAL APPLICATIONS

A study has shown an almost complete curative effect on UTIs caused by *Pseudomonas aeruginosa* and *Klebsiella pneumoniae* in children treated with probiotics in combination with traditional antibiotic prophylaxis. Clinical trial revealed alleviated reoccurrence of UTI on treatment with *Lactobacillus rhamnosus* or *L. fermentum* via vaginal route in women (New et al. 2022, Beyitler and Kavukcu 2017).

In other case studies, significantly decreased incidences of UTI in women were observed when treated with suppositories containing strains of *Lactobacillus crispatus*, with no adverse effects (Shoureshi, Niino, and Eilber 2023).

As mentioned above, pathogens have developed drug-resistance ability, so the same case has happened with *Helicobacter pylori*-associated infection showing resistance toward traditional antibiotics like macrolide and some quinolones. However, a study has shown a remarkable positive result when probiotics are given in synergism with standard treatment, minimizing the adverse effects (Chen et al. 2019).

Different studies have reported that AAD has been successfully overcome by using probiotics in combination with traditional antibiotics. AAD is mostly associated with broad-spectrum antibiotics like clindamycin, beta-lactams, and third-generation cephalosporins. However, probiotics of specific strains like *Lactobacillus rhamnosus* and *Saccharomyces boulardii* are to be used to get the maximum results with reduced adverse complications (Kopacz and Phadtare 2022).

Traditional antibiotics have been associated with dysbiosis, as stated above, so clinical trials have shown significant results in mitigating gut microbiota imbalance, and restoring and promoting the growth of beneficial microorganisms was evident. This reduces the risk of long-term gastrointestinal issues and other infections (Vitetta et al. 2014).

Traditional antibiotics are associated with impaired function of the immune system, resulting in frequent attacks by other pathogenic microbes. Co-administration with probiotics modulates and enhances the immune function in the human body, combating infection more effectively and potentiating the efficacy of antibiotics. It is also useful in cancer treatment as an immunotherapy (Singh, Alexander, and Martin 2023).

Traditional antibiotics have been associated with fungal infection of mouth and vagina as adverse effect, and reduction in the microbiome resulting in candida infection. Successful treatment was attained when probiotic combined with conventional antibiotics was given, restoring pH and microbiota balance and increasing their colonization (Wu et al. 2022). Another study showed significant treatment of vaginal infection of bacterial vaginosis and recurrent vulvovaginal candidiasis in women with probiotics (*Lactobacillus rhamnosus* and *Lactobacillus gasseri*) combined with traditional antibiotics or anti-fungal (Russo et al. 2019).

In one of the clinical studies, antimicrobial-resistant *Pseudomonas aeruginosa* from burn wounds has shown a better inhibitory effect against traditional antibiotics when combined with probiotics like *L. salivarius* (Oryan et al. 2018). Another clinical application revealed that the topical use of probiotics like *S. cerevisiae* in addition to traditional antibiotics increased the healing effect of burn wounds (Gaziano et al. 2020).

Probiotics have revealed attractive results in treating inflammatory bowel disease (IBD) and associated gut issues like bloating, nausea, etc. Particularly for pouchitis and ulcerative colitis, probiotics-antibiotic therapy has been quite appraisable in suppressing the disease by enhancing the gut microbiome environment (Santacroce, Charitos, and Bottalico 2019).

One placebo study has proved positive results in obese post-menopausal women by altering their abnormal glucose level, lipid profile, waist circumference, visceral fat, and serum uric acid level (Szulińska et al. 2018).

Probiotics have also been found to be effective in post-operative infection and have promising results from the number of trials that were conducted. *Pediococcus pentosaceus, Leuconostoc mesenteroides, L. paracasei ssp. paracasei, L. plantarum,* along with fibers (inulin, oat bran, pectin, starch) when given in patients post-surgery showed promising efficacy results against infection, however, the usage varies on the type of surgery (Li et al. 2019).

Overall, the synergistic strategy of probiotics and traditional antibiotics in biomedicine is proving to be among the first and effective alternatives in combating and counteracting pathogenic microorganisms. Adding to the efficacy, minimizing the adverse effects, and effectiveness against drug-resistant bacteria, all these properties make this approach very useful and efficacious in medicine.

8.6 CHALLENGES AND CONSIDERATIONS

Probiotics are found to be highly strain-specific in terms of their action. The effect on specific bacteria depends on the sensitivity of particular probiotic strains against them. This property specificity among probiotic strains is because of the vast differences in genomic makeup among them, thereby varying their mechanisms of action, susceptibility against particular microorganisms, and interface with the host. As the usage and clinical application is growing among many countries, choosing the right probiotics for a particular bacterial infection is quite a tedious and challenging task for both medical personnel and patients. Not all probiotics are effective and useful for all pathogenic microbes; selectivity of specific probiotic strains for particular infectious bacteria is needed. Probiotics are both strain-specific as well disease-specific, which if ignored, leads to harmful and lethal consequences on human health. Use of the wrong probiotics for any particular disease or infection causes severe side effects or unwanted outcomes and even affects the other antibiotics or medication negatively; this makes it crucial to take the right probiotics for any infection or disease to ensure safety and effectiveness (Sniffen et al. 2018).

Choosing the probiotics that are not the choice for particular infections or diseases have a reverse impact on human health like probiotic-associated dysbiosis, GIT complications, immune malfunction, risk of developing drug-resistant bacteria, interference in traditional antibiotics or other medication, and even psychological impact.

Probiotics of selective strains of *Lactobacillus rhamnosus* and *Bifidobacterium infantis* have shown to be noteworthy in their sensitivity against AAD and traveler's diarrhea. It also proved to be effective in treating irritable bowel syndrome (IBS) symptoms, including bloating, flatulence, and abdominal pain (McFarland, Karakan, and Karatas, 2021).

Selective strains of *Lactobacillus casei Shirota* and *Bifidobacterium animalis subsp. Lactis* have been known to boost immune system responses, specifically in enhancing the activity of NK cells, which are the main component in infection eradication, including respiratory and against tumor (Hori, Matsuda, and Oishi 2020).

Strain-specific *Lactobacillus reuteri* and *L. rhamnosus* proved to be a good choice against respiratory tract infections, including flu and cold, mitigating their effect in both children and adults (Pico-Monllor et al. 2021). Various strains have proven useful in treating infections, diseases, or disorders and are mentioned in Table 8.1.

Appropriate and adequate doses and formulations are requisite for the synergism and maximum efficacy of probiotics. Several studies have shown that probiotics do possess dose-dependent response and also formulation-dependent efficacy (Sniffen et al. 2018). Probiotics must be formulated in such a way to protect them from destruction due to gastric attack and bile destruction. Similarly, appropriate doses for specific probiotics are important for effective results for specific

TABLE 8.1
Some Selected Strains Proved Useful in Infections/Diseases

Probiotics	Indications	References
Bifidobacterium infantis, L. acidophilus, Bac. coagulans, L. plantarum, L. salivarius, S. boulardii	Irritable bowel syndrome	Ritchie and Romanuk (2012)
L casei, L bulgaricus, and *S thermophilus*	Antibiotic-associated diarrhea	
Saccharomyces boulardii and *L. rhamnosus*	Traveler's diarrhea	Aponte, Murru, and Shoukat (2020)
S. boulardii and *Lactobacillus johnsonii*	*H. pylori* infection	Ritchie and Romanuk (2012)
Lactobacillus paracasei, L. casei	Respiratory infection	Aponte, Murru, and Shoukat (2020)
Lactobacillus crispatus, L. acidophilus, L. rhamnosus, Bifidobacterium bifidum, and *Bifidobacterium lactis*	UTI	
S. cerevisiae, Clostridium butyricum,	Bacterial vaginosis	
L. rhamnosus, L. acidophilus, L. casei, B. bifidum, and *Lactobacillus fermentum*	Immunity disorders	
L. casei, L acidophilus, L. crispatus B. bifidum	Anti-cancer effects	
B. longum, B. breve, and *B. animalis, B. bifidum*	Obesity	
B. longum, L. acidophilus lysates, C. butyricum, some species of Streptococcus	Neurodegenerative disorders	

TABLE 8.2
Formulations/Dosage Forms of Different Probiotics

Probiotics	Dose	Formulations
Lactobacillus reuteri	100–200 million CFU	Chewable tablets, drops, lozenges
Streptococcus thermophilus	10 billion CFU	Yogurt
Bifidobacterium lactis	Up to 3.5 billion CFU	Capsules, powder
L. salivarius	5 billion CFU	Powder
Saccharomyces boulardii	250 mg	Powder
Sterptococcus thermophilus	10 billion CFU	Yogurt
B. infantis	1 billion CFU	Capsule
L. sporogenes	3 billion CFU	Chewable tablets
B. animalis	1.5 billion CFU	Capsules
L. rhamnosus	Less than 0.5–3.5 billion CFU	Capsules
L. plantarum	3 billion CFU	Chewable tablets

complications having a specific threshold. One meta-analysis has proven that AAD treatment with probiotic therapy among children showed varied results upon varying the dose. Studies have shown a specific dose of probiotics is effective against IBS (10–20 billion CFUs per day of a specific strain) (Goldenberg et al. 2017). A list of probiotics, along with dose and the formulations, is given in Table 8.2.

As the diversity and usage of probiotics have spread throughout the countries for their therapeutic benefits, the concern for safety and proper monitoring remains an issue. Apart from using them as dietary supplements, probiotics are also used as therapy in medical complications, which leads to differences in regulatory rules across the world. The U.S. FDA has skipped the pre-market approval requirements for probiotics used as supplements, making it a concern about the safety, efficacy, viability, and efficacy of probiotics and label claiming (Spacova et al. 2023). Additional safety concerns include the compatibility of probiotic strains with traditional antibiotics or other medications, the risk of serious complications in comorbid patients (such as those who are immunocompromised), and the potential for probiotics to harbor harmful microbes. Furthermore, there is a lack of standardized, precise data and comprehensive safety evaluations, which limits our understanding of the optimal use of probiotics. Key gaps include their role as a synergistic therapy with antibiotics, the selection of specific strains for particular conditions, and the long-term effects of probiotic use. Keeping in view all the complications and concerns related to the use and introduction of probiotics in the medical field, several clinical trials, meta-analyses, and research are underway to evaluate the accuracy, safety, and best possible benefits of probiotics. This advancement in probiotics can lead to novel, innovative approaches to deal with the different conditions for a better and healthier society.

8.7 FUTURE DIRECTIONS

The synergistic approach of probiotics and traditional antibiotics is a vast and evolving area of research in the medical field and is going to be the focal point of the future. It has the potential of serving the healthcare in a very efficient and promising way, increasing the therapeutic efficacy not only against infectious diseases but in other complications too with diminished side effects.

Development in more precise and enhanced microbiome mapping and molecular genomics of probiotics will pave the way for new advancement in personalized probiotic and traditional antibiotics synergistic combinations. This strategy will enhance the targeted health issues with better patient compliance and decrease side effects.

Evaluation of the precise selectivity of probiotics in combination with traditional antibiotics is needed in future research to target the drug-resistant microorganisms and to inhibit the pathogens in developing resistance against drugs. The prevention of dysbiosis and traditional antibiotic-related GIT issues by selecting more specific probiotic strains will be a remarkable feat in health sciences. This approach will restore gut flora after antibiotic use.

Probiotic-antibiotic synergy has the scope for the development of new targeted delivery systems like nanotechnology studies, microencapsulation, and allowing the medication to be more efficacious and risk-free.

The use of probiotics and prebiotics (synbiotics) in combination with traditional antibiotics has a scope to further augment the therapeutic effect and elevate the recovery of gut microbiota and boost the patient recovery. Also, the synergistic approach of probiotic and traditional antibiotics is being investigated for complications other than GIT, like UTI, respiratory infection, and even therapy for cancer treatment as well (Balta et al. 2021). This innovative research could enhance the overall health sector for better care.

The development of age- and dose-specific regimens of probiotics in combination with traditional antibiotics is another impressive novel achievement in this area of research. Children and infants are more prone to antibiotic-related issues, and the introduction of probiotics at specific doses could be a boost for improving such complications (Sniffen et al. 2018). However, there is still a scope for more precise and innovative techniques that are still being developed.

More research and studies on a mass scale are the need of the hour and are even going on at present, establishing more accurate information related to efficacy, compatibility of specific probiotics with antibiotics, adverse effects, appropriate dose formulation, etc. All such studies will become a baseline and ease the treatment of medical conditions properly and efficiently.

The growing popularity of probiotics, along with their synergistic approach, indicates the potential for their therapeutic use. This will attract the researchers to evaluate more promising results and offer promising health benefits in patients and address the global challenges like drug resistance.

8.8 CONCLUSION

According to all studies, it can be concluded that the synergistic approach of probiotics and traditional antibiotics is truly proving to be a rare foot in the field of health sciences and technology. Most of the research has shown promising results with positive feedback. Activity against infectious diseases has been appraised so far using the synergistic approach of probiotics-antibiotics, whether it's against AAD or gut issues. Still, there is a scale for improvement in the approach to additional innovations, like specificity in strains, combinations, patient compliance, minimizing side effects, etc. The extraction of bioactive substances from probiotics and evaluating their potency could be yet another scope to research, which could be another way of resolving the health issues. Finally, this novel approach is a boost for medical science but still needs tremendous and dedicated research to prove wider and efficient benefits.

REFERENCES

Acharjee, M., Hasan, F., Islam, T., Nur, I.T., Begum, N., Mazumder, C., Lubna, M.A., Zerin, N., Shahriar, A. and Mahmud, M.R., 2022. In-vitro antibacterial activity of commercially available probiotics on food-borne pathogens along with their synergistic effects with synthetic drugs. *Metabolism Open*, *14*, p. 100187.

Akgül, T. and Karakan, T., 2018. The role of probiotics in women with recurrent urinary tract infections. *Turkish Journal of Urology*, *44*(5), p. 377.

Aponte, M., Murru, N. and Shoukat, M., 2020. Therapeutic, prophylactic, and functional use of probiotics: A current perspective. *Frontiers in Microbiology*, *11*, p. 562048.

Azad, M. A. K., Sarker, M. and Wan, D., 2018. Immunomodulatory effects of probiotics on cytokine profiles. *BioMed Research International*, *2018*(1), p. 8063647.

Balta, I., Butucel, E., Mohylyuk, V., Criste, A., Dezmirean, D.S., Stef, L., Pet, I. and Corcionivoschi, N., 2021. Novel insights into the role of probiotics in respiratory infections, allergies, cancer, and neurological abnormalities. *Diseases*, *9*(3), p. 60.

Batta, K., Thakur, M. and Meghwal, M., 2024. Vitamins production from probiotic bacteria. In S. A. Ashraf and M. Kuddus (Eds.), *Microbial Vitamins and Carotenoids in Food Biotechnology* (pp. 149–177). Academic Press.

Beyitler, I. and Kavukcu, S., 2017. Probiotics for prophylaxis and treatment of urinary tract infections in children. *Iranian Journal of Pediatrics*, 27(2), p. e7695. https://doi.org/10.5812/ijp.7695

Bhowmick, A., Oishi, T.S. and Aishy, R.I., 2022. Current antibiotic-resistant crisis and initiatives to combat antimicrobial resistance: A review from global perspective (*Doctoral dissertation*, Brac University).

Chen, Y.-H., Tsai, W. H., Wu, H.-Y., Chen, C. Y., Yeh, W. L., Chen, Y. H., Hsu, H.Y., et al. 2019. Probiotic Lactobacillus spp. act against Helicobacter pylori-induced inflammation. *Journal of Clinical Medicine*, 8(1), p. 90.

Chinemerem Nwobodo, D., Ugwu, M.C., Oliseloke Anie, C., Al-Ouqaili, M.T., Chinedu Ikem, J., Victor Chigozie, U. and Saki, M., 2022. Antibiotic resistance: The challenges and some emerging strategies for tackling a global menace. *Journal of Clinical Laboratory Analysis*, *36*(9), p. e24655.

Elshaghabee, F.M. and Rokana, N., 2022. Mitigation of antibiotic resistance using probiotics, prebiotics and synbiotics. A review. *Environmental Chemistry Letters*, *20*(2), pp. 1295–1308.

Eslami, M., Bahar, A., Keikha, M., Karbalaei, M., Kobyliak, N. M. and Yousefi, B., 2020. Probiotics function and modulation of the immune system in allergic diseases. *Allergologia et Immunopathologia*, *48*(6), pp. 771–788.

Falagas, M. E., Betsi, G. I., Tokas, T., and Athanasiou, S., 2006. Probiotics for prevention of recurrent urinary tract infections in women: A review of the evidence from microbiological and clinical studies. *Drugs*, *66*, pp. 1253–1261.

Fernández, J., Redondo-Blanco, S., Gutiérrez-del-Río, I., Miguélez, E.M., Villar, C.J. and Lombo, F., 2016. Colon microbiota fermentation of dietary prebiotics towards short-chain fatty acids and their roles as anti-inflammatory and antitumour agents: A review. *Journal of Functional Foods*, *25*, pp. 511–522.

Galdeano, C.M., De Moreno De Leblanc, A., Vinderola, G., Bonet, M.B. and Perdigón, G., 2007. Proposed model: Mechanisms of immunomodulation induced by probiotic bacteria. *Clinical and Vaccine Immunology*, *14*(5), pp. 485–492.

Gaziano, R., Sabbatini, S., Roselletti, E., Perito, S. and Monari, C., 2020. Saccharomyces cerevisiae-based probiotics as novel antimicrobial agents to prevent and treat vaginal infections. *Frontiers in Microbiology*, *11*, p. 718.

Goldenberg, J. Z., Yap, C., Lytvyn, L., Lo, C. K. F., Beardsley, J., Mertz, D. and Johnston, B. C., 2017. Probiotics for the prevention of Clostridium difficile-associated diarrhea in adults and children. *Cochrane Database of Systematic Reviews*, *12*(12), p. CD006095. https://doi.org/10.1002/14651858.CD006095.pub4

Hori, T., Matsuda, K. and Oishi K., 2020. Probiotics: A dietary factor to modulate the gut microbiome, host immune system, and gut–brain interaction. *Microorganisms*, *8*(9), p. 1401.

Hutchings, M. I., Truman, A. W. and Wilkinson, B., 2019. Antibiotics: Past, present and future. *Current Opinion in Microbiology*, *51*, pp. 72–80.

Javanshir, N., Hosseini, G.N.G., Sadeghi, M., Esmaeili, R., Satarikia, F., Ahmadian, G. and Allahyari, N., 2021. Evaluation of the function of probiotics, emphasizing the role of their binding to the intestinal epithelium in the stability and their effects on the immune system. *Biological Procedures Online*, *23*, pp. 1–17.

Ji, J. and Yang, H., 2020. Using probiotics as supplementation for Helicobacter pylori antibiotic therapy. *International Journal of Molecular Sciences*, *21*(3), p. 1136.

Kang, M.J., Baek, K.R., Lee, Y.R., Kim, G.H. and Seo, S.O., 2022. Production of vitamin K by wild-type and engineered microorganisms. *Microorganisms*, *10*(3), p. 554.

Kesavelu, D. and Jog, P., 2023. Current understanding of antibiotic-associated dysbiosis and approaches for its management. *Therapeutic Advances in Infectious Disease*, *10*, p. 20499361231154443.

Kopacz, K. and Phadtare, S., 2022. Probiotics for the prevention of antibiotic-associated diarrhea. *Healthcare*, *10*(8), p. 1450.

Korpela, K., Salonen, A., Virta, L.J., Kekkonen, R.A., Forslund, K., Bork, P. and De Vos, W.M., 2016. Intestinal microbiome is related to lifetime antibiotic use in Finnish pre-school children. *Nature Communications*, *7*(1), p. 10410.

Kwoji, I. D., Aiyegoro, O. A., Okpeku, M. and Adeleke, M. A., 2021. Multi-strain probiotics: Synergy among isolates enhances biological activities. *Biology*, *10*(4), p. 322.

Li, C., Niu, Z., Zou, M., Liu, S., Wang, M., Gu, X., Lu, H., Tian, H. and Jha, R., 2020. Probiotics, prebiotics, and synbiotics regulate the intestinal microbiota differentially and restore the relative abundance of specific gut microorganisms. *Journal of Dairy Science*, *103*(7), pp. 5816–5829.

Li ,C., Liu, L., Gao, Z., Zhang, J., Chen, H., Ma, S., Liu, A., Mo, M., Wu, C., Chen, D., Liu, S., Xie, J., Huang, Y., Qiu, H., Yang, Y., 2021. Synbiotic Therapy Prevents Nosocomial Infection in Critically Ill Adult Patients: A Systematic Review and Network Meta-Analysis of Randomized Controlled Trials Based on a Bayesian Framework. Front Med (Lausanne). Jul 15; *8* (693188). doi: 10.3389/fmed.2021.693188.

Madden-Fuentes, R. J., Arshad, M., Ross, S. S. and Seed, P. C., 2015. Efficacy of fluoroquinolone/probiotic combination therapy for recurrent urinary tract infection in children: A retrospective analysis. *Clinical Therapeutics*, *37*(9), pp. 2143–2147.

Maier, L., Goemans, C. V., Wirbel, J., Kuhn, M., Eberl, C., Pruteanu, M., Müller, P., et al., 2021. Unravelling the collateral damage of antibiotics on gut bacteria. *Nature*, *599*(7883), 120–124.

Markowiak-Kopeć, P. and Śliżewska, K., 2020. The effect of probiotics on the production of short-chain fatty acids by human intestinal microbiome. *Nutrients*, *12*(4), p. 1107.

McDonnell, L., Gilkes, A., Ashworth, M., Rowland, V., Harries, T.H., Armstrong, D. and White, P., 2021. Association between antibiotics and gut microbiome dysbiosis in children: Systematic review and meta-analysis. *Gut Microbes*, *13*(1), p. 1870402.

McFarland, L.V., 2017. Common organisms and probiotics: Saccharomyces boulardii. In M. H. Floch, Y. Ringel and W. A. Walker (Eds.), *The Microbiota in Gastrointestinal Pathophysiology* (pp. 145–164). Academic Press.

McFarland, L.V., Karakan, T. and Karatas, A., 2021. Strain-specific and outcome-specific efficacy of probiotics for the treatment of irritable bowel syndrome: A systematic review and meta-analysis. *EClinicalMedicine*, *41*, p. 101154. https://doi.org/10.1016/j.eclinm.2021.101154

Moghadam, S.S., Khodaii, Z., Zadeh, S.F., Ghooshchian, M., Aghmiyuni, Z.F. and Shabestari, T.M., 2018. Synergistic or antagonistic effects of probiotics and antibiotics-alone or in combination-on antimicrobial-resistant Pseudomonas aeruginosa isolated from burn wounds. *Archives of Clinical Infectious Diseases*, *13*(3), p. e63121. https://doi.org/10.5812/archcid.63121

New, F.J., Theivendrampillai, S., Juliebø-Jones, P. and Somani, B., 2022. Role of probiotics for recurrent UTIs in the twenty-first century: A systematic review of literature. *Current Urology Reports*, *23*(2), pp. 19–28.

Oryan, A., Jalili, M., Kamali, A. and Nikahval, B., 2018. The concurrent use of probiotic microorganism and collagen hydrogel/scaffold enhances burn wound healing: An in vivo evaluation. *Burns*, *44*(7), pp. 1775–1786.

Patangia, D.V., Anthony Ryan, C., Dempsey, E., Paul Ross, R. and Stanton, C., 2022. Impact of antibiotics on the human microbiome and consequences for host health. *Microbiologyopen*, *11*(1), p. e1260.

Pendharkar, S., Skafte-Holm, A., Simsek, G. and Haahr, T., 2023. Lactobacilli and their probiotic effects in the vagina of reproductive age women. *Microorganisms*, *11*(3), 636.

Pico-Monllor, J.A., Ruzafa-Costas, B., Nunez-Delegido, E., Sanchez-Pellicer, P., Peris-Berraco, J. and Navarro-Lopez, V., 2021. Selection of probiotics in the prevention of respiratory tract infections and their impact on occupational health: Scoping review. *Nutrients*, *13*(12), p. 4419.

Potrykus, M., Czaja-Stolc, S., Stankiewicz, M., Kaska, Ł. and Małgorzewicz, S., 2021. Intestinal microbiota as a contributor to chronic inflammation and its potential modifications. *Nutrients*, *13*(11), p. 3839.

Raheem, A., Liang, L., Zhang, G. and Cui, S., 2021. Modulatory effects of probiotics during pathogenic infections with emphasis on immune regulation. *Frontiers in Immunology*, *12*, p. 616713.

Reid, G. 2016. Probiotics: Definition, scope and mechanisms of action. *Best Practice & Research Clinical Gastroenterology*, *30*(1), 17–25.

Ritchie, M.L. and Romanuk, T.N., 2012. A meta-analysis of probiotic efficacy for gastrointestinal diseases. *PLoS One*, *7*(4), p. e34938.

Russo, R., Superti, F., Karadja, E. and De Seta, F., 2019. Randomised clinical trial in women with recurrent Vulvovaginal Candidiasis: Efficacy of probiotics and lactoferrin as maintenance treatment. *Mycoses*, *62*(4), pp. 328–335.

Sagar, S., Kaistha, S., Das, A. J. and Kumar, R., 2019. *Antibiotic Resistant Bacteria: A Challenge to Modern Medicine*. Springer Singapore.

Santacroce, L., Charitos, I. A. and Bottalico, L., 2019. A successful history: Probiotics and their potential as antimicrobials. *Expert Review of Anti-Infective Therapy*, *17*(8), pp. 635–645.

Shafaghi, A., Pourkazemi, A., Khosravani, M., Asl, S.F., Maafi, A.A., Roshan, Z.A. and Rahimabad, J.A., 2016. The effect of probiotic plus prebiotic supplementation on the tolerance and efficacy of Helicobacter pylori eradication quadruple therapy: A randomized prospective double blind controlled trial. *Middle East Journal of Digestive Diseases*, *8*(3), p. 179.

Shoureshi, P. S., Niino, C. and Eilber, K. S., 2023. Can vaginal lactobacillus suppositories help reduce urinary tract infections? *International Urogynecology Journal*, *34*(11), pp. 2713–2718.

Singh, A., Alexander, S. G. and Martin, S., 2023. Gut microbiome homeostasis and the future of probiotics in cancer immunotherapy. *Frontiers in Immunology*, *14*, 1114499.

Sniffen, J. C., McFarland, L. V., Evans, C. T. and Goldstein, E. J. C., 2018. Choosing an appropriate probiotic product for your patient: An evidence-based practical guide. *PLoS One*, *13*(12), e0209205.

Spacova, I., Binda, S., Haar, J. A. T., Henoud, S., Legrain-Raspaud, S., Dekker, J., Espadaler-Mazo, J. et al., 2023. Comparing technology and regulatory landscape of probiotics as food, dietary supplements and live biotherapeutics. *Frontiers in Microbiology*, *14*, p. 1272754.

Stavropoulou, E., and Bezirtzoglou, E., 2020. Probiotics in medicine: A long debate. *Frontiers in Immunology*, *11*, 2192.

Szulińska, M., Łoniewski, I., Van Hemert, S., Sobieska, M. and Bogdański, P., 2018. Dose-dependent effects of multispecies probiotic supplementation on the lipopolysaccharide (LPS) level and cardiometabolic profile in obese postmenopausal women: A 12-week randomized clinical trial. *Nutrients*, *10*(6), p. 773.

Vitetta, L., Briskey, D., Alford, H., Hall, S. and Coulson, S., 2014. Probiotics, prebiotics and the gastrointestinal tract in health and disease. *Inflammopharmacology*, *22*, pp.135–154.

Wu, Y., Hu, S., Wu, C., Gu, F. and Yang, Y., 2022. Probiotics: Potential novel therapeutics against fungal infections. *Frontiers in Cellular and Infection Microbiology*, *11*, p. 793419.

Yassour, M., Vatanen, T., Siljander, H., Hämäläinen, A.-M., Härkönen, T., Ryhänen, S. J., Franzosa, E. A. et al., 2016. Natural history of the infant gut microbiome and impact of antibiotic treatment on bacterial strain diversity and stability. *Science Translational Medicine*, 8(343), pp. 343ra81.

Yousefi, B., Eslami, M., Ghasemian, A., Kokhaei, P., Salek Farrokhi, A. and Darabi, N., 2019. Probiotics importance and their immunomodulatory properties. *Journal of Cellular Physiology*, *234*(6), pp. 8008–8018.

9 Probiotics as Biotechnological Tools in Personalized Medicine
Opportunities for Africa

Abigarl Ndudzo[1,2], Thubelihle Ngwenya[1], Nicole Kanenungo[1], Simangaliso Mhlanga[1], Kimberly Elliot[1], Senzeni Sibanda[1], Petronella Danga[1], Vigilant Sibanda[1], Sakhile Ndlovu[1], Angela Sibanda Makuvise[3], and Stabbin Maluphaso[1]

[1]Faculty of Environmental and Life Sciences, Department of Biotechnology, Lupane State University, Bulawayo, Zimbabwe

[2]Department of Molecular Biology and Biotechnology, Pan African University Institute of Basic Sciences and Technology, Jomo Kenyatta University of Agriculture and Technology, Juja, Nairobi, Kenya

[3]Faculty of Applied Science, Department of Applied Biology and Biochemistry, National University of Science and Technology, Bulawayo, Zimbabwe

9.1 INTRODUCTION

Probiotics, extensively studied in developed countries, are recognized for conferring health benefits and are commonly found in dairy milk and fermented products (Ukeyima et al., 2010). In Africa, fermented foods are acknowledged as rich sources of beneficial microorganisms, presenting an opportunity to harness probiotics' potential for promoting health (Obafemi et al., 2022). The emergence of probiotics in personalized medicine, particularly in Africa, is facilitated by advancements in genetic engineering tools for probiotic lactic acid bacteria (Mugwanda et al., 2023). Additionally, the escalating concern of antimicrobial resistance (AMR) globally, including in Africa, underscores the need for alternative strategies. Probiotics offer promise in combating infectious diseases and reducing antibiotic reliance by modulating gut microbiota and enhancing host immunity (Murugaiyan et al., 2022; Petrariu et al., 2024; Nyathi et al., 2024). However, the potential of probiotics as biotechnological tools in personalized medicine, specifically in Africa, remains largely unexplored. This review aims to comprehensively examine the current applications of probiotic foods in Africa and explore probiotics' potential as biotechnological tools in personalized medicine, focusing on opportunities and challenges within the African context. By evaluating the impact of intestinal microbiota on health, the unique microbial landscape of African fermented foods, and considering implications for AMR, this review seeks to illuminate the prospects of leveraging probiotics for advancing personalized medicine in Africa.

9.2 LITERATURE REVIEW

9.2.1 Health Care Status in Africa

Africa faces significant challenges in combating infectious and non-infectious diseases, primarily due to socioeconomic factors, low incomes, and inadequate healthcare infrastructure (Obande et al., 2021). Infectious diseases, such as HIV/AIDS, tuberculosis (TB), malaria, and more recently,

DOI: 10.1201/9781032676951-9

COVID-19, pose substantial burdens on Africa's economies and healthcare systems (WHO: HIV & AIDS, 2023; Ojo et al., 2022; Cabore et al., 2022). TB remains a significant challenge in Africa, with approximately 2.5 million new cases and 417,000 deaths reported in 2020. The disease burden is exacerbated by factors such as poor sanitation and overcrowding. (World Economic Forum, 2023). Similarly, malaria remains prevalent in Africa, accounting for the majority of global cases, reflecting widespread impoverishment and limited access to preventive measures (WHO: Malaria, 2023).

Non-infectious diseases, including cancer, hypertension, and diabetes mellitus, also contribute significantly to Africa's disease burden (Rebbeck, 2020). Cancer patterns vary across regions, with breast, bladder, and liver cancers prevalent in Northern Africa, while prostate, lung, and colorectal cancers predominate in Southern Africa (Niohuru, 2023). However, challenges such as limited medical supplies, research funding, and epidemiological knowledge contribute to rising death rates from prostate and cervical cancers in Southern Africa (Sharma et al., 2022). Hypertension, a leading cause of cardiovascular-related mortality globally, affects around 30.8% of Africa's population and is particularly prevalent in rural communities and low-income households (Yuyun et al., 2023). With aging populations, increased urbanization, and lifestyle changes, the prevalence of hypertension is expected to rise further in Africa (Niohuru, 2023).

Diabetes mellitus poses another significant health challenge, with an estimated 24 million undiagnosed cases in Africa, and the burden projected to increase substantially by 2045 (WHO, 2022). Limited access to quality healthcare exacerbates the challenges faced by individuals with non-communicable diseases like diabetes, resulting in delayed diagnosis and increased mortality rates, particularly among younger age groups (IDF, 2017; WHO, 2022). Addressing these challenges requires concerted efforts to strengthen healthcare systems, improve access to essential medical supplies and services, and implement of effective prevention and management strategies tailored to Africa's unique healthcare landscape.

9.2.2 Food Security and Nutrition

Malnutrition and food insecurity are known to perpetuate the effects of both communicable and non-communicable diseases in Africa (Atukunda et al., 2021). Food security is achieved when everyone, at all times, has physical, social, and economic access to enough, safe, and nutritious food that satisfies their dietary needs and food preferences for an active and healthy life (FAO, IFAD, UNICEF, WFP, WHO, 2019; FAO, 2021). The population of Africa was 1.38 billion in 2021 and is projected to rise to 1.71 billion by 2030 (Mohajan, 2021). For leaders, feeding this population while using the same land resources would be the major challenge (Atukunda et al., 2021). Because most people in Africa have poor incomes and opt for cheap, less nutrient dense meals, hinger and malnutritrion are widespread. (United Nations, 2019). Alarmingly, while adult overweight and obesity rates continue to rise across all areas (about 12.8% in Africa), they are increasing for children in African nations (FAO, 2020). This raises the likelihood of diet-related illnesses (United Nations, 2021). Additionally, a number of African nations, including Somalia, South Sudan, Chad, Central African Republic, and the Democratic Republic of the Congo, have seen an increase in hunger and food insecurity due to poor governance, climate change and variability, a lack of land resources, political conflict, wars, and uprisings (FAO, 2020; Otekunrin et al., 2020; Atukunda et al., 2021).

9.2.3 Antimicrobial Resistance

In Africa, AMR has become a growing concern for microorganisms that cause diseases like AIDS, gonorrhea, meningitis, typhoid, and malaria, to name a few (Elton et al., 2020). Infections become more difficult to treat as antimicrobials lose their effectiveness, increasing the risk of disease spread, life-threatening sickness, and death (Nwobodo et al., 2022; Nyathi et al., 2024). AMR increases medical expenses since it causes prolonged illnesses, leads to longer hospital stays, and the need for more expensive medications (WHO: AMR Fact Sheet, 2023).

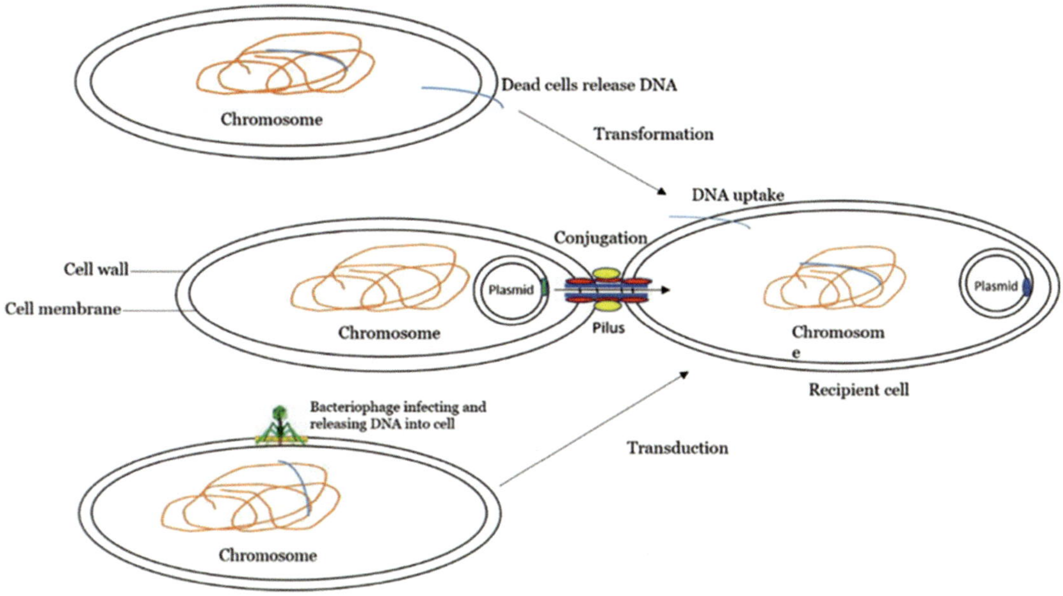

FIGURE 9.1 Mechanisms of horizontal gene transfer. (Adapted from Ndudzo et al., 2022.)

Inappropriate use of antimicrobials in contemporary agricultural practices (where they are employed to prevent diseases and as growth promoters), as well as globalization and poor hygiene, enhance the likelihood that AMR may emerge. AMR develops naturally in different ways (Reygaert, 2018). Horizontal gene transfer, a process in which an organism transfers genetic material to another organism that is not its progeny, can result in AMR. It permits the emergence of fresh variants without causing that variant to mutate. The genetic material enters the cell either through a transposon or a DNA plasmid. Transformation (the absorption of genetic material from a dead organism); transduction (involving the transfer of a DNA fragment from one bacteria to another by a bacteriophage) (Reygaert, 2018); and conjugation (the exchange of genetic material between two living organisms) are the primary ways by which horizontal gene transfer takes place as depicted in Figure 9.1.

Several studies have shown that (Nwobodo et al., 2022; Michaelis & Grohmann, 2023), microorganisms develop resistance to antibiotics through a variety of mechanisms, including intrinsic resistance (drug uptake restriction, active drug efflux) and acquired resistance (drug target modification, acquisition of antibiotic resistance genes like lactamase coding genes from the environment and/or other microorganisms) as shown in Figure 9.2.

These mechanisms have rendered most of the clinically important microorganisms insensitive to antimicrobial control thus putting the medical systems under massive strain.

9.3 PERSONALIZED MEDICINE

Personalized medicine, also known as precision or individualized medicine, tailors treatments to individuals based on their genetics, lifestyle, and environment (Ayivi et al., 2020; Vaz & Kumar, 2021; Eid et al., 2019). This approach aims to replace the conventional one-size-fits-all method, which has often failed to address high death rates, particularly in developing nations (Akil et al., 2021). By delivering unique medications at optimal dosages, personalized medicine seeks to enhance health outcomes (Dash et al., 2019; Parmar et al., 2021). The benefits of personalized medicine include its ability to predict, prevent, and diagnose diseases, provide tailored therapies, increase safety, reduce adverse drug reactions, and lower costs associated with trial-and-error approaches (Akil et al., 2021; Neznanov, 2021). Next-generation sequencing, along with diagnostic

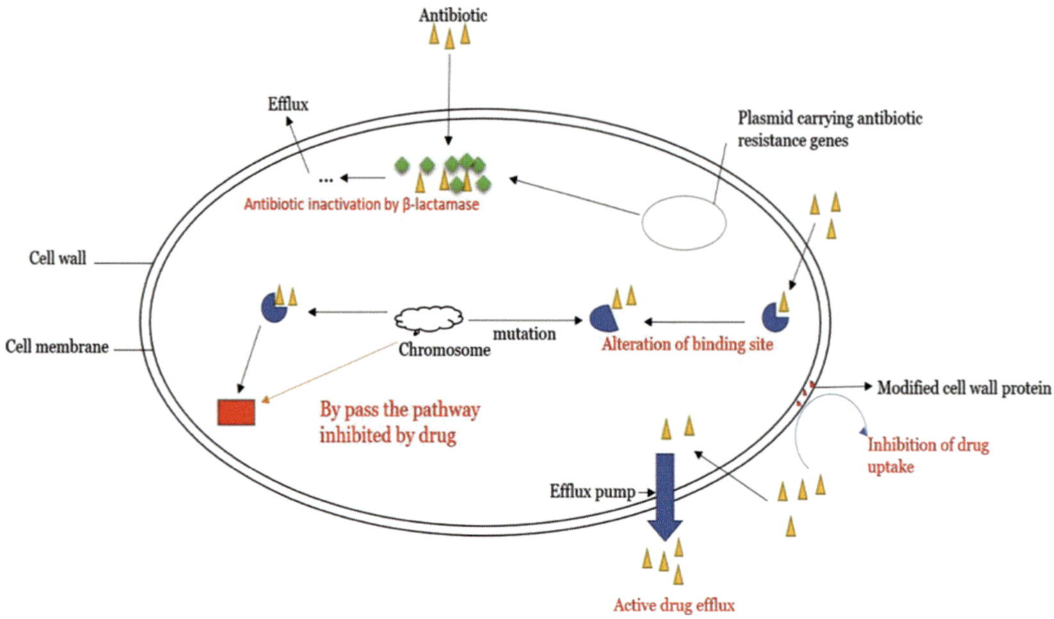

FIGURE 9.2 Mechanisms of antimicrobial resistance. (Adapted from Ndudzo et al., 2022.)

tests and medical history, guides personalized medicine by identifying genetic and microbiological characteristics to develop individualized treatments (Akhoon, 2021; Parmar et al., 2021; Eid et al., 2019). This involves sequencing and analyzing a person's entire genome using various molecular profiling techniques, enabling the identification of genetic mutations influenced by environmental and lifestyle factors (Pravin et al., 2021).

Doctors predict patient responses to medications based on their DNA and biomarkers, facilitating disease identification, prevention, and treatment (Akhoon, 2021). By understanding the genetic makeup of diseases, personalized medicine simplifies diagnosis and aims to deliver optimal care by considering genetic, clinical, and environmental factors (Mukherjee et al., 2022). The transition from conventional to personalized medicine is crucial for considering individual differences (Akil et al., 2021). Personalized medicine has shown efficacy in various areas, including Mendelian disorders, cancer therapy, and diagnostics, where tailored medications and dosages improve treatment outcomes (Dugger et al., 2018).

9.3.1 Status of Personalized Medicine in Africa

Africa relies heavily on traditional medicine, which, though efficient, lacks the accuracy and effectiveness of individualized treatment, posing concerns for the continent's high mortality rates (Pereira et al., 2021). With Africa's diverse population, there is significant potential to contribute to genetic data and reduce disease-related mortality (Chikowore et al., 2021). However, Africa lags behind Europe and Asia in personalized medicine adoption due to limited resources and infrastructure (Mathuba et al., 2021). The existing genetic data, predominantly from European and Asian populations, are not suitable for African populations due to environmental factors influencing treatment efficacy (Lumaka et al., 2022; Chikowore et al., 2021). This disparity highlights the need to address gaps in African population genetics and advance medical research (Pereira et al., 2021). Despite initiatives to involve African nations in global genome data collection, significant disparities remain (Magavern et al., 2021; Wonkam, 2021).

Efforts are underway to integrate African nations into personalized medicine initiatives to bridge these gaps and reduce health disparities (Omotoso et al., 2022). Projects like the EU-Africa PerMed seek to include Africa in the International Consortium for Personalized Medicine, aiming to strengthen bilateral relationships and support African health-related groups (EU+ Africa PerMed, 2021). By involving Africa in personalized medicine initiatives, there is potential to lessen health disparities between industrialized and developing countries (Chakravarthy et al., 2020). Africa's adoption of personalized medicine lags behind other regions due to limited resources and infrastructure. Efforts to include African nations in global genome data collection and personalized medicine initiatives are underway, with the goal of reducing health disparities and advancing medical research on the continent.

9.3.2 The Microbiome

The microbiome encompasses all microorganisms' collective genetic makeup in a specific environment, including symbiotic and pathogenic microbiota found in various ecosystems such as soil, human skin, and the gut (Gilbert et al., 2018; Sinkkonen, 2022). It comprises fungi, viruses, bacteria, and protozoa, playing essential roles in immune regulation, digestion, and protection against harmful microorganisms (Silva et al., 2020; de Vos et al., 2022). Factors like environment and diet influence microbiome transmission from parents to children, impacting health outcomes such as allergies (Gilbert et al., 2018; Bao et al., 2021). Dysbiosis, or microbiome disruption, leads to disease by enabling colonization by pathogenic microorganisms (Gilbert et al., 2018; de Vos et al., 2022). Given its significance in health maintenance, microbiome investigations should be integrated into personalized treatment approaches.

9.3.3 The Significance of the Gut Microbiome in Personalized Medicine

The microbiome serves as a critical biomarker in predicting disease phenotypes and treatment responses in precision medicine, enabling targeted therapeutic interventions (Yan et al., 2022). Disruption in the interactions between microbiota and hosts can lead to various conditions, including cirrhosis, colorectal cancer, and diabetes mellitus (Alsayed et al., 2023; Chadha et al., 2021). Understanding the microbiome's impact on health and therapeutic outcomes is essential in personalized medicine, where treatments are tailored to individual patients based on their unique microbiome and genetic makeup (Pereira et al., 2021; Vijay & Valdes, 2022). This personalized approach considers the gut microbiota's significant influence on medication response and interactions, aiming to maximize efficacy while minimizing side effects (Consortium, 2019; Wu et al., 2021).

The microbiome's role extends beyond disease prediction and treatment response to shaping the effectiveness of medicinal compounds and antibiotic interventions (Sharpton et al., 2021; Alpuim et al., 2021). Antibiotics, while eliminating pathogens, also impact beneficial microorganisms, highlighting the need for tailored therapies that preserve beneficial bacteria while targeting pathogens (Alsayed et al., 2023). By leveraging DNA sequencing, researchers identify a patient's unique microbiome profile, which informs treatment decisions and ensures therapies are compatible with the patient's microbial composition (Chadha et al., 2021; Pereira et al., 2021).

Studying the intricate relationship between personalized medicine and the microbiome is crucial for diagnosing diseases accurately and delivering targeted therapies that selectively target pathogenic bacteria while preserving beneficial ones (Consortium, 2019; McCoubrey et al., 2021). This understanding also facilitates the development of potent probiotics that can restore a healthy microbiome, ultimately improving patient outcomes (Shanahan et al., 2021). Therefore, integrating microbiome analysis into personalized medicine enhances treatment efficacy, minimizes adverse effects, and holds promise for advancing precision healthcare.

9.4 PROBIOTICS IN MEDICINE

9.4.1 Characteristics of Probiotics

Probiotics, including bacteria, yeasts, fungi, and viruses, are essential components of the body's natural microflora, particularly in the intestines where they offer various health benefits (Valdes et al., 2018). These microorganisms, like *Lactobacillus rhamnosus* GG, are identified by their genus, species, and strain, as each strain may confer different health advantages (Sanders et al., 2018; Swanson et al., 2020). Common probiotic genera include *Lactobacillus*, *Bifidobacterium*, and *Saccharomyces*, among others (Das et al., 2022).

Probiotics must possess certain characteristics to be effective and safe for use:

- They should colonize and spread in their intended location to be tolerated by the immune system (Gao et al., 2022).
- Probiotics must be non-pathogenic, non-allergenic, and non-mutagenic/carcinogenic, meeting the "Generally Regarded As Safe" (GRAS) criteria (Gao et al., 2022).
- They should thrive in real-world conditions, including low pH and high bile acid concentrations (Sanders et al., 2018).
- Probiotics should maintain the sensory qualities of food and perform well during food processing (Swanson et al., 2020).
- Delivery of probiotics should ensure that they are alive, at an effective dose, and provide health benefits (Das et al., 2022).

These characteristics are crucial for ensuring the efficacy and safety of probiotics in various applications, including food and medicine.

9.4.2 Sources of Probiotics

Probiotics are present in various forms including food, dietary supplements, and pharmaceuticals, with dietary sources like fermented foods providing additional health benefits beyond their nutritional value (Vera-Santander et al., 2023; Sanders et al., 2018). Fermented foods contain live microorganisms that support gastrointestinal and overall health, enhancing vitamin levels and inactivating toxic substances (Valdes et al., 2018; Vera-Santander et al., 2023).

Lactobacillus and *Bifidobacterium* species are commonly found in fermented foods, with *Lactobacillus* being fermentative, Gram-positive spore-forming bacteria that aid in nutrient metabolism (Ayivi et al., 2020). Anaerobic *Bifidobacterium* species in the large intestine produce health-promoting short-chain fatty acids and vitamins (Khushboo & Malik, 2023). *Streptococcus thermophilus* and *Lactobacillus bulgaricus* in yogurt metabolize lactose and exhibit antibacterial activity (Ayivi et al., 2020). *Saccharomyces boulardii*, non-colonizing yeast, secretes compounds that break down bacterial toxins and boost vitamin synthesis (Moré & Vandenplas, 2018).

Probiotic supplements are available in various forms such as tablets, powders, capsules, and liquids, offering an alternative for probiotics that cannot withstand food processing (Wang et al., 2022; Ayivi et al., 2020). Regular consumption of probiotic-rich foods and prebiotics like fiber supports a healthy digestive system (Singh et al., 2022). Overall, probiotics provide various health benefits through their diverse mechanisms of action as shown in Table 9.1.

9.4.3 African Fermented Foods as a Source of Probiotics

African fermented foods are a rich source of probiotics, containing abundant potential probiotic microorganisms such as lactic acid bacteria (LAB), *Saccharomyces*, and *Bacillus* as shown in Table 9.2. These microorganisms produce health-promoting compounds, such as antimicrobials, essential nutrients, molecules with antioxidant activity, and probiotic strains. Traditional African

TABLE 9.1
Health Benefits of Probiotics

Health Benefits	Mode of Action	References
Maintains a proper human microflora balance	Preserving the equilibrium between bacteria and pathogens that is required for an organism to operate normally	Ukeyima et al. (2010)
	Restoration of natural microbiota following antibiotic-induced dysbiosis	
Inhibits growth of pathogenic organisms	Preventing food poisoning by limiting the activity of pathogenic intestinal microbiota acquired from tainted food and the environment, like *Clostridium perfringens*, *Salmonella enteritidis*, and *Escherichia coli*	Feng et al. (2018)
Maintains proper gut physiology	Restores gut composition, prevents gut inflammation, and other intestinal problems which occur as a result of environmental stress such as heat, cold, or noise	Lobionda et al. (2019)
Modulates the immunity system	By interacting with immune-competent cells, lactic acid bacteria in probiotics strengthen the immune system and change how cytokines are produced.	Mazzantini et al. (2021)
Regulation of gut motility	Regulating gut motility promotes communication between the central nervous system and the bacterial flora in the intestines, as the CNS alters the gut's permeability and motility, which in turn influences the microbiota.	Li et al.(2020)
Prevention of oral infections	Lactic acid bacteria interact with dental biofilms or plaques destroy the causative agent with its antimicrobial activity	Lobionda et al. (2019)
Preventing nutrient deficiencies	*Lactobacillus plantarum*, *L. reuteri*, *Bifidobacterium adolescentis*, and *Bifidobacterium pseudocatenulatum* are natural producers of B vitamins	Ukeyima et al. (2010)
	Improve the immune system's performance and the body's ability to absorb vitamins and mineral components in addition to promoting the production of organic acids and amino acids.	Das et al. (2022)
Production of enzymes and cofactors	Enzymes, such as esterase, lipase, and coenzymes A, Q, NAD, and NADP, are produced by gut microbiota	Pravin et al. (2021)
	Probiotic products with antibacterial, anticancer, and immunosuppressive characteristics include acidophiline, bacitracin, and lactacin.	Wonkam (2021)
Treatment of inflammatory bowel disease	Cytokines can be delivered to target areas by lactic acid bacteria and *Escherichia coli*.	Campos et al. (2023)
Prevention of obesity and protection against colon cancer	*E. coli*, *S. aureus*, and *Bifidobacterium* are important non-pathogen inducers of apoptosis in cancer cells.	Dey et al. (2022)
Facilitation of healing process	Probiotics speed up the recovery of infected skin wounds and intestinal ulcers.	de Simone (2019)
Maintenance of a healthy skin	Skin microbiota serves as a protective barrier and controls the skin's inflammatory response to epidermal injury by encouraging the creation of cytokines, which affects the pathogen's direct killing, the relocation of pathogenic bacteria, and the stimulation of fibroblasts.	Mukherjee et al. (2022)
Treatment of diarrhea	Application of *Saccharomyces boulardii* yeast results in cure and reduced frequency of complaints; probiotics increase amount of IgA antibodies resulting in arrest of viral diarrhea; *S. boulardii* and *L. rhamnosus* GG reduce the development of antibiotic-associated diarrhea	Mohajan (2021)
Prenatal immunization	Pregnant women are given probiotic supplements to help the developing fetus avoid autoimmune disorders and syndromes, effectively immunizing the child.	Nwobodo et al. (2022)
Adjuncts to antibiotic therapy	In addition to medicines, *Lactobacillus*-containing probiotics have been shown to improve *Helicobacter pylori* eradication in patients with peptic ulcers.	Idrees et al. (2022)
Probiotics as adjuvants	Since probiotics have antidiabetic and anti-inflammatory characteristics, they are essential for both preventing and treating type II diabetes mellitus.	Ayesha et al. (2023)

TABLE 9.2
Diversity of African Traditional Fermented Foods and Beverages

Substrate	Fermented Product	Microorganisms	Region	References
Maize, sorghum, millet	Ogi, Uji	*Lactobacillus plantarum*, *L. fermentum*, *S. cerevisae*, *Candida krusei*, *Corynebacterium* spp., *Acetobacter* spp., *L. cellobiosus*, *L. buchneri*, *Geotrichum candidum*, *Geotrichum fermentum*, *Rhodotorula graminis*, *Pediococcus acidilactici*	West Africa	Okafor et al. (2018); ShiferwaTerefe and Augustin (2020)
Maize, sorghum	Mahewu (non-alcoholic beverage)	*Lactobacillus bulgaricus*, *L. brevis*, *L. delbrueckii* subsp. *delbrueckii*, *Leuconostoc* spp., Heterofermentative lactobacilli, *Candida kefyr*, *S. cerevisae*	Southern Africa	Gadaga et al. (1999); ShiferwaTerefe and Augustin (2020); Obafemi et al. (2023)
Maize	Munkoyo (chibwantu)	*Lactobacillus* spp., *Weisella*	Southern Africa	Schoustra et al. (2013)
Millet, sorghum	Kunu zaki	*L. fermentum*, *Ped. pentosaceus*, *Weisella confusa*, *Enterococcus faecalis*	West Africa	Ezekiel et al. (2019); Obafemi et al. (2022)
Sorghum	Chibuku (alcoholic beverage)	*Lactobacillus* spp., *S. cerevisae*	Southern Africa	Gadaga et al. (1999); Mokoena et al. (2016)
Maize, sorghum	Umqomboti (alcoholic beverage)	*Lactobacillus* spp., *S. cerevisae*	Southern Africa	Mokoena et al. (2016)
Cassava	Fufu	*Ped. pentosaceus*, *L. fermentum*, *L. plantarum*	West Africa	Obafemi et al. (2022)
Cassava	Agbelima	*L. plantarum*, *L. brevis*, *L. fermentum*, *Leuconostoc mesenteroides*, *Bacillus* spp., *C. tropicalis*, *G. candidum*, *Penicillium* spp.	West Africa	Obafemi et al. (2022)
Cassava	Gari	*L. fermentum*, *L. plantarum*, *W. confusa*, *S. cerevisae*, *Kluyveromyces marxianus*, *Hansenia guilliermondii*, *Leuconostoc fallax*, *Bacillus subtilis*	West Africa	Mokoena et al. (2016)
Cassava, maize, sorghum, millet	Togwa	*Lactobacillus* spp., *Ped. pentosaceus*, *W. confusa*, *Issatchenkia orientalis*, *S. cerevisae*, *Candida pelliculosa*, *C. tropicalis*	East Africa	Mugula et al. (2003); Mokoena et al. (2016)
Milk	Amasi/ Mabisi/Kule naoto/Ergo	*Lactococcus*, *Lactobacillus* and *Streptococcus* spp., *Leuconostoc* spp.	Southern Africa/ East Africa	Doutoum et al. (2013); Nyambane et al. (2014); Jans et al. (2017); Osvik et al. (2013); Obafemi et al. (2022)

(*Continued*)

TABLE 9.2 (*Continued*)
Diversity of African Traditional Fermented Foods and Beverages

Substrate	Fermented Product	Microorganisms	Region	References
Cow Milk	Amabere amaruranu	*Streptococcus thermophilus*, *Lactobacillus plantarum*, *Leuconostoc mesenteroides*, Yeasts, *L. fermentum*, *Lb. brevis*, *Lc. lactis* subsp. *cremoris*, *L. paracasei*, *L. acidophilus*, *Lc. lactis* subsp. *lactis*	Southern Africa	Zheng et al. (2020); Obafemi et al. (2022)
Cow Milk	Leben	*Lc. lactis* subsp. *lactis*, *Lc. lactis* subsp. *lactis biovardiacetylactis*, *Kluyveromyces lactis*, *S. cerevisae*, *E. faecalis*, *L. brevis*, *Leuconostoc mesenteroides* subsp. *dextranicum*	North Africa	Jans et al. (2017); Obafemi et al. (2022)
African locust bean	Iru	*Bacillus*, *Staphylococcus* spp., *Leuconostoc mesenteroides*, *Lactobacillus plantarum*	West Africa	Mokoena et al. (2016)
Amarula fruit	Marula/buganu	Yeasts	Southern Africa	Gadaga et al. (1999); Mokoena et al. (2016)

fermented foods play a significant role in the diet of African populations, contributing to food security, nutrition, and overall health (Mokoena et al., 2016). The probiotic microbes associated with fermented foods are hugely beneficial, degrading antinutritional substances like mycotoxins and phytic acid, improving digestion and the production of nutrients, inhibiting the growth of food spoilers, and preventing and treating diarrhea (Obafemi et al., 2022). The use of probiotics in personalized medicine is an emerging field with significant potential, particularly in the African context.

Several studies have demonstrated the beneficial roles of probiotic bacteria associated with traditional fermented foods such as *L. casei*, *L. plantarum*, *Bifidobacterium bifidum*, and *Bacillus subtilis* (Obafemi et al., 2022; Mgbodile and Nwagu, 2023). Recent research has supported the potential of African fermented foods in the management of viral infections including SARS-CoV-2 cases (Mgbodile and Nwagu, 2023). African fermentation techniques have been practiced for centuries, offering valuable insights into the health benefits of fermented foods and their potential for personalized medicine (Girgis et al., 2022).

9.4.4 Mechanism of Action of Probiotics

Probiotics work through four different processes, which are depicted in Figure 9.3. The mechanisms play a crucial role in prevention, infection therapy, and gut microbiota maintenance for the host.

9.4.4.1 Competition with Pathogens for Nutrients and Adhesion to the Epithelium

Probiotic strains' capacity to cluster together creates a barrier that guards against harmful germs invading the epithelium (Sanders et al., 2018). Probiotic bacteria might be able to stick to epithelial cells and prevent pathogens from entering. Additionally, this has a significant impact on the host's health, setting off a signaling cascade that modulates the immune system (Campos et al., 2023; Wang et al., 2022). This action is crucial for the prevention and treatment of infectious infections as well as for chronic gastrointestinal tract inflammation. Additionally, research posits that probiotics may help eradicate cancer cells (Singh et al., 2022).

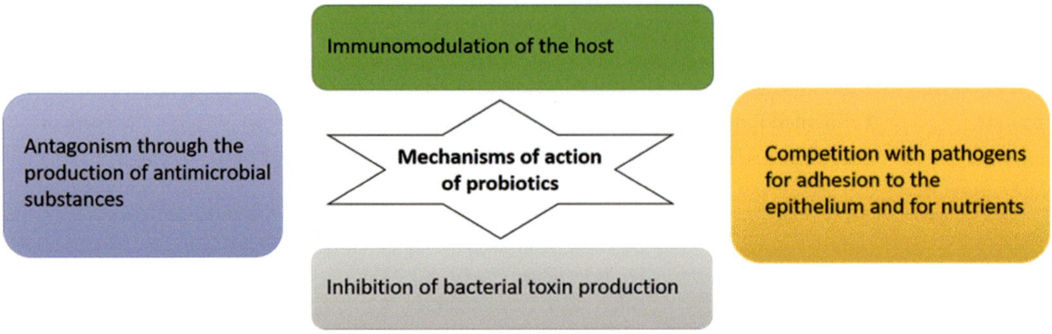

FIGURE 9.3 Mechanisms of action of probiotics.

9.4.4.2 Antagonism through Production of Antimicrobial Substances

Deconjugated bile acids, which are derivatives of bile acids, are produced by probiotic bacteria and have a higher antibacterial impact than the bile salts made by their host (Prete et al., 2020).

9.4.4.3 Immunomodulation of the Host

The metabolic processes of the host are significantly influenced by the gut bacteria. Recent metagenomic studies have shown that they have an impact on the development and physiology of the host and are engaged in the control of the host immune system (Terpou et al., 2019). Specific species and strain characteristics, including cellular structure, cell surface size, metabolic characteristics, and chemicals released by microbes, play a significant impact in how well probiotics work. Three mechanisms underlie the gut microbiota's ability to modulate the immune system, including probiotic bacteria (Prete et al., 2020). These include the development and maintenance of the immune system's capacity to tolerate environmental antigens, the production, and management of immune responses against bacterial and viral infections, and the inhibition of auto-aggressive and allergic responses (Wiertsema et al., 2021). Increased production of immunoglobulins, increased activity of macrophages and lymphocytes, and stimulation of the production of interferon all contribute to probiotic-induced immunological stimulation (Li et al., 2020). Through metabolites, cellular wall components, and DNA identified by specialized host cells, probiotics affect the congenital and acquired immune systems (Idrees et al., 2022).

Intestinal, immune, and epithelial cells are the main host cells required for the immune response. Cell wall components of LAB boost the activity of macrophages, which quickly eliminate microorganisms by increasing the synthesis of lysosomal enzymes and free oxygen radicals (Liu et al., 2022). Additionally, probiotics have the ability to promote the cytokine production of immune-competent cells in the gastro intestinal tract (Wang et al., 2022). Glucans, which are found in the cellular walls of yeast, are linked to the immunological activity of the organism; they prompt the reticuloendothelial system to respond (Vera-Santander et al., 2023).

9.4.4.4 Inhibition of Bacterial Toxin Production

This is based on actions that result in toxin inactivation. Adsorption can lead to detoxification; however, microbial metabolism of mycotoxins can also have this effect (Liu et al., 2022). Not all probiotics will have a detoxifying impact since detoxifying abilities vary depending on the strain. The use of a probiotic blend, which combines bacteria with different modes of action, may enhance the protection provided by a biotherapeutic therapy (synergistic action) (Das et al., 2022).

9.4.5 Roles of Probiotics in Personalized Medicine

Probiotics, such as *Lactobacillus* and *Bifidobacterium* strains, play a significant role in personalized medicine by leveraging the genetic makeup of an individual and microbiome variability to enhance therapeutic effectiveness. Unlike prescribing the same drug dosage universally, personalized medicine recognizes individual differences in drug response based on genetic variations and microbial composition, particularly in the gut (Athiyyah et al., 2019). Probiotics like *Lactobacillus* stimulate immunity based on individual health status and assist in repairing colonic CD4+ T cells, thereby reducing AIDS-induced diarrhea (Ceccarelli et al., 2019).

Genome mapping has facilitated the understanding of individual genetic variability, enabling targeted therapies in personalized medicine to improve treatment efficacy (Cuevas-Sierra et al., 2019). This approach is particularly effective in managing immunity-related conditions such as inflammation, asthma, musculoskeletal disorders, and various chronic diseases (Marques et al., 2020). Probiotics, found in food, have immense potential in both general and personalized preventive medicine, with their effects influenced by host-related characteristics such as age and dietary habits (Valdes et al., 2018).

Lactobacillus and *Bifidobacterium* strains, commonly found in fermented foods, contribute to immunological tolerance and prevent disease phenotypes like inflammatory bowel syndrome and disorder, particularly prevalent in the elderly (Gao et al., 2022). Additionally, they enhance the integrity of the intestinal barrier and modify intestinal immunity responses (Chen et al., 2020). *Akkermansia muciniphila* (AM), a non-motile bacterium primarily found in the intestinal mucus layer, regulates host immune system homeostasis, enhances metabolism, and exhibits antiaging and anticancer activities (Depommier et al., 2019). AM is beneficial in treating metabolic syndrome and enhancing glucose metabolism, making it a potential therapeutic option for type II diabetes (Rodrigues et al., 2022). In conjunction with native *Escherichia coli Nissle 1917* (EcN), AM controls cancer by specifically colonizing solid tumors and inducing apoptosis in colon cancer cells (Dey & Chaudhuri, 2022). AM's role in the brain-gut microbiota axis and its use in managing radiation-induced diarrhea further underscore its therapeutic potential (Hiippala et al., 2018; Xu et al., 2020).

Probiotics like *Clostridium butyricum* aid in preventing diabetes by improving intestinal community balance (de Melo Pereira et al., 2018). During the COVID-19 pandemic, probiotics such as *Bifidobacterium* and *Lactobacillus* showed efficacy in enhancing the immune response against viral infections and reducing lung damage (Dey & Chaudhuri, 2022). Probiotics are also effective in treating dietary issues like cholestasis, contributing to the development of non-alcoholic fatty liver disease (Meroni et al., 2019). Major probiotics including *Bifidobacterium* and *Faecalibacterium* reduce the severity of acute diarrheal symptoms by preventing pathogenic species from attaching to epithelial cells and releasing bacteriocins (Iqbal et al., 2021). Additionally, probiotics have positive effects on mental health by enhancing mental agility and reducing stress, making them suitable for treating conditions like Alzheimer's disease (Dey & Chaudhuri, 2022). *Faecalibacterium prausnitzii*, for example, reduces the brain buildup of lipopolysaccharides and amyloid associated with Alzheimer's disease (Moir et al., 2018).

AMR poses a serious health threat, with the gut microbiome acting as a reservoir for antibiotic resistance genes (resistome) that can spread to pathogens and contribute to the formation of drug-resistant bacteria (Theophilus and Taft, 2023). The overuse of antibiotics contributes to resistome growth, highlighting the importance of probiotics in restoring microbiome equilibrium and preventing further resistome expansion (Radovanovic et al., 2023; Nwobodo et al., 2022). Probiotics play a crucial role in personalized medicine by leveraging an individual's genetic makeup and microbiome variability to enhance therapeutic effectiveness across various health conditions, from metabolic disorders to infectious diseases and cancer. Their multifaceted benefits underscore their potential as integral components of personalized preventive and therapeutic strategies.

9.5 CHALLENGES ASSOCIATED WITH THE USE OF PROBIOTICS IN PERSONALIZED MEDICINE

Probiotics, living bacteria that promote host health, are increasingly popular, yet concerns regarding their safety, particularly in vulnerable populations like the elderly and infants, have been raised (Siciliano et al., 2021; Zhu et al., 2023). Ensuring the safety of probiotics is paramount, especially considering potential harmful side effects, particularly in at-risk groups such as surgical patients and premature newborns (Shripada et al., 2020). Probiotic cultures must be engineered to survive processing, storage, and gastrointestinal transit while maintaining viability and reaching the target site in the host in sufficient numbers (Zucko et al., 2020). Studies on their safety in specific populations, including infants and the elderly, are necessary as probiotic supplements are regulated as foods, not medications (Kothari et al., 2019). Although probiotics can cause mild side effects like gas or bloating, there is also a risk of infections requiring antibiotic treatment, especially in immunocompromised individuals (Ale & Binetti, 2021). Ensuring the safety of genetically altered probiotic strains is crucial to prevent environmental contamination and the transmission of genetic material (Plavec & Berlec, 2020).

While dairy remains the most popular food source for probiotics, incorporating probiotics into non-dairy products presents challenges due to issues with survival, viability, stability, and storage (Feng et al., 2018; Pimentel et al., 2021). Probiotic cultures in non-dairy products often struggle to grow, affecting their stability, particularly concerning variables like water activity, oxygen tension, and temperature (Küçükgöz &Trząskowska, 2022; Feng et al., 2018). Storage at room temperature, common for many non-dairy products, can compromise probiotic stability (Wang et al., 2022). Additionally, antibiotic resistance genes have been found in common probiotic species, raising concerns about long-term immunological responses and competition with commensal gut microbiota (Shripada et al., 2020; Das et al., 2022).

To address these concerns, there is growing interest in non-viable probiotic preparations, known as ghost probiotics, which offer benefits without the risks associated with live bacteria (Siciliano et al., 2021). Ghost probiotics consist of non-viable microbial cells or crude cell extracts that, when provided in sufficient concentrations, confer benefits to consumers orally or topically. These preparations aim to mitigate the negative effects associated with live probiotics while still providing health benefits (Siciliano et al., 2021).

9.6 ADVANCEMENTS IN GENETIC TOOLS FOR ENGINEERING LACTIC ACID BACTERIA

Recent advancements in genetic engineering and gene editing technologies have opened up new avenues for modifying probiotics, holding significant implications for personalized medicine, especially in the African context. Genetic engineering tools allow for the modification of bacterial genomes through gene knock-outs, gene insertions, or introducing mutations, thereby tailoring probiotics for therapeutic applications (Mugwanda et al., 2023; Yadav & Shukla, 2019). These modifications have been explored in various therapeutic areas, including oncology, inflammatory bowel disease, and obesity, showcasing the potential of genetically engineered probiotics in improving health outcomes (Obafemi et al., 2022). However, the effectiveness of genetic engineering tools is often strain or species-specific, necessitating the development of standardized engineering tool sets to ensure broad applicability (Mugwanda et al., 2023). Moreover, safety concerns surrounding bioengineered probiotics require careful consideration, with particular attention to avoid the disruption of essential genes or the production of harmful metabolites. The utilization of essential biotherapeutic molecules derived from fiber-rich indigenous diets in Africa holds promise for disease downregulation (Obafemi et al., 2022).

Various genetic engineering techniques have been employed to modify probiotics, each with its advantages and limitations. Non-replicative plasmid-mediated homologous recombination and temperature-sensitive plasmid-mediated homologous recombination techniques have been utilized, albeit with inefficiencies and constraints (Zuo et al., 2019; Mu et al., 2022). Prophage recombinase-assisted dsDNA or ssDNA recombinase and CRISPR-Cas9 systems have shown promise in achieving stable mutations and precise editing in certain bacterial strains (Zuo et al., 2019; Hidalgo-Cantabranaet al., 2019; Wang et al., 2022). However, challenges such as editing escape and strain-dependent targeting need to be addressed to enhance the efficiency and applicability of these techniques (Mu et al., 2022; Vento et al., 2019). The development of inducible plasmid self-destruction (IPSD) assisted genome engineering offers a more adaptable and widespread technique, enabling gene deletion, insertion, and replacement in Lactobacilli and Bifidobacteria (Mu et al., 2022). This technology has been utilized to investigate probiotic-host cell adaptation and engineer probiotics with improved robustness (Zuo et al., 2019; Mu et al., 2022). Genetic engineering and gene editing technologies hold promise for modifying probiotics to provide personalized and targeted treatments for various diseases in Africa.

9.7 CURRENT ISSUES SURROUNDING THE USE OF PROBIOTICS IN PERSONALIZED MEDICINE

Probiotics have garnered attention for their potential health benefits, but concerns regarding their safety, regulation, and quality control persist. Regulation of probiotics varies globally, with national regulatory standards playing a crucial role. In the USA, the Food and Drug Administration (FDA) evaluates probiotics based on their intended use, safety, potency, and efficacy (Salminen et al., 2021; Dwyer et al., 2018). Probiotics fall into different categories, including biological products, medications, or dietary supplements, each requiring specific regulatory testing and approval processes (Claudio, 2019). The FDA assesses probiotics for premarket approval as food additives or for Generally Recognized As Safe (GRAS) status before marketing (Ivan et al., 2019). In South Africa, the South African Health Products Regulatory Authority (SAHPRA) oversees probiotic registration and use (Happel et al., 2017). However, many over-the-counter probiotics lack formal review or clinical trials. Limited claims for functional food are permitted, with only four species of probiotic bacteria allowed labeling as probiotics (Happel et al., 2020). Europe and Brazil also have regulatory frameworks for probiotics. The European Food Safety Authority (EFSA) identifies microorganisms deemed safe for food use, while Brazil established standards for probiotic assessment in 1999, updating regulations as knowledge evolves (Swanson et al., 2020; Mukherjee et al., 2022).

Quality control is an ethical issue in the probiotic industry, with concerns about mislabeling, contamination, and adherence to regulatory standards (Mazzantini et al., 2021). Studies have reported discrepancies between labeled and actual microbial strains in probiotic products, raising questions about their efficacy and safety (de Simone, 2019). The potential for lateral or horizontal gene transfer among probiotics is another safety concern, especially with increasing probiotic therapy alongside antibiotics. Although no proof of gene transfer has been established, it remains a concern in the scientific community. Harmonized regulatory standards and rigorous quality control measures are essential to ensure the safety and effectiveness of probiotic products globally.

9.8 CONCLUSION AND FUTURE PERSPECTIVES

The exploration of probiotics as biotechnological tools in personalized medicine holds immense promise, particularly for the unique health challenges faced by populations in Africa. The intricate interplay between an individual's microbiome and their response to diseases and treatments underscores the significance of personalized medicine in tailoring healthcare solutions. Probiotics, with their ability to modulate the gut microbiota and influence various aspects of health, emerge as pivotal players in this paradigm. The existing research underscores the potential applications of probiotics in addressing infectious diseases, non-communicable diseases, metabolic disorders, and even mental health issues. The focus on African fermented foods as rich sources of beneficial microorganisms presents an exciting avenue for harnessing probiotics in the region, aligning with local dietary practices and preferences. However, it is crucial to approach the integration of probiotics in personalized medicine with caution, considering safety issues, especially in vulnerable populations.

Future perspectives should involve rigorous clinical studies and trials to validate the efficacy and safety of probiotics in diverse African populations. Collaboration between researchers, healthcare practitioners, and biotechnological experts is essential for the development of tailored probiotic interventions that consider regional variations in microbiota and genetic makeup. Furthermore, efforts should be directed toward optimizing the production and storage of probiotics in non-dairy products, ensuring their accessibility and feasibility in resource-constrained settings. The fight against infectious diseases, prevalent in Africa, could benefit significantly from probiotic interventions that enhance immunity and combat AMR. Additionally, the potential of probiotics in addressing non-communicable diseases, such as diabetes and cardiovascular disorders, requires comprehensive investigations to establish specific strains and formulations that yield optimal outcomes.

In the future, probiotics could become an integral part of public health initiatives in Africa, contributing to preventive strategies and personalized treatment approaches. As advancements in biotechnology continue, there is an exciting opportunity to delve deeper into the molecular mechanisms underlying probiotic action, paving the way for precision probiotics designed for individual health needs.

REFERENCES

Akhoon N. Precision medicine: A new paradigm in therapeutics. *International Journal of Preventive Medicine*. 2021; 12. https://doi.org/10.4103/ijpvm.IJPVM_375_19

Akil AA, Yassin E, Al-Maraghi A, Aliyev E, Al-Malki K, Fakhro KA. Diagnosis and treatment of type 1 diabetes at the dawn of the personalized medicine era. *Journal of Translational Medicine*. 2021; 19(1): 137. https://doi.org/10.1186/s12967-021-02778-6

Ale EC, Binetti AG. Role of probiotics, prebiotics, and synbiotics in the elderly: Insights into their applications. *Frontiers Microbiology*. 2021; 12: 631254. https://doi.org/10.3389/fmicb.2021.631254

Alpuim CD, Nobre JG, Batista MV, et al. Human microbiota and breast cancer-Is there any relevant link?: A literature review and new horizons toward personalised medicine. *Frontiers Microbiology*. 2021; 12: 584332. https://doi.org/10.3389/fmicb.2021.584332

Alsayed AR, Abed A, Khader HA, et al. Molecular accounting and profiling of human respiratory microbial communities: Toward precision medicine by targeting the respiratory microbiome for disease diagnosis and treatment. *International Journal of Molecular Science*. 2023; 24: 4086. https://doi.org/10.3390/ijms24044086

Athiyyah AF, Brahmantya H, Dwiastuti S, et al. Effect of *Lactobacillus plantarum* IS-10506 on blood lipopolysaccharide level and immune response in HIV-infected children. *Iranian Journal of Microbiology*. 2019; 11(2): 137–144. https://www.ncbi.nlm.nih.gov/pmc/articles/PMC6635306/

Atukunda P, Eide WB, Kardel KR, Iversen PO, Westerberg AC. Unlocking the potential for achievement of the UN Sustainable Development Goal 2 – 'Zero Hunger' – in Africa: Targets, strategies, synergies and challenges. *Food & Nutrition Research*. 2021; 65. https://doi.org/10.29219/fnr.v65.7686

Ayesha IE, Monson NR, Klair N, Patel U, Saxena A, Patel D, Venugopal S. Probiotics and Their Role in the Management of Type 2 Diabetes Mellitus (Short-Term Versus Long-Term Effect): A Systematic Review and Meta-Analysis. *Cureus*. 2023 Oct 9; 15(10): e46741. https://doi.org/10.7759/cureus.46741

Ayivi RD, Gyawali R, Krastanov A, et al. Lactic acid bacteria: Food safety and human health applications. *Dairy*. 2020; 1(3): 202–232. https://doi.org/10.3390/dairy1030015

Bao R, Hesser LA, He Z, Zhou X, Nadeau KC, Nagler CR. Fecal microbiome and metabolome differ in healthy and food-allergic twins. *Journal of Clinical Investigations*. 2021; 131(2): e141935. https://doi.org/10.1172/JCI141935

Cabore JW, Karamagi HC, Kipruto HK, et al. COVID-19 in the 47 countries of the WHO African region: A modelling analysis of past trends and future patterns. *The Lancet Global Health*. 2022; 10(8): e1099–e1114. https://doi.org/10.1016/S2214-109X(22)00233-9

Campos GM, Américo MF, dos Santos Freitas A, et al. *Lactococcus lactis* as an interleukin delivery system for prophylaxis and treatment of inflammatory and autoimmune diseases. *Probiotics & Antimicrobial Proteins*. 2023; 7: 1–15. https://doi.org/10.1007/s12602-023-10041-1

Campos LM, Hernández-Chiñas U, Moreno-Guerrero SS, Ramírez-Pacheco A, Eslava CA.. Probiotic properties and immunomodulatory activity of *Lactobacillus* strains isolated from dairy products. *Microorganisms*. 2021; 9(4). https://doi.org/10.3390/microorganisms9040825

Ceccarelli G, Statzu M, Santinelli L, et al. Challenges in the management of HIV infection: Update on the role of probiotic supplementation as a possible complementary therapeutic strategy for cART treated people living with HIV/AIDS. *Expert Opinion Biological Therapy*. 2019; 19 (9): 949–965. https://doi.org/10.1080/14712598.2019.1638907

Chadha J, Nandi D, Atri Y, Nag A. Significance of human microbiome in breast cancer: Tale of an invisible and an invincible. *Seminars in Cancer Biology*. 2021; 70: 112–127. https://doi.org/10.1016/j.semcancer.2020.07.010

Chakravarthy R, Stallings SC, Williams M, et al. Factors influencing precision medicine knowledge and attitudes. *PLoS One*. 2020; 15(11): e0234833. https://doi.org/10.1371/journal.pone.0234833

Chen D, Chen G, Chen C, Zeng X, Ye H. Prebiotics effects in vitro of polysaccharides from tea flowers on gut microbiota of healthy persons and patients with inflammatory bowel disease. *International Journal of Biological Macromolecules*. 2020; 158: 968–976. https://doi.org/10.1016/j.ijbiomac.2020.04.248

Chikowore T, Kamiza AB, Oduaran OH, Machipisa T, Fatumo S. Non-communicable diseases pandemic and precision medicine: Is Africa ready? *EBioMedicine*. 2021; 65: 103260. https://doi.org/10.1016/j.ebiom.2021.103260

Claudio S. The unregulated probiotic market. *Clinical Gastroenterology and Hepatology*. 2019; 17(5): 809–817. https://doi.org/10.1016/j.cgh.2018.01.018

Consortium RN. The integrative human microbiome project. *Nature*. 2019; 569(7758): 641–648. https://doi.org/10.1038/s41586-019-1238-8

Cuevas-Sierra A, Ramos-Lopez O, Riezu-Boj JI, Milagro FI, Martinez JA. Diet, gut microbiota, and obesity: Links with host genetics and epigenetics and potential applications. *Advanced Nutrition*. 2019; 10 (Supplement 1): S17–S30. https://doi.org/10.1093/advances/nmy078

Das TK, Pradhan S, Chakrabati S, Mondal KM, Ghosh K. Current status of probiotic and related health benefits. *Applied Food Research*. 2022; 2: 100185. https://doi.org/10.1016/j.afres.2022.100185

Dash S, Shakyawar SK, Sharma M, Kaushik S. Big data in healthcare: Management, analysis and future prospects. *Journal of Big Data*. 2019; 6(1): 1–25. https://doi.org/10.1186/s40537-019-0217-0

Depommier C, Everard A, Druart C, et al. Supplementation with *Akkermansia muciniphila* in overweight and obese human volunteers: A proof-of-concept exploratory study. *Nature Medicine*. 2019; 25(7): 1096–1103. https://doi.org/10.1038/s41591-019-0495-2

Dey P, Ray Chaudhuri S. Cancer-associated microbiota: From mechanisms of disease causation to microbiota-centric anti-cancer approaches. *Biology*. 2022; 11(5): 757. https://doi.org/10.3390/biology11050757

Doutoum AA, Tidjani A, Abdelaziz Arada I, et al. Identification of the lactic microorganism responsible for milk coagulation in Abeche (Chad). *African Journal of Food Science*. 2013; 7: 103–106. https://doi.org/10.5897/AJFS12.161

Dugger SA, Platt A, Goldstein DB. Drug development in the era of precision medicine. *Nature Reviews Drug Discovery*. 2018; 17(3): 183–196. https://doi.org/10.1038/nrd.2017.226

Dwyer JT, Coates PM, Smith MJ. Dietary supplements: Regulatory challenges and research resources. *Nutrients*. 2018; 10(1): 41. https://doi.org/10.3390/nu10010041

Eid A, Mhatre I, Richardson JR. Gene-environment interactions in Alzheimer's disease: A potential path to precision medicine. *Pharmacology & Therapeutics*. 2019; 199: 173–187. https://doi.org/10.1016/j.pharmthera.2019.03.005.

Elton L, Thomason MJ, Tembo J, Velavan TP, Pallerla SR, Arruda LB. Antimicrobial resistance preparedness in sub-Saharan African countries. *Antimicrobial Resistance and Infection Control*. 2020; 9: 145. https://doi.org/10.1186/s13756-020-00800-y

Ezekiel CN, Ayeni KI, Ezeokoli OT, et al. High-throughput sequence analyses of bacterial communities and multi-mycotoxin profiling during processing of different formulations of *Kunu*, a traditional fermented beverage. *Frontiers in Microbiology*. 2019; 9: 3282. https://doi.org/10.3389/fmicb.2018.03282

EU+ Africa PerMed. Participation of NACOSTI in EU-Africa PerMed. EU-Africa Permed Launching Event: Building Links between Europe and Africa in Personalised Medicine. 2021. https://www.nacosti.go.ke/2021/06/21/participation-of-nacosti-in-eu-africa-permed-project/ Accessed 24 April 2023

FAO, IFAD, UNICEF, WFP, WHO. *The State of Food Security and Nutrition in the World 2019. Safeguarding Against Economic Slowdowns and Downturns*. FAO, Rome. 2019. https://www.fao.org/3/ca5162en/ca5162en.pdf

FAO, IFAD, UNICEF, WFP, WHO. *The State of Food Security and Nutrition in the World 2021. Transforming Food Systems for Food Security, Improved Nutrition and Affordable Healthy Diets for All*. FAO, Rome. 2021. https://doi.org/10.4060/cb4474en

FAO. *The State of Food Security and Nutrition in the World. Transforming Food Systems for Affordable Healthy Diets*. Food and Agriculture Organization (FAO) of the United Nations, Rome. 2020.

Feng P, Ye Z, Kakade A, Virk AK, Li X, Liu P. A review on gut remediation of selected environmental contaminants: Possible roles of probiotics and gut microbiota. *Nutrients*. 2018; 11(1): 22. https://doi.org/10.3390/nu11010022

Gadaga TH, Mutukumira AN, Narvhus JA, Feresu SB. A review of traditional fermented foods and beverages of Zimbabwe. *International Journal of Food Microbiology*. 1999; 53(1): 1–11. https://doi.org/10.1016/s0168-1605(99)00154-3

Gao X, Zhao J, Zhang H, Chen W, Zhai Q. Modulation of gut health using probiotics: The role of probiotic effector molecules. *Journal of Future Foods*. 2022; 2 (1): 1–12. https://doi.org/10.1016/j.jfutfo.2022.03.011

Gilbert JA, Blaser MJ, Caporaso JG, Jansson JK, Lynch SV, Knight R. Current understanding of the human microbiome. *Nature Medicine*. 2018; 24(4): 392–400. https://doi.org/10.1038/nm.4517

Girgis E, Galicia G, Salatiel D. Traditional fermented foods and beverages from around the world and their health benefits. *Microorganisms*. 2022; 10(6): 1151. https://doi.org/10.3390/microorganisms10061151

Happel AU, Jaumdally SZ, Pidwell T, et al. Probiotics for vaginal health in South Africa: What is on retailers' shelves? *BMC Womens Health*. 2017; 17(7). https://doi.org/10.1186/s12905-017-0362-6

Happel AU, Singh R, Mitchev N, et al. Testing the regulatory framework in South Africa – a single-blind randomized pilot trial of commercial probiotic supplementation to standard therapy in women with bacterial vaginosis. *BMC Infectious Diseases*. 2020; 20: 491. https://doi.org/10.1186/s12879-020-05210-4

Hidalgo-Cantabrana C, Goh YJ, Pan M, Sanozky-Dawes R, Barrangou R. Genome editing using the endogenous type I CRISPR-Cas system in *Lactobacillus crispatus. Proceedings of the National Academy of Sciences of the United States of America*. 2019; 116: 15774–15783. https://doi.org/10.1073/pnas.1905421116

Hiippala K, Jouhten H, Ronkainen A, et al. The potential of gut commensals in reinforcing intestinal barrier function and alleviating inflammation. *Nutrients*. 2018; 10: 988. https://doi.org/10.3390/nu10080988

International Diabetes Federation: *IDF Diabetes Atlas*, 8th Edition. 2017. https://www.diabetesatlas.org

Idrees M, Imran M, Atiq N, et al. Probiotics, their action modality and the use of multi-omics in metamorphosis of commensal microbiota into target-based probiotics. *Frontiers in Nutrition*. 2022; 9: 959941. https://doi.org/10.3389/fnut.2022.959941

Iqbal Z, Ahmed S, Tabassum N, Bhattacharya R, Bose D. Role of probiotics in prevention and treatment of enteric infections: A comprehensive review. *Biotechnology*. 2021; 11(5): 242. https://doi.org/10.1007/s13205-021-02796-7

Ivan W. Regulation of Probiotics in the USA: Food. 2019. Retrieved from https://www.nutraingredients-usa.com/Article/2016/04/25/Regulation-of-Probiotics-in-the-USA-Food

Jans C, Meile L, Kaindi DWM, et al. African fermented dairy products–Overview of predominant technologically important microorganisms focusing on African Streptococcus infantarius variants and potential future applications for enhanced food safety and security. *International Journal of Food Microbiology*. 2017; 250: 27–36. https://doi.org/10.1016/j.ijfoodmicro.2017.03.012

Kothari D, Patel S, Kim KM. Probiotic supplements might not be universally-effective and safe: A review. *Biomedicine & Pharmacotherapy*. 2019; 111: 537–47. https://doi.org/10.1016/j.biopha.2018.12.104

Küçükgöz K, Trząskowska M. Nondairy probiotic products: Functional foods that require more attention. *Nutrients*. 2022; 14(4): 753. https://doi.org/10.3390/nu14040753

Khushboo KA, Malik T. Characterization and selection of probiotic lactic acid bacteria from different dietary sources for development of functional foods. *Frontiers in Microbiology*. 2023; 14. https://doi.org/10.3389/fmicb.2023.1170725

Li XJ, You XY, Wang CY, Li XL, Sheng YY, Zhuang PW. Bidirectional Brain-gut-microbiota Axis in increased intestinal permeability induced by central nervous system injury. *CNS Neuroscience & Therapeutics*. 2020; 26(8): 783–790. https://doi.org/10.1111/cns.13401

Liu Y, Wang J, Wu, C. Modulation of gut microbiota and immune system by probiotics, pre-biotics, and post-biotics. *Frontiers in Nutrition*. 2022; 8: 634897. https://doi.org/10.3389/fnut.2021.634897

Lobionda S, Sittipo P, Kwon HY, Lee YK. The role of gut microbiota in intestinal inflammation with respect to diet and extrinsic stressors. *Microorganisms*. 2019; 7(8): 271. https://doi.org/10.3390/microorganisms7080271

Lumaka A, Carstens N, Devriendt K, et al. Increasing African genomic data generation and sharing to resolve rare and undiagnosed diseases in Africa: A call-to-action by the H3Africa rare diseases working group. *Orphanet Journal of Rare Diseases*. 2022; 17: 230. https://doi.org/10.1186/s13023-022-02391-w

Mathuba B, Koromina M, Mitropoulou C, Patrinos, GP. Catalyzing clinical implementation of pharmacogenomics and personalized medicine interventions in Africa. *Pharmacogenomics*. 2021; 22(2): 115–122. https://doi.org/10.2217/pgs-2020-0101

Magavern EF, Gurdasani D, Ng FL, Soo-Jin Lee S. Health equality, race and pharmacogenomics. *British Journal of Clinical Pharmacology*. 2021; 88(1): 27. https://doi.org/10.1111/bcp.14983

Marques AM, Sarandy MM, Novaes RD, Gonçalves RV, Freitas MB. Preclinical relevance of probiotics in type 2 diabetes: A systematic review. *International Journal of Experimental Pathology*. 2020; 101(3–4): 68–79. https://doi.org/10.1111/iep.12359

Mazzantini D, Calvigioni M, Celandroni F, Lupetti A, Ghelardi E. Spotlight on the compositional quality of probiotic formulations marketed worldwide. *Frontiers in Microbiology*. 2021; 12. https://doi.org/10.3389/fmicb.2021.693973

McCoubrey LE, Gaisford S, Orlu M, Basit AW. (2021). Predicting drug-microbiome interactions with machine learning. *Biotechnology Advances*. 2021; 54: 107797. https://doi.org/10.1016/j.biotechadv.2021.107797

de Melo Pereira G V, de Oliveira Coelho B, Magalhães Júnior AI, Thomaz-Soccol V, Soccol, CR. How to select a probiotic? A review and update of methods and criteria. *Biotechnology Advances*. 2018; 36(8): 2060–2076. https://doi.org/10.1016/j.biotechadv.2018.09.003

Meroni M, Longo M, Dongiovanni P. The role of probiotics in nonalcoholic fatty liver disease: A new insight into therapeutic strategies. *Nutrients*. 2019; 11(11): 2642. https://doi.org/10.3390/nu11112642

Mgbodile FC, Nwagu TNT. Probiotic therapy, African fermented foods and food-derived bioactive peptides in the management of SARS-CoV-2 cases and other viral infections. *Biotechnology Reports (Amsterdam, Netherlands)*. 2023; 38: e00795. https://doi.org/10.1016/j.btre.2023.e00795

Michaelis C, Grohmann E. Horizontal gene transfer of antibiotic resistance genes in biofilms. *Antibiotics*. 2023; 12: 328. https://doi.org/10.3390/antibiotics12020328

Mohajan HK. Food insecurity and malnutrition of Africa: A combined attempt can reduce them. *Journal of Economic Development, Environment and People*. 2021; 11(1): 24–34. https://doi.org/10.26458/jedep.v1i1.716

Moir RD, Lathe R, Tanzi RE. The antimicrobial protection hypothesis of Alzheimer's disease. *Alzheimers Dement*. 2018; 14(12): 1602–1614. https://doi.org/10.1016/j.jalz.2018.06.3040

Mokoena MP, Mutanda T, Olaniran AO. Perspectives on the probiotic potential of lactic acid bacteria from African traditional fermented foods and beverages. *Food & Nutrition Research*. 2016; 60(1): 29630. https://doi.org/10.3402/fnr.v60.29630

Moré MI, Vandenplas Y. *Saccharomyces boulardii* CNCM I-745 improves intestinal enzyme function: A trophic effects review. Clinical medicine insights. *Gastroenterology*. 2018; 11: 1179552217752679. https://doi.org/10.1177/1179552217752679

Mu Y, Zhang C, Li T, et al. Development and applications of CRISPR/Cas9-based genome editing in *Lactobacillus*. *International Journal of Molecular Science*. 2022; 23: 12852. https://doi.org/10.3390/ijms232112852

Mugula JK, Nnko SA, Narvhus JA, Sørhaug T. Microbiological and fermentation characteristics of togwa, a Tanzanian fermented food. *International Journal of Food Microbiology*. 2003; 80(3): 187–199. https://doi.org/10.1016/s0168-1605(02)00141-1

Mugwanda K, Hamese S, Van Zyl WF, et al. Recent advances in genetic tools for engineering probiotic lactic acid bacteria. *Bioscience Reports*. 2023; 43(1): BSR20211299. https://doi.org/10.1042/BSR20211299

Murugaiyan J, Kumar PA, Rao GS, et al. Progress in alternative strategies to combat antimicrobial resistance: Focus on antibiotics. *Antibiotics (Basel)*. 2022; 11(2):200. https://doi.org/10.3390/antibiotics11020200

Mukherjee AME, Kenny JG, Cotter PD. Global regulatory frameworks for fermented foods: A review. *Frontiers in Nutrition*. 2022; 9. https://doi.org/10.3389/fnut.2022.902642

Ndudzo A, Ndlovu S, Nyathi N, Makuvise-Sibanda, A. Unlocking the potential of ghost probiotics in combating anti-microbial resistance. In *The Global Antimicrobial Resistance Epidemic: Innovative Approaches and Cutting Edge solutions*. IntechOpen. 2022. https://doi.org/10.5772/intechopen.104126

Neznanov N. A paradigm shift to treat psychoneurological disorders. *Personalized Psychiatry and Neurology*. 2021; 1(1): 1–2.

Niohuru I. Disease burden and mortality. In: *Healthcare and Disease Burden in Africa. SpringerBriefs in Economics*. Springer, Cham. 2023. https://doi.org/10.1007/978-3-031-19719-2_3

Nwobodo DC, Ugwu MC, Anie CO, et al. Antibiotic resistance: The challenges and some emerging strategies for tackling a global menace. *Journal Clinical Laboratory Analysis*. 2022; 36: e24655. https://doi.org/10.1002/jcla.24655

Nyambane B, Thari WM, Wangoh J, Njage PM. Lactic acid bacteria and yeasts involved in the fermentation ofamabere amaruranu, a Kenyan fermented milk. *Food Science & Nutrition*. 2014; 2(6):692–699. https://doi.org/10.1002/fsn3.162

Nyathi N, Ndhlovu D, Rabvukwa E and Ndudzo A. *Improving Antimicrobial Stewardship in Human Health Using Probiotics and Their Derivatives*. IntechOpen. 2024. https://doi.org/10.5772/intechopen.112497

Obafemi YD, Oranusi SU, Ajanaku KO, Akinduti PA, Leech J, Cotter PD. African fermented foods: Overview, emerging benefits, and novel approaches to microbiome profiling. *NPJ Science of Food*. 2022; 6(1): 15. https://doi.org/10.1038/s41538-022-00130-w

Obande GA, Bagudo AI, Mohamad S, et al. Current state of COVID-19 pandemic in Africa: Lessons for today and the future. *International Journal of Environmental Research and Public Health*. 2021; 18(19): 9968. https://doi.org/10.3390/ijerph18199968

Ojo T, Ruan C, Hameed T, et al. HIV, tuberculosis, and food insecurity in Africa-A syndemics-based scoping review. *International Journal of Environmental Research and Public Health*. 2022; 19(3): 1101. https://doi.org/10.3390/ijerph19031101

Okafor UI, Omemu AM, Obadina AO, Bankole MO, Adeyeye SA. Nutritional composition and antinutritional properties of maize ogi co-fermented with pigeon pea. *Food Science & Nutrition*. 2018; 6: 424–439.

Omotoso OE, Teibo JO, Atiba FA, Oladimeji T, Adebesin AO, Babalghith, K. Bridging the genomic data gap in Africa: Implications for global disease burdens. *Global Health*. 2022; 18: 103. https://doi.org/10.1186/s12992-022-00898-2

Osvik RD, Sperstad S, Breines EM, et al. Bacterial diversity of amasi, a South African fermented milk product, determined by clone library and denaturing gradient gel electrophoresis analysis. *African Journal of Microbiology Research*. 2013; 7: 4146–4148. https://doi.org/10.5897/AJMR12.2317

Otekunrin OA, Sawicka B, Ayinde IA. Three Decades of Fighting against Hunger in Africa: Progress, Challenges and Opportunities. *World Nutrition*. 2020; 11(3): 86–111. https://doi.org/10.26596/wn.202011386-111

Parmar PH, Patel A, Mishra MM, Parmar K. Personalized Medicine: Time for One Person, One Medicine. *International Journal of Pharmaceutics and Drug Analysis*. 2021; 9(2): 86–92. https://doi.org/10.47957/ijpda.v9i2.462

Pereira L, Mutesa L, Tindana P, Ramsay M. African genetic diversity and adaptation inform a precision medicine agenda. *Nature Reviews Genetics*. 2021; 22(5): 284–306. https://doi.org/10.1038/s41576-020-00306-8

Petrariu O-A, Barbu IC, Niculescu A-G, Constantin M, Grigore GA, Cristian R-E, Mihaescu G and Vrancianu CO. (2024) Role of probiotics in managing various human diseases, from oral pathology to cancer and gastrointestinal diseases. *Frontiers in Microbiology*. 2024; 14: 1296447. https://doi.org/10.3389/fmicb.2023.1296447

Pimentel TC, Costa WKAD, Barão CE, Rosset M, Magnani M. Vegan probiotic products: A modern tendency or the newest challenge in functional foods. *Food Research International*. 2021; 140: 110033. https://doi.org/10.1016/j.foodres.2020.110033

Plavec TV, Berlec A. Safety aspects of genetically modified lactic acid bacteria. *Microorganisms*. 2020; 8(2): 297. https://doi.org/10.3390/microorganisms8020297

Pravin B, Kishor O, Ashwini B. Personalized medicine: The future of modern medicine. *Authorea Preprints*. 2021. https://doi.org/10.22541/au.163940692.29299498/v1

Prete R, Long SL, Gallardo AL, Gahan CG, Corsetti A, Joyce SA. Beneficial bile acid metabolism from *Lactobacillus plantarum* of food origin. *Scientific Reports*. 2020; 10: 1165. https://doi.org/10.1038/s41598-020-58069-5

Radovanovic M, Kekic D, Gajic I, et al. (2023). Potential influence of antimicrobial resistance gene content in probiotic bacteria on the gut resistome ecosystems. *Frontiers in Nutrition*. 2023; 10. https://doi.org/10.3389/fnut.2023.1054555

Rebbeck TR. Cancer in Sub-Saharan Africa. *Science*. 2020; 367(6473):27–28. https://doi.org/10.1126/science.aay4743

Reygaert WC. An overview of the antimicrobial resistance mechanisms of bacteria. *AIMS Microbiology*. 2018; 4(3): 482–501. https://doi.org/10.3934/microbiol.2018.3.482

Rodrigues VF, Elias-Oliveira J, Pereira ÍS, et al. *Akkermansia muciniphila* and gut immune system: A good friendship that attenuates inflammatory bowel disease, obesity, and diabetes. *Frontiers in Immunology*. 2022; 13: 934695. https://doi.org/10.3389/fimmu.2022.934695

Salminen S, Collado MC, Endo A, et al. The International Scientific Association of Probiotics and Prebiotics (ISAPP) consensus statement on the definition and scope of postbiotics. *Nature Reviews Gastroenterology & Hepatology*. 2021; 18(9): 649–667. https://doi.org/10.1038/s41575-021-00440-6

Sanders ME, Benson A, Lebeer S, Merenstein DJ, Klaenhammer TR. Shared mechanisms among probiotic taxa: Implications for general probiotic claims. *Current Opinion in Biotechnology*. 2018; 49: 207–216. https://doi.org/10.1016/j.copbio.2017.09.007

Schoustra SE, Kasase C, Toarta C, Kassen R, Poulain AJ. Microbial community structure of three traditional Zambian fermented products: Mabisi, chibwantu and munkoyo. *PLoS One*. 2013; 8(5):e63948. https://doi.org/10.1371/journal.pone.0063948

Shanahan F, Ghosh TS, O'Toole PW. The healthy microbiome—What is the definition of a healthy gut microbiome? *Gastroenterology*. 2021; 160(2): 483–494. https://doi.org/10.1053/j.gastro.2020.09.057.

Sharma R, Aashima NM, Fronterre C, et al. (2022). Mapping cancer in Africa: A comprehensive and comparable characterization of 34 cancer types using estimates from GLOBOCAN 2020. *Frontiers in Public Health*. 2022; 10: 839835. https://doi.org/10.3389/fpubh.2022.839835

Sharpton SR, Schnabl B, Knight R, Loomba R. (2021). Current concepts, opportunities, and challenges of gut microbiome-based personalized medicine in nonalcoholic fatty liver disease. *Cell Metabolism*. 2021; 33(1): 21–32. https://doi.org/10.1016/j.cmet.2020.11.010

Shiferaw Terefe N, Augustin MA. Fermentation for tailoring the technological and health related functionality of food products. *Critical Reviews in Food Science and Nutrition*. 2020; 60: 2887–2913.

Shripada R, Gayatri AJ, Sanjay P. Paraprobiotics. In Faintuch J, Faintuch, S., Eds.; *Precision Medicine for Investigators, Practitioners and Providers* (pp. 39–49). Academic Press, Cambridge, MA. 2020.

Siciliano RA, Reale A, Mazzeo MF, et al. Paraprobiotics: A new perspective for functional foods and nutraceuticals. *Nutrients*. 2021; 13(4): 1225. https://doi.org/10.3390/nu13041225

Silva YP, Bernardi A, Frozza RL. The role of short-chain fatty acids from gut microbiota in gut-brain communication. *Frontiers in Endocrinolology*. 2020; 11: 25. https://doi.org/10.3389/fendo.2020.00025

de Simone C. The unregulated probiotic market. *Clinical Gastroenterology and Hepatology*. 2019; 17(5): 809–817. https://doi.org/10.1016/j.cgh.2018.01.018

Singh RP, Shadan A, Ma Y. Biotechnological applications of probiotics: A multifarious weapon to disease and metabolic abnormality. *Probiotics and Antimicrobial Proteins*. 2022; 14: 1184–1210. https://doi.org/10.1007/s12602-022-09992-8

Sinkkonen A. (2022). Distortion of the microbiota of the natural environment by human activities. In *Evolution, Biodiversity and a Reassessment of the Hygiene Hypothesis* (pp. 221–242). Springer, Cham. 2022. https://doi.org/10.1007/978-3-030-91051-8_8

Swanson KS, Gibson GR, Hutkins R, et al. (2020). The International Scientific Association for Probiotics and Prebiotics (ISAPP) consensus statement on the definition and scope of synbiotics. *National Reviews of Gastroenterology and Hepatology*. 2020; 17: 687–701. https://doi.org/10.1038/s41575-020-0344-2

Terpou A, Papadaki A, Lappa IK, Kachrimanidou V, Bosnea LA, Kopsahelis N. Probiotics in food systems: Significance and emerging strategies towards improved viability and delivery of enhanced beneficial value. *Nutrients*. 2019; 11(7). https://doi.org/10.3390/nu11071591

Theophilus RJ, Taft DH. Antimicrobial Resistance Genes (ARGs), the gut microbiome, and infant nutrition. *Nutrients*. 2023; 15(14): 3177. https://doi.org/10.3390/nu15143177

Ukeyima MT, Enujiugha VN, Sanni TA. Current applications of probiotic foods in Africa. *African Journal of Biotechnology*. 2010; 9(4): 394–401. https://www.academicjournals.org/AJB

United Nations. Economic and Social Council; United Nations. Economic Commission for Africa; United Nations. Economic and Social Council (2021-01). Building forward better: Towards a resilient and green Africa to achieve the 2030 Agenda and Agenda 2063. UN. ECA African Science, Technology and Innovation Forum (3rd session: 2021, Feb. 25–26: Brazzaville, Congo). https://repository.uneca.org/handle/10855/47770

United Nations. No Time to Wait: Securing the Future from Drug-Resistant Infections. Report to the Secretary-General of the United Nations (Interagency Coordination Group on Antimicrobial Resistance. 2019. https://news.un.org/en/story/2019/04/1037471

Vaz VM, Kumar L. 3D printing as a promising tool in personalized medicine. *AAPS PharmSciTech*. 2021; 22(1): 1–20. https://doi.org/10.1208/s12249-020-01905-8

Valdes AM, Walter J, Segal E, Spector TD. Role of the gut microbiota in nutrition and health. Science and politics of nutrition. *BMJ*. 2018; 361: k2179. https://doi.org/10.1136/bmj.k2179

Vera-Santander VE, Hernandez-Figueroa RH, Himenez-Manguia MT, Mani-Lopez E, Lopez-Malo A. Health benefits of consuming foods with bacterial probiotics, postbiotics, and their metabolites: A review. *Molecules*. 2023; 28(3): 1230. https://doi.org/10.3390/molecules28031230

Vento JM, Crook N, Beisel C L. Barriers to genome editing with CRISPR in bacteria. *Journal of Industrial Microbiology and Biotechnology*. 2019; 46: 1327–1341. https://doi.org/10.1007/s10295-019-02195-1

Vijay A, Valdes, AM. Role of the gut microbiome in chronic diseases: A narrative review. *European Journal of Clinical Nutrition*. 2022; 76: 489–501. https://doi.org/10.1038/s41430-021-00991-6

de Vos WM, Tilg H, Van Hul M, Cani DP. Gut microbiome and health: Mechanistic insights. *Gut*. 2022; 71: 1020–1032. https://doi.org/10.1136/gutjnl-2021-326789

Wang X, Zheng W, Zhou H, et al. Improved dsDNA recombineering enables versatile multiplex genome engineering of kilobase-scale sequences in diverse bacteria. *Nucleic Acids Research*. 2022; 50(3): e15. https://doi.org/10.1093/nar/gkab1076

Wiertsema SP, van Bergenhenegouwen J, Garssen J, Knippels LMJ. The interplay between the gut microbiome and the immune system in the context of infectious diseases throughout life and the role of nutrition in optimizing treatment strategies. *Nutrients*. 2021; 13(3): 886. https://doi.org/10.3390/nu13030886

Wonkam A. Sequence three million genomes across Africa. *Nature*. 2021; 590 (7845): 209–211. https://doi.org/10.1038/d41586-021-00313-7

World Economic Forum.World TB Day: Tuberculosis is on the rise globally – Here's how we can end it for good. 2023. https://www.weforum.org/agenda/2023/03/tuberculosis-is-on-the-rise-globally-heres-how-we-can-end-tb-for-good/. Accessed 19 May 2023.

World Health Organisation. Africa region tops world undiagnosed diabetes: WHO analysis. 2022. https://www.afro.who.int/news/african-region-tops-world-undiagnosed-diabetes-who-analysis

World Health Organisation. HIV and AIDS. 2023. https://www.who.int/news-room/fact-sheets/detail/hiv-aids. Accessed 24 April 2023

World Health Organisation. Antimicrobial resistance: Fact sheet. 2023. https://www.who.int/news-room/fact-sheets/detail/antimicrobial-resistance. Accessed 4 April 2024.

World Health Organisation. Malaria. 2023. https://www.who.int/news-room/fact-sheets/detail/malaria. Accessed 24 April 2023.

Wu G, Zhao N, Zhang C, Lam YY, Zhao L. Guild-based analysis for understanding gut microbiome in human health and diseases. *Genome Medicine*. 2021; 13(1): 1–12. https://doi.org/10.1186/s13073-021-00840-y

Xu X, Liv J, Guo F, Li J, Jia Y, Jiang D. Gut microbiome influences the efficacy of PD-1 antibody immunotherapy on MSS-type colorectal cancer via metabolic pathway. *Frontiers in Microbiology*. 2020; 11. https://doi.org/10.3389/fmicb.2020.00814

Yadav M, Shukla P. Recent systems biology approaches for probiotics use in health aspects: A review. 3 Biotech. 2019; 9(12): 448. https://doi.org/10.1007/s13205-019-1980-5

Yan J, Shin TS, Kim JS, Jee YK, Kim YK. A new horizon of precision medicine: Combination of the microbiome and extracellular vesicles. *Experimental & Molecular Medicine*. 2022; 54: 466–482. https://doi.org/10.1038/s12276-022-00748-6

Yuyun MF, Sliwa K, Kengne AP, Mocumbi AO, Bukhman G. Cardiovascular diseases in sub-Saharan Africa compared to high-income countries: An epidemiological perspective. *Global Heart*. 2023; 15(1): 15. https://doi.org/10.5334/gh.403

Zheng J, Wittouck S, Salvetti E, et al. A taxonomic note on the genus *Lactobacillus*: Description of 23 novel genera, emended description of the genus *Lactobacillus* Beijerinck 1901, and union of *Lactobacillaceae* and *Leuconostocaceae*. *International Journal of Systematic and Evolutionary Microbiology*. 2020; 70(4): 2782–2858. https://doi.org/10.1099/ijsem.0.004107

Zhu J, Pitre T, Ching C, Zeraatkar D, Gruchy S. Safety and efficacy of probiotic supplements as adjunctive therapies in patients with COVID-19: A systematic review and meta-analysis. *PLoS One*. 2023; 18(3): e0278356. https://doi.org/10.1371/journal.pone.0278356

Zucko J, Starcevic A, Diminic J, Oros D, Mortazavian AM, Putnik P. Probiotic – Friend or foe? *Current Opinions in Food Science*. 2020; 32: 45–49. https://doi.org/10.1016/j.cofs.2020.01.007

Zuo F, Zeng Z, Hammarström L, Marcotte H. Inducible Plasmid Self-Destruction (IPSD) assisted genome engineering in *Lactobacilli* and *Bifidobacteria*. *ACS Synthetic Biology*. 2019; 8(8): 1723–1729. https://doi/10.1021/acssynbio.9b00114

10 Exploring the Potential of Probiotics in Enhancing Antituberculosis Treatment

Maryam Meskini[1,2,3], *Mina Rezghi Rami*[4], *Mohammed A. E. Elmakki*[5,6], and *Seyyed Mohammad Amin Mousavi-Sagharchi*[7]

[1]Microbiology Research Center (MRC), Pasteur Institute of Iran, Tehran, Iran

[2]Mycobacteriology & Pulmonary Research Department, Pasteur Institute of Iran, Tehran, Iran

[3]Faculty of Natural and Agricultural Sciences, Department of Genetics, University of the Free State, Bloemfontein, South Africa

[4]Department of Chemistry, K. N. Toosi University of Technology, Tehran, Iran

[5]Department of Chemistry, Omdurman Islamic University, Omdurman, Sudan

[6]Department of Chemistry, University of the Free State, Bloemfontein, South Africa

[7]Department of Microbiology, College of Basic Sciences, Shahr-e-Qods Branch, Islamic Azad University, Tehran, Iran

10.1 INTRODUCTION

Tuberculosis (TB), caused by *Mycobacterium tuberculosis* (MTB), remains a major global health concern (Meskini et al., 2023; Mousavi-Sagharchi et al., 2024). A World Health Organization (WHO) estimate shows that in 2022, an estimated 10.6 million people were diagnosed with TB (95% UI: 9.9–11.4 million). It increased from 10.3 million in 2021 to 10.0 million in 2020 (Bagcchi, 2023). As a result, the downward trend that had been sustained over the last many years has reversed, continuing the trend of reverse. Globally, the burden of TB is further aggravated by the emergence of multidrug-resistant (MDR) and extensively drug-resistant (XDR) strains of this disease, which pose significant challenges to the development of effective treatments (Madadi-Goli et al., 2024). Even though the standard treatment regimen for drug-susceptible TB is effective if it is combined with antibiotics such as isoniazid (INH), rifampicin (RIF), pyrazinamide (PZA), and ethambutol (EMB), many of these medicines have adverse effects, having an impact on patient compliance and resulting in suboptimal outcomes (World Health Organization, 2020).

Also note that adverse events such as gastrointestinal disturbances and hepatotoxicity are significant challenges in the treatment of TB, which is one of the major issues in the field. Consequently, patients may fail to adhere to their treatment due to these side effects, resulting in poor adherence (Forget & Menzies, 2006; Meskini, Zamani, et al., 2024). It is becoming increasingly apparent that adjunctive therapies can enhance the efficacy of treatments and minimize adverse effects associated with anti-TB medications, which is essential due to the necessity of prolonged treatment and the potential side effects associated with these medications (Mishra et al., 2020; Saghi et al., 2015).

DOI: 10.1201/9781032676951-10

A recent research has highlighted that the microbiome plays an extremely important role in human health and disease. A microbiome is a collection of microbes that reside in various parts of the body, including the gastrointestinal tract, the respiratory system, and the skin, all of which form a complex community (Young, 2017). Particularly the gut microbiome, which is present in every single human being, plays an important role in modulating immune responses, affecting the metabolism of drugs, and generally maintaining a healthy lifestyle. It has been shown that anti-TB drugs, particularly RIF and PZA, may disrupt the gut microbiome, which can lead to dysbiosis, a condition characterized by an imbalance in microbial composition and function due to the disruption of the gut microbiome (Khaledi et al., 2024). It has been shown that dysbiosis is associated with adverse effects on the gastrointestinal tract, reduced immunity, and the development of antibiotic-resistant bacteria as well as other adverse effects. These disruptions may result in decreased compliance on the part of the patient, resulting in an increased risk of relapse or treatment failure on their part (Yu et al., 2024). Probiotics can play a crucial role in mitigating these negative effects by maintaining gut health and improving immune function since they have the ability to improve gut health and strengthen the immune system (Mazziotta, Tognon, Martini, Torreggiani, & Rotondo, 2023).

To modify immune responses, probiotics provide a promising avenue of intervention due to their ability to modify the composition of the gut microbiota, which has emerged as an important determinant of systemic immunity, as well as to modulate immune responses (Wang, Zhang, & Zhang, 2021). As part of its role in regulating immune function, the gut microbiota has a pivotal role to play, including the ability to produce short-chain fatty acids (SCFAs), modulate the integrity of the mucosal barrier, and interrelate with gut-associated lymphoid tissue (GALT) v(Rooks & Garrett, 2016). Inflammatory and infectious diseases, such as TB, have been linked to dysbiosis of the gut microbiota, a condition that disrupts the intestinal flora. There is a compelling reason for researching probiotics as an adjunctive therapy in the treatment of TB based on the intricate interplay between host immunity and the pathogenesis of MTB, the causative agent of TB disease (Hrncir, 2022). In order to evade immune surveillance and establish chronic infections within the host, MTB has evolved sophisticated mechanisms. In the context of a dysregulation of the immune response, especially in the context of the balance between Th1/Th2 cytokines, disease progression and the effect of treatment can be negatively influenced. Consequently, it is of paramount importance to ensure immune homeostasis is restored as well as to strengthen the immune system to treat TB (Zhuang, Yang, Li, Ye, & Gong, 2024).

As a result of several preclinical studies, there has been compelling evidence that probiotics can have immunomodulatory effects in patients with TB. The administration of probiotics in animal models of TB has been shown to enhance Th1 immune responses, which are characterized by increased production of pro-inflammatory cytokines such as interferon gamma (IFN-γ) and tumor necrosis factor (TNF), which are crucial for the management of the infection. Furthermore, probiotics have been demonstrated in experimental models to be effective in mitigating TB-induced lung pathology and improving survival outcomes when administered to mice with TB (Pavlidou, Fasoulas, Mantzorou, & Giaginis, 2022).

Probiotics may exert antimicrobial activity against MTB by regulating the immune system through a variety of mechanisms in addition to their direct effect on immune function. It has been shown *in vitro* that certain probiotic strains produce antimicrobial peptides and secondary metabolites that exhibit inhibitory effects against MTB (Silva et al., 2020). Moreover, probiotics can alter the gut environment in such a way as to create conditions that are unfavorable for the survival and spread of MTB. The antimicrobial properties of these probiotics complement the immunomodulatory effects of probiotics, potentially improving the overall efficacy of anti-TB therapies by enhancing the immune system (Wang et al., 2021). Despite the absence of clinical trials on the application of probiotics in adjunctive therapy among TB treatment, there has been an increase in the number of clinical trials. There have been encouraging reports from initial clinical trials, such as increased rates of sputum conversion, resolution of radiological lesions, and a decrease in inflammatory markers from the start of the trial. Probiotics have also been said to be less likely to cause undesirable effects than conventional anti-tuberculosis treatment, and that they might play a role in preventing treatment-induced toxicity from tuberculosis treatment (Guo, Li, Cai, & Kang, 2024).

Probiotics are currently being used for TB control but there are still many challenges and questions that need to be addressed. There needs to be further research to determine the best probiotic strain selection, dose regimens, and timing of supplementation when using probiotics (Sanders et al., 2013). Also, it is essential that the potential interaction of probiotics with antiTB drugs and their impact on the pharmacokinetics and effectiveness of medicines is carefully examined. In addition, there is sparse information on how host factors such as nutritional status, comorbidities, and genetic predisposition affect the effectiveness of probiotics in the treatment of TB (Purdel, Ungurianu, Adam-Dima, & Margină, 2023). The study of probiotics as adjuvant therapy in TB is promising for the augmentation of antiTB treatment by affecting host immunity and having antimi-crobal activity against MTB by modulation of host immunity. It has been found to have beneficial impacts on immune response, lung lesion, and cure rate in preclinical models, and results of preliminary clinical trials have been promising. To optimize probiotic therapy, address safety concerns, and elucidate mechanisms of action against TB, even more research must be conducted (Otdushkina, Zakharova, Kholodov, & Pyanzova, 2023; Rabetafika, Razafindralambo, Ebenso, & Razafindralambo, 2023). Through accessing the possibility of probiotics, it is perhaps possible to create new strategies of improving TB control and reducing the global burden of this revolting illness as well as lowering the rate of death (O'Toole, Marchesi, & Hill, 2017). We clarify in this chapter why probiotics should be included in antiTB treatment regimens and further elaborate on the potential mechanisms whereby probiotics may contribute to the therapeutic efficacy in anti-TB drugs.

10.2 PROBIOTICS: BASICS AND MECHANISMS

10.2.1 Definition and Types of Probiotics

A living bacterium is what the word "probiotic" has come to mean when administered to a host in adequate quantities and has the ability to enhance the health of that host. Although bacteria make up the vast majority of probiotics, some yeasts are also capable of performing the same function. Some examples of fermented foods that include probiotics are yogurt, kefir, kimchi, and sauerkraut. Other examples include sauerkraut. Nutrient supplements are another possible source of these nutrients (FAW/WHO, 2006; Salminen et al., 2021). "Probiotic" means "for life" in ancient Greek. According to Markowiak et al, Ferdinand Vergin most likely originated the expression "probiotic" during his 1954 essay ("Anti-und Probiotika"). In the essay, Vergin contrasted the gut microbiota and how antibiotics and other antibacterial medications have impact on it, both positively and negatively (Maftei et al., 2024; Markowiak & Slizewska, 2017).

Lilly and Stillwell used the term "probiotics" in 1965 to characterize chemicals produced by one kind of bacteria that promote the expansion of another kind of bacteria (Lilly & Stillwell, 1965). In 1989, Fuller began the contemporary age of probiotics by stating that food fermentation uses live active cultures and that certain microbial strains are introduced to food goods. In addition to their use in food supplements and topical medications, Fuller noted that probiotics have found their way into cosmetics such as gels and soaps (Fuller, 1989). "Live microorganisms which, when administered in adequate amounts, confer a health benefit on the host" is the current scientific consensus regarding probiotics, according to a 2002 proposal by the WHO and the Food and Agriculture Organization of the United Nations. The Argentinian government sought this definition; it resulted from a joint FAO/WHO Expert Consultation that took place amid an international trade dispute over powdered milk that included lactic acid bacteria (LAB) (FAW/WHO, 2006). Even among members of the same species of microbes, the effects and advantages of probiotics are dose-dependent, as stated in the FAO/WHO Expert Consultation (Britton, Hoffmann, & Khoruts, 2021). The ISAPP, an international organization that studies probiotics and prebiotics, likewise agreed with that description (Hill et al., 2014). On the other hand, probiotics are live microbial feed additives that help maintain a healthy gut flora in the host animal. The original aim of probiotics was to alter the gut flora to promote human and animal health. At this time, several *Bifidobacteria* and *Lactobacilli* strains have been investigated extensively that people may use to cure or prevent gastrointestinal (GI) symptoms. Dietary probiotics provide several health benefits, including regulating the microbiota in the gut, boosting the immune system, increasing the bioavailability and synthesis of nutrients, and providing relief for those with lactose intolerance while also decreasing the likelihood of certain diseases (Thilagavathi, 2020).

Although many different types of probiotics have been found to improve human health, the bulk of these benefits have been related to specific strains of probiotics. Probiotics are derived from a range of organisms, including yeasts, *Bacillus*, *Lactobacillus*, *Bifidobacterium*, and *Pediococcus* (Soemarie, Milanda, & Barliana, 2021). Evidence suggests that some probiotic strains belonging to the following genera may be beneficial to human health: *Enterococcus*, *Saccharomyces*, *Bifidobacterium*, *Leuconostoc*, *Bacillus*, *Streptococcus*, *Pediococcus*, and *Escherichia coli* (Fijan, 2014).

Lactobacillus: Lactobacillus belongs to the family of many rod-shaped, facultatively anaerobic, or microaerophilic Gram-positive bacteria. Their ability to convert hexose carbs to lactic acid creates an acidic environment that inhibits the growth of several harmful bacteria. Among the lactic acid bacteria (LAB) family, they are an important member, which also includes *Enterococcus*, *Leuconostoc*, *Pediococcus*, *Streptococcus*, and *Lactobacillus* (Makarova et al., 2006).

Bifidobacterium: Anaerobic bacteria, many of which are Gram-positive, are prevalent inside the genus *Bifidobacterium*. Numerous animal gastric tracts and vaginas, including humans, are home to these endosymbiotic bacteria. Because of their various resistance mechanisms to bile salts, probiotic microorganisms from the species Bifidobacterium are also often used. The presence of this bodily fluid is crucial for the formation of probiotic bacteria, which are responsible for their beneficial effects (Fijan, 2014).

Saccharomyces: There are many different kinds of yeasts in the genus *Saccharomyces*. Some examples include *Saccharomyces cerevisiae*, which is used to make bread, beer, and wine; both *Saccharomyces bayanus* and *Saccharomyces boulardii* have medicinal and oenological uses. Kefir is also made by Saccharomyces yeasts in symbiotic matrices with bacteria (Witthuhn, Schoeman, & Britz, 2005). *S. boulardii* has excellent probiotic properties, including the ability to thrive in the stomach's acidic environment and improve the microbiota in the intestines. It is the most common and widely examined species exhibiting these characteristics. There is new evidence that employing probiotic yeast in food creation is a feasible approach. This has opened up new possibilities for a more integrated approach that may leverage *S. boulardii* to create innovative functional food products (Souza et al., 2022). One common probiotic yeast used to treat gastrointestinal issues, such as diarrhea, is *S. boulardii*. It has been argued for a long time whether it is the organism in question is either a separate species or a genetic variety of the model yeast *S. cerevisiae*, since it has a significant resemblance in its genetic composition. How well the probiotic can adapt to the host's physiological circumstances is determined by the primary genetic difference between *S. boulardii* and S. cerevisiae. The optimal growth temperature, gastric resistance, and viability at low pH are some of the distinguishing morphological and physiological traits of *S. boulardii* that contribute to its probiotic effectiveness. Increased gut barrier function, pathogen competitive exclusion, antimicrobial peptide production, immunological modulation, trophic effects, and other mechanisms have all contributed to its probiotic effectiveness (Pais, Almeida, Yilmaz, & Teixeira, 2020).

Lactococcus: A family of Gram-positive LAB commonly used often to make fermented dairy products. Acidification is key in inhibiting the formation of spoilage microorganisms in milk products (Fijan, 2014). *Lactococcus lactis* is a kind of bacterium that makes lactic acid.

It is deemed safe for human consumption and is widely utilized in the food sector, notably in the preparation of fermented dairy products. Several investigations have looked into lactococci's technical and probiotic qualities. However, the probiotic qualities of the strains of *L. lactis* often found in dairy products have received surprisingly little attention in research. Researchers Chiara et al. studied the probiotic characteristics of *L. lactis* strains isolated from natural whey starter cultures. These characteristics included their ability to survive in gastric juice simulations, tolerance of bile salts, hydrophobicity, and auto- and co-aggregation. The results revealed that several strains under investigation have strong probiotic properties, as shown by their high levels of hydrophobicity and auto-aggregation, as well as their low co-aggregation with the tested pathogenic strains. To determine which of four highly desirable strains might prevent *Salmonella typhimurium* and enteroinvasive *E. coli* (EIEC) from adhering to epithelial cells, a co-cultured epithelial cell model was used. Strains A3, A5, I4, and I7 of *L. lactis* were identified and capable of both pathogen competition and providing protection to cells that had suffered from a history of *E. coli* or *Salmonella* infections. To create ground breaking functional meals, it is necessary to identify new probiotic LAB strains in dairy products (Chiara et al., 2024).

Streptococcus* and *Enterococcus: *Streptococcus pyogenes*, *Streptococcus pneumoniae*, and vancomycin-resistant *Enterococcus faecium* are only a few strains of the LAB that are associated with major healthcare-associated infections (Hossain, 2014). The researchers Ayyash et al. used M17 agar to extract two species of LAB—*Streptococcus* and *Enterococcus*—from raw camel milk. These bacteria displayed significant probiotic properties, rendering them viable contenders for the production of functional foods that may offer health advantages. In their investigation, eight LAB strains were extracted from camel milk. These strains demonstrated notable efficacy in removing cholesterol, the ability to inhibit pathogen growth, susceptibility to antibiotics without experiencing hemolysis, and a modest tolerance for acid and bile (Ayyash et al., 2018).

Bacillus: *Bacillus subtilis*, *Bacillus cereus*, and *B. subtilis* coagulans are members of the aerobic Gram-positive spore-forming or facultative aerobic classes of the genus *Bacillus* that are purported to possess probiotic properties. In conjunction with other microorganisms, *Bacillus coagulans* has demonstrated the greatest efficacy in the management of antibiotic-related diarrhea and its prevention (Fijan, 2014; Hempel et al., 2012). In their investigation, Lee et al. examined the probiotic properties of species of Bacillus, both in their spore and vegetative states. These forms exhibit remarkable resistance to environmental stresses, including heat, gastric conditions, and moisture. *B. subtilis*, *B. cereus*, *B. clausii*, *B. coagulans*, *B. licheniformis*, and *B. polyfermenticus* are the commercial Bacillus probiotic strains in use. In addition to their antibacterial and anticancer capabilities, these strains also demonstrate antioxidant and vitamin-producing capabilities. On the other hand, probiotics produced by Bacillus may transmit genes that make bacteria resistant to antibiotics and produce toxins and biogenic amines; thus, their safety is a concern. Research utilizing probiotic Bacillus strains to examine the microbiome in humans is restricted. Research on the microbiome has predominantly focused on chickens, rodents, and pigs. As probiotic carriers, certain Bacillus probiotics are utilized as fermentation initiates in soybeans and plants, as well as dietary supplements in baked goods. A synopsis of the characterization of Bacillus species concerning human probiotics, such as their security, microbiota, and probiotic transporter, is provided in their study (Lee, Kim, & Paik, 2019). In their investigation, Soren et al. explored the impacts of *B. subtilis* probiotics on a multitude of poultry production parameters, including carcass characteristics, gut histomorphology, immunity, and gut microbiota. Supplementation with Bacillus subtilis probiotics may enhance growth performance, alter the gastrointestinal microbiome, and stimulate the immune system, according to the findings; the carcass trait was not affected. Probiotic effectiveness in poultry production is contingent on the condition of the birds' intestines, the rate of probiotic incorporation, the quality of the forage, and water supplied (Soren, Mandal, Roy, Samanta, & Hansda, 2023).

Escherichia: *E. coli* is a commensal in the gut microbiota, yet it and other Enterobacteriaceae members are known to be pathogenic. *E. coli* colonizes first and stays permanently, and certain strains secrete antimicrobial colicin. The best-studied beneficial strain is 1917-isolated *E. coli* Nissle. Adaptability to withstand varied circumstances and easy genetic alterations allowed *E. coli* strains to provide antioxidant, anti-inflammatory, and anticancer compounds. As synbiotics, genetically edited *E. coli* strains that converted sucrose and fructose into insulin and mannitol had a strong impact on avoiding metabolic illness. *E. coli* is becoming a powerful probiotic platform for generating strains that may manage metabolic and multifactorial disorders, including cancer (Gattupalli & Gattupalli, 2021).

10.2.2 Mechanisms of Action in the Gastrointestinal Tract

The primary obstacle related to probiotics in the food business is their vulnerability to processing conditions and their sensitivity to gastrointestinal stressors. Nevertheless, when it comes to health advantages, consumers consistently exhibit a strong preference for probiotic products (Konuray & Erginkaya, 2018; Meskini, Amanzadeh, et al., 2024). The field of probiotics has made tremendous strides, but a pivotal discovery regarding the elucidation of their mechanism of action is still pending. Probiotics may have a beneficial effect on the human body by competitively excluding harmful bacteria, enhancing the function of the intestinal barrier, modulating the immune system, and producing neurotransmitters. Competing with pathogens for nutrients and receptor-binding sites, probiotics make it harder for pathogens to survive in the gut (Plaza-Diaz, Ruiz-Ojeda, Gil-Campos, & Gil, 2019). Chemicals like SCFAs are produced by probiotics, which provide them with antibacterial properties, organic acids, and hydrogen peroxide (Ahire, Jakkamsetty, Kashikar, Lakshmi, & Madempudi, 2021). In addition, probiotics regulate the gastrointestinal immune response, modulate the activity of tight junction proteins like occludin and claudin 1, and promote mucin protein production, all of which contribute to the intestinal barrier's improved function (Bu et al., 2022; Chang et al., 2021; Ma, Son, Yoo, Joo, & Kim, 2022). Probiotics have the ability to control and adjust the natural and acquired immunological reaction by influencing lymphocytes (both B and T), dendritic cells (DC), and macrophages. Probiotics enhance the production of anti-inflammatory cytokines by engaging with the cells that line the intestines and attracting immune cells called macrophages and mononuclear cells. Moreover, probiotics may create neurotransmitters in the gastrointestinal system via the gut–brain axis. Specific probiotic strains have the ability to regulate the levels of serotonin, gamma-aminobutyric acid, and dopamine, which in turn impact mood, behavior, gastrointestinal movement, and pathways connected to stress (Gangaraju, Raghu, & Siddalingaiya Gurudutt, 2022; Sajedi, Shabani, & Elmieh, 2021).

There are more than a hundred trillion microbes in the human digestive system, including bacteria, viruses, fungi, and protozoa, representing at least a thousand different species (Wang, Yao, Lv, Ling, & Li, 2017). A minimal number of bacteria are present in the stomach and duodenum due to the inhibitory or eradication impacts of pancreatic secretions, bile, and stomach acid. The population of microorganisms inside the gut gradually rises as one moves farther down, with the jejunum containing around 10^4 cells per gram and the distal ileum containing around 10^9 cells per gram. The colon is the most densely inhabited area of the gastrointestinal system, mostly by anaerobic microorganisms. It harbors up to 10^{12} cells per gram of intestinal material (Guarner et al., 2012). The local chemical environment, oxygen levels, nutrients in the host tissue, and immune system participation are the precise circumstances that bacteria need to establish themselves in the human body. Consequently, the effects of probiotics are intricate and can vary depending on the specific strain. Understanding the intricate factors that bacteria rely on to thrive within the human body involves examining the local chemical environment, oxygen levels, nutrients in the host tissue, and the role of the immune system, either individually or in various combinations. Research has shown that certain probiotic microorganisms may enhance resilience and restore balance to the gut microbiota after disruption caused by antibiotics or other stress-inducing factors. However, scientists have challenges in precisely defining the characteristics of a healthy microbiome (Jenkins & Mason, 2022).

The populations of invading microbes varied between healthy and sick individuals, according to Wang et al. Hence, racial/ethnic background, age, gender, and location all have a role in the microbiome. In light of the above, there is a considerable variation in the variety of gut microbes among individuals. Varieties of genes, the amount of microbes introduced into the body at birth, and dietary habits all contribute to the unique microbiome that each person has (Wang et al., 2017).

Upon entering the intestine, probiotics engage in interactions with intestinal microorganisms. The intestinal mucosa mainly acts as a physical barrier to keep harmful chemicals out of the gut and the intestinal lumen. In reaction to bacteria that have made it into the intestines, probiotics strengthen the four barriers: mechanical, chemical, biological, and immunological (Cheng, Pan, Chang, Jiang, & Sang, 2020). Upon entering the intestine, these substances engage in interactions with intestinal cells in an effort to return normal function to the intestinal mechanical barrier, promote mucosal regeneration, increase mucus production, maintain mucosal barrier integrity, and restore intestinal permeability (Toumi et al., 2013).

Considerable advancements have been noted in the domain of probiotic research in recent times, particularly with regard to the identification and attributes of distinct probiotic cultures, as well as their effects on people's well-being. This is because probiotics may exhibit unique patterns of temporary colonization in the human gut, their mode of action against diseases is understood to be mediated by a variety of mechanisms. These mechanisms vary depending on the microbiota of the individual, the GI tract area, and the probiotic strain or strains (Maftei et al., 2024; Zmora et al., 2018).

According to published data, probiotics have been found to exert their main effects by working on the epithelial barrier, improving adhesion to the intestinal mucosa, producing antibacterial substances, competing with pathogens, inhibiting bacterial adhesion, and influencing the immune system (Devendra Singh & Agarwal, 2021; Kang et al., 2023; Singh, Singh, Bhushan Mishra, Sharma, & Agarwal, 2022).

Probiotics disrupt the development of pathogenic microorganisms within the gastrointestinal ecosystem. This is achieved, for instance, by expediting the passage of food through the gastrointestinal tract and impeding the capacity of pathogenic bacteria to adhere to and colonize the lining of the digestive tract. Another way probiotics work is by lowering the colon's pH by increasing the synthesis of bioactive metabolites such as SCFAs. Enzyme activity, bile salt metabolism, vitamin production in the GI tract, and toxin neutralization are other action pathways (Guo et al., 2019).

The mechanism by which it inhibits the development of harmful bacteria by directly attaching to Gram-negative bacteria, enhances the intestinal electrolyte absorption, and modulates lipid metabolism is described here. The improvement of the function of the intestinal barrier and the reduction of the production of pro-inflammatory indicators (such as cytokines) result from the biochemical signaling mechanisms by which probiotics interact with cells both within and outside the gastrointestinal tract (Milajerdi et al., 2020).

Specific cell types make up the mucous layer and the intestinal epithelium, both of which act as physical barriers against intestinal microorganisms. For instance, upon contact with intestinal bacteria, Paneth cells synthesize and secrete antimicrobial peptides; Saccharomyces cerevisiae cells contribute to mucus production; enterocytes absorb molecules from the intestinal lumen; and endocrine cells regulate hormonal signaling. Together, these cells form the structure and function of the intestinal epithelium (Burgueno & Abreu, 2020; Wu et al., 2019).

The main tasks of the intestinal epithelium mucus layer are to provide a barrier against the harmful luminal environment, allow food move through, and prevent the attachment of disease-causing substances to the intestinal wall. The intestinal mucosa is the initial component that functions as a physical barrier, ensuring that harmful compounds throughout the digestive tract remain at a safe distance from the intestine. Once probiotics infiltrate the intestine, they engage in interactions with intestinal cells and microbes to improve their physical, biological, chemical, and immunological defenses (Cheng et al., 2020). The major goals include increasing mucus production, restoring intestinal permeability, encouraging mucosal regeneration, and preserving the integrity of the mucosal barrier and proper mechanical barrier function of the intestinal mucosa (Toumi et al., 2013). By means of these mechanisms, probiotics are capable of exerting a significantly more substantial impact on human health and diseases, encompassing both children and adults (Maftei et al., 2024).

10.2.3 Immunomodulatory Effects of Probiotics

Compensatory microorganisms and immune cells are able to develop harmonious relationships in the intestinal environment (Zhang, Tang, Chen, Xie, & Tao, 2019). The progressive development of this heterogeneous ecosystem has been facilitated by repeated encounters with various antigens. Constantly in proximity to intestinal commensal microorganisms, the cellular and molecular components constituting gut immunity contribute to understanding the control of immune responses and the delicate balance of gut homeostasis (Mazziotta et al., 2023). Dysbiosis can occur when there are modifications with regard to the intestinal microbiota composition. This has the potential to set off a cascade of events that are linked to the maturation, specialization, and proliferation of gut immune cells. Metabolic disorders (Martinez, Leone, & Chang, 2017), chronic inflammation (Wang, Chen, & Wang, 2020), cancer (Sharma et al., 2021), and neurodegenerative diseases mediated through the gut–brain axis (Rutsch, Kantsjo, & Ronchi, 2020) are all associated with the emergence of severe health problems (Wang et al., 2019) when there is a disruption in this intricate microbial community. With regard to the restoration of intestinal microbial diversity, probiotics have been proposed as a potentially effective therapeutic approach (Plaza-Diaz et al., 2019).

In addition to producing bioactive compounds and competitive exclusion and inhibition of detrimental invading pathogens, probiotics exert their immunomodulatory effects on the host through colonization of the disrupted mucosal barrier; bacteriocins have recently attracted a lot of interest as antibacterial agents. Bacteriocins have long been used to extend the shelf life of food. Their potential as immune-modulating agents and therapeutic antimicrobials, however, is attracting more and more attention. Antibiotic-resistant bacterial infections have been identified as a worldwide public health problem. There is hope that bacteriocins, which may either kill all bacteria or only those that are resistant to antibiotics, can end this worldwide epidemic. Notably, despite its significance in ensuring the safety of food as a natural substitute for chemical preservatives, nisin is still the only bacteriocin that regulatory bodies have officially authorized as a food preservative. Furthermore, inadequate evidence on the safety and toxicity of bacteriocins provides a barrier to the more widespread use of bacteriocins in the food and medical industries (Soltani et al., 2021).

Probiotics' immunomodulatory activities have received a lot of interest as a therapy for degenerative and other disorders caused by pathogenic microbes. Probiotics improve innate immunity by exhibiting numerous antiviral activities. Furthermore, studies have shown that probiotics may activate B cells and affect cytokine production, therefore improving the gut barrier's function. Consequently, the body's adaptive mechanisms are set into motion. Nevertheless, enough information regarding the precise mechanisms by which probiotics exert immunomodulatory benefits in inflammatory therapy is currently lacking. Understanding cytokine release by Th2 cells, DCs, monocytes, B cells, and Tregs is also critical for developing novel probiotic strains (Azad, Sarker, & Wan, 2018).

In the human intestine, immune cells and commensal microbes have a constant line of communication and interaction, working together to ensure a stable and healthy immune system. To fully grasp the immune system's relationship with microbiota, one must have an in-depth familiarity with the complex mechanisms that regulate immunogenicity and tolerance. Probiotic bacteria have the ability to interact with and stimulate immune cells in the intestines, as well as the microflora that naturally reside there. This interaction helps to regulate and maintain specific immune functions and overall immune balance. Probiotic microorganisms have remarkable immunomodulatory and health-promoting characteristics, according to mounting data. Therefore, probiotics may provide a potential strategy to enhance immune system activities. Research on the effects of probiotics on the immune system is still in its infancy. On the other hand, their metabolic and nutritional qualities have been the subject of a great deal of research. Because of this gap in our understanding, we do not yet know how probiotics interact with the immune cells of the host. The latest findings and their implications of the interaction between probiotic bacteria and immune cells in strengthening immune functions were investigated by Mazziotta et al. Therefore, their research provided a synopsis of the current understanding of how probiotic bacteria modulate the immune system of their hosts (Mazziotta et al., 2023).

In their research, Al-Najjar et al. emphasized the important role of multi-species probiotics in boosting immune function. The immunostimulatory effects shown in the research are probably caused by the secretions and metabolites of different species of bacteria that are included in the probiotics. Unfortunately, the precise mechanisms behind these immunomodulatory effects remain a mystery until a thorough chemical investigation of probiotic secretions is conducted. Finding out what components make up probiotic secretions and what bioactive qualities they possess, such as phenolic and flavonoid chemicals, should be a top priority for future study. This may provide a deeper understanding of their medicinal possibilities. When designing functional foods and supplements to improve health and wellness, it may be very beneficial to have a thorough grasp of the connection between probiotic secretions and immunological activity (Al-Najjar et al., 2024).

When it comes to controlling dendritic cells, using the probiotic strain *L. rhamnosus* JB-1 has several benefits. This is accomplished by triggering the synthesis of heme oxygenase and activating pattern recognition receptors such as DC-SIGN and TLR-2. This research emphasizes that L. rhamnosus JB-1 may have anti-inflammatory properties. By stimulating DCs generated from human monocytes, this strain can block the expression of co-stimulatory molecules, cytokine synthesis and maturation, and TH1/TH17. It is well established that *L. rhamnosus* JB-1 possesses immunomodulatory effects, including Foxp3 expression and IL-10 induction. Cell wall components of probiotic bacterial strains influence the immunomodulation of DCs. Dendritic cells' TLR-2 receptors are activated by capsular polysaccharide, which in turn triggers T-helper cells to produce IL-10. Consequently, it helps alleviate the inflammatory reaction associated with colitis (Siddique, Akram, Alghamdi, Arshad, & Siddique, 2022).

Improved gut microbial population, increased mucus production, and protection of tight junction proteins against lipopolysaccharides (LPSs) are all benefits of probiotics. As a consequence of LPS binding to endothelial cell toll-like receptors (TLR 2, 4), DCs, and macrophages are activated, leading to an increase in inflammatory markers. Additionally, probiotic treatment may reduce dysbiosis in the gut and intestinal permeability, which in turn may forestall the development of inflammatory biomarkers and needless immune system activation. To better target Th2 cells, probiotics may enhance T-cell differentiation and promote the production of Th2 cytokines including IL-4 and IL-10. From a broad biological plausibility to specific *in vitro* and *in vivo* evidence that probiotics are beneficial, this review delves into the interplay between gut microbiota and the immune system. Additionally, it delves into the possible uses of state-of-the-art treatment methods (Cristofori et al., 2021).

Complex probiotics composed of various strains of bacteria have shown a more pronounced impact on immune regulation. As an illustration, VSL#3, (Cheng et al., 2020) this product, classified as a medical food, contains a total of eight live strains of bacteria. These include four strains of Lactobacillus, three strains of Bifidobacterium, and one strain of Streptococcus. The probiotic qualities of *B. lactis* XLTG11 were found in the intestines of healthy newborns. Immune system regulation, inflammation reduction, intestinal barrier strengthening, and gut microbiota balancing are all under its purview (Xu et al., 2022). The Lactobacillus plantarum P8 strain may thrive in the human gut and alter the microbiome in a beneficial way. Additionally, it may help improve the barrier function of the gut flora, which in turn helps reduce the presence of dangerous bacteria and encourages the development of good bacteria. Furthermore, studies have shown that it positively affects humoral immunity, cellular immunity, and the immunological response of lymphoid tissues in the intestines, among other parts of the immune system (L. Wang et al., 2014). Research has shown that *L. casei* Zhang can boost the Th1 to Treg ratio and promote antibody class switching in activated B cells during the B cell immune response. Through the expression and production of IgG, IgA, or IgE, the organism can exert immunomodulatory effects, much like a neuroscientist studying the intricacies of the brain (Fu et al., 2020; Wang et al., 2013).

In their study, Ma et al. examined the immunomodulatory effects of a complex probiotic on mice that had been immunosuppressed using cyclophosphamide (CTX). Thus, in this experiment, a complex probiotic powder was created by combining three strains known for their exceptional probiotic properties. The aim was to investigate the impact of this powder on the host's immune system. Complex probiotics shown potential in enhancing splenic and thymus function and intestinal structure, according to the research. Mice given the complex probiotics showed an increase in cell counts (both white and red) and a rise in the proportion of lymphocyte subpopulations in the spleen. Furthermore, after the complex probiotic administration, an upregulation of cytokine levels (IL-6, IL-10, IL-1β, and IFN-γ) was seen. These advanced probiotics have the potential to rebalance the disrupted gut microbiota. This study offers theoretical backing for the immune-boosting capabilities of a combination of microbial strains, including *B. lactis* XLTG11, *L. casei* Zhang, and L. plantarum P8. It also provides a foundation for the future development and utilization of this preparation from a pharmacological standpoint (Ma et al., 2023).

10.3 TB TREATMENT LANDSCAPE

Anti-TB medications have been arranged according to their clinical efficacy and tolerability. These medications, which target both drug-sensitive (DS) and drug-resistant (DR) strains of the illness, are the cornerstone of treatment regimens for TB. It is essential to comprehend the characteristics and classification of these medications to create efficient treatment plans and reduce the prevalence of TB worldwide (Ramachandran & Swaminathan, 2015).

First-line medications, such as RIF, INH, PZA, EMB, and streptomycin, make up the main drug arsenal for treating DS-TB (Figure 10.1). All these medications target distinct parts of the physiology of the TB bacteria, which together help to eradicate the infection (Xavier, Sharumathi, Parthasarathy, Mani, & Mohanasundaram, 2023). While RIF interferes with RNA synthesis, INH inhibits the synthesis of mycolic acid, which is necessary for the development of cell walls. EMB interferes with the development of cell walls, streptomycin prevents the synthesis of proteins, and PZA targets dormant bacteria. These medications work well together to treat DS-TB with the least amount of toxicity, which is why they are the mainstay of TB treatment regimens worldwide (Zade & Raut, 2023). TB control efforts are confronted with substantial obstacles due to the growth of DR strains, such as XDR and MDR-TB. In these situations, first-line medications are frequently ineffective, and second-line medications, sometimes referred to as reserve medications, must be used. These medications, which include ciprofloxacin, amikacin, and capreomycin, have different modes of action but are often more harmful and less effective than first-line medications. Second-line medications are essential for treating DR-TB and stopping its spread, despite several disadvantages (Monde et al., 2023). When all other TB treatment options have failed, third-line medications are the last choice. Certain antibiotics, such as amoxicillin/clavulanate and linezolid, are frequently used in combination with other medications as salvage therapy because of their dubious efficacy against TB (Koehler et al., 2023). They nevertheless play a significant role in managing patients that are challenging to treat and filling in the gaps in the current treatment regimens, even though their role in TB treatment is restricted.

The WHO has taken the lead in creating treatment guidelines for TB, periodically revising them to account for developments in the field (de Lima Corvino, Shrestha, Hollingshead, & Kosmin, 2023). Based on variables like medication resistance patterns and treatment histories, these guidelines provide treatment plans customized to patients' individual requirements. There have been attempts in the past few years to decrease treatment durations and simplify TB treatment regimens, especially for MDR-TB. An important development in the therapy of TB is the introduction of shorter treatment regimens lasting 9–12 months for certain individuals. The WHO's recommendation to shorten

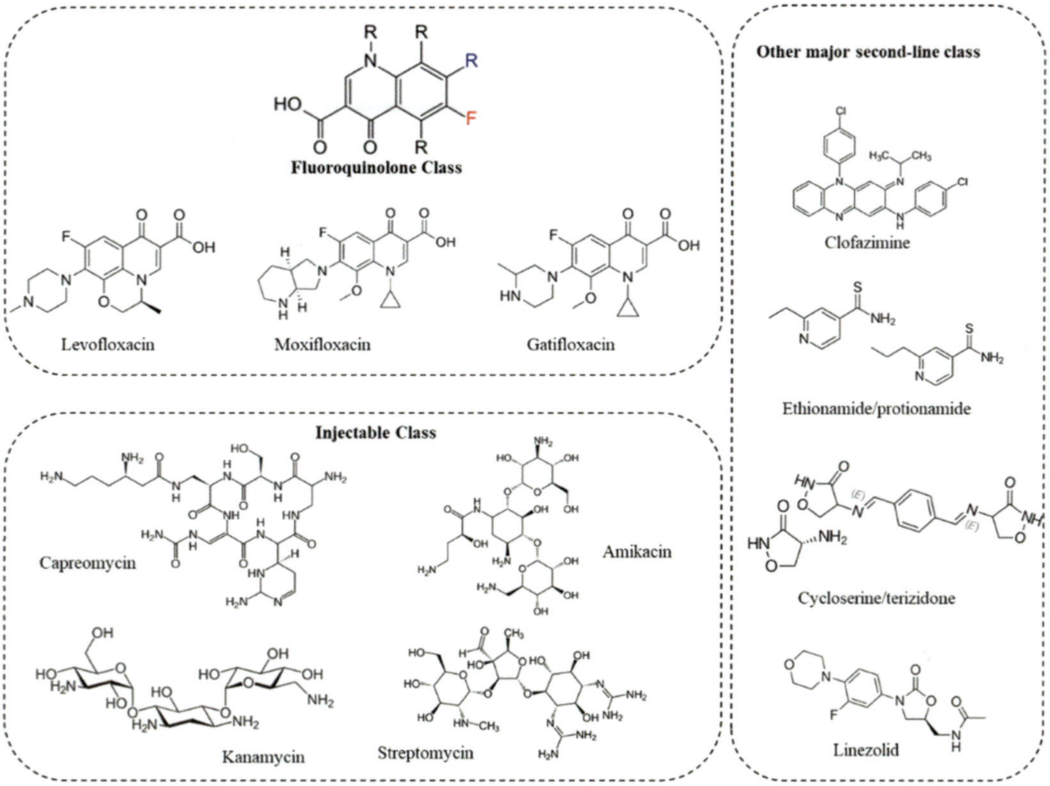

FIGURE 10.1 Common drugs for treating TB.

the course of MDR-TB therapy to a duration of 9–12 months was a key component of their plan. The goal of this recommendation was to promote patient compliance, expedite treatment procedures, and eventually improve MDR-TB cases' treatment outcomes. Healthcare professionals could maximize resources, lower the chance of medication-related side effects, and lessen the financial strain of lengthy treatment plans by cutting the length of treatment. By following these regimens, treatment outcomes can be improved and the impact of MDR-TB on healthcare systems and impacted populations can be lessened (Motta et al., 2024). Even with these developments, there are still issues with TB therapy optimization and treatment success.

Current research is concentrated on creating novel medications and TB treatment strategies to overcome these issues. The creation of new antimicrobial agents and the repurposing of current medications show promise in terms of increasing treatment options and addressing antibiotic resistance (Ahmed et al., 2023). Moreover, initiatives to enhance diagnostic equipment and fortify healthcare infrastructure are critical to the early detection and efficient treatment of TB (Roy et al., 2024). As a result, the field of TB treatment is intricate and multidimensional, offering physicians a wide range of medications and treatment plans. Comprehending the categorization and characteristics of anti-TB medications is crucial in formulating efficacious therapeutic approaches and addressing the worldwide TB outbreak. We can endeavor to eliminate TB as a worldwide health hazard by tackling obstacles including medication resistance, treatment duration, and toxicity.

The mainstay of TB treatment for a long time has been first-line medications, which mainly target DS-TB strains. These medications were also suggested for decreasing treatment durations and strengthening treatment regimens for MDR-TB cases, indicating the need for more potent treatment alternatives. The main objective of anti-TB therapy is to give medical professionals safe and efficient drugs that can cure all strains of TB while also lowering treatment costs and duration. Modifying currently available anti-TB medications is one way to accomplish this, and it offers a simple and affordable path for the creation of new therapy alternatives (World Health Organization, 2019).

For more than 60 years, INH and its variants have been the essential first-line medications in the TB treatment. INH, an isonicotinic acid hydrazide, is the most effective and powerful first-line medication. Its bactericidal concentration for DS strains of *Mycobacterium bovis* and MTB usually varies between 0.02 and 0.2 µg/mL. INH is a prodrug that needs the mycobacterial cell's catalase–peroxidase enzymes to activate it. Once activated, INH targets multiple MTB genes, with inhA being especially important, to deliver its antimicrobial effects (Natalia, Syah, Retnoningrum, & Kusuma, 2016). INH mainly acts by inhibiting the formation of mycolic acid in the bacterial cell wall, which damages the integrity of the cell wall and ultimately results in the death of the bacteria. Despite its amazing effectiveness, INH has faced difficulties. As a result of incorrect and continuous administration over time, resistance to the medicine has developed in over 10% of TB patients worldwide.

One notable derivative of INH, known as Sudoterb or LL-3858, features a pyrrole core in addition to the INH moiety. This derivative has demonstrated promising activity against both DS and MDR strains of MTB, with minimal inhibitory concentrations (MICs) values ranging from 0.06 to 0.5 µg/mL (Hu et al., 2017). Developed by Lupin Ltd., Sudoterb is currently undergoing clinical trials, highlighting its potential as a future treatment option (Tandon & Nath, 2017). Among a series of N-alkyl-2-isonicotinoylhydrazine-1-carboxamides, the N-methyl-substituted derivative emerged as the most active, exhibiting efficacy comparable to that of INH but with lower toxicity. This derivative shares a similar mechanism of action with INH, targeting InhA and inhibiting mycolic acid synthesis within MTB cells (Rychtarčíková et al., 2014).

Furthermore, condensation products of INH with fatty acids, including saturated ones, have shown significantly improved antimycobacterial properties against DS and DR-TB compared to INH and RIF. Enhanced lipophilicity plays a crucial role in augmenting the antimycobacterial activity of these derivatives. Their mechanism of action involves the inhibition of enoyl-ATP reductase (InhA), similar to INH, thereby blocking mycolic acid biosynthesis (Rodrigues et al., 2013). Additionally, certain INH derivatives containing 1,2,3-triazole fragments have exhibited promising antimycobacterial activity against MTB strains, with one derivative demonstrating particularly high efficacy *in vitro*. Similarly, linked products of INH derivatives and 1,3,5-triazine with substituted 2-aminopyridines have shown potent inhibitory effects against MTB growth (Kumar et al., 2014; Raval, Patel, Patel, & Patel, 2011).

Another promising avenue in drug development involves furoxans, which generate nitric oxide (NO) during biotransformation. NO has been shown to possess antimicrobial properties, capable of damaging bacterial DNA and inducing macrophage apoptosis. Phenylsulfonyl derivatives of INH and furoxans, with INH fragments strategically positioned, have exhibited highly promising activity against both DS and DR-TB (dos Santos Fernandes et al., 2016). Modifications to the hydrazine part of INH have been explored to block acetylation, a predominant metabolic pathway that can reduce INH's antimycobacterial activity. Hydrazide-hydrazone derivatives of INH have shown excellent antimycobacterial activity, offering potential as future treatment options for TB (dos Santos Fernandes et al., 2017).

Moving beyond INH, RIF, and its derivatives have also been extensively studied for their efficacy against MTB. RIF, a key first-line drug, exerts potent bactericidal activity and is often used in combination with INH for DS-TB treatment. Several RIF derivatives, such as rifalazil (RLZ), rifabutin (RBT), and rifapentine (RPT), have been synthesized to enhance activity and prolong metabolic half-life (Naveen Kumar et al., 2014).

PZA has emerged as a significant player in the treatment of TB since its introduction as an anti-TB drug in 1980. It holds a unique position as a sterilizing agent, playing a crucial role in shortening the duration of anti-TB therapy from 9 to 12 months to just 6 months (Hameed et al., 2018).

Functioning as a prodrug, PZA is transformed by the bacterial enzyme pyrazinamidase into pyrazinoic acid (POA) within the mycobacterial cell. POA then acts to inhibit fatty acid synthesis in mycobacteria, ultimately leading to bacterial death. This mechanism of action underscores PZA's efficacy in targeting TB bacteria and highlights its inclusion in TB treatment regimens featuring new drugs currently undergoing clinical trials (Kwon, 2017).

Exploration into PZA derivatives has yielded intriguing findings, with synthetic analogs such as 5-chloropyrazinamide (5-ClPZA) exhibiting activity against MTB, albeit with varying degrees of effectiveness. Similarly, derivatives based on 5-chloro-N-phenylpyrazine-2-carboxamide anilides have demonstrated high activity against various strains of MTB, including MTB H37Rv, *M. kansasii*, and *M. avium*. These derivatives exhibit promising antimycobacterial properties, with some compounds showing a broad spectrum of activity against test strains (Ahmad et al., 2012; Zitko et al., 2013). Furthermore, derivatives based on 3-chloropyrazine-2-carboxamide and N-alkyl modifications of PZA have shown considerable potential in inhibiting MTB growth. Compounds such as N-(2-morpholinoethyl) pyrazine-2-carboxamide display notable bacteriostatic effects against MTB strains, highlighting their efficacy as potential anti-TB agents (Jandourek et al., 2017; Zhou, Yang, & Huang, 2017).

RIF and its derivatives play a crucial role in TB treatment, with RIF being a cornerstone first-line drug due to its potent bactericidal activity. When combined with INH, RIF forms the backbone of anti-TB therapy for DS-TB cases. RIF, a semisynthetic derivative, exhibits high effectiveness against MTB, with MICs, typically ranging from 0.1 to 0.2 μg/mL. As an ansamycin, RIF belongs to a class of lipophilic macrocyclic antibiotics known for their ability to readily penetrate cell membranes (Anand, Upadhyaya, & Tripathi, 2015; Kayukova & Berikova, 2020). Notably, RIF demonstrates the capability to cross the blood–brain barrier, achieving concentrations sufficient to eradicate MTB within the central nervous system (Prideaux et al., 2015).

In addition to RIF, several derivatives have been synthesized to enhance efficacy and extend metabolic half-life. These derivatives include RLZ, also known as KRM-1648), RBT, and RPT. RBT, RPT, and rifalazil, along with RIF, belong to the ansamycin class of antibiotics. Given their effectiveness in combating MTB, especially in cases involving patients co-infected with HIV, these RIF derivatives are crucial components of comprehensive TB treatment regimens (Figure 10.2) (Nusrath Unissa, Hanna, & Swaminathan, 2016).

CGP-7040, another RIF derivative, has shown promising activity against both DS and DR-TB, outperforming RIF and RBT *in vitro*. This compound exhibits enhanced stability and activity

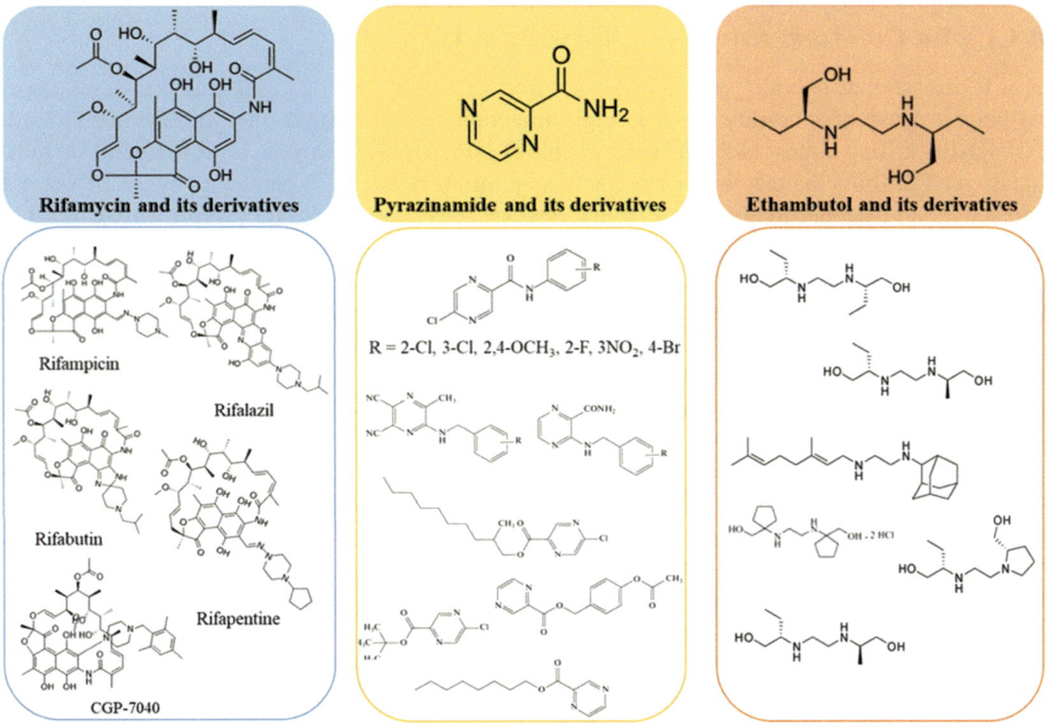

FIGURE 10.2 Various anti-TB drugs and their derivatives.

against MTB, positioning it as a potential candidate for further development as a TB treatment option (Kayukova & Berikova, 2020).

EMB, also known as (S,S)-ethambutol dihydrochloride, stands as a cornerstone in TB treatment regimens. First synthesized in 1961, EMB's efficacy is closely linked to its stereospecificity, with the (S,S)-isomer exhibiting significantly higher anti-TB activity compared to its counterpart, the R,R-isomer (Kumar & Narsaiah, 2014). Efficient synthesis of the (S,S)-EMB isomer, achieved asymmetrically with high enantiomeric purity, underscores its importance in TB therapy. The mixture of EMB stereoisomers demonstrates potent activity against DS-TB strains, with an MIC of 0.5 μg/mL (Asif, 2012). However, resistance in MTB strains with mutations in the gene embB codon 306 significantly elevates the MIC to 16–32 μg/mL (Safi et al., 2013). Efforts to enhance EMB's efficacy have led to the exploration of second-generation derivatives through combinatorial chemical methods. A library of over 63,000 compounds, primarily lipophilic asymmetric diamines with bulky hydrophobic fragments, was synthesized and screened. Among these compounds, SQ7, SQ37, and SQ125 demonstrated promising *in vitro* activity against MTB strains, with MIC values comparable to or greater than EMB (Kurosu & Denny, 2016; R. E. Lee et al., 2003). Notably, N-geranyl-N′-(2-adamantyl)ethane-1,2-diamine, known as SQ109, emerged as a particularly potent derivative, exhibiting a 14- to 35-fold increase in activity against DS and DR-TB strains compared to EMB. With an MIC of 0.9 μg/mL against EMB-resistant strains, SQ109 demonstrated bactericidal activity and enhanced the efficacy of other anti-TB drugs like INH, RIF, and TMC207. This synergistic effect, coupled with its ability to penetrate the bacterial cell wall and prevent mycolic acid synthesis, holds promise for shortening treatment duration (Machado, Girardini, Viveiros, & Pieroni, 2018). SQ109 and its derivatives have undergone extensive screening against MTB strains, with compounds such as Compound 12 and SQ109, exhibiting MIC values in the range 0.5–1 μM. Notably, Compound 16, featuring a long alkene chain, demonstrated exceptional activity with MIC values as low as 0.25–0.5 μM (Onajole et al., 2010).

10.4 THE RATIONALE FOR PROBIOTIC INTEGRATION

10.4.1 The Gut–Lung Axis and Its Relevance in TB

TB is a disease that is both preventable and cured. However, in 2022, TB was the world's second biggest cause of mortality from a single infectious agent, behind only coronavirus disease (COVID-19), killing almost twice as many people as HIV/AIDS. Each year, almost 10 million individuals get TB. Immediate action is needed to stop the worldwide TB pandemic by 2030, a target endorsed by all Member States of the United Nations (UN) and the WHO (WHO, 2023). MTB is the bacillus that causes TB. It spreads by coughing and expulsion of germs. It is believed that over a quarter of the worldwide population has been infected with TB (Houben & Dodd, 2016). The anti-TB medication regimen, which consists of four kinds of medicines administered for 6 months, is the primary approach to TB management. Prolonged use of the anti-TB medication regimen may induce a decline in patient adherence, which may also influence therapeutic effectiveness and lead to MTB resistance to the anti-TB medicine (AlSahafi et al., 2019). The incidence of resistance to the first-line anti-TB treatment, known as multiple MDR-TB, and resistance to the first and second line, known as XDR-TB, is increasing over time and is a worldwide health concern today (Simmons et al., 2018). Over time, decreasing patient adherence might lead to resistance (AlSahafi et al., 2019). This underlines the need to find novel treatment strategies that may tackle the problem of resistance to anti-TB medicine (Sanjiwani, Setiawan, Putra, & Darwinata, 2021).

Recent years have seen the presentation of therapeutic innovation ideas for TB therapy, with the gut–lung axis being one of the potential paths being revealed. An intricate system that explains how two faraway organs, the lungs and the intestines, work together is called the gut–lung axis. Problems with the gut microbiota's equilibrium may have an effect on the lungs' health and vice versa (He et al., 2017). Research has shown that the maturation of the respiratory tract is influenced

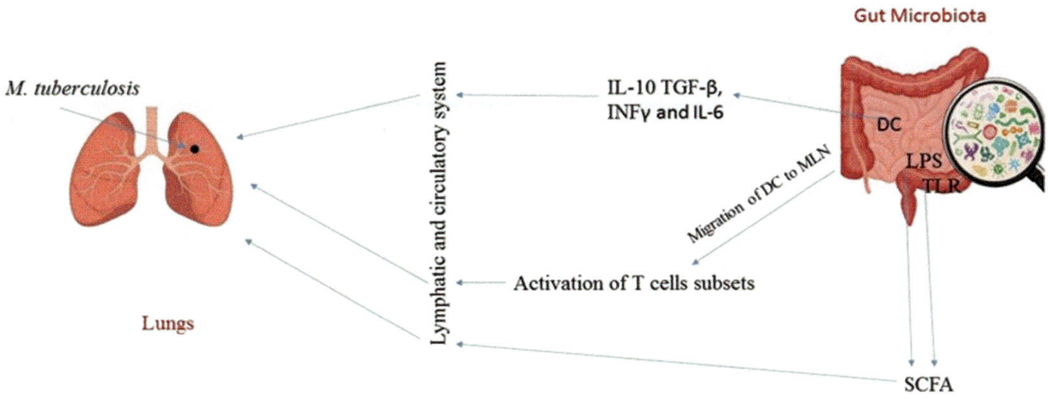

FIGURE 10.3 The immunological response to MTB involves the gut–lung axis mechanism (Sanjiwani et al., 2021).

by the control of gut microbiota colonization (Li, Zhu, Liao, Wang, & Wan, 2019). People with TB have changes in the balance of the microbiota in their gut. However, it is unclear if these changes are caused by the illness, the treatment, or a genetic trait that can make pathogens grow faster. A number of studies have shown that gut bacteria play a part in treating TB. A decrease in inflammation and a boost in defense are some of the processes that are set off (Dumas et al., 2018). However, the full process of how gut bacteria protect against TB is still not fully known. One possible way is by turning on Granulocyte-macrophage colony-stimulating factor (GM-CSF), which is a granulocyte-macrophage colony-stimulating factor. GM-CSF is released by the body as part of the immune system and to control myelopoiesis. It plays a major role in the inflammatory process. GM-CSF has been shown in the past to lower the production of MTB inside cells and control inflammation in granulomas (Rothchild et al., 2017).

There is hope for a TB treatment that modulates the gut–lung axis now that we know the gut microbiota may inhibit GM-CSF activation. The use of probiotics is one strategy for influencing the gut–lung axis. Previous research has shown that probiotics, which are beneficial microorganisms, may both reduce inflammation in TB and protect the liver from the harmful effects of anti-TB drugs (Horvath et al., 2016; Sanjiwani et al., 2021).

Figure 10.3 illustrates the mechanism of immunity control through the gut–lung axis. Similar to an ecologist, the intestinal microbiota triggers alterations in signaling and prompts the migration of DCs to the lymphatic tract (Negi, Pahari, Bashir, & Agrewala, 2019). DCs are involved in the activation of different T cells in the mesenteric lymph nodes and the stimulation of cytokine production, including IL-10, TGF-β, INF-γ, and IL-6 (Geva-Zatorsky et al., 2017). When T cells are activated, they bind to their receptors. In the case of an infection in the respiratory tract, cells from the MLN and GALT will move to the lung mucosa to offer protection and reduce inflammation. Microbiota products, such as LPS, have the ability to bind to TLR on intestinal epithelial cells and macrophages. This binding can then trigger the production of cytokines and chemokines. In addition to its products, the microbiota metabolite in the form of SCFA has been discovered to have a role in the gut–lung axis, as it can potentially move to the lungs (Samuelson, Welsh, & Shellito, 2015).

Several ways modulate the gut–lung axis. Probiotics are a simple and well-studied approach. Probiotics activate GM-CSF to treat TB. Table 10.1 summarizes the results of many studies that looked at the impact of probiotic supplementation on cytokine and interleukin levels in relation to TB and lung infections (Sanjiwani et al., 2021).

There are a number of health advantages associated with these species, as stated by the International Scientific Association for Probiotics Consensus. Among them are the following: the production of SCFAs, the regulation of intestinal motility, the outcompetition of dangerous

TABLE 10.1
Multiple Investigations Have Been Conducted on the Administration of Probiotics and Their Impact on Cytokines and Interleukins Associated with Lung Infections

Interference	Impacts	References
Vitamin B and a multi-strain probiotic	Increased IFN-γ and IL-12	Suprapti et al. (2018)
CNCM I-1518 probiotic strain of *Lactobacillus paracasei*	Increased IL-10 in lung tissues	Robak et al. (2018)
Probiotic of NOD2-stimulating bacteria	Reduced TNF-α and IL-1β in BALF	Belkacem et al. (2017)
Probiotic *Lactobacillus fermentum*-1	Increased IgA Reduced IL-6 and TNF-α	Youn et al. (2012)
Probiotic *Lactobacillus rhamnosus* GG	Reduced IL-6 and TNF-α	Khailova, Baird, Rush, McNamee, and Wischmeyer (2013)
Probiotic *Lactobacillus rhamnosus* GG	Increased IL-10 in serum and lung tissues Increased T-cell regulator and neutrophil infiltration in lung, Reduced expression of TLR2 gene and Myd88	Martens et al. (2018)
Longum 51A probiotic Bifidobacterium	Reduced TNF-α and IL-6 in BALF Increased IL-10 in lung tissues, macrophage phagocytosis ability	Vieira et al. (2016)

microbes, and the restoration of the balance of gut bacteria. The protective effect of probiotics may be linked to alveolar NK and macrophage activation. Research on mice supports the idea that *Lactobacillus pentosus* treatment boosts NK cell activity and IFN-γ production in the spleen. CD11c+ DCs produce IL-12 after lactobacilli contact with them via TLR2 and TLR4. Lactobacilli strains vary in their capacity to increase IL-12 and IFN-γ levels. Other inflammatory cells affected by probiotics include lung Treg and Th17 lymphocytes. Th17s eliminate pathogens and Tregs regulate immune response in humans and rats (Comberiati et al., 2021).

Researchers found three different kinds of Lactobacillus in the feces of wild boars: *L. casei*, *L. plantarum*, and *L. salivarius*. When tested against MTC complex members *M. bovis* and MTB, these species showed inhibitory efficacy against bovine TB. Their effects on circulating phagocytes' ability to absorb *M. bovis* were also pH-dependent. All lactobacilli killed *M. bovis* at low pH, but only *L. plantarum* and *L. casei* did so at neutral pH. Bacteriocins with an antimycobacterial and immunomodulating collagen adhesion protein were found in the latter genomes. Lactobacillus plantarum also inhibited macrophage uptake via antagonistic competition. The scientists hypothesized that oral lactobacilli with antimycobacterial activity may diminish *M. bovis* intestine concentration and transmission between domestic and wild animals (Bravo et al., 2019).

The principal infectious agent responsible for causing TB, the leading infectious disease killer and a major public health problem, is MTB. About 1.5 million people die from TB each year, while an estimated 2–3 billion individuals have latent MTB infection. A new wave of MDR, XDR, and TDR strains has emerged, prompting researchers to seek other treatments. Probiotics may be an anti-TB medication candidate in these cases, and thus, we developed them. Here, we take a look at the anti-tubercular capabilities of *L. rhamnosus* PMC203, a strain taken from the vaginal microbiota of generally healthy women. PMC203 successfully eradicated DS and DR-MTB from RAW 264.7 murine macrophage cells without damaging the cells themselves. Also hindered was the formation of MTB in broth culture media. Neither body weight nor clinical symptoms were altered by PMC203, according to the findings of a 2-week repeated oral dose toxicity test in guinea pigs. We found that PMC203 upregulates the expression of known autophagy gene markers, which promotes autophagy in a dose-dependent way. This suggests that there is an intrinsic mechanism for cell death (Rahim et al., 2022).

Probiotic treatment is essential for boosting the immune system when someone is infected with TB. It does this via many processes, including controlling the development of MTB, controlling inflammation in granulomas, and facilitating the restoration of lung tissue. Similar to a biochemist, probiotics have the ability to not only boost immunity against TB but also provide protection to the liver and alleviate gastrointestinal issues caused by anti-TB medications. Further investigation is required to fully understand the potential of probiotic-based therapy as an additional treatment for TB, since its effectiveness and direct impact have not yet been confirmed (Sanjiwani et al., 2021).

10.4.2 Potential Synergies between Probiotics and Anti-TB Drugs

It was Lederberg and McCray first suggested the idea of human microbiota, which refers to the populations of microbes found in many parts of the human body, such as the skin, vagina, mouth, stomach, and lower respiratory system (Human Microbiome Project, 2012). The microbiota is sometimes referred to as a "forgotten organ" due to its complex and dynamic nature. It serves a crucial function in several physiological processes, such as digestion, nutrition absorption, and immune system control (Cani, 2018). Both the systemic and pulmonary immune systems benefit from a well-balanced gut microbiome. It does this via modulating immunological cell migration and death, T-cell growth, inflammation, and toll-like receptor signaling activation (Samuelson et al., 2015). Important functions of the pulmonary microbiota include infection defense, Th1/Th2 immune response regulation, and inflammation management (Balcells, Yokobori, Hong, Corbett, & Cervantes, 2019). Hence, the human microbiota may have a significant impact in combating MTB infection. Microbiota changes may have a role in the development of TB; therefore, it is important to study the composition of the human microbiota at each stage of MTB infection. This data has the potential to aid in the realization of the End TB Strategy's (WHO) stated objectives (Tarashi et al., 2018).

Changes to the microbiota in the gut have the potential to interfere with macrophage activity and the activation of immune cells that are supposed to eradicate MTB infection. Consequently, to improve the effectiveness of anti-TB medications and the host immune response against MTB, it may be useful to take supplements that regulate the gut microbiota and keep things in balance (Liu, Wang, & Wu, 2021).

Metchnikoff postulated in the early 1900s that the beneficial microbes in fermented milk were responsible for the unusually long lifespans seen among Bulgarian peasants. Consequently, the term "probiotic" was first proposed (Gibson et al., 2017). A probiotic was defined in 2014 as having certain components, an appropriate viable count that lasts beyond the expiration date, and enough evidence of health benefits. Hill also looked into the safety of probiotics (Hill et al., 2014). Research has shown that probiotics alter microbiota composition by encouraging the growth of beneficial bacteria and reducing the activity of unwanted ones (Azad, Sarker, Li, & Yin, 2018). Some probiotic bacteria showed antibacterial characteristics against TB in an *in vitro* investigation. These bacteria were *Lactobacillus brevis*, *Lactobacillus plantarum*, and *L. fermentum* (Gavrilova et al., 2014). *B. bovis* Calmette-Guerin (BCG) was the target of potent antibacterial action in an additional *in vitro* study including specific probiotic bacterial strains. The metabolites produced by these bacteria, which contain genes that code for class II bacteriocins and bacteriolysins, are believed to be the cause of the observed antimycobacterial effect. In addition, phagocyte BCG absorption was significantly reduced by *L. plantarum*, dramatically increased by *L. casei*, and unaffected by *L. salivarius* (Bravo et al., 2019). In a study conducted by Negi et al. (2019), the usage of antibiotics resulted in a reduction in Bacteroidetes and Firmicutes, as shown by an *in vivo* animal model, while increasing the presence of proteobacteria. The expression of the macrophage-inducible Ca^{2+}-dependent lectin receptor (mincle) decreased due to the shift in bacterial composition. Mincle is an essential pattern recognition receptor that connects to carbohydrate structures in several diseases, including MTB. Consequently, this decline in mincle expression had an impact on the activation of the innate immune response (Schierwagen et al., 2020). Furthermore, changes in gut microbiota result in a higher load of MTB, a reduced population of effector and memory T cells, and an increase in regulatory T cells in the lungs. Nevertheless, the addition of *Lactobacillus plantarum* MTCC 2621 probiotics could potentially reverse the expression of mincle and MHC-II on lung dendritic cells. Therefore, this may lead to a reduction in MTB lung loads, a decrease in regulatory T cells, and an increase in CD4T cells with a CD44hi phenotype and a CD62LloCD44hi phenotype that are activated and effector memory (Negi et al., 2019). A study conducted by Liu et al. found that *L. plantarum*, a probiotic, may enhance the function of lung DCs and T cells in dysbiotic mice, particularly in relation to their reaction to MTB. Probiotics may modulate the immune system and have antagonistic effects on MTB, according to the little research on the topic. Based on these findings, probiotics may be a new option for treating TB (Liu et al., 2021).

Acute or antibiotic diarrhea in children is treated with probiotics. *L. casei*, a popular probiotic, regulates intestinal microbiota, and relieves academic stress-related gastrointestinal discomfort. Zebrafish intestinal pro-inflammatory cytokines can be reduced by *L. casei* ATCC334's anti-inflammatory capabilities. Recent research on old mice found that the intestinal barrier might be efficiently protected by combining dietary fiber with *L. casei*. The combination also caused an upregulation of genes associated with SCFA receptors in the small intestine and an elevation in SCFA levels. One research found that *L. casei* supplementation increased SCFAs, specifically propionic and butyric acids, in antibiotic-treated rats (Li et al., 2022); According to the available data, probiotics and anti-TB drugs may change the gut microbiota composition, which might affect the intestinal barriers. Our earlier clinical experiment showed that consuming *L. casei* drinks may decrease the occurrence of overall gastrointestinal side effects linked to anti-TB medications (Lin et al., 2020).

The purpose of this 6-week animal study was to investigate the potential protective effects of *L. casei* against intestinal damage caused by anti-TB medications. According to our study, the usage of anti-TB medications negatively affected the intestinal walls. Alterations to mucus layer composition, gut microbiota structure disturbance, and decreased metabolism of SCFAs were all part of this. However, we also discovered that the introduction of *L. casei* ATCC334 had a positive impact on the intestinal barrier function, mitigating these damages. Similar to a microbiologist, one way to achieve this improvement effect is by increasing the number of goblet cells, the content of MUC-2, and the levels of immune factors like IL-10 and IgA. Additionally, enhancing the presence of beneficial microbes in the intestines and improving the metabolism of SCFAs can also contribute to this effect. Similar to a microbiologist, one would observe that the mucus system of the colon is connected to the surface of epithelial cells. The main structural component, known as MUC-2, is secreted by goblet cells (Carlson, Lock, & Carrier, 2018). Li et al. discovered *L. casei* mitigated the intestinal barrier damage caused by anti-TB medications in rats. Our research examined SCFAs as metabolites to understand their process. Limitations were the unknown mechanisms of intestinal flora or SCFAs that cause intestinal damage. Whether these gut alterations are connected to endocrine metabolism needs additional study. *L. casei* alone can only partially repair intestinal injury, demonstrating its efficacy and limits. They proposed further studies to reduce anti-TB drug-induced intestinal damage. *L. casei* mildly alleviated anti-TB drug-induced intestinal damage in this investigation. Gut microorganisms and SCFAs may affect metabolic pathways. The aforementioned findings raised concerns regarding antibiotic overuse and preventative actions. However, anti-TB therapy without medications proved impossible. Correct and appropriate use of anti-TB medications and gastrointestinal tract protection were important considerations. Our research guided the quest for better preventative or treatment strategies (Li et al., 2022).

10.5 EVIDENCE FROM PRECLINICAL STUDIES

Investigating the composition of the microbiome in TB patients and hosts infected with MTB using animal models has become a focal point in recent research endeavors. This area of study has attracted considerable interest due to its potential to uncover critical insights into the interplay between the host's microbial communities and MTB infection. By employing animal models, researchers aim to simulate and understand the complex dynamics between the pathogen and the host microbiota, shedding light on how alterations in microbial populations influence disease progression, treatment outcomes, and host immune responses. The findings from these studies hold promise for the development of novel therapeutic strategies targeting the microbiome to enhance the efficacy of existing TB treatments and improve patient outcomes (Hong et al., 2016; Naidoo et al., 2019; Namasivayam, Sher, Glickman, & Wipperman, 2018; Wood, Elaine, & Mehta, 2017).

Mori et al. (2021) offered a comprehensive overview of insights gained from examining the colonic (fecal) and lung microbiota in both human subjects and animal models following MTB infection, juxtaposed with noninfected controls. This study not only consolidated existing knowledge but also uncovered novel associations and patterns within the microbiome landscape in response to MTB challenge. By analyzing both human and animal models, the research broadened our understanding of how MTB infection has impact on microbial communities across different anatomical sites, elucidating potential systemic effects on the microbiota. These findings underscore the intricate interplay between MTB and the host microbiome, paving the way for further investigations to delineate the mechanisms underlying microbiota alterations and their implications for TB pathogenesis and treatment (Mori, Morrison, & Blumenthal, 2021).

In general, analyses of fecal microbiota profiles among various groups, including treatment-adolescent, new-onset, and recurrent TB patients, consistently reveal a notable reduction in bacterial diversity compared to control individuals. This decline in microbial diversity highlights a common trend associated with TB infection and suggests a potential dysbiosis within the gut microbiota. However, while the decrease in bacterial diversity is a shared observation across different patient categories, there may exist nuanced variations in the specific microbial taxa affected or the extent of dysbiosis experienced by each group. Understanding these variations could provide valuable insights into the dynamic interplay between TB infection and the gut microbiome, offering potential avenues for targeted interventions or therapeutic strategies aimed at restoring microbial balance and improving patient outcomes. Further exploration into the underlying mechanisms driving these alterations in microbial diversity, such as immune modulation or metabolic shifts induced by TB infection, is warranted to comprehensively decipher the intricate relationship between the gut microbiota and TB disease progression (Li et al., 2019).

These studies reveal changes in the relative abundances of bacterial lineages affiliated with families such as *Ruminococcaceae* and *Lachnospiraceae*. Despite varied observations, these bacterial families, predominant in the human colon, are recognized for their fermentation capabilities, yielding SCFAs like acetate and butyrate (Louis & Flint, 2009; Rinninella et al., 2019). Butyrate, with its immunomodulatory properties, alongside other potentially anti-inflammatory factors produced by these bacterial lineages, underscores their relevance in host responses to MTB infection (Atarashi et al., 2013; Atarashi et al., 2011).

Furthermore, changes in non-spore-forming Gram-negative bacterial lineages, including *Prevotella* and Bacteroides from the phylum *Bacteroidetes*, have been documented in individuals infected with MTB. Alongside these alterations, there is an observed increase in the relative abundance of proteobacteria. Understanding the dynamics of these bacterial lineages in response to MTB infection is crucial for unraveling the complex interactions between the pathogen and the host microbiota. Moreover, exploring the functional roles of these microbial changes could provide insights into the mechanisms underlying TB pathogenesis and may open avenues for novel therapeutic interventions aimed at modulating the gut microbiota to enhance host resistance against MTB infection (Sokol et al., 2008). These bacterial groups, which are known for their capacity to produce SCFAs and other metabolites such as propionate and acetate, play a significant role in the dysbiosis observed in the gut during MTB infection. The dysbiosis refers to an imbalance or disruption in the composition of the gut microbiota. The production of SCFAs and other metabolites by these bacterial groups not only influences the local gut environment but also has systemic effects on the host. Understanding the specific roles of these metabolites and their producers in the context of MTB infection could provide valuable insights into the complex interplay between the gut microbiota and the host immune response during TB. Further investigation into the mechanisms underlying these interactions is essential for elucidating the pathogenesis of TB and developing novel therapeutic strategies targeting the gut microbiota (Ward, 2015).

Research conducted in murine models suggests that infection with MTB triggers notable alterations in the composition of the colonic microbiota within a few days. These changes are characterized by a reduction in the relative abundances of bacterial families such as *Lachnospiraceae* and *Ruminococcaceae*, even in the absence of detectable mycobacterial DNA in fecal samples. These findings underscore the intricate and rapid interplay between MTB infection and the gut microbiota, indicating the potential systemic effects of pulmonary TB on microbial communities residing in distant anatomical sites. Further exploration into the mechanisms underlying these microbial shifts could provide valuable insights into the pathogenesis of TB and its systemic implications for host–microbiota interactions (Winglee et al., 2014). This indicates that shifts in the gut microbiota might originate from initial alterations in mucosal immune responses triggered by lung infection.

Various animal models serve as indispensable tools for conducting preclinical testing of drugs or chemical compounds, as well as specific toxicity studies. Among these models, rodents, including rats and mice, stand out as the most commonly utilized species. However, other animal species such as guinea pigs, rabbits, and hamsters are also employed, albeit less frequently (Zwierzyna & Overington, 2017). In addition to vertebrate models, invertebrate models play a crucial role in assessing various biological activities. Notable examples include *Drosophila melanogaster*, commonly known as the fruit fly, and *Caenorhabditis elegans*, a type of nematode worm (Badyal & Desai, 2014). Furthermore, zebrafish emerge as a prominent vertebrate model due to its rapid development and suitability for toxicity testing. This aquatic species has demonstrated exceptional efficacy as a model organism in toxicity studies (Zwierzyna & Overington, 2017).

10.6 MICROBIOTA AND PROBIOTICS IN ANTI-TB PRECLINICAL AND CLINICAL STUDIES

Enhancing anti-TB drug effects in the modern era with MTB drug resistance is of major importance in the clinical treatment of patients. Besides, prevention of anti-TB drug side effects is a crucial matter in dealing with MTB. Probiotics are the best option to play these roles simultaneously (Liu et al., 2021). The gut–lung nervous axis and gut immune system relation clarify the effects of probiotics and microbiota on human immune system regulation and promotion. On the other hand, probiotics have some voice in pharmacokinetics and absorption of medicine in the gut, leading to the emergence of pharmacomicrobiomics sciences (Doestzada et al., 2018).

Different studies have proven that individuals and societies with a deficiency of microbiota are so vulnerable to different diseases due to the lack of a practical regulated immune system. In these people, prescription of probiotics with specific treatment is necessary for the proper outcome of medicine and treatment (Diallo et al., 2021). Fighting TB like any other pathogenic disease needs effective immune system function. Probiotics by boosting and regulating the immune system can meet our needs in TB-related immune deficiency. On the other hand, second-line anti-TB treatment drugs may lead to changes and decrease in the population of normal flora and microbiota due to antibiotic activities (C. Wu et al., 2023). The prescription of probiotics in TB treatments is essential to compensate for this change (Figure 10.4). Further, as mentioned, probiotics are an effective factor in the enhancement of the immune system and resulting in TB treatment.

Probiotics are generally recognized as safe and consumption is approved by the Food and Drug Administration (FDA). Our gut is full of different bacterial strains with a beneficial role in digestion and absorption to immune responses induction for active immunity. Common and experimented probiotics contain *Lactobacillus, Saccharomyces, Enterococcus, Bifidobacterium, Streptococcus,* and *Bacillus*. Studies based on the immune activation and regulation effects of probiotics demonstrate that the consumption of distinct and a cocktail of probiotics is more effective than probiotics. For example, stimulatory effects of *Saccharomyces boulardii, L. casei* LBC80R, *Lactobacillus*

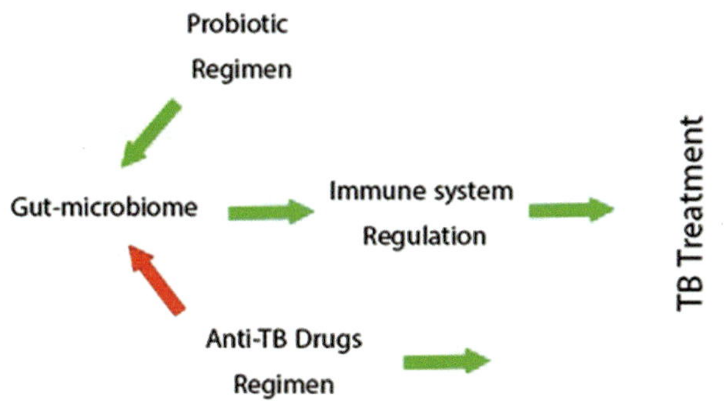

FIGURE 10.4 Effect of probiotics and anti-TB on the gut microbiome and relation with TB treatment.

acidophilus CL1285, *Lactobacillus delbrueckii* subsp. *bulgaricus*, *Streptococcus salivarius,* and *Bifidobacterium bifidum* may provide better results than a single probiotic. The mentioned probiotics are able to regulate immunity by different functions in intestinal lumina and mucosa (Hoffmann et al., 2014).

Probiotics and beneficial bacteria can expose their antigens (Ags) to antigen-presenting cells (APCs). APCs activate the immune system by presenting Ags to immune cells (T-helper and regulatory T cells). After presentation and maturation, differentiation into types of immune cells will occur (Mazziotta et al., 2023; Taverniti & Guglielmetti, 2011). In the second step, these immune cells produce cytokines and chemokines. Furthermore, gut microbiota and probiotics by their secondary metabolite are effective in the activation of innate immune system (Eribo et al., 2020). The activated immune system is able to deal with TB and other diseases better than the nonactivated mode. CD4+ cells are the main cells in the adaptive immune system that are usually present in the lumina of the gut. Probiotic bacteria are presented by DCs directly to $CD4^+$ and $CD8^+$ (Mazziotta et al., 2023). Microbiota like *Clostrodia* in the gut can induce the $CD4^+$ to activate and differentiate to their subtypes such as Th1, Th2, Th17, and T regulator (Atarashi et al., 2011; Wu & Wu, 2012).

Prescription of probiotics with anti-TB drugs in different animal studies was observed. Probiotics can improve and promote the host's immune system and may have a voice in the absorption of drugs in the lumina gut (Liu et al., 2021). However, the main observed object is the immunoregulatory effect of probiotics *in vivo* and *in vitro* studies. Yu et al. (2024) assessed the relationship between microbiota and SCFAs in rat gut with anti-TB drug prescriptions. They figured out that CSTAs may change in anti-TB regimens, and this change is effective on anti-TB drug absorption. Permeability of the gut was increased after the probiotic prescription. This change is effective in the absorption of more amount of anti-TB drugs in the gut. Another important factor is interactions between anti-TB drugs and microbiota in preclinical and clinical studies. Table 10.2 shows the preclinical and basic studies about the effect of anti-TB drugs on gut microbiota. In comprehensive investigation of all these studies, we can conclude that bacterial species *Lactobacillus, Bifidobacterium, Bacteroides, Campylobacter Butyricicoccus, Ruminococcus, Acetivibrio, Peptococcus, Alkaliphilus, Bacteroidota,* and *Bacillota* found to be decreased in mice. *Enterococcus, Erysipelatoclostridium, Pseudomonadota, Clostrodia,* and *Verrucomicrobia* increased after the use of anti-TB drugs (Figure 10.5). The reduction in bacterial abundance is related to the antibiotic effects of anti-TB drugs, whereas the increases result from the depletion of other microbiota and the availability of more prebiotics and nutrients in addition to their resistance to anti-TB.

After reviewing the clinical studies on probiotics and microbiota measurement in patients prescribed anti-TB drugs, we can describe the interaction between them. Anti-TB drugs have effects on the gut microbiota of humans. Stool samples are collected and investigated by different methods such as culture and biochemical tests. *Prevotella, Bifidobacterium, Bacteroides, Hemophilus parainfluenzae, Roseburia hominis, R. inulinivorans, Bifidobacteriaceae, Ruminococcaceae, F. prausnitzii,*

TABLE 10.2
Studies on the Effect of anti-TB Drugs on Microbiota in Animals

Animal	Bacteria Change	References
Mouse	Depletion of *Lactobacillus, Bifidobacterium, Bacteroides, Campylobacter* and increased *Enterococcus*	Khan and Agrewala (2016)
Mouse	*Butyricicoccus, Peptococcus Ruminococcus, Acetivibrio,* and *Alkaliphilus* were decreased, and *Erysipelatoclostridium* was increased.	Namasivayam et al. (2017)
Mouse	*Bacillota* and *Bacteroidota* were reduced, and *Pseudomonadota* enhanced	Dumas, Peixoto, Hudrisier, and Neyrolles (2018)
Mouse	*Bacillota* was reduced and Verrucomicrobia and *Bacteroidetes* were enhanced	Khan et al. (2019)
Mouse	Abundance of *Campylobacter, Bifidobacterium,* and *Lactobacillus* was reduced, and the abundance of *Bacteroides* and *Enterococcus* enhanced	Negi et al. (2019)

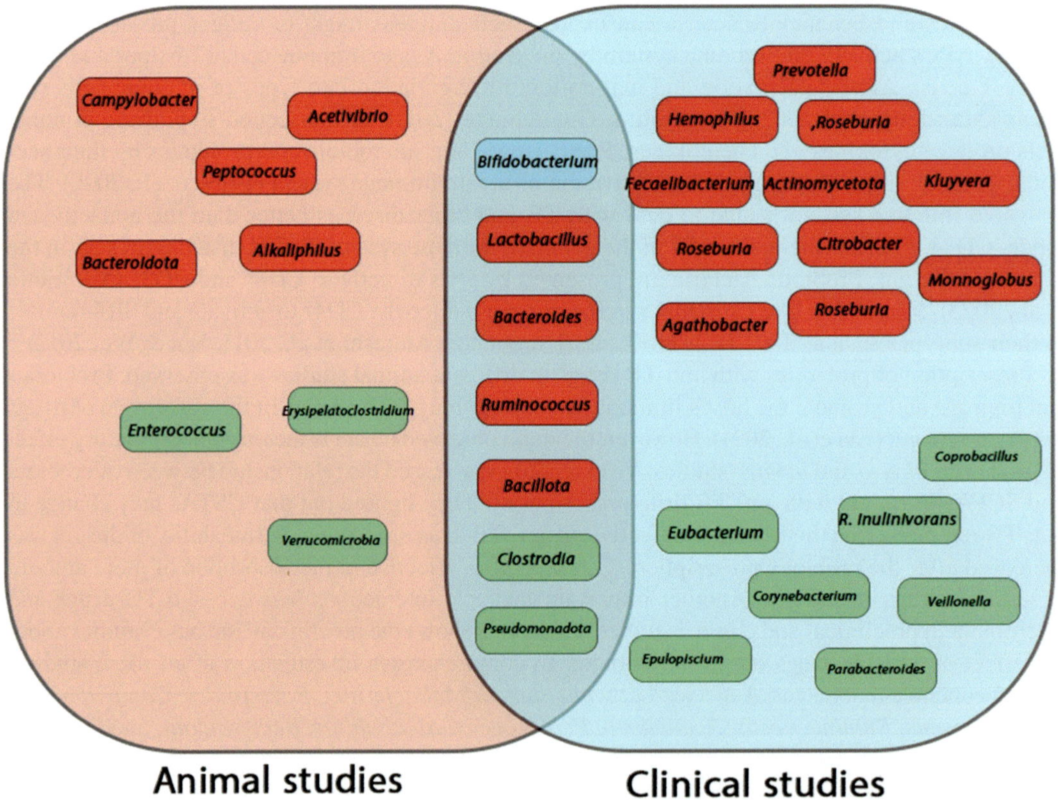

FIGURE 10.5 Clinical and preclinical studies on microbiota changes after anti-TB regimens. Red backgrounded bacteria are decreased. Green backgrounded bacteria are increased and *Bifidobacterium* has no specific change (in some studies increased and in other studies decreased).

Ruminococcus, *Fecaelibacterium*, *Bacillota*, *Actinomycetota*, *Kluyvera*, *Roseburia*, *Citrobacter*, *Agathobacter Lactobacillus*, *Faecalibacterium*, *Roseburia*, *Dorea*, *Monnoglobus*, *L. delbrueckii*, *L. casei*, *L. acidophilus,* and *L. plantarum* were found to be decreased in human stools. Moreover, *Roseburia inulinivorans*, *Pseudomonadota*, *Ruminococcus*, *Faecalibacterium*, *Eubacterium*,

TABLE 10.3
Systematic Table of Clinical Studies, Interaction between Anti-TB Drugs and Microbiota

Country	Microbiota	No. of Patients	References
India	*Eubacterium rectale, Faecalibacterium prausnitzii,* and *Roseburia inulinivorans* were increased, *Bifidobacterium* and *Prevotella were* decreased.	6	Maji et al. (2018)
Haiti	*Faecalibacterium Ruminococcus* and *Eubacterium* were increased, and *Bacteroides* was reduced.	19	Wipperman et al. (2017)
China	*Bacteroidota* was reduced, and *Pseudomonadota* and *Actinomycetota* were increased.	37	Luo et al. (2017)
China	*Clostridium bolteae* and *Coprobacillus* were increased, and *Roseburia hominis, Hemophilus parainfluenzae,* and *R. inulinivorans* were decreased.	46	Hu et al. (2019b)
China	*Ruminococcaceae Bifidobacteriaceae,* and *F. prausnitzii* were decreased.	18	Li et al. (2019)
Taiwan	*Bacillota* was decreased.	80	Huang et al. (2019)
China	*Bacillota* and *Actinomycetota* were decreased.	76	Jinyu Wang and Ma (2020)
China	*Roseburia, Kluyvera,* and *Citrobacter* were decreased, and *Epulopiscium, Corynebacterium,* and *Atopobium* were increased.	156	Shi et al. (2021)
China	*Bacteroides* and *Bifidobacterium* were increased, *Faecalibacterium,* and *Roseburia* were decreased.	142	Wang et al. (2022)
China	*Parabacteroides, Bacteroides,* and *Veillonella* were increased, *Agathobacter Lactobacillus, Bifidobacterium, Faecalibacterium, Dorea, Monnoglobus,* and *Roseburia* were decreased.	132	Ye et al. (2022)
	Bifidobacterium animalis and *B. bifidum* were increased and *Lactobacillus casei, L. delbrueckii, L. plantarum, and L. acidophilus* were decreased.	30	Otdushkina et al. (2023)

Actinomycetota, Clostridium bolteae, Coprobacillus, Corynebacterium, Epulopiscium, Atopobium Bacteroides, Parabacteroides, and *Veillonella* were increased (Table 10.3). The increases are due to the abundance of foods and prebiotics that support microbiota resistant to anti-TB drugs.

Basic and preclinical studies on this object demonstrate bright and hopeful results. Studies on rats and other animals regarding the probiotics and their effects on immune system regulation and antibacterial peptides designed the path to noninvasive dealing with diseases such as TB. The next step is the investigation and precipitation of probiotics in patients with TB or other similar respiratory diseases. Clinical studies and observations show the great potential of probiotics in enhancing anti-TB in the human body. Studies on the clinical prescription of probiotics with specific drugs and the effect of gut-microbiota condition in respiratory infected patients are designed. Table 10.4 shows and describes some of the research on the effect of probiotics on the respiratory treatment process by immunoregulatory effect. IL-12, IFN-Gamma, IL10, Treg, and neutrophils, are increased

TABLE 10.4
Administration of Probiotics in Patients with Respiratory Diseases

Country	Probiotic	Immunological Effect	References
Indonesia	Multistarin	Increased IL-12 and IFN-Gamma	Suprapti et al. (2018)
Colorado	*L. rhamnosus GG*	Decreased IL-6 and TNF-α	Khailova et al. (2013)
Belgium	*L. actobacillus rhamnosus GG*	Increased IL-10 in serum and lung tissues Increased T-cell regulator and neutrophil Infiltration in the lung, and decreased expression of *TLR2* gene and *Myd88*	Martens et al. (2018)
Brazil	*Bifidobacterium longum 51A*	TNF-α and IL-6 in BALF decreased IL-10 in lung tissues Increased.	Vieira et al. (2016)
China	*L. casei*	TNF-α, IL-6, IL-10, IL-12 decreased	Jiang et al. (2022)

immune factors after probiotic regimens. These factors are directly effective in responses against MTB. On the other hand, some factors such as IL-6, TNF, TLR-2, and Myd88 are decreased. These researches demonstrate the new roadmap of TB drug regimens for treatment. Safe treatment of TB without side effects of TB drugs is possible with probiotics prescription. Having a rich and diverse microbiota in the gut is essential for a strong immune system in the host. While this is important for everyone, it is even more necessary for patients with TB.

10.7 MECHANISTIC INSIGHTS

The human intestinal environment is a bustling metropolis, teeming with a vibrant community of bacterial inhabitants, surpassing a grand total of 1,000 distinct species and spanning over 50 diverse genera (Onajole et al., 2010). Among the various segments of the gastrointestinal tract, the large intestine contains the highest concentration of bacteria, estimated at approximately 10^{11} to 10^{12} cells per gram (Fujimura et al., 2016). This intricate network of gut microbiota serves as a cornerstone in upholding host well-being by engaging in a myriad of functions, including nutrient metabolism, energy regulation, immune modulation, and safeguarding against pathogens (Takiishi, Fenero, & Câmara, 2017). Hence, preserving a balanced composition of gut microbiota is crucial for overall well-being.

A wealth of research has illuminated the positive impacts of interventions focused on fostering balance within the gut microbiota, encompassing an array of strategies such as prebiotics, probiotics, and postbiotic metabolites. These interventions, characterized by their ability to promote harmony within the intestinal microbial community, serve to strengthen the integrity of the gut barrier and modulate immune responses, as depicted in Figure 10.6. By bolstering the body's defenses against pathogenic intruders and alleviating the susceptibility to various conditions, ranging from obesity and type 2 diabetes to inflammatory bowel disease, cancer, cardiovascular disorders, liver ailments, and central nervous system disorders, these interventions contribute significantly to overall host health and well-being (Tsai et al., 2019).

Over the last 10 years, progress in high-throughput sequencing technologies has sparked a notable increase in microbiome investigations, facilitating extensive profiling of microbiota composition across a spectrum of medical conditions (Belkaid & Hand, 2014; Levy, Kolodziejczyk, Thaiss, & Elinav, 2017; Rodríguez et al., 2015). While the definitive causative link between changes in the microbiome and specific disease conditions remains elusive, it is widely acknowledged as a conceivable factor across various ailments. In the context of TB, the microbiome may exert influence in multiple ways: first, by influencing the subsets or functionality of immune cells, thereby potentially having impact on susceptibility to TB or response to treatment, whether locally within the lung or systemically via the intestinal microbiota; second, by modifying drug absorption during TB treatment; and third, by generating antimicrobial or immune-regulating compounds that could directly impede the growth of MTB (Rooks & Garrett, 2016).

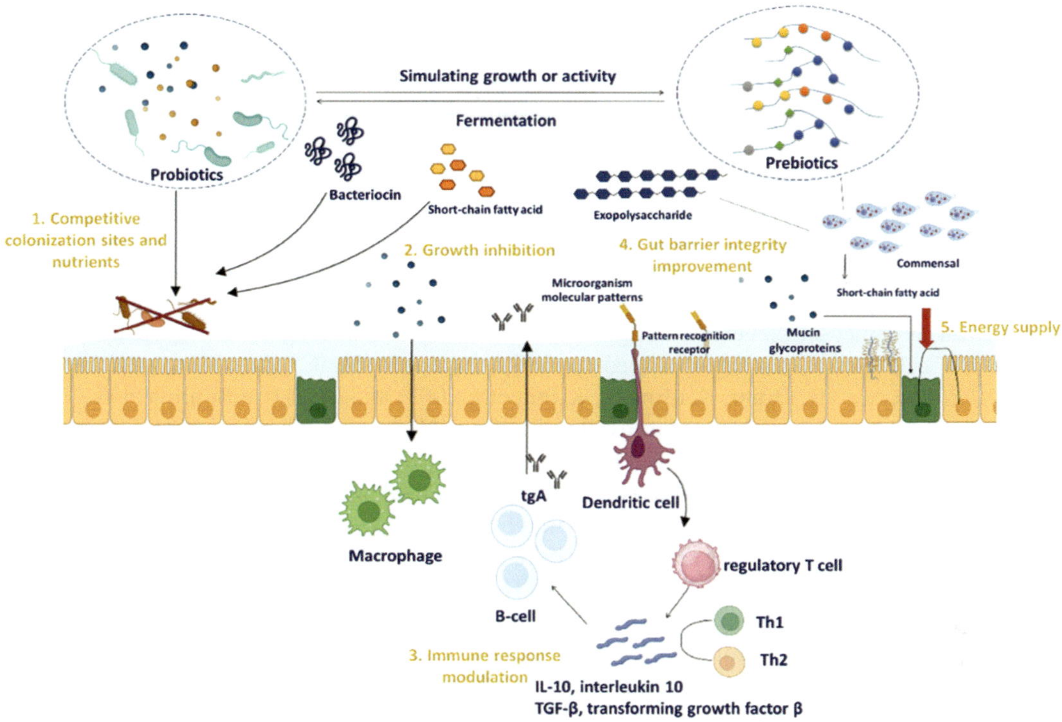

FIGURE 10.6 Contribution of prebiotics, probiotics, and postbiotics to intestinal microbiota balance, gut barrier integrity, and immune regulation.

The transmission of TB occurs through the inhalation of aerosols-carrying bacteria, which are expelled by individuals with active TB through coughing. Upon inhalation, the host may develop latent TB infection (LTBI), which can be detected using tests such as the tuberculin skin test (TST) or interferon gamma release assay (IGRA). However, some exposed individuals are able to clear the infection before LTBI is established, leading to negative results for TST/IGRA (Gury-BenAri et al., 2016). Microbiota differences might influence the initial clearance of infection through innate immune mechanisms. Studies comparing normal and germ-free mice have revealed differences in the transcriptional profiles of innate lymphoid cells, myeloid cell function (Khosravi et al., 2014), and the formation of mucus layers (Johansson et al., 2015), highlighting the microbiome's role in immune defense and resistance against pathogens. Certain commensal bacteria and their antimicrobial products may enhance resistance to pathogens through various mechanisms (Ubeda, Djukovic, & Isaac, 2017), such as niche competition (Kamada et al., 2012) and bacteriolytic activity (Satoh-Takayama et al., 2008). Nevertheless, there remains a scarcity of comprehensive data regarding the microbiota's involvement in the mediation of initial TB resistance. Investigations employing broad-spectrum antibiotic treatments in mice and those introducing dysbiosis-inducing commensals indicate a plausible influence of the microbiota on the initial host resistance mechanism, potentially affecting the susceptibility to TB infection among exposed human populations (Khan et al., 2016). Exploring the microbiome compositions of household contacts equally exposed to TB, but with varying results in IFN-γ release assay (IGRA) tests, could offer valuable insights into potential mechanisms of host protection (Majlessi et al., 2017).

Numerous mechanisms have been suggested to understand the shift from LTBI to active TB disease, with emerging evidence indirectly implicating the microbiota in disease progression. Notably, research has shown that LTBI patients harboring *Helicobacter pylori* in their gut flora had a 50% lower likelihood of developing active TB disease, mirroring findings that link specific intestinal bacteria to susceptibility to respiratory syncytial virus infection and pediatric asthma (Fujimura et al., 2014). Recent research has suggested a potential link between the transition from LTBI to active TB (Segal et al., 2017). In HIV-infected individuals undergoing antiretroviral therapy (ART), alterations in the lung microbiota composition are noteworthy. Elevated levels of SCFAs like propionate and butyrate in the bloodstream were linked to an increased susceptibility to active TB. These SCFAs were found to be associated with higher levels of anaerobic bacteria, including *Prevotella*, which are known to produce these lipid molecules. Furthermore, these SCFAs were observed to inhibit the production of IFN-γ and IL-17A, essential cytokines crucial for T-cell effector function, suggesting a potential mechanism through which certain pulmonary anaerobes might contribute to TB progression. This highlights the significance of microbial metabolic activity at mucosal surfaces as a contributing factor for the development of active TB (Dodd et al., 2017; Negatu et al., 2018).

Moreover, commensal metabolites might have an impact on the progression of TB by activating innate T-cell subsets via major histocompatibility complex (MHC) class I-like proteins such as CD1 (Ulrichs, Moody, Grant, Kaufmann, & Porcelli, 2003) and MR1 (Le Bourhis et al., 2010). These proteins activate invariant natural killer T cells (iNKT cells) (Chackerian, Alt, Perera, & Behar, 2002), germ line-encoded mycolyl lipid-reactive (GEM) T cells (Van Rhijn et al., 2013), and mucosa-associated invariant T (MAIT) cells (Gold et al., 2010), speculated to be involved in host resistance to MTB. Interestingly, MAIT cells are absent in germ-free mice (Treiner et al., 2003), implying that the microbiota could potentially have an impact on their development and function, thereby potentially influencing the progression of TB. There is a pressing need for prospective studies to directly uncover the links between microbiota composition and the advancement of LTBI in humans. Such investigations could unveil correlations between transcriptional signatures in the blood and the composition of the microbiota, offering invaluable insights for both diagnosis and comprehension of the mechanisms driving the development of active TB (Zak et al., 2016).

The changes occurring in the microbiota during active TB were explored in mouse models. Using inbred specific-pathogen-free (SPF) mice, which have a uniform intestinal microbiota among individuals, researchers observed slight alterations in taxa after exposure to aerosol MTB. These changes primarily affected the relative abundance of the Clostridiales order (Winglee et al., 2014). Numerous investigations into the microbiome highlight variations both among individuals and across different geographical regions concerning MTB infection or TB disease. These studies delve into the microbiota present in stool, sputum, or bronchoalveolar lavage (BAL) fluid obtained from afflicted patients. For example, recent studies have unveiled heightened diversity and quantities of actinobacteria and proteobacteria in individuals experiencing recurrent TB, alongside diminished levels of *Prevotella* and members of the Clostridiales order, in contrast to those deemed healthy (Luo et al., 2017).

Assessment of changes in the lung microbiome induced by active MTB infection relies on studies utilizing sputum samples (Cheung et al., 2013; Cui et al., 2012; Wu et al., 2013) and BAL fluid(Cui et al., 2012). The analysis of the lung microbiota presents a complex endeavor, primarily due to practical hurdles in acquiring uncontaminated samples and recruiting suitable healthy controls for bronchoalveolar lavage (BAL) fluid extraction. Additionally, the transient nature of the lung microbiome, coupled with its lower biomass, adds to the intricacies of this analysis (Huffnagle, Dickson, & Lukacs, 2017). Despite encountering these obstacles, specific changes in both the diversity and makeup of the lung microbiota have been associated with various stages of TB in humans, including new MTB infections, recurring TB diseases, and instances of treatment failures. Interestingly, while it might be anticipated that the presence of mycobacteria would lead to a reduction in diversity, this decrease is not consistently observed across studies (Dickson, Erb-Downward, & Huffnagle, 2013). Moreover, our comprehension of the interplay among MTB infection, TB disease, and the pulmonary microbiome is impeded by the dearth of data derived from animal models. This deficiency is, in part, attributable to the diminished abundance of microbiota within the lung tissue. However, current investigations into experimental MTB infection in nonhuman primates present a promising avenue for bridging this gap in understanding. By leveraging nonhuman primate models, researchers can explore the dynamics of microbiota within the lungs during TB infection, thereby enhancing our insight into the complex relationship between microbial communities and disease progression in the respiratory system.

10.8 CHALLENGES AND FUTURE DIRECTIONS

The development of probiotics as a tool for curing TB is a promising but challenging research avenue. In order to avoid these challenges, which are outlined below, rigorous scientific research, as well as clinical trials, will need to be undertaken in order to unlock the full potential of probiotics in the treatment of TB.

Variability in the Efficacy of Probiotic Strains: In the context of adding probiotics to antiTB (TB) treatment, one of the largest problems is that there is high variability in the efficacy among different strains of probiotics. Notice that certain strains of bacteria have different effects on gut health, immune function, and patient outcomes. It is of paramount importance that future research focuses on identifying the strains that will deliver the most benefits to the cure of TB in the future. A large array of probiotics is presently being rigorously investigated *in vitro* and *in vivo* to establish their immunomodulatory and gut-protective properties *in vitro* and *in vivo* (Rahim et al., 2022).

Optimum Dosage and Delivery Systems: The second challenge is discovering the optimum dosage and delivery system for probiotics in the treatment of TB. It is important that the probiotic dose should be sufficient enough to impart a benefit and not cause any harm to the human body, and the delivery system should be able to carry the probiotics to the target location in the gut. Additional studies are necessary in order to create standardized dosage regimens, as well as modes of delivery, such as capsules or probiotics incorporated into food, which will provide assurance regarding the viability of the probiotics during storage and when being digested (Ezike et al., 2023).

Understanding Probiotic and Antibiotic Interactions: There should be clear knowledge of the interaction between anti-TB antibiotics and probiotics. There are certain antibiotics that suppress the growth of probiotics, while there are some antibiotics which are able to stimulate it. There should be proper understanding of these interactions so that drugs can be coadministered. Future studies must explore how different antibiotics administered in the management of TB (e.g., INH, RIF, EMB) affect probiotic functions and if some probiotics can possibly prevent antibiotic side effects such as dysbiosis or intestinal distress (Cunningham et al., 2021).

Clinical Trials and Patient Diversity: Another matter that must be attended to in order to determine whether probiotics are beneficial to TB treatment is carrying out good clinical trials. These trials must be done in a way that covers patient diversity, age, gender, comorbidities, and severity of the TB. To guarantee that the results can be generalized, future studies should include a wide range of patients and take into account a variety of demographic factors. Largescale, multicenter trials with rigorous control and randomization are needed to achieve this goal (Willis, Isaacs, & Khunti, 2021).

Mechanistic Insights into Probiotic Actions: Despite our best efforts, we still do not fully understand the mechanisms through which probiotics may enhance anti-TB treatment. To better understand how probiotics influence cytokine production, immune cell activation, and gut microbiome composition, more research into the immunomodulatory effects of probiotics is needed. Future work should be focused on elucidating these mechanisms at a molecular level, using techniques such as metagenomics, transcriptomics, and proteomics to obtain a deeper understanding of these mechanisms (PlazaDiaz et al., 2019).

Regulatory and Safety Considerations: Regulation and safety concerns are the barriers to the clinical adoption of probiotics as an adjunct to anti-TB treatments, and this is a significant hurdle to their adoption. One needs to ensure that probiotics can be consumed long-term safely without any side effects or unwanted interaction with other medications and that they are safe for long-term use. A rigorous safety assessment in the future must be conducted to dispel these fears and to develop guidelines for the clinical use of probiotics in TB therapy in collaboration with regulatory bodies (Rahim et al., 2022).

Integration with Personalized Medicine: As the world is shifting towards more personalized medicine, one of the most potential fields of future research would be the integration of probiotics into personalized TB treatment routines. Under this approach, probiotic therapy would be tailored to one's personal composition of the microbiome and immune system. Whether personalized probiotic therapy can increase the efficacy of anti-TB therapy and reduce side effects needs to be ascertained with more research (Pritchard, Hulick, & Wells, 2021).

10.9 CONCLUSION

Probiotics are of immense potential as an adjuvant treatment for TB, aiming at problems of microbiome imbalance and immune disruption caused by anti-TB drugs. Probiotics have been found to modulate the gut microbiota, enhance intestinal barrier function, and stimulate favorable immune responses, suggesting their potential to neutralize side effects of anti-TB treatments and augment therapeutic efficacy.

Immunity is essential in governing and healing MTB infection, and probiotics have already been demonstrated to be able to modulate immune response in a positive manner. Probiotics are able to strengthen the efficacy of macrophages and natural killer cells, induce anti-inflammatory cytokines, and suppress pro-inflammatory indicators. By such action, probiotics have the potential to boost the host immune response against MTB, making anti-TB therapy more effective as well as reducing the inflammation and tissue damage caused by the disease or drugs. Moreover, probiotics have been shown to neutralize antibioticassociated diarrhea, a common side effect of TB therapy, by increasing the beneficial bacteria in the gut while suppressing the pathogens. They are also helpful to the liver in modulating oxidative stress and inflammation and, potentially, reducing the risk of drug-induced hepatotoxicity. The growing evidence highlights the potential of probiotics to enhance anti-TB treatment efficacy and quality of life among patients.

By sustaining microbiome balance, boosting immune response, and alleviating deleterious drug effects, probiotics offer a novel complementary treatment strategy to conventional TB therapies. Additional research is needed to establish the optimal strains, dosages, and treatment regimens to further maximize their applications in clinical settings. Such advances could bring novel solutions to the battle against TB, one of the globe's most stubborn and destructive infectious diseases.

ACKNOWLEDGMENT

We utilized AI-powered tools to check grammar and enhance the academic quality of the text, which was primarily writer by the authors.

REFERENCES

Ahire, J. J., Jakkamsetty, C., Kashikar, M. S., Lakshmi, S. G., & Madempudi, R. S. (2021). In vitro evaluation of probiotic properties of Lactobacillus plantarum UBLP40 isolated from traditional indigenous fermented food. *Probiotics Antimicrob Proteins, 13*(5), 1413–1424. doi:10.1007/s12602-021-09775-7

Ahmad, Z., Tyagi, S., Minkowski, A., Almeida, D., Nuermberger, E. L., Peck, K. M., … Grosset, J. H. (2012). Activity of 5-chloro-pyrazinamide in mice infected with Mycobacterium tuberculosis or Mycobacterium bovis. *Indian Journal of Medical Research, 136*(5), 808–814.

Ahmed, S., Ahmed, M. Z., Rafique, S., Almasoudi, S. E., Shah, M., Jalil, N. A. C., & Ojha, S. C. (2023). Recent approaches for downplaying antibiotic resistance: Molecular mechanisms. *BioMed Research International, 2023*, 5250040. doi:10.1155/2023/5250040.

Al-Najjar, M. A. A., Abdulrazzaq, S. B., Alzaghari, L. F., Mahmod, A. I., Omar, A., Hasen, E., … Barakat, M. (2024). Evaluation of immunomodulatory potential of probiotic conditioned medium on murine macrophages. *Scientific Reports, 14*(1), 7126. doi:10.1038/s41598-024-56622-0

AlSahafi, A. J., Shah, H. B. U., AlSayali, M. M., Mandoura, N., Assiri, M., Almohammadi, E. L., … AlGarni, F. (2019). High non-compliance rate with anti-tuberculosis treatment: A need to shift facility-based directly observed therapy short course (DOTS) to community mobile outreach team supervision in Saudi Arabia. *BMC Public Health,* 19(1), 1168. doi:10.1186/s12889-019-7520-8

Anand, N., Upadhyaya, K., & Tripathi, R. (2015). Drug development pipeline for the treatment of tuberculosis: Needs, challenges, success and opportunities for the future. *Chemistry and Biology Interface, 5*, 84–127.

Asif, M. (2012). Mini review on Ethmabutol and its some analogues as antitubercular agents. *Elixir Pharm, 53*, 11774–11778.

Atarashi, K., Tanoue, T., Oshima, K., Suda, W., Nagano, Y., Nishikawa, H., … Hase, K. (2013). Treg induction by a rationally selected mixture of Clostridia strains from the human microbiota. *Nature, 500*(7461), 232–236.

Atarashi, K., Tanoue, T., Shima, T., Imaoka, A., Kuwahara, T., Momose, Y., … Ohba, Y. (2011). Induction of colonic regulatory T cells by indigenous Clostridium species. *Science, 331*(6015), 337–341.

Ayyash, M., Abushelaibi, A., Al-Mahadin, S., Enan, M., El-Tarabily, K., & Shah, N. (2018). In-vitro investigation into probiotic characterisation of Streptococcus and Enterococcus isolated from camel milk. *LWT, 87*, 478–487. doi:10.1016/j.lwt.2017.09.019

Azad, M. A. K., Sarker, M., Li, T., & Yin, J. (2018). Probiotic species in the modulation of gut microbiota: An overview. *Biomed Research International,* 2018, 9478630. doi:10.1155/2018/9478630

Azad, M. A. K., Sarker, M., & Wan, D. (2018). Immunomodulatory effects of probiotics on cytokine profiles. *Biomed Research International,* 2018, 8063647. doi:10.1155/2018/8063647

Badyal, D. K., & Desai, C. (2014). Animal use in pharmacology education and research: The changing scenario. *Indian Journal of Pharmacology, 46*(3), 257–265.

Bagcchi, S. (2023). WHO's global tuberculosis report 2022. *The Lancet Microbe, 4*(1), e20.

Balcells, M. E., Yokobori, N., Hong, B. Y., Corbett, J., & Cervantes, J. (2019). The lung microbiome, vitamin D, and the tuberculous granuloma: A balance triangle. *Microbial Pathogenesis, 131*, 158–163. doi:10.1016/j.micpath.2019.03.041

Belkacem, N., Serafini, N., Wheeler, R., Derrien, M., Boucinha, L., Couesnon, A., … Bourdet-Sicard, R. (2017). Lactobacillus paracasei feeding improves immune control of influenza infection in mice. *PLoS One, 12*(9), e0184976. doi:10.1371/journal.pone.0184976

Belkaid, Y., & Hand, T. W. (2014). Role of the microbiota in immunity and inflammation. *Cell, 157*(1), 121–141.

Bravo, M., Combes, T., Martinez, F. O., Cerrato, R., Rey, J., Garcia-Jimenez, W., … Gutierrez-Merino, J. (2019). Lactobacilli isolated from wild boar (Sus scrofa) Antagonize Mycobacterium bovis Bacille Calmette-Guerin (BCG) in a species-dependent manner. *Frontiers in Microbiology, 10*, 1663. doi:10.3389/fmicb.2019.01663

Britton, R. A., Hoffmann, D. E., & Khoruts, A. (2021). Probiotics and the microbiome-how can we help patients make sense of probiotics? *Gastroenterology, 160*(2), 614–623. doi:10.1053/j.gastro.2020.11.047

Bu, Y., Liu, Y., Liu, Y., Wang, S., Liu, Q., Hao, H., & Yi, H. (2022). Screening and probiotic potential evaluation of bacteriocin-producing Lactiplantibacillus plantarum in vitro. *Foods, 11*(11). doi:10.3390/foods11111575

Burgueno, J. F., & Abreu, M. T. (2020). Epithelial toll-like receptors and their role in gut homeostasis and disease. *Nature Reviews Gastroenterology & Hepatology, 17*(5), 263–278. doi:10.1038/s41575-019-0261-4

Cani, P. D. (2018). Human gut microbiome: Hopes, threats and promises. *Gut, 67*(9), 1716–1725. doi:10.1136/gutjnl-2018-316723

Carlson, T. L., Lock, J. Y., & Carrier, R. L. (2018). Engineering the mucus barrier. *Annual Review of Biomedical Engineering, 20*, 197–220. doi:10.1146/annurev-bioeng-062117-121156

Chackerian, A., Alt, J., Perera, V., & Behar, S. M. (2002). Activation of NKT cells protects mice from tuberculosis. *Infection and Immunity, 70*(11), 6302–6309.

Chang, Y. H., Jeong, C. H., Cheng, W. N., Choi, Y., Shin, D. M., Lee, S., & Han, S. G. (2021). Quality characteristics of yogurts fermented with short-chain fatty acid-producing probiotics and their effects on mucin production and probiotic adhesion onto human colon epithelial cells. *Journal of Dairy Science, 104*(7), 7415–7425. doi:10.3168/jds.2020-19820

Cheng, F. S., Pan, D., Chang, B., Jiang, M., & Sang, L. X. (2020). Probiotic mixture VSL#3: An overview of basic and clinical studies in chronic diseases. *World Journal of Clinical Cases, 8*(8), 1361–1384. doi:10.12998/wjcc.v8.i8.1361

Cheung, M. K., Lam, W. Y., Fung, W. Y. W., Law, P. T. W., Au, C. H., Nong, W., … Tsui, S. K. W. (2013). Sputum microbiota in tuberculosis as revealed by 16S rRNA pyrosequencing. *PLoS One, 8*(1), e54574.

Chiara, I. D., Marasco, R., Gala, M. D., Fusco, A., Donnarumma, G., & Muscariello, L. (2024). Probiotic properties of Lactococcus lactis strains isolated from natural whey starter cultures. *Foods, 13*(6), 1–14. doi:10.3390/foods13060957

Comberiati, P., Cicco, M. D., Paravati, F., Pelosi, U., Gangi, A. D., Arasi, S., … Licari, A. (2021). The role of gut and lung microbiota in susceptibility to tuberculosis. *International Journal of Environmental Research and Public Health, 18*(22), 12220. doi:10.3390/ijerph182212220

Cristofori, F., Dargenio, V. N., Dargenio, C., Miniello, V. L., Barone, M., & Francavilla, R. (2021). Anti-inflammatory and immunomodulatory effects of probiotics in gut inflammation: A door to the body. *Frontiers in Immunology, 12*, 578386. doi:10.3389/fimmu.2021.578386

Cui, Z., Zhou, Y., Li, H., Zhang, Y., Zhang, S., Tang, S., & Guo, X. (2012). Complex sputum microbial composition in patients with pulmonary tuberculosis. *BMC Microbiology, 12*, 1–8.

Cunningham, M., Azcarate-Peril, M. A., Barnard, A., Benoit, V., Grimaldi, R., Guyonnet, D., … Obis, D. (2021). Shaping the future of probiotics and prebiotics. *Trends in Microbiology, 29*(8), 667–685.

de Lima Corvino, D., Shrestha, S., Hollingshead, C., & Kosmin, A. (2023). *Tuberculosis screening*. Treasure Island, FL: StatPearls Publishing. https://www.ncbi.nlm.nih.gov/books/NBK448205/

Devendra Singh, D., & Agarwal, V. (2021). Screening of antimicrobial, anti-quorum sensing activity and cytotoxicity of origanumoil against Gram-positive and Gram-negative bacteria. *Biomedicine, 41*(3), 599–603. doi: 10.51248/.v41i3.1200

Diallo, D., Somboro, A. M., Diabate, S., Baya, B., Kone, A., Sarro, Y. S., … Diakite, M. (2021). Antituberculosis therapy and gut microbiota: Review of potential host microbiota directed-therapies. *Frontiers in Cellular and Infection Microbiology, 11*, 673100.

Dickson, R. P., Erb-Downward, J. R., & Huffnagle, G. B. (2013). The role of the bacterial microbiome in lung disease. *Expert Review of Respiratory Medicine, 7*(3), 245–257.

Dodd, D., Spitzer, M. H., Van Treuren, W., Merrill, B. D., Hryckowian, A. J., Higginbottom, S. K., … Fischbach, M. A. (2017). A gut bacterial pathway metabolizes aromatic amino acids into nine circulating metabolites. *Nature, 551*(7682), 648–652.

Doestzada, M., Vila, A. V., Zhernakova, A., Koonen, D. P. Y., Weersma, R. K., Touw, D. J., … Fu, J. (2018). Pharmacomicrobiomics: A novel route towards personalized medicine? *Protein Cell, 9*(5), 432–445. doi:10.1007/s13238-018-0547-2

dos Santos Fernandes, G. F., de Souza, P. C., Marino, L. B., Chegaev, K., Guglielmo, S., Lazzarato, L., … Dos Santos, J. L. (2016). Synthesis and biological activity of furoxan derivatives against Mycobacterium tuberculosis. *European Journal of Medicinal Chemistry, 123*, 523–531.

dos Santos Fernandes, G. F., De Souza, P. C., Moreno-Viguri, E., Santivañez-Veliz, M., Paucar, R., Pérez-Silanes, S., … Fruttero, R. (2017). Design, synthesis, and characterization of N-oxide-containing heterocycles with in vivo sterilizing antitubercular activity. *Journal of Medicinal Chemistry, 60*(20), 8647–8660.

Dumas, A., Corral, D., Colom, A., Levillain, F., Peixoto, A., Hudrisier, D., … Neyrolles, O. (2018). The host microbiota contributes to early protection against lung colonization by Mycobacterium tuberculosis. *Frontiers in Immunology, 9*, 2656. doi:10.3389/fimmu.2018.02656

Eribo, O. A., du Plessis, N., Ozturk, M., Guler, R., Walzl, G., & Chegou, N. N. (2020). The gut microbiome in tuberculosis susceptibility and treatment response: Guilty or not guilty? *Cellular and Molecular Life Sciences, 77*, 1497–1509.

Ezike, T. C., Okpala, U. S., Onoja, U. L., Nwike, P. C., Ezeako, E. C., Okpara, J. O., … Odoh, E. C. (2023). Advances in drug delivery systems, challenges and future directions. *Heliyon, 9*(6), e17488. doi: 10.1016/j.heliyon.2023.e17488

FAW/WHO. (2006). Probiotics in food, Health and nutritional properties and guidlines for evaluation. *FAO Food and Nutrition Paper, 85*, 1–50.

Fijan, S. (2014). Microorganisms with claimed probiotic properties: An overview of recent literature. *International Journal of Environmental Research and Public Health, 11*(5), 4745–4767. doi:10.3390/ijerph110504745

Forget, E. J., & Menzies, D. (2006). Adverse reactions to first-line antituberculosis drugs. *Expert Opinion on Drug Safety, 5*(2), 231–249.

Fu, L., Xie, M., Wang, C., Qian, Y., Huang, J., Sun, Z., … Wang, Y. (2020). Lactobacillus casei zhang alleviates shrimp tropomyosin-induced food allergy by switching antibody isotypes through the NF-kappaB-dependent immune tolerance. *Molecular Nutrition & Food Research, 64*(10), e1900496. doi:10.1002/mnfr.201900496

Fujimura, K. E., Demoor, T., Rauch, M., Faruqi, A. A., Jang, S., Johnson, C. C., … Lukacs, N. W. (2014). House dust exposure mediates gut microbiome Lactobacillus enrichment and airway immune defense against allergens and virus infection. *Proceedings of the National Academy of Sciences, 111*(2), 805–810.

Fujimura, K. E., Sitarik, A. R., Havstad, S., Lin, D. L., Levan, S., Fadrosh, D., … Lukacs, N. W. (2016). Neonatal gut microbiota associates with childhood multisensitized atopy and T cell differentiation. *Nature Medicine, 22*(10), 1187–1191.

Fuller, R. (1989). Probiotics in man and animals. *Journal of Applied Bacteriology, 66*, 365–378. doi:10.1111/j.1365-2672.1989.tb05105.x

Gangaraju, D., Raghu, A. V., & Siddalingaiya Gurudutt, P. (2022). Green synthesis of γ-aminobutyric acid using permeabilized probiotic Enterococcus faecium for biocatalytic application. *Nano Select, 3*(10), 1436–1447. doi:10.1002/nano.202200059

Gattupalli, N. k., & Gattupalli, A. (2021). Potential of Escherichia coli probiotics for improved health and disease management. In M. S. Erjavec (Ed.), *Escherichia coli – Old and New Insights* (pp. 1–14). IntechOpen. doi: 10.5772/intechopen.100380

Gavrilova, N. N., Ratnikova, I. A., Sadanov, A. K., Bayakisheva, K., Tourlibaeva, Z. J., & Belikova, O. A. (2014). Application of probiotics in complex treatment of tuberculosis. *International Journal of Engineering Research and Applications, 4*(11), 13–18.

Geva-Zatorsky, N., Sefik, E., Kua, L., Pasman, L., Tan, T. G., Ortiz-Lopez, A., … Kasper, D. L. (2017). Mining the human gut microbiota for immunomodulatory organisms. *Cell, 168*(5), 928–943 e911. doi:10.1016/j.cell.2017.01.022

Gibson, G. R., Hutkins, R., Sanders, M. E., Prescott, S. L., Reimer, R. A., Salminen, S. J., … Reid, G. (2017). The International Scientific Association for Probiotics and Prebiotics (ISAPP) consensus statement on the definition and scope of prebiotics. *Nature Reviews: Gastroenterology & Hepatology, 14*, 491–502. doi:10.1038/nrgastro.2017.75

Gold, M. C., Cerri, S., Smyk-Pearson, S., Cansler, M. E., Vogt, T. M., Delepine, J., … Yu, Y. Y. (2010). Human mucosal associated invariant T cells detect bacterially infected cells. *PLoS Biology, 8*(6), e1000407.

Guarner, F., Sanders, M. E., Eliakim, R., Fedorak, R., Gangl, A., Garisch, J., … Mair, A. L. (2012). World gastroenterology organisation global guidelines: probiotics and prebiotics october 2011. *Journal of Clinical Gastroenterology, 46*(6), 468–481. doi: 10.1097/MCG.0b013e3182549092

Guo, J., Li, L., Cai, Y., & Kang, Y. (2024). The development of probiotics and prebiotics therapy to ulcerative colitis: A therapy that has gained considerable momentum. *Cell Communication and Signaling: CCS, 22*(1), 268. doi: 10.1155/2019/5796491

Guo, S., Chen, S., Ma, J., Ma, Y., Zhu, J., Ma, Y., … Pan, Y. (2019). Escherichia coli Nissle 1917 protects intestinal barrier function by inhibiting NF-kappaB-mediated activation of the MLCK-P-MLC signaling pathway. *Mediators of Inflammation,* 2019, 5796491. doi:10.1155/2019/5796491

Gury-BenAri, M., Thaiss, C. A., Serafini, N., Winter, D. R., Giladi, A., Lara-Astiaso, D., … David, E. (2016). The spectrum and regulatory landscape of intestinal innate lymphoid cells are shaped by the microbiome. *Cell, 166*(5), 1231–1246. e1213.

Hameed, H. A., Islam, M. M., Chhotaray, C., Wang, C., Liu, Y., Tan, Y., … Yew, W. W. (2018). Molecular targets related drug resistance mechanisms in MDR-, XDR-, and TDR-Mycobacterium tuberculosis strains. *Frontiers in Cellular and Infection Microbiology, 8*, 114.

He, Y., Wen, Q., Yao, F., Xu, D., Huang, Y., & Wang, J. (2017). Gut-lung axis: The microbial contributions and clinical implications. *Critical Reviews in Microbiology, 43*(1), 81–95. doi:10.1080/1040841X.2016.1176988

Hempel, S., Newberry, S. J., Maher, A. R., Wang, Z., Miles, J. N. V., Shanman, R., … Shekelle, P. G. (2012). Probiotics for the prevention and treatment of antibiotic-associated diarrhea. *JAMA,* 307(18), 1959–1969. doi:10.1001/jama.2012.3507

Hill, C., Guarner, F., Reid, G., Gibson, G. R., Merenstein, D. J., Pot, B., … Sanders, M. E. (2014). Expert consensus document. The International Scientific Association for Probiotics and Prebiotics consensus statement on the scope and appropriate use of the term probiotic. *Nature Reviews Gastroenterology & Hepatology, 11*(8), 506–514. doi:10.1038/nrgastro.2014.66

Hoffmann, D. E., Fraser, C. M., Palumbo, F., Ravel, J., Rowthorn, V., & Schwartz, J. (2014). Probiotics: Achieving a better regulatory fit. *Food Drug Law Journal, 69*(2), 237–272, ii.

Hong, B.-Y., Maulén, N. P., Adami, A. J., Granados, H., Balcells, M. E., & Cervantes, J. (2016). Microbiome changes during tuberculosis and antituberculous therapy. *Clinical Microbiology Reviews, 29*(4), 915–926.

Horvath, A., Leber, B., Schmerboeck, B., Tawdrous, M., Zettel, G., Hartl, A., … Stadlbauer, V. (2016). Randomised clinical trial: The effects of a multispecies probiotic vs. placebo on innate immune function, bacterial translocation and gut permeability in patients with cirrhosis. *Alimentary Pharmacology & Therapeutics, 44*(9), 926–935. doi:10.1111/apt.13788

Hossain, Z. (2014). Bacteria: Streptococcus. In *Encyclopedia of Food Safety* (pp. 535–545). Academic Press. doi:10.1016/b978-0-12-378612-8.00116-5

Houben, R. M., & Dodd, P. J. (2016). The global burden of latent tuberculosis infection: A re-estimation using mathematical modelling. *PLoS Medicine, 13*(10), e1002152. doi:10.1371/journal.pmed.1002152

Hrncir, T. (2022). Gut microbiota dysbiosis: Triggers, consequences, diagnostic and therapeutic options. *Microorganisms, 10*(3), 578. doi: 10.3390/microorganisms10030578

Hu, Y., Feng, Y., Wu, J., Liu, F., Zhang, Z., Hao, Y., … Lv, N. (2019a). The gut microbiome signatures discriminate healthy from pulmonary tuberculosis patients. *Frontiers in Cellular and Infection Microbiology, 9*, 90.

Hu, Y., Yang, Q., Liu, B., Dong, J., Sun, L., Zhu, Y., … Chen, X. (2019b). Gut microbiota associated with pulmonary tuberculosis and dysbiosis caused by anti-tuberculosis drugs. *Journal of Infection, 78*(4), 317–322.

Hu, Y.-Q., Zhang, S., Zhao, F., Gao, C., Feng, L.-S., Lv, Z.-S., … Wu, X. (2017). Isoniazid derivatives and their anti-tubercular activity. *European Journal of Medicinal Chemistry, 133*, 255–267.

Huang, S. F., Yang, Y. Y., Chou, K. T., Fung, C. P., Wang, F. D., & Su, W. J. (2019). Systemic proinflammation after Mycobacterium tuberculosis infection was correlated to the gut microbiome in HIV-uninfected humans. *European Journal of Clinical Investigation, 49*(5), e13068.

Huffnagle, G., Dickson, R., & Lukacs, N. (2017). The respiratory tract microbiome and lung inflammation: A two-way street. *Mucosal Immunology, 10*(2), 299–306.

Human Microbiome Project, C. (2012). Structure, function and diversity of the healthy human microbiome. *Nature, 486*(7402), 207–214. doi:10.1038/nature11234

Jandourek, O., Tauchman, M., Paterova, P., Konecna, K., Navratilova, L., Kubicek, V., … Dolezal, M. (2017). Synthesis of novel pyrazinamide derivatives based on 3-chloropyrazine-2-carboxamide and their antimicrobial evaluation. *Molecules, 22*(2), 223.

Jenkins, G., & Mason, P. (2022). The role of prebiotics and probiotics in human health: A systematic review with a focus on gut and immune health. *Food & Nutrition Journal, 7*(1), 1–19. doi:10.29011/2575-7091.100245

Jiang, L., Wang, J., Xu, L., Cai, J., Zhao, S., & Ma, A. (2022). Lactobacillus casei modulates inflammatory cytokines and metabolites during tuberculosis treatment: A post hoc randomized controlled trial. *Asia Pacific Journal of Clinical Nutrition, 31*(1), 66–77. doi:10.6133/apjcn.202203_31(1).0008

Johansson, M. E., Jakobsson, H. E., Holmén-Larsson, J., Schütte, A., Ermund, A., Rodríguez-Piñeiro, A. M., … Bäckhed, F. (2015). Normalization of host intestinal mucus layers requires long-term microbial colonization. *Cell Host & Microbe, 18*(5), 582–592.

Kamada, N., Kim, Y.-G., Sham, H. P., Vallance, B. A., Puente, J. L., Martens, E. C., & Núñez, G. (2012). Regulated virulence controls the ability of a pathogen to compete with the gut microbiota. *Science, 336*(6086), 1325–1329.

Kang, M. J., Jeong, H., Kim, S., Shin, J., Song, Y., Lee, B. H., … Park, Y. S. (2023). Structural analysis and prebiotic activity of exopolysaccharide produced by probiotic strain Bifidobacterium bifidum EPS DA-LAIM. *Food Science and Biotechnology, 32*(4), 517–529. doi:10.1007/s10068-022-01213-w

Kayukova, L., & Berikova, E. (2020). Modern anti-tuberculosis drugs and their classification. Part I: First-line drugs. *Pharmaceutical Chemistry Journal, 54*(6), 555–563.

Khailova, L., Baird, C. H., Rush, A. A., McNamee, E. N., & Wischmeyer, P. E. (2013). Lactobacillus rhamnosus GG improves outcome in experimental pseudomonas aeruginosa pneumonia: Potential role of regulatory T cells. *Shock, 40*(6), 496–503. doi:10.1097/SHK.0000000000000066

Khaledi, M., Poureslamfar, B., Alsaab, H. O., Tafaghodi, S., Hjazi, A., Singh, R., … Sameni, F. (2024). The role of gut microbiota in human metabolism and inflammatory diseases: A focus on elderly individuals. *Annals of Microbiology, 74*(1), 1.

Khan, N., & Agrewala, J. N. (2016). Alteration in the gut microbiota provokes susceptibility to tuberculosis. *Frontiers in Immunology, 7*, 218883.

Khan, N., Mendonca, L., Dhariwal, A., Fontes, G., Menzies, D., Xia, J., … King, I. L. (2019). Intestinal dysbiosis compromises alveolar macrophage immunity to Mycobacterium tuberculosis. *Mucosal Immunology, 12*(3), 772–783.

Khan, N., Vidyarthi, A., Nadeem, S., Negi, S., Nair, G., & Agrewala, J. N. (2016). Alteration in the gut microbiota provokes susceptibility to tuberculosis. *Frontiers in Immunology, 7*, 218883.

Khosravi, A., Yáñez, A., Price, J. G., Chow, A., Merad, M., Goodridge, H. S., & Mazmanian, S. K. (2014). Gut microbiota promote hematopoiesis to control bacterial infection. *Cell host & Microbe, 15*(3), 374–381.

Koehler, N., Andres, S., Merker, M., Dreyer, V., John, A., Kuhns, M., … Zur Wiesch, P. A. (2023). Pretomanid-resistant tuberculosis. *Journal of Infection, 86*(5), 520–524.

Konuray, G., & Erginkaya, Z. (2018). Potential use of Bacillus coagulans in the food industry. *Foods, 7*(6). doi:10.3390/foods7060092

Kumar, D., Khare, G., Kidwai, S., Tyagi, A. K., Singh, R., & Rawat, D. S. (2014). Synthesis of novel 1, 2, 3-triazole derivatives of isoniazid and their in vitro and in vivo antimycobacterial activity evaluation. *European Journal of Medicinal Chemistry, 81*, 301–313.

Kumar, J. K., & Narsaiah, A. V. (2014). Stereoselective synthesis of tuberculostatic agent (S, S)-ethambutol. *Organic Communications, 7*(1), 28.

Kurosu, M., & Denny, W. (2016). *Advances in Tuberculosis Medicinal Chemistry*. Future Medicine.

Kwon, Y.-S. (2017). Clinical implications of new drugs and regimens for the treatment of drug-resistant tuberculosis. *Chonnam Medical Journal, 53*(2), 103.

Le Bourhis, L., Martin, E., Péguillet, I., Guihot, A., Froux, N., Coré, M., … Premel, V. (2010). Antimicrobial activity of mucosal-associated invariant T cells. *Nature Immunology, 11*(8), 701–708.

Lee, N. K., Kim, W. S., & Paik, H. D. (2019). Bacillus strains as human probiotics: Characterization, safety, microbiome, and probiotic carrier. *Food Science and Biotechnology, 28*(5), 1297–1305. doi:10.1007/s10068-019-00691-9

Lee, R. E., Protopopova, M., Crooks, E., Slayden, R. A., Terrot, M., & Barry, C. E. (2003). Combinatorial lead optimization of [1, 2]-diamines based on ethambutol as potential antituberculosis preclinical candidates. *Journal of Combinatorial Chemistry, 5*(2), 172–187.

Levy, M., Kolodziejczyk, A. A., Thaiss, C. A., & Elinav, E. (2017). Dysbiosis and the immune system. *Nature Reviews Immunology, 17*(4), 219–232.

Li, W., Zhu, Y., Liao, Q., Wang, Z., & Wan, C. (2019). Characterization of gut microbiota in children with pulmonary tuberculosis. *BMC Pediatrics, 19*(1), 445. doi:10.1186/s12887-019-1782-2

Li, Y., Zhao, L., Hou, M., Gao, T., Sun, J., Luo, H., … Cai, J. (2022). Lactobacillus casei improve anti-tuberculosis drugs-induced intestinal adverse reactions in rat by modulating gut microbiota and short-chain fatty acids. *Nutrients, 14*(8). doi:10.3390/nu14081668

Lilly, D. M., & Stillwell, R. H. (1965). Probiotics: Growth promoting factors produced by microorganisms. *Science, 147*, 747–748. doi:10.1126/science.147.3659.747

Lin, S., Zhao, S., Liu, J., Zhang, J., Zhang, C., Hao, H., … Ma, A. (2020). Efficacy of proprietary Lactobacillus casei for anti-tuberculosis associated gastrointestinal adverse reactions in adult patients: A randomized, open-label, dose-response trial. *Food & Function, 11*(1), 370–377. doi:10.1039/c9fo01583c

Liu, Y., Wang, J., & Wu, C. (2021). Microbiota and tuberculosis: A potential role of probiotics, and postbiotics. *Frontiers in Nutrition, 8*, 626254. doi:10.3389/fnut.2021.626254

Louis, P., & Flint, H. J. (2009). Diversity, metabolism and microbial ecology of butyrate-producing bacteria from the human large intestine. *FEMS Microbiology Letters, 294*(1), 1–8.

Luo, M., Liu, Y., Wu, P., Zheng, H., Pandol, S. J., & Han, Y.-P. (2017). Alternation of gut microbiota in patients with pulmonary tuberculosis. *Frontiers in Physiology, 8*, 283746.

Ma, W., Li, W., Yu, S., Bian, H., Wang, Y., Jin, Y., … Huang, L. (2023). Immunomodulatory effects of complex probiotics on the immuno-suppressed mice induced by cyclophosphamide. *Frontiers in Microbiology, 14*, 1055197. doi:10.3389/fmicb.2023.1055197

Ma, X. Y., Son, Y. H., Yoo, J. W., Joo, M. K., & Kim, D. H. (2022). Citation: Tight junction protein expression-inducing probiotics alleviate TNBS-induced cognitive impairment with colitis in mice. *Nutrients, 14*(14). doi:10.3390/nu14142975

Machado, D., Girardini, M., Viveiros, M., & Pieroni, M. (2018). Challenging the drug-likeness dogma for new drug discovery in tuberculosis. *Frontiers in Microbiology, 9*, 383242.

Madadi-Goli, N., Ahmadi, K., Meskini, M., Vaziri, F., Fateh, A., & Siadat, S. D. (2024). Tuberculosis prevention, diagnosis, and treatment financial profile during 2006–2021: Part B. *Journal of Preventive, Diagnostic and Treatment Strategies in Medicine, 3*(1), 37–43.

Maftei, N. M., Raileanu, C. R., Balta, A. A., Ambrose, L., Boev, M., Marin, D. B., & Lisa, E. L. (2024). The potential impact of probiotics on human health: An update on their health-promoting properties. *Microorganisms, 12*(2). doi:10.3390/microorganisms12020234

Maji, A., Misra, R., Dhakan, D. B., Gupta, V., Mahato, N. K., Saxena, R., … Singh, Y. (2018). Gut microbiome contributes to impairment of immunity in pulmonary tuberculosis patients by alteration of butyrate and propionate producers. *Environmental Microbiology, 20*(1), 402–419. doi:10.1111/1462-2920.14015

Majlessi, L., Sayes, F., Bureau, J.-F., Pawlik, A., Michel, V., Jouvion, G., … Peano, C. (2017). Colonization with Helicobacter is concomitant with modified gut microbiota and drastic failure of the immune control of Mycobacterium tuberculosis. *Mucosal Immunology, 10*(5), 1178–1189.

Makarova, K., Slesarev, A., Wolf, Y., Sorokin, A., Mirkin, B., Koonin, E., … Mills, D. (2006). Comparative genomics of the lactic acid bacteria. *Proceedings of the National Academy of Sciences of the USA, 103*(42), 15611–15616. doi:10.1073/pnas.0607117103

Markowiak, P., & Slizewska, K. (2017). Effects of probiotics, prebiotics, and synbiotics on human health. *Nutrients, 9*(9). doi:10.3390/nu9091021

Martens, K., Pugin, B., De Boeck, I., Spacova, I., Steelant, B., Seys, S., … Hellings, P. (2018). Probiotics for the airways: Potential to improve epithelial and immune homeostasis. *Allergy, 73*(10), 1954–1963. doi:10.1111/all.13495

Martinez, K. B., Leone, V., & Chang, E. B. (2017). Western diets, gut dysbiosis, and metabolic diseases: Are they linked? *Gut Microbes, 8*(2), 130–142. doi:10.1080/19490976.2016.1270811

Mazziotta, C., Tognon, M., Martini, F., Torreggiani, E., & Rotondo, J. C. (2023). Probiotics mechanism of action on immune cells and beneficial effects on human health. *Cells, 12,184*, 1–33. doi:10.3390/cells12010184

Meskini, M., Amanzadeh, A., Salehi, F., Bouzari, S., Karimipoor, M., Fuso, A., … Siadat, S. D. (2024). A protocol to isolate and characterize pure monocytes and generate monocyte-derived dendritic cells through FBS-Coated flasks. *Scientific Reports, 14*(1), 23956.

Meskini, M., Madadi, N., Ahmadi, K., Vaziri, F., Fateh, A., & Siadat, S. D. (2023). Tuberculosis prevention, diagnosis, and treatment financial profile during 2006–2021: PART A. *Cost Effectiveness and Resource Allocation, 21*(1), 68.

Meskini, M., Zamani, M. S., Amanzadeh, A., Bouzari, S., Karimipoor, M., Fuso, A., … Siadat, S. D. (2024). Epigenetic modulation of cytokine expression in Mycobacterium tuberculosis-infected monocyte derived-dendritic cells: Implications for tuberculosis diagnosis. *Cytokine, 181*, 156693.

Milajerdi, A., Mousavi, S. M., Sadeghi, A., Salari-Moghaddam, A., Parohan, M., Larijani, B., & Esmaillzadeh, A. (2020). The effect of probiotics on inflammatory biomarkers: A meta-analysis of randomized clinical trials. *European Journal of Nutrition, 59*(2), 633–649. doi:10.1007/s00394-019-01931-8

Mishra, R., Krishan, S., Siddiqui, A. N., Kapur, P., Khayyam, K. U., & Sharma, M. (2020). Potential role of adjuvant drugs on efficacy of first line oral antitubercular therapy: Drug repurposing. *Tuberculosis, 120*, 101902.

Monde, N., Munyeme, M., Chongwe, G., Wensman, J. J., Zulu, M., Siziya, S., … Malama, S. (2023). First and second-line anti-tuberculosis drug-resistance patterns in pulmonary tuberculosis patients in Zambia. *Antibiotics, 12*(1), 166.

Mori, G., Morrison, M., & Blumenthal, A. (2021). Microbiome-immune interactions in tuberculosis. *PLoS Pathogens, 17*(4), e1009377. doi:10.1371/journal.ppat.1009377

Motta, I., Boeree, M., Chesov, D., Dheda, K., Günther, G., Horsburgh Jr, C. R., … McIlleron, H. M. (2024). Recent advances in the treatment of tuberculosis. Clinical Microbiology and Infection, *30*(9), 1107–1114. doi: 10.1016/j.cmi.2023.07.013

Mousavi-Sagharchi, S. M. A., Afrazeh, E., Seyyedian-Nikjeh, S. F., Meskini, M., Doroud, D., & Siadat, S. D (2024). New insight in molecular detection of Mycobacterium tuberculosis. *AMB Express, 14*(1), 1–20.

Naidoo, C. C., Nyawo, G. R., Wu, B. G., Walzl, G., Warren, R. M., Segal, L. N., & Theron, G. (2019). The microbiome and tuberculosis: State of the art, potential applications, and defining the clinical research agenda. *The Lancet Respiratory Medicine, 7*(10), 892–906.

Namasivayam, S., Maiga, M., Yuan, W., Thovarai, V., Costa, D. L., Mittereder, L. R., … Trinchieri, G. (2017). Longitudinal profiling reveals a persistent intestinal dysbiosis triggered by conventional anti-tuberculosis therapy. *Microbiome, 5*, 1–17.

Namasivayam, S., Sher, A., Glickman, M. S., & Wipperman, M. F. (2018). The microbiome and tuberculosis: Early evidence for cross talk. *MBio, 9*(5), doi:10.1128/mbio.01420-01418.

Natalia, D., Syah, Y., Retnoningrum, D., & Kusuma, H. (2016). Mutation of katG in a clinical isolate of Mycobacterium tuberculosis: Effects on catalase-peroxidase for isoniazid activation. *The Ukrainian Biochemical Journal, 88*(5), 71–81.

Naveen Kumar, H., Parumasivam, T., Jumaat, F., Ibrahim, P., Asmawi, M. Z., & Sadikun, A. (2014). Synthesis and evaluation of isonicotinoyl hydrazone derivatives as antimycobacterial and anticancer agents. *Medicinal Chemistry Research, 23*, 269–279.

Negatu, D. A., Liu, J. J., Zimmerman, M., Kaya, F., Dartois, V., Aldrich, C. C., … Dick, T. (2018). Whole-cell screen of fragment library identifies gut microbiota metabolite indole propionic acid as antitubercular. *Antimicrobial Agents and Chemotherapy, 62*(3), doi:10.1128/aac.01571-01517.

Negi, S., Pahari, S., Bashir, H., & Agrewala, J. N. (2019). Gut microbiota regulates mincle mediated activation of lung dendritic cells to protect against Mycobacterium tuberculosis. *Frontiers in Immunology, 10*, 1142. doi:10.3389/fimmu.2019.01142

Nusrath Unissa, A., Hanna, L. E., & Swaminathan, S. (2016). A note on derivatives of isoniazid, Rifampicin, and pyrazinamide showing activity against resistant Mycobacterium tuberculosis. *Chemical Biology & Drug Design, 87*(4), 537–550.

O'Toole, P. W., Marchesi, J. R., & Hill, C. (2017). Next-generation probiotics: The spectrum from probiotics to live biotherapeutics. *Nature Microbiology, 2*(5), 1–6.

Onajole, O. K., Govender, P., van Helden, P. D., Kruger, H. G., Maguire, G. E., Wiid, I., & Govender, T. (2010). Synthesis and evaluation of SQ109 analogues as potential anti-tuberculosis candidates. *European Journal of Medicinal Chemistry, 45*(5), 2075–2079.

Otdushkina, L. Y., Zakharova, Y. V., Kholodov, A. A., & Pyanzova, T. V. (2023). Microbiological evaluation of probiotic therapy in patients with pulmonary tuberculosis. *Russian Journal of Infection and Immunity, 13*(3), 517–525.

Pais, P., Almeida, V., Yilmaz, M., & Teixeira, M. C. (2020). Saccharomyces boulardii: What makes it tick as successful probiotic? *Journal of Fungi (Basel), 6*(2). doi:10.3390/jof6020078

Pavlidou, E., Fasoulas, A., Mantzorou, M., & Giaginis, C. (2022). Clinical evidence on the potential beneficial effects of probiotics and prebiotics in cardiovascular disease. *International Journal of Molecular Sciences, 23*(24), 15898.

Plaza-Diaz, J., Ruiz-Ojeda, F. J., Gil-Campos, M., & Gil, A. (2019). Mechanisms of action of probiotics. *Advances in Nutrition, 10*, S49–S66. doi:10.1093/advances/nmy063

Prideaux, B., Via, L. E., Zimmerman, M. D., Eum, S., Sarathy, J., O'brien, P., … Chen, P.-Y. (2015). The association between sterilizing activity and drug distribution into tuberculosis lesions. *Nature Medicine, 21*(10), 1223–1227.

Pritchard, D., Hulick, P. J., & Wells, C. J. (2021). The integration of personalized medicine into health systems: Progress and a path forward. *Personalized Medicine, 18*(6), 527–531. doi: 10.2217/pme-2021-0102

Purdel, C., Ungurianu, A., Adam-Dima, I., & Margină, D. (2023). Exploring the potential impact of probiotic use on drug metabolism and efficacy. *Biomedicine & Pharmacotherapy, 161*, 114468.

Rabetafika, H. N., Razafindralambo, A., Ebenso, B., & Razafindralambo, H. L. (2023). Probiotics as antibiotic alternatives for human and animal applications. *Encyclopedia, 3*(2), 561–581.

Rahim, M. A., Seo, H., Kim, S., Tajdozian, H., Barman, I., Lee, Y., … Song, H.-Y. (2022). In vitro antituberculosis effect of probiotic Lacticaseibacillus rhamnosus PMC203 isolated from vaginal microbiota. *Scientific Reports, 12*(1), 8290. doi:10.1038/s41598-022-12413-z

Ramachandran, G., & Swaminathan, S. (2015). Safety and tolerability profile of second-line anti-tuberculosis medications. *Drug Safety, 38*, 253–269.

Raval, J. P., Patel, N. H., Patel, H. V., & Patel, P. S. (2011). In vitro antimycobacterial activity of novel N′-(4-(substituted phenyl amino)-6-(pyridin-2-ylamino)-1, 3, 5-triazin-2-yl) isonicotinohydrazide. *Medicinal Chemistry Research, 20*, 274–279.

Rinninella, E., Raoul, P., Cintoni, M., Franceschi, F., Miggiano, G. A. D., Gasbarrini, A., & Mele, M. C. (2019). What is the healthy gut microbiota composition? A changing ecosystem across age, environment, diet, and diseases. *Microorganisms, 7*(1), 14.

Robak, O. H., Heimesaat, M. M., Kruglov, A. A., Prepens, S., Ninnemann, J., Gutbier, B., … Opitz, B. (2018). Antibiotic treatment-induced secondary IgA deficiency enhances susceptibility to Pseudomonas aeruginosa pneumonia. *Journal of Clinical Investigation, 128*(8), 3535–3545. doi:10.1172/JCI97065

Rodrigues, M. O., Cantos, J. B., D'Oca, C. R. M., Soares, K. L., Coelho, T. S., Piovesan, L. A., … D'Oca, M. G. M. (2013). Synthesis and antimycobacterial activity of isoniazid derivatives from renewable fatty acids. *Bioorganic & Medicinal Chemistry, 21*(22), 6910–6914.

Rodríguez, J. M., Murphy, K., Stanton, C., Ross, R. P., Kober, O. I., Juge, N., … Jenmalm, M. C. (2015). The composition of the gut microbiota throughout life, with an emphasis on early life. *Microbial Ecology in Health and Disease, 26*(1), 26050.

Rooks, M. G., & Garrett, W. S. (2016). Gut microbiota, metabolites and host immunity. *Nature Reviews Immunology, 16*(6), 341–352.

Rothchild, A. C., Stowell, B., Goyal, G., Nunes-Alves, C., Yang, Q., Papavinasasundaram, K., … Behar, S. M. (2017). Role of granulocyte-macrophage colony-stimulating factor production by T Cells during Mycobacterium tuberculosis infection. *MBio, 8*(5). doi:10.1128/mBio.01514-17

Roy, P., Das, A., Panda, A., Shivya, T., Priyadarshani, A., & Patro, B. K. (2024). India marching towards TB elimination: How far we are. *Indian Journal of Tuberculosis, 71*(2), 213–218. doi:10.1016/j.ijtb.2023.06.019

Rutsch, A., Kantsjo, J. B., & Ronchi, F. (2020). The gut-brain axis: How microbiota and host inflammasome influence brain physiology and pathology. *Frontiers in Immunology, 11*, 604179. doi:10.3389/fimmu.2020.604179

Rychtarčíková, Z., Krátký, M., Gazvoda, M., Komlóová, M., Polanc, S., Kočevar, M., … Vinšová, J. (2014). N-Substituted 2-isonicotinoylhydrazinecarboxamides—New antimycobacterial active molecules. *Molecules, 19*(4), 3851–3868.

Safi, H., Lingaraju, S., Amin, A., Kim, S., Jones, M., Holmes, M., … Fleischmann, R. (2013). Evolution of high-level ethambutol-resistant tuberculosis through interacting mutations in decaprenylphosphoryl-β-D-arabinose biosynthetic and utilization pathway genes. *Nature Genetics, 45*(10), 1190–1197.

Saghi, H., Bahador, A., Dastjerdi, F. A., Asadolahi, H., Neyshaboori, M., Mohammad, E., … Esmaeili, D. (2015). Antibacterial effects of herbal compounds against Acinetobacter baumannii isolated from Hospital of Tehran, Iran. *Global Journal of Infectious Diseases and Clinical Research, 1*(1), 018–020.

Sajedi, D., Shabani, R., & Elmieh, A. (2021). Changes in leptin, serotonin, and cortisol after eight weeks of aerobic exercise with probiotic intake in a cuprizone-induced demyelination mouse model of multiple sclerosis. *Cytokine, 144*, 155590. doi:10.1016/j.cyto.2021.155590

Salminen, S., Collado, M. C., Endo, A., Hill, C., Lebeer, S., Quigley, E. M. M., … Vinderola, G. (2021). The International Scientific Association of Probiotics and Prebiotics (ISAPP) consensus statement on the definition and scope of postbiotics. *Nature Reviews Gastroenterology & Hepatology, 18*(9), 649–667. doi:10.1038/s41575-021-00440-6

Samuelson, D. R., Welsh, D. A., & Shellito, J. E. (2015). Regulation of lung immunity and host defense by the intestinal microbiota. *Frontiers in Microbiology, 6*. doi:10.3389/fmicb.2015.01085

Sanders, M. E., Guarner, F., Guerrant, R., Holt, P. R., Quigley, E. M., Sartor, R. B., … Mayer, E. A. (2013). An update on the use and investigation of probiotics in health and disease. *Gut, 62*(5), 787–796.

Sanjiwani, M. I. D., Setiawan, N. B. W., Putra, A. I. Y. D., & Darwinata, A. E. (2021). Probiotic-based therapy for active tuberculosis infection: The role of gut-lung axis and granulocyte macrophage-colony stimulating factor. *Jurnal Respirasi, 7*(2), 93. doi:10.20473/jr.v7-I.2.2021.93-99

Satoh-Takayama, N., Vosshenrich, C. A., Lesjean-Pottier, S., Sawa, S., Lochner, M., Rattis, F., … Mandelboim, O. (2008). Microbial flora drives interleukin 22 production in intestinal NKp46+ cells that provide innate mucosal immune defense. *Immunity, 29*(6), 958–970.

Schierwagen, R., Uschner, F. E., Ortiz, C., Torres, S., Brol, M. J., Tyc, O., … Klein, S. (2020). The role of macrophage-inducible C-type lectin in different stages of chronic liver disease. *Frontiers in Immunology, 11*, 1352. doi:10.3389/fimmu.2020.01352

Segal, L. N., Clemente, J. C., Li, Y., Ruan, C., Cao, J., Danckers, M., … Diaz, P. (2017). Anaerobic bacterial fermentation products increase tuberculosis risk in antiretroviral-drug-treated HIV patients. *Cell Host & Microbe, 21*(4), 530–537. e534.

Sharma, V. R., Singh, M., Kumar, V., Yadav, M., Sehrawat, N., Sharma, D. K., & Sharma, A. K. (2021). Microbiome dysbiosis in cancer: Exploring therapeutic strategies to counter the disease. *Seminars in Cancer Biology, 70*, 61–70. doi:10.1016/j.semcancer.2020.07.006

Shi, W., Hu, Y., Ning, Z., Xia, F., Wu, M., Hu, Y. O., … Xu, B. (2021). Alterations of gut microbiota in patients with active pulmonary tuberculosis in China: A pilot study. *International Journal of Infectious Diseases, 111*, 313–321.

Siddique, F., Akram, K., Alghamdi, E. S., Arshad, Q., & Siddique, A. (2022). The immunomodulatory role of probiotics. In E. F. Robles (Ed.), *Prebiotics and Probiotics – From Food to Health*. IntechOpen. doi:10.5772/intechopen.98839

Silva, D. R., Sardi, J. d. C. O., de Souza Pitangui, N., Roque, S. M., da Silva, A. C. B., & Rosalen, P. L. (2020). Probiotics as an alternative antimicrobial therapy: Current reality and future directions. *Journal of Functional Foods, 73*, 104080.

Simmons, J. D., Stein, C. M., Seshadri, C., Campo, M., Alter, G., Fortune, S., … Hawn, T. R. (2018). Immunological mechanisms of human resistance to persistent Mycobacterium tuberculosis infection. *Nature Reviews Immunology, 18*(9), 575–589. doi:10.1038/s41577-018-0025-3

Singh, D., Singh, V., Bhushan Mishra, S., Sharma, D., & Agarwal, V. (2022). Evaluation of anti-biofilm, anti-quorum, anti-dysenteric potential of designed polyherbal formulation: In vitro and in vivo study. *Journal of Applied Biomedicine, 20*(1), 7–14. doi:10.32725/jab.2022.005

Soemarie, Y. B., Milanda, T., & Barliana, M. I. (2021). Fermented foods as probiotics: A review. *Journal of Advanced Pharmaceutical Technology and Research, 12*(4), 335–339. doi:10.4103/japtr.japtr_116_21

Sokol, H., Pigneur, B., Watterlot, L., Lakhdari, O., Bermúdez-Humarán, L. G., Gratadoux, J.-J., … Corthier, G. (2008). Faecalibacterium prausnitzii is an anti-inflammatory commensal bacterium identified by gut microbiota analysis of Crohn disease patients. *Proceedings of the National Academy of Sciences, 105*(43), 16731–16736.

Soltani, S., Hammami, R., Cotter, P. D., Rebuffat, S., Said, L. B., Gaudreau, H., … Fliss, I. (2021). Bacteriocins as a new generation of antimicrobials: Toxicity aspects and regulations. *FEMS Microbiology Reviews, 45*(1). doi:10.1093/femsre/fuaa039

Soren, S., Mandal, G. P., Roy, B., Samanta, I., & Hansda, R. N. (2023). Assessment of Bacillus subtilis based probiotics on health and productive performance of poultry: A review. *Indian Journal of Animal Health*. doi:10.36062/ijah.2023.spl.00923

Souza, H. F. d., Carosia, M. F., Pinheiro, C., Carvalho, M. V. d., Oliveira, C. A. F. d., & Kamimura, E. S. (2022). On probiotic yeasts in food development: Saccharomyces boulardii, a trend. *Food Science and Technology, 42*. doi:10.1590/fst.92321

Suprapti, B., Suharjono, S., Raising, R., Yulistiani, Y., Izzah, Z., Nilamsari, W. P., … Bachtiar, A. (2018). Effects of probiotics and vitamin B supplementation on IFN-γ and IL-12 levels during intensive phase treatment of tuberculosis. *Indonesian Journal of Pharmacy, 29*(2), 80.

Takiishi, T., Fenero, C. I. M., & Câmara, N. O. S. (2017). Intestinal barrier and gut microbiota: Shaping our immune responses throughout life. *Tissue Barriers, 5*(4), e1373208.

Tandon, R., & Nath, M. (2017). Tackling drug-resistant tuberculosis: Current trends and approaches. *Mini Reviews in Medicinal Chemistry, 17*(6), 549–570.

Tarashi, S., Ahmadi Badi, S., Moshiri, A., Nasehi, M., Fateh, A., Vaziri, F., & Siadat, S. D. (2018). The human microbiota in pulmonary tuberculosis: Not so innocent bystanders. *Tuberculosis, 113*, 215–221. doi:10.1016/j.tube.2018.10.010

Taverniti, V., & Guglielmetti, S. (2011). The immunomodulatory properties of probiotic microorganisms beyond their viability (ghost probiotics: proposal of paraprobiotic concept). *Genes & Nutritions, 6*(3), 261–274. doi:10.1007/s12263-011-0218-x

Thilagavathi, T. (2020). Probiotics, prebiotics, synbiotics and its health benefits. *International Journal of Current Microbiology and Applied Sciences, 9*(11), 497–511. doi:10.20546/ijcmas.2020.911.061

Toumi, R., Abdelouhab, K., Rafa, H., Soufli, I., Raissi-Kerboua, D., Djeraba, Z., & Touil-Boukoffa, C. (2013). Beneficial role of the probiotic mixture ultrabiotique on maintaining the integrity of intestinal mucosal barrier in DSS-induced experimental colitis. *Immunopharmacology and Immunotoxicology, 35*(3), 403–409. doi:10.3109/08923973.2013.790413

Treiner, E., Duban, L., Bahram, S., Radosavljevic, M., Wanner, V., Tilloy, F., … Lantz, O. (2003). Selection of evolutionarily conserved mucosal-associated invariant T cells by MR1. *Nature, 422*(6928), 164–169.

Tsai, Y.-L., Lin, T.-L., Chang, C.-J., Wu, T.-R., Lai, W.-F., Lu, C.-C., & Lai, H.-C. (2019). Probiotics, prebiotics and amelioration of diseases. *Journal of Biomedical Science, 26*, 1–8.

Ubeda, C., Djukovic, A., & Isaac, S. (2017). Roles of the intestinal microbiota in pathogen protection. *Clinical & Translational Immunology, 6*(2), e128.

Ulrichs, T., Moody, D. B., Grant, E., Kaufmann, S. H., & Porcelli, S. A. (2003). T-cell responses to CD1-presented lipid antigens in humans with Mycobacterium tuberculosis infection. *Infection and Immunity, 71*(6), 3076–3087.

Van Rhijn, I., Kasmar, A., De Jong, A., Gras, S., Bhati, M., Doorenspleet, M. E., … De Jager, W. (2013). A conserved human T cell population targets mycobacterial antigens presented by CD1b. *Nature Immunology, 14*(7), 706–713.

Vieira, A. T., Rocha, V. M., Tavares, L., Garcia, C. C., Teixeira, M. M., Oliveira, S. C., … Nicoli, J. R. (2016). Control of Klebsiella pneumoniae pulmonary infection and immunomodulation by oral treatment with the commensal probiotic Bifidobacterium longum 5(1A). *Microbes and Infection, 18*(3), 180–189. doi:10.1016/j.micinf.2015.10.008

Wang, B., Yao, M., Lv, L., Ling, Z., & Li, L. (2017). The human microbiota in health and disease. *Engineering, 3*(1), 71–82. doi:10.1016/j.eng.2017.01.008

Wang, G., Huang, S., Wang, Y., Cai, S., Yu, H., Liu, H., … Qiao, S. (2019). Bridging intestinal immunity and gut microbiota by metabolites. *Cellular and Molecular Life Sciences, 76*(20), 3917–3937. doi:10.1007/s00018-019-03190-6

Wang, J., Chen, W. D., & Wang, Y. D. (2020). The relationship between gut microbiota and inflammatory diseases: The role of macrophages. *Frontiers in Microbiology, 11*, 1065. doi:10.3389/fmicb.2020.01065

Wang, J., & Ma, A. (2020). Long-term effects of multi-drug-resistant tuberculosis treatment on gut microbiota and its health consequences. *Frontiers in Microbiology, 11*, 500124.

Wang, L., Zhang, J., Guo, Z., Kwok, L., Ma, C., Zhang, W., … Zhang, H. (2014). Effect of oral consumption of probiotic Lactobacillus planatarum P-8 on fecal microbiota, SIgA, SCFAs, and TBAs of adults of different ages. *Nutrition, 30*(7–8), 776–783 e771. doi:10.1016/j.nut.2013.11.018

Wang, X., Zhang, P., & Zhang, X. (2021). Probiotics regulate gut microbiota: An effective method to improve immunity. *Molecules, 26*(19), 6076.

Wang, Y., Deng, Y., Liu, N., Chen, Y., Jiang, Y., Teng, Z., … Xiang, Y. (2022). Alterations in the gut microbiome of individuals with tuberculosis of different disease states. *Frontiers in Cellular and Infection Microbiology, 12*, 836987.

Wang, Y., Xie, J., Wang, N., Li, Y., Sun, X., Zhang, Y., & Zhang, H. (2013). Lactobacillus casei Zhang modulate cytokine and toll-like receptor expression and beneficially regulate poly I:C-induced immune responses in RAW264.7 macrophages. *Microbiology and Immunology, 57*(1), 54–62. doi:10.1111/j.1348-0421.516.x

Ward, B. (2015). Bacterial energy metabolism. In Y.-W. Tang, D. Liu, & J. Schwartzman (Eds.), *Molecular Medical Microbiology* (pp. 201–233). Academic Press.

World Health Organization (2019). *WHO Consolidated Guidelines on Drug-Resistant Tuberculosis Treatment.* World Health Organization.

World Health Organization (2020). *WHO Operational Handbook on Tuberculosis: Module 4: Treatment: Drug-Resistant Tuberculosis Treatment.* World Health Organization

WHO. (2023). Global Tuberculosis Report (pp. 1–99). https://www.who.int/publications/i/item/9789240083851

Willis, A., Isaacs, T., & Khunti, K. (2021). Improving diversity in research and trial participation: The challenges of language. *The Lancet Public Health, 6*(7), e445–e446.

Winglee, K., Eloe-Fadrosh, E., Gupta, S., Guo, H., Fraser, C., & Bishai, W. (2014). Aerosol Mycobacterium tuberculosis infection causes rapid loss of diversity in gut microbiota. *PLoS One, 9*(5), e97048.

Wipperman, M. F., Fitzgerald, D. W., Juste, M. A. J., Taur, Y., Namasivayam, S., Sher, A., … Glickman, M. S. (2017). Antibiotic treatment for Tuberculosis induces a profound dysbiosis of the microbiome that persists long after therapy is completed. *Scientific Reports, 7*(1), 10767. doi:10.1038/s41598-017-10346-6

Witthuhn, R. C., Schoeman, T., & Britz, T. J. (2005). Characterisation of the microbial population at different stages of Kefir production and Kefir grain mass cultivation. *International Dairy Journal, 15*(4), 383–389. doi:10.1016/j.idairyj.2004.07.016

Wood, M. R., Elaine, A. Y., & Mehta, S. (2017). The human microbiome in the fight against tuberculosis. *The American Journal of Tropical Medicine and Hygiene, 96*(6), 1274.

Wu, C., Yi, H., Hu, Y., Luo, D., Tang, Z., Wen, X., … Wu, S. (2023). Effects of second-line anti-tuberculosis drugs on the intestinal microbiota of patients with rifampicin-resistant tuberculosis. *Frontiers in Cellular and Infection Microbiology, 13*, 326.

Wu, H. J., & Wu, E. (2012). The role of gut microbiota in immune homeostasis and autoimmunity. *Gut Microbes, 3*(1), 4–14. doi:10.4161/gmic.19320

Wu, J., Liu, W., He, L., Huang, F., Chen, J., Cui, P., … Zhang, Y. (2013). Sputum microbiota associated with new, recurrent and treatment failure tuberculosis. *PLoS One, 8*(12), e83445.

Wu, Y., Tang, L., Wang, B., Sun, Q., Zhao, P., & Li, W. (2019). The role of autophagy in maintaining intestinal mucosal barrier. *Journal of Cellular Physiology, 234*(11), 19406–19419. doi:10.1002/jcp.28722

Xavier, R. M., Sharumathi, S., Parthasarathy, A. K., Mani, D., & Mohanasundaram, T. (2023). Limited sampling strategies for therapeutic drug monitoring of anti-tuberculosis medications: A systematic review of their feasibility and clinical utility. *Tuberculosis*, 102367.

Xu, B., Liang, S., Zhao, J., Li, X., Guo, J., Xin, B., … Ma, W. (2022). Bifidobacterium animalis subsp. lactis XLTG11 improves antibiotic-related diarrhea by alleviating inflammation, enhancing intestinal barrier function and regulating intestinal flora. *Food & Function, 13*(11), 6404–6418. doi:10.1039/d1fo04305f

Ye, S., Wang, L., Li, S., Ding, Q., Wang, Y., Wan, X., … Li, X. (2022). The correlation between dysfunctional intestinal flora and pathology feature of patients with pulmonary tuberculosis. *Frontiers in Cellular and Infection Microbiology, 12*, 1090889.

Youn, H. N., Lee, D. H., Lee, Y. N., Park, J. K., Yuk, S. S., Yang, S. Y., … Song, C. S. (2012). Intranasal administration of live Lactobacillus species facilitates protection against influenza virus infection in mice. *Antiviral Research, 93*(1), 138–143. doi:10.1016/j.antiviral.2011.11.004

Young, V. B. (2017). The role of the microbiome in human health and disease: An introduction for clinicians. *BMJ, 356*. doi:10.1136/bmj.j831

Yu, Y.-J., Liu, X.-D., Liao, C., Yu, R., Wang, X., Li, M., & Wang, Y. (2024). Targeting gut microbiota for immunotherapy of diseases. *Archives of Toxicology*, *98*(8), 2429–2439. doi:10.1007/s00204-024-03770-x

Zade, D., & Raut, P. (2023). Hepatotoxicity induced by anti-tuberculosis medications: Exploring causes, risk factors, and potential mitigation strategies: Hepatotoxicity in anti-tuberculosis therapy: Causes, risks, and mitigation. *Journal of Drug Delivery and Biotherapeutics, 1*(01), 1–10.

Zak, D. E., Penn-Nicholson, A., Scriba, T. J., Thompson, E., Suliman, S., Amon, L. M., … Hussey, G. D. (2016). A blood RNA signature for tuberculosis disease risk: a prospective cohort study. *The Lancet, 387*(10035), 2312–2322.

Zhang, Z., Tang, H., Chen, P., Xie, H., & Tao, Y. (2019). Demystifying the manipulation of host immunity, metabolism, and extraintestinal tumors by the gut microbiome. *Signal Transduction and Targeted Thera py, 4*, 41. doi:10.1038/s41392-019-0074-5

Zhou, S., Yang, S., & Huang, G. (2017). Design, synthesis and biological activity of pyrazinamide derivatives for anti-Mycobacterium tuberculosis. *Journal of Enzyme Inhibition and Medicinal Chemistry, 32*(1), 1183–1186.

Zhuang, L., Yang, L., Li, L., Ye, Z., & Gong, W. (2024). Mycobacterium tuberculosis: Immune response, biomarkers, and therapeutic intervention. *MedComm, 5*(1), e419.

Zitko, J., Servusová, B., Paterová, P., Mandíková, J., Kubíček, V., Kučera, R., … Doležal, M. (2013). Synthesis, antimycobacterial activity and in vitro cytotoxicity of 5-chloro-N-phenylpyrazine-2-carboxamides. *Molecules, 18*(12), 14807–14825.

Zmora, N., Zilberman-Schapira, G., Suez, J., Mor, U., Dori-Bachash, M., Bashiardes, S., … Elinav, E. (2018). Personalized gut mucosal colonization resistance to empiric probiotics is associated with unique host and microbiome features. *Cell, 174*(6), 1388–1405 e1321. doi:10.1016/j.cell.2018.08.041

Zwierzyna, M., & Overington, J. P. (2017). Classification and analysis of a large collection of in vivo bioassay descriptions. *PLoS Computational Biology, 13*(7), e1005641.

11 Cascades and Implications of Antimicrobial Resistance on Human Health

Lamar Al-Muslem[1], *Ayman M. Al-Qaaneh*[2,3], *Zainab Al-Salem*[4], *Mohammad Azam Ansari*[3], *and Suriya Rehman*[4*]

[1]Mawhiba Research Enrichment Program, Institute for Research and Medical Consultations (IRMC), Imam Abdulrahman Bin Faisal University, Dammam, Saudi Arabia

[2]Faculty of Allied Medical Sciences, Al-Balqa Applied University (BAU), Al-Salt, Jordan

[3]Department of Pharmaceutical Technology, Faculty of Pharmacy, Jordan University of Science and Technology (JUST), Irbid, Jordan

[4]Department of Epidemic Diseases Research, Institute for Research and Medical Consultations (IRMC), Imam Abdulrahman Bin Faisal University, Dammam, Saudi Arabia

*Corresponding author: surrehman@iau.edu.sa

11.1 INTRODUCTION

Antimicrobial resistance (AMR) has emerged as an urgent global health challenge, with profound implications for human and animal health. This phenomenon is characterized by the increasing resistance of microorganisms, including bacteria, viruses, and parasites, to the effects of antimicrobial drugs, such as antibiotics and antivirals. The ramifications of AMR extend far beyond the realm of medicine, having impact on the foundations of modern healthcare and necessitating immediate and concerted global action (World Health Organization, 2014). AMR is driven by a complex interplay of factors, each contributing to the growing resistance of microorganisms to antimicrobial drugs. A significant driver is the overuse and misuse of these drugs in both human healthcare and agriculture. The widespread and often inappropriate use of antimicrobials has created a breeding ground for resistance to flourish (Ventola, 2015). Furthermore, inadequate infection prevention and control practices in healthcare settings allow resistant pathogens to persist and spread, further exacerbating the problem (European Centre for Disease Prevention and Control, 2020). In our interconnected world, global travel and trade facilitate the rapid dissemination of drug-resistant microbes across borders, making AMR a truly global concern (Rehman, 2023) (Figure 11.1).

The consequences of AMR are profound and multifaceted. Prolonged illnesses are a hallmark of AMR as infections become increasingly challenging to treat effectively. This leads to increased mortality rates, particularly among vulnerable populations, and places immense pressure on healthcare systems (O'Neill et al., 2016). The economic burden of AMR is significant, encompassing escalating healthcare costs, prolonged hospital stays, and lost productivity due to resistant infections. Recognizing the gravity of the situation, governments, international organizations, and the scientific community have taken measures to address AMR.

DOI: 10.1201/9781032676951-11

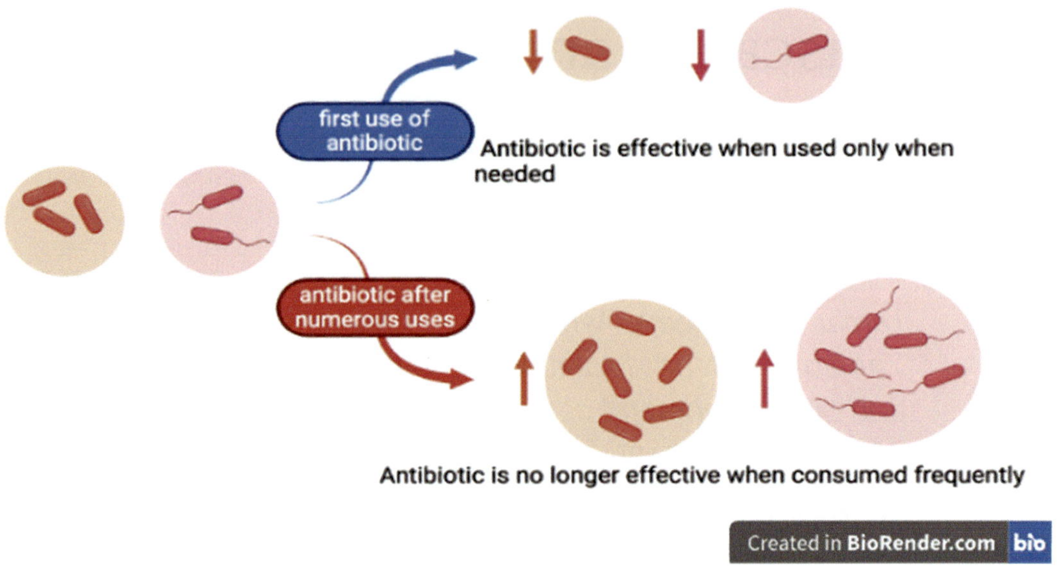

FIGURE 11.1 Contribution of misuse and overuse of antibiotics on microbial drug resistance.

Enhanced surveillance systems have been established to monitor resistance patterns, enabling the development of targeted interventions (World Health Organization, 2014). Antibiotic stewardship programs in healthcare settings aim to promote responsible antimicrobial use and reduce overprescription, thus curbing the development of resistance (CDC, 2024). Moreover, research and development efforts focus on discovering new antimicrobial agents and alternative treatment approaches to combat resistant infections (Ventola, 2015). In conclusion, AMR presents a global challenge with far-reaching implications for public health, necessitating immediate and coordinated action at local, national, and international levels. The insights from authoritative sources in this introduction set the stage for a comprehensive examination of AMR, its root causes, far-reaching consequences, and potential solutions to mitigate its impact and protect the health of future generations.

11.2 ANTIBIOTIC RESISTANCE AND THE RISE OF MULTIDRUG RESISTANCE

Ciba foundation symposium (London, 1996) shed light on the escalating challenge of antibiotic resistance, which has developed into a global health crisis. It traced the history of antibiotics, noting that their discovery, particularly penicillin in the 1940s, which represented a significant milestone in medical science and drastically reduced the number of deaths from infectious diseases (Wilkins, 1996). However, the initial triumph seemed to be short-lived. By the late 1960s, the emergence of antibiotic-resistant bacteria posed a new threat, and the first cases of resistance were detected in Japan in the mid-1950s. Antibiotic resistance, fueled by Darwinian selection, has now spread across the globe, leading to an increase in mortality rates and taking a heavy financial toll on healthcare systems. The symposium discussed how the misuse of antibiotics in different environments contributes to resistance in soil and water as well as in patients. It was suggested that resistance genes originally from soil microorganisms have spread through processes such as conjugation and transformation (Figure 11.2). The conference highlighted how the spread of resistance is influenced by factors such as regional differences in antibiotic use and hospital regulations (Fatima et al., 2023).

Resistance develops in a stepwise manner, whereby low-level resistance can escalate to high-level resistance through a series of mutations. Population genetic studies indicate that once resistance has developed, it usually persists. This may be due to subsequent mutations that compensate for an initial reduction in the bacterium's survival advantage. In addition, the symposium examined

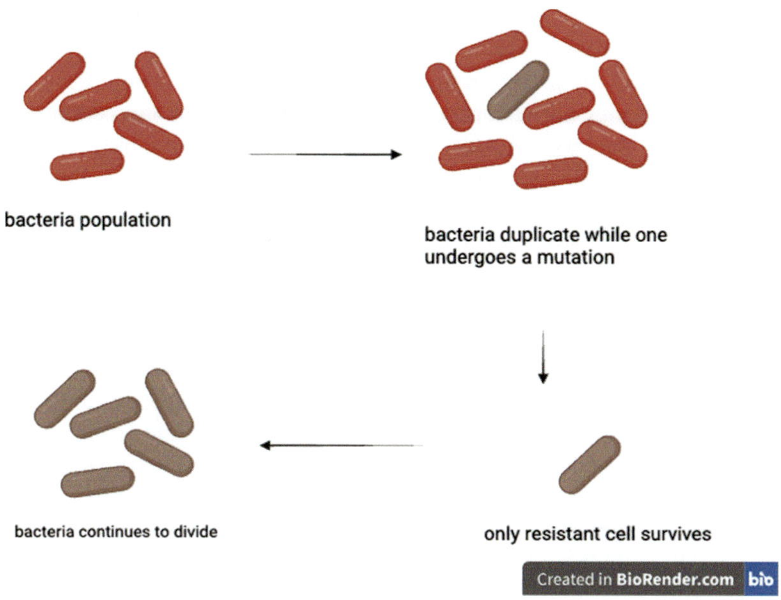

FIGURE 11.2 Natural spreading of resistant bacteria.

molecular mechanisms of resistance, including the evolution of beta-lactamases and the function of multidrug resistance plasmids. Also, societal factors such as global travel and urbanization have accelerated the spread of resistance, underscoring the complexity of this modern medical dilemma (Wilkins, 1996).

Microorganisms can either naturally resist antibiotics or develop resistance after exposure. This resistance can occur through mutations or the acquisition of resistance genes, which can spread between bacteria by conjugation, transformation, or transduction. The spread of resistance traits is unpredictable; for example, the β-lactamase gene is widespread in *Staphylococcus aureus* but not in enterococci. Fortunately, vancomycin resistance genes remain rare in *Staphylococcus aureus* despite being present in enterococci (Sabtu et al., 2015). Large-scale studies are necessary to understand how antimicrobial prescribing patterns contribute to AMR and to develop targeted guidelines for AMR stewardship. Integrated surveillance across human and animal populations, using standardized methods, is crucial for identifying transmission routes. Sequencing-based approaches can aid in tracing pathogen lineages. Individual-level data on patient outcomes in response to antimicrobial use are lacking, but electronic health records (EHRs) offer potential for such studies. However, there are disparities in EHR systems, especially in low- and middle-income countries, which require improved infrastructure and capacity building. Expansion of surveillance programs to include animal infections and veterinary practices is also necessary. Sharing data through a centralized repository could enable broader assessments of the AMR relationship on a national or regional scale (Allcock et al., 2017).

11.3 DRUG RESISTANCE IN BACTERIA

The study of bacterial adaptation to environmental changes has revealed an intrinsic property of microorganisms and microorganisms to evolve rapidly and efficiently. This inherent plasticity is reflected in the development of increased pathogenicity and the ability to thrive in media that are not traditionally conducive to their growth. These adaptations are not just random; they are the mechanisms by which bacteria can give rise to new strains, often exhibiting significant differences in their characteristics, including but not limited to their sensitivity to pharmacological interventions. The development of bacterial strains that are resistant to the effects of drugs used to combat them is a natural consequence of this adaptability, representing an enormous challenge in the fight against infectious diseases. The ability of bacteria to evade the mechanisms of action of antibiotics not only complicates treatment strategies but also poses a significant risk to public health as it can lead to the spread of more virulent and less controllable infections. Research into the mechanisms underlying acquired drug resistance is therefore of paramount importance, aiming to unravel the complex interplay between bacterial genetic mutations, the selection pressure exerted using antibiotics, and the environmental factors that can favor the horizontal transmission of resistance genes. Focusing on research analyzing the multiple processes that lead to the acquisition of drug resistance will help to understand not only the biological basis of resistance but also the epidemiological trends characterizing its spread. The implications of this research are far-reaching, providing a framework on which new strategies for antibiotic development can be built and emphasizing the need for strict, responsible use of existing antibiotics to prevent the further spread of resistant strains. It underscores the importance of a multidisciplinary approach, combining insights from molecular biology, genetics, pharmacology, and public health to develop comprehensive measures that can effectively counter the growing threat of drug-resistant bacteria. In summary, the emergence of drug-resistant bacteria because of their adaptive evolution is a critical area of research essential to maintain the utility of antibiotics (Al-Ghamdi et al., 2023) (Figure 11.3).

The researchers' work is an integral part of the global health community's ongoing efforts to understand and combat the persistence and spread of these resistant strains, ensuring the effectiveness and efficacy of antibiotic therapies for future generations (Yudkin, 1953). A previous study showed that with the serious issue of antimicrobial-resistant bacterial infections, some experts and groups are now keeping tabs on how common different drug-resistant organisms are. They use methods like multilocii sequence typing, which gives a detailed picture, and whole genome sequencing, which provides comprehensive information, to classify antimicrobial-resistant bacteria at a molecular level. For Gram-positive bacteria, the main ones that are resistant to antibiotics

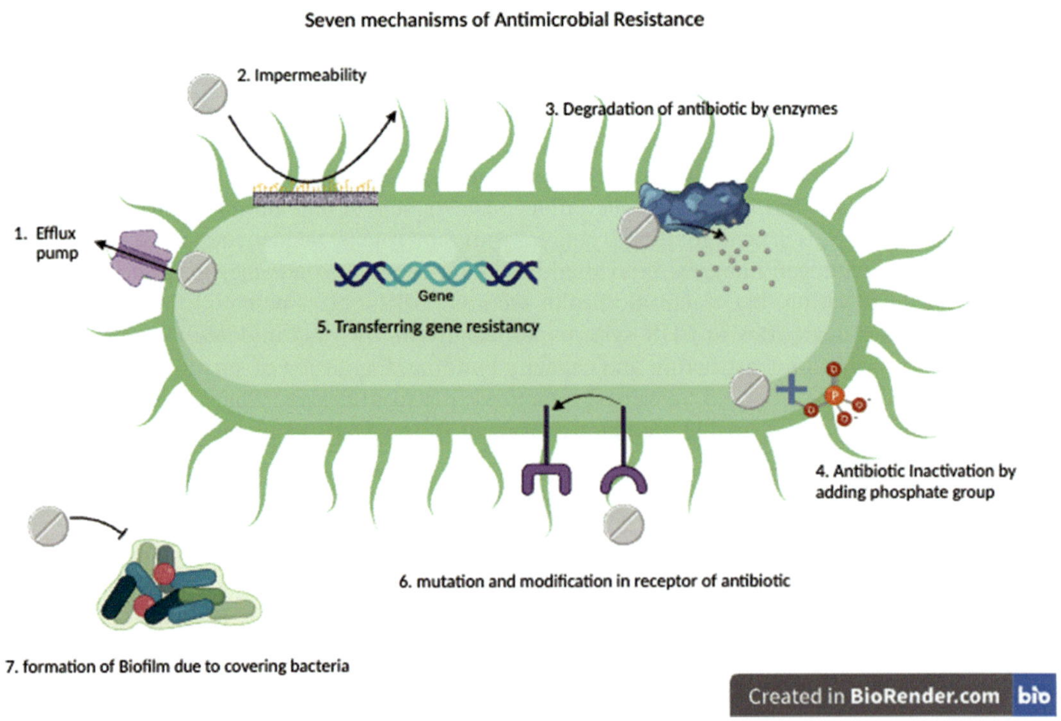

FIGURE 11.3 Mechanisms of antimicrobial resistance in bacteria.

include methicillin-resistant *Staphylococcus aureus* (MRSA), vancomycin-resistant enterococcus (VRE), and penicillin-resistant *Streptococcus pneumoniae* (PRSP). MRSA is found all over the world, especially in places like South America, North America, and Japan, with rates of over 40%. PRSP is particularly challenging for Europe. When it comes to VRE, *Enterococcus faecium* is more widespread than *Enterococcus faecalis*. In the case of Gram-negative bacteria, the main ones that are resistant to antibiotics include *Escherichia coli* and *Klebsiella pneumoniae* (*K. pneumoniae*), which are resistant to cephalosporin and/or carbapenem, as well as *Acinetobacter baumannii* (*A. baumannii*) and *Pseudomonas aeruginosa* (*P. aeruginosa*), which are resistant to carbapenem. Since the 2000s, there's been a worrying global spread of Extended-Spectrum Beta-Lactamases (ESBL) producing bacteria in communities, posing a threat to public health. In recent years, there's been an increase in the use of carbapenem antibiotics, leading to a rise in carbapenem resistance among clinical strains, including *A. baumannii*, *P. aeruginosa*, and *K. pneumoniae*. Multidrug-resistants (MDRs) *A. baumannii* and *P. aeruginosa* may become a major clinical challenge in the future (Zhu et al., 2022). Another research discussed the escalating threat of antibiotic resistance in bacteria, particularly multidrug-resistant foodborne pathogens prevalent in developing nations like Africa and Asia, which poses a significant global health challenge. The horizontal transfer of resistance genes among bacterial species intensifies this risk, necessitating vigilant monitoring and global containment efforts. Diverse bacterial profiles in livestock, influenced by geographical and socioeconomic factors, further complicate this issue. Immediate action is needed to develop alternative control strategies beyond antibiotics, with innovative approaches such as bacteriophages and combination therapies showing promise. Preventive measures, including sanitation improvements and vaccine development, are critical in curbing infections, highlighting the need for a comprehensive approach to combat antibiotic resistance and safeguard human and animal health (Urban-Chmiel et al., 2022).

11.4 DRUG RESISTANCE IN FUNGI

The protracted duration of therapy for mycoses, particularly systemic infections, significantly heightens the likelihood of antifungal drug resistance emerging, posing substantial clinical hurdles. While literature indicates that innate resistance among human pathogenic fungi is uncommon, the generation of resistant mutants remains challenging compared to bacteria (Fatima et al., 2021). Notably, traditional antifungals such as amphotericin B have not been extensively linked with resistance issues, presenting a relatively favorable scenario. However, the advent of flucytosine has drawn attention to the potential risks of fungal resistance. Unlike its counterparts, flucytosine has exhibited a relatively high and frequent occurrence of resistance, both in laboratory settings and clinical environments. This development has spurred concerns within the medical mycology community, highlighting the imperative of meticulously addressing resistance concerns when introducing novel antifungal medications. The emergence of secondary resistance, particularly in immunocompromised patients, poses a significant clinical threat. It is imperative to promptly and appropriately select treatment regimens to mitigate this risk. When necessary, the substitution of one antifungal agent with another from a different class, possessing greater potency and reduced toxicity, becomes imperative. Combination therapy, especially incorporating 5-flucytosine and amphotericin B, has demonstrated promise in forestalling secondary resistance. The administration of systemic 5-flucytosine for deep-seated infections raises apprehensions regarding its profound impact on microorganisms and the subsequent development of resistance. This underscores the necessity for a cautious approach in prescribing antifungal medications. Correlations between primary and secondary resistance exhibit variability, with 5-flucytosine showing a notable association. The infrequency of resistance issues with most antifungals, excluding 5-flucytosine, does not guarantee immunity from future challenges, particularly with recently developed drugs and those with limited historical usage. The comprehension of fungal resistance entails evaluating Minimum inhibitory concentrations (MIC), yet definitions, terminology, and methodologies differ, underscoring the need for standardization. Strain-specific variances and inaccuracies in experimental conditions may yield artificially elevated MIC values, potentially underestimating the genuine risk of resistance. The emergence of resistance, even in isolated instances, underscores the continual necessity for vigilant research and monitoring to address the threats of fungal resistance in clinical settings. Persistent endeavors are indispensable to augment our understanding of fungal resistance mechanisms and devise strategies to mitigate its impact on antifungal therapies (Iwata, 1992). Fungal pathogens represent a significant public health threat, exacerbated by AMR and a dearth of novel treatments. The World Health Organization (WHO) underscores the severity of this issue through its priority pathogen list. Effective prevention, particularly in hospital settings, necessitates rigorous disinfection protocols. Enhanced diagnostic methods for fungal infections and judicious use of antifungals are imperative. Additionally, understanding and mitigating biofilm-forming fungi on medical devices, as well as preventing mycotoxins in food, are crucial. Adopting a OneHealth approach, which integrates human, animal, and environmental health, is essential to address this complex challenge (Garvey & Rowan, 2023; Gow et al., 2022). Antifungal susceptibility testing (AFST) is pivotal in contemporary medical practice, given the mounting challenge of fungal infections and the consequential rise of drug-resistant strains. Its significance lies in guiding clinicians toward the most effective antifungal treatments tailored to individual patients while concurrently monitoring the evolution of drug resistance patterns. Despite the existence of established methodologies like broth microdilution, which are considered the gold standard due to their reliability, their prolonged turnaround time poses a notable limitation in clinical settings where timely intervention is crucial. In response to this challenge, researchers and clinicians have explored alternative AFST assays that offer expedited results. However, these rapid methods often trade off precision for speed, raising concerns about their reliability in clinical decision-making. To address this gap, recent advancements have introduced innovative approaches, such as a low-cost agar screening method specifically designed to assess echinocandin susceptibility in Aspergillus species. This method not only accelerates the testing process but also enhances cost-effectiveness, making it particularly promising for resource-limited healthcare settings.

Nevertheless, interpreting AFST results remains a complex endeavor, especially in the absence of universally accepted clinical breakpoints. This ambiguity can impede treatment decisions and compromise patient outcomes. To mitigate this challenge, emerging technologies such as matrix-assisted laser desorption/ionization time-of-flight mass spectrometry (MALDI-TOF MS) and molecular-based resistance detection methods provide the potential for rapid and precise assessment of antifungal susceptibility profiles. However, their widespread adoption faces hurdles related to standardization, validation, and interpretation, highlighting the need for further research and refinement; while AFST plays an indispensable role in the management of fungal infections, there is an urgent need for standardized and expedited methodologies to address the escalating threat of drug resistance. By harnessing innovative technologies and fostering collaboration between researchers, clinicians, and regulatory bodies, the field of AFST can continue to evolve, ensuring optimal patient care in the face of evolving fungal pathogens (Arastehfar et al., 2020). Research across multiple studies has consistently pointed to the significant role of extracellular matrix (ECM) in fungal biofilms, particularly those predominantly composed of β-1, 3 glucans, in fostering resistance to antifungal agents. This matrix serves as a formidable barrier, impeding the free movement of drugs within the biofilm structure. Consequently, the effectiveness of antifungal treatments is compromised, as the drugs struggle to penetrate the biofilm. The composition and abundance of ECM emerge as pivotal determinants influencing drug diffusion dynamics and setting the threshold for resistance within the biofilm community under scrutiny (Seneviratne et al., 2008). Moreover, extensive investigations have revealed a compelling association between the levels of glucans and the development of multidrug resistance in biofilms. This correlation underscores the intricate interplay between ECM composition, drug diffusion, and the evolution of resistance mechanisms within fungal biofilms. As such, understanding the dynamics of ECM and its constituent components becomes imperative in devising strategies to combat biofilm-associated infections effectively (Prasad et al., 2014). Each fungal biofilm has its own unique way of forming and building up. For example, the development of *Candida albicans* biofilm can be broken down into four main stages: first, there's adhesion, where the fungus attaches itself; then proliferation, when it starts to multiply; after that, there's maturation and dispersion, where the biofilm becomes more organized and spreads out; and finally, as the biofilm matures, it becomes more resistant to drugs. This makes treating fungal infections tricky and contributes significantly to the severity of these diseases. Scientists are still working to fully understand how biofilms function. But they've identified some potential ways to tackle them: firstlyl, by either preventing their formation or breaking down their structure once they've formed; secondly, by using a combination of different antibiotics or trying out new antifungal drugs to combat biofilms that have already developed; and thirdly, by continuing to explore new targets for drugs that could disrupt biofilm formation or weaken existing ones (Cui et al., 2022).

11.5 CONSEQUENCES OF ANTIBIOTIC RESISTANCE

The escalating global crisis of bacterial resistance to antibiotics is a multifaceted issue intricately linked to the extensive use of antibiotics and their ubiquitous presence in the environment. Over time, a historical lack of decisive action has significantly contributed to the severity of the current crisis. One pivotal factor highlighted in the text is the misuse of antibiotics, which has played a major role in fostering bacterial drug resistance. As susceptible strains diminish, a fertile environment for the proliferation of resilient bacterial flora is created, elucidating the emergence of the resistance problem (Dadgostar, 2019).

In recent years, the use of antibiotics in veterinary medicine has been declining worldwide. However, the overall use of antibiotics in human healthcare remains high, with overuse being a significant issue, particularly in low- and middle-income countries. While most countries have developed national action plans to fight antibiotic resistance, the specific approaches and strategies they use can differ greatly. The COVID-19 pandemic has worsened the issue of antibiotic resistance. A significant number of patients with COVID-19 were initially given antibiotics (Al-Qaaneh et al., 2022). A study from a hospital in China revealed that 71% of hospitalized COVID-19 patients received antibiotic treatment. Additionally, data shows that children who were treated with antibiotics for COVID-19 had a higher risk of severe outcomes. This is one of the most serious dangers of antibiotic resistance in COVID-19 patients (Al-Qaaneh & Al-Ghamdi, 2021; Nadgir & Biswas, 2023). An exploration into the components of drug resistance reveals the complex interplay between the antibiotic agent and resistance genes, both of which are indispensable for the manifestation of clinical resistance. This dual requirement underscores the complexity of the issue at hand. The pervasive use of antibiotics across diverse sectors, including human healthcare, animal farming, and agriculture, is underscored, with particular emphasis on the substantial utilization of antibiotics in agriculture, especially on fruit trees. Moreover, the impact of antibiotic resistance extends beyond the total quantity employed; the way antibiotics are utilized becomes crucial. Low-level, prolonged usage is identified as optimal conditions for the selection of bacterial resistance. Introducing the concept of selection density, the text accentuates the ecological nuances of resistance emergence. The frequency of antibiotic-resistant bacteria is intricately linked to a number of individuals receiving the drug and the spatial confines of the treatment area. Geographical distribution of antibiotics is identified as a pivotal factor influencing the prevalence of antibiotic-resistant bacteria. Additionally, understanding the "life after treatment" of antibiotics is crucial, drawing attention to their stability in the environment and proposing that environmental residues may significantly contribute to the selection and propagation of resistant strains. Mechanisms of resistance, ranging from direct attacks on the drug to alterations in its target or transport, are explored. Environmental antibiotic residues emerge as a potentially dominant force in the selection and perpetuation of resistant strains, further underscoring the urgency of addressing this multifaceted crisis through comprehensive strategies targeting antibiotic use and environmental management. Furthermore, collaborative efforts on a global scale are imperative to tackle this crisis effectively. This involves not only promoting responsible antibiotic usage in healthcare and agriculture but also implementing stringent regulations to limit environmental contamination. Additionally, investing in research and development to discover novel antibiotics and alternative treatments is crucial for combating resistance in the long term. By addressing the various facets of antibiotic resistance comprehensively, society can mitigate its impact and safeguard the effectiveness of antibiotics for future generations (Calam, 2008). Other consequences also include public funding that has been allocated for research on antibiotic resistance has historically remained insufficient. In many industrialized nations, this critical issue has been perceived merely as an inevitable consequence of antibiotic utilization, relegating epidemiological and societal considerations to the periphery while the research agenda has predominantly been shaped by the pharmaceutical industry. Regrettably, this approach has engendered significant knowledge gaps, particularly regarding the health and economic ramifications of antibiotic resistance, especially within community settings. Articulating the public health ramifications of antibiotic resistance presents a formidable challenge due to the diverse array of pathogens involved, each with unique modes of transmission and disease manifestations. The adverse effects on patients, such as prolonged illness or heightened mortality attributed to antibiotic resistance, are obfuscated amidst a myriad of clinical syndromes. Moreover, since antibiotic resistance does not manifest as a discernible disease entity, it remains largely imperceptible, rendering it unfamiliar to individuals outside the medical domain. In locales where new antibacterial agents are economically accessible, therapeutic regimens can be transitioned to newer antibiotics in response to escalating resistance levels to older counterparts. Conversely, this adaptive measure is unfeasible in impoverished regions where second- and third-line therapies for drug-resistant infections are often absent. Consequently, the scarcity of antibiotics in such regions amplifies the deleterious impact of resistance to first-line antibiotics. Notably, industrialized nations are also grappling with this issue as the dearth of new drug developments has diminished the efficacy of empiric antibiotic treatment, thereby exposing patients to infections caused by bacteria resistant to all available antibiotics (Cars & Nordberg, 2005). As argued, the imperative to combat antibiotic resistance involves two key strategies: developing new antibiotics and exploring alternatives like vaccines and probiotic therapy. Advances in understanding bacterial genomes are facilitating the identification of potential drug targets, alongside progress in combinatorial chemistry. However, there's a sense of urgency as this may be the final opportunity for antibiotic innovation, emphasizing the need for prudent use of existing drugs. Additionally, efforts to create effective vaccines against bacterial infections show promise, although their realization may take time (Rao, 1998).

11.6 PROBLEMS MILITATING NEW ANTIBIOTIC DEVELOPMENT/PRODUCTION

Addressing the complexities of troubleshooting fermentation processes is vital, especially in the antibiotic industry's early stages where infections cause significant production interruptions. The need for new antibiotics arises from emerging opportunistic pathogens, notably affecting immunocompromised individuals due to factors such as AIDS and aggressive cancer treatments. While existing treatments can help, many common antibiotics may not work against these pathogens, and antibiotic resistance is a growing concern (Silver & Bostian, 1993).

However, concerted collaborative efforts over time have significantly ameliorated plant operations, facilitating extended periods of uninterrupted production. The spectrum of challenges presented includes fluctuations in yield, ranging from moderate declines to severe plunges, each posing its own set of obstacles. Particularly severe fluctuations, indicative of substantial disturbances within the process, present formidable hurdles. Effectively addressing issues such as infections and yield variations necessitates a concerted team effort involving engineers, plant personnel, and sometimes microbiologists. A meticulous and thorough diagnosis becomes paramount, especially considering the potential interplay between these two critical challenges. Infections, often stemming from batch contamination, introduce a layer of uncertainty into production and process development. Detection methods encompass a range of techniques including direct inspection, microscopic examination, and culture tests. Regular testing of plant samples becomes imperative for confirming infections. Common culprits include bacteria and yeasts, with Gram-negative rods being particularly prevalent. Identifying the specific organism sheds light on the source of infection, frequently traced back to issues such as leaks, inadequate sterilization procedures, or operational shortcomings. Navigating sterilization challenges demands seasoned judgment in determining the appropriate time and temperature parameters. A guideline of sterilization at 125°C for 45 minutes is generally considered suitable, though complexities arise when dealing with intricate plant components. Employing a control-chart approach to investigate infections aids in pinpointing problem areas, often involving a thorough review of records, particularly in the seed-culture domain. The narrative underscores the critical importance of logical analysis, effective management of sterilization challenges, thorough investigation of infections, and proactive measures to address batch variations in ensuring plant safety. Troubleshooting fermentation process issues highlight the need to consider both major and minor factors in process development investigations. Methods for assessing culture and diagnosing batch conditions emerge as pivotal components, emphasizing early detection, systematic approaches, and the importance of further investigation and standardization. In serial production settings, control charts emerge as invaluable tools for detecting changes and troubleshooting production variations. Delving into the metabolic aspects of fluctuations underscores the importance of early observation and detection. Evaluation approaches for assessing culture quality encompass a diverse array of methods, providing valuable insights into growth dynamics and metabolic considerations. In contemporary clinical and scientific contexts, advancements in drug development are pivotal for therapeutic efficacy and addressing diverse challenges within the therapeutic domain. A significant focus lies on the advancement of antibiotics to combat the pressing issue of antibiotic resistance, which poses a substantial threat to public health. Efforts are directed toward targeting both biofilm and planktonic infections, aiming to disrupt essential bacterial processes such as DNA and RNA synthesis or cell wall formation. Many medications are derived from compounds historically employed by microbes in their own evolutionary battles. However, bacterial evolution and their consequent evasion of conventional antibiotics have necessitated the exploration of more potent measures, occasionally requiring the utilization of last-resort antibiotics or high dosages to counter multidrug-resistant microorganisms. In instances where bacteria form biofilms, their heightened resistance often demands aggressive intervention techniques, coupled with intensive antibiotic therapy (Zhang et al., 2022). This, in turn, escalates the risks associated with adverse effects, resistance development, and treatment failure, thereby exacerbating the financial challenges inherent in drug development and treatment protocols. Despite concerted efforts toward the development of new antibiotics, reports indicate that the existing pipeline may not be sufficiently robust to meet current clinical demands. Merely a fraction of antibiotics in the development phase are projected to secure market entry, with many lacking innovative mechanisms of action and efficacy against well-validated resistant pathogens. This is further compounded by the waning interest from major pharmaceutical conglomerates, which prioritize investments in more financially lucrative chronic illness medications. Apprehensions regarding antibiotic overuse and consequent resistance have prompted a cautious approach among medical practitioners. Newer antibiotics are often reserved for severe cases, while older counterparts are predominantly utilized in less critical scenarios. While this conserves newer antibiotics for exigent circumstances, it also diminishes their return on investment. However, due to the stochastic nature of microbial evolution, forecasting the emergence of resistance remains challenging. The process of developing novel antibiotics is characterized by its substantial financial investment and time-intensive nature. A single drug entails considerable costs and an extensive timeframe before reaching commercial availability. Given the urgency in combating antibiotic resistance, complementary strategies such as preventative measures are being explored alongside antibiotic therapy (Muteeb et al., 2023). The European Medicines Agency's guideline on environmental risk assessment for medicinal products doesn't currently address the risk of antibiotic resistance from environmental contamination by antibiotics.

Antibiotics can promote the emergence of antibiotic-resistant bacteria, even at concentrations considered safe. An addendum is urgently needed to assess the risks of antibiotics inducing antibiotic resistance, incorporating molecular and culture-based methods to determine the concentrations at which resistance is likely to occur. This addition is crucial for establishing a reliable risk assessment framework for antibiotics in the environment (Prasad et al., 2014).

11.7 EMERGING STRATEGIES FOR TACKLING ANTIBIOTIC RESISTANCE

The urgent global need for comprehensive solutions to combat the growing menace of antibiotic resistance and its associated challenges is imperative. A recent report from the WHO underscores the alarming reality that, despite the ongoing development of new antibiotics, none are projected to effectively combat the most severe forms of antibiotic-resistant bacteria (Alharbi et al., 2025). This situation necessitates a shift in focus from merely discovering new antibiotics to developing multifaceted strategies that can limit or circumvent resistance development. One promising avenue is the exploration of immuno-antibiotics, which delves into the intricate interactions between antibiotics and the host's innate immunity (Saba et al., 2025). These immuno-antibiotics could offer indirect effects, enhancing bacterial clearance and reducing the likelihood of resistance emergence. However, there is a pressing need for further research to deepen our understanding of how antibiotics interact with innate immune components (Nahvi et al., 2025). Antibiotics traditionally target essential bacterial activities, and resistance can arise through various mechanisms. To combat the evolution of antibiotic resistance, it is imperative to address two critical factors: the emergence of resistance genes in pathogens and the reproductive advantage these genes acquire from antibiotic use. This can be effectively managed through antibiotic stewardship, which entails using antibiotics only when necessary and selecting narrow-spectrum antibiotics whenever possible. Additionally, preventing infections through vaccines, stringent hygiene practices, and proper sanitation significantly reduces the need for antibiotic use (Smith et al., 2015). A groundbreaking development in the fight against antimicrobial resistance is the introduction of dual-acting immuno-antibiotics (DAIAs). These DAIAs target specific bacterial pathways, demonstrating efficacy in preclinical models of Gram-negative bacterial infections and marking a potential watershed moment in the global battle against antibiotic resistance. In parallel, researchers are exploring biochemical approaches to enhance antibiotic susceptibility without the necessity of developing entirely new antibiotics. By neutralizing natural resistance defense mechanisms, this innovative approach aims to potentiate the effectiveness of existing antibiotics. The complex process of antibiotic-mediated bacterial killing involves the production of reactive oxygen species (ROS), contributing to biochemical, molecular, and ultrastructural changes in bacteria. The Fenton-like reaction and the role of hydroxyl radicals in antibiotic-induced bacterial death have become focal points of investigation. Understanding these intricate processes is crucial for developing alternative antibiotic therapies. Researchers are delving into targeting the SOS response, an inducible DNA repair process linked to bacterial adaptation, pathogenesis, and antibiotic resistance. Inhibiting the SOS response is considered a potential strategy to prevent antibiotic resistance and enhance the activities of bactericidal antibiotics. Another intriguing area of study involves the role of hydrogen sulfide (H_2S) in generalized antibiotic resistance in bacteria. Endogenous synthesis of H_2S is believed to inhibit the cellular generation of ROS, enhancing bacterial resilience to oxidative stress and antibiotics. Inhibiting the H_2S biosynthetic pathway is being explored as a means to increase antibiotic sensitivity and counteract resistance. Distinguishing between the effects of endogenous and exogenous H_2S on bacteria is crucial. While endogenous H_2S production protects bacteria, exogenous H_2S may have cytotoxic effects. Nevertheless, H_2S-releasing compounds are being investigated for their potential to improve antibiotic efficacy and reverse resistance in bacteria that do not naturally produce H_2S. This multifaceted approach, encompassing immuno-antibiotics, biochemical strategies, and targeted inhibition, represents a comprehensive strategy to address antibiotic resistance and pave the way for more effective treatments against infectious diseases (Ding et al., 2023). The utilization of natural bioactive substances as potential remedies for infections constitutes a rich vein of human ingenuity and scientific inquiry. Over centuries, civilizations worldwide have drawn upon the therapeutic properties of plants, harnessing their intrinsic capabilities to combat various ailments. This longstanding tradition, coupled with contemporary scientific rigor, has engendered a renewed interest in botanical compounds as a reservoir of novel antibiotics. An eminent characteristic of plant-derived compounds lies in their accessibility. Nature generously furnishes an abundance of bioactive molecules, often found in commonplace herbs, fruits, and other botanical sources. This accessibility, paired with affordability and relatively low-risk profiles, underscores the potential of these compounds as viable alternatives to traditional antibiotics (Rehman et al., 2024). Historical exemplars abound, such as the utilization of honey for wound healing and Echinacea for its immunomodulatory effects. These time-honored remedies stand as a testament to the efficacy of plant-derived compounds in combating infections. Nevertheless, with the escalating challenge of bacterial resistance to conventional antibiotics, the imperative for novel therapeutic strategies has become increasingly urgent. In response, scientists have delved deeply into the molecular intricacies of botanical compounds, unveiling their diverse mechanisms of action against microbial pathogens. From inhibiting efflux pumps to disrupting bacterial membranes, impeding biofilm formation, and interfering with DNA and protein synthesis, these compounds manifest a multifaceted armamentarium against infectious agents. However, the transition from laboratory discovery to clinical application is beset with challenges. While promising in vitro studies offer insights into efficacy, translating these findings into tangible treatments necessitates navigation of complex pharmacokinetic and pharmacodynamic terrains. Factors such as tissue penetration, bioavailability, and potential interactions with host physiology profoundly influence the therapeutic potential of natural molecules. Notwithstanding these hurdles, the pursuit of innovative antimicrobial treatments derived from botanical sources perseveres. A nuanced comprehension of the pharmacological attributes of plant-derived compounds, conjoined with advancements in drug delivery technologies, holds promise for ushering in a new epoch of antimicrobial therapy (Jubair et al., 2021). Addressing AMR demands a multifaceted approach tailored to the specific context of pathogens, whether human or zoonotic, and the environments where they thrive, such as hospitals or communities. For instance, combating multidrug-resistant tuberculosis (MDR-TB) emphasizes patient adherence to treatment regimens. In hospital settings, a comprehensive strategy intertwining infection control measures and antimicrobial stewardship programs, led by interdisciplinary teams including infectious disease physicians, clinical pharmacists, and microbiologists, is essential. These initiatives are pivotal in halting the spread of antimicrobial-resistant microorganisms and safeguarding the effectiveness of available antimicrobials. Expanding antimicrobial stewardship to encompass family physicians in community healthcare settings is crucial due to significant antibiotic consumption there. Achieving this involves a range of measures, from educating consumers to training healthcare professionals, improving diagnostic capabilities, formulating treatment guidelines, and conducting rigorous prescription audits. These efforts combat antibiotic misuse and unnecessary prescriptions. In veterinary medicine, urgent action is needed to monitor antimicrobial use in food animals, as highlighted on the 2011 World Health Day. Remedial interventions should include enforcing regulations on antimicrobial use in food-producing animals, enhancing surveillance and monitoring systems, and reducing reliance on antimicrobials through improved animal husbandry practices. By addressing AMR through tailored strategies across different contexts, we can mitigate its impact and preserve the effectiveness of antimicrobials (Prestinaci et al., 2015).

11.8 CONCLUSION

The challenge of AMR presents a multifaceted and urgent global concern, with profound implications for healthcare, mortality rates, and the stability of healthcare systems. AMR emerges from a complex interplay of factors, including inappropriate antimicrobial usage in both healthcare and agriculture, inadequate infection control measures, and the international spread of resistant microbes through travel. To effectively address AMR, immediate and concerted action is necessary, encompassing heightened surveillance, judicious antimicrobial stewardship, and sustained research efforts to develop new treatment strategies.

Interdisciplinary collaboration in understanding bacterial adaptation and resistance mechanisms is crucial. Additionally, the examination of fungal resistance emphasized the clinical challenges involved and stressed the importance of carefully selecting antifungal agents. The impact of antibiotic resistance extends beyond healthcare, affecting various sectors and ecosystems. Addressing antibiotic resistance requires tackling inherent challenges within these processes.

Promising emerging strategies, such as immuno-antibiotics and targeted modulation of bacterial responses, offer comprehensive approaches to combat resistance. However, combating AMR demands global cooperation, responsible practices, and innovative solutions to safeguard public health effectively. A holistic approach is essential to preserve the efficacy of antibiotics for future generations. By deploying multifaceted strategies and fostering international collaboration, we can mitigate the threat of antimicrobial resistance and ensure the continued effectiveness of these crucial treatments.

CONFLICT OF INTEREST

None to declare.

All the authors have read and approved the manuscript.

REFERENCES

Alharbi, L. N. F., Rehman, S., Azmi, S., Alamri, A., Alnimr, A., & Ansari, M. A. (2025). Novel circular antimicrobial peptides to combat a critical listed bacterial pathogen multi drug resistant Acinetobacter baumannii. Microbial Pathogenesis, *203*, 107448.

Al-Ghamdi, F. H., Al-Qaaneh, A. M., & Al-Shehri, B. A. (2023). Effect of antibiotics' electromagnetic spectrums on the bacterial growths: Beyond drug-receptor interaction. *Journal of Medical Pharmaceutical and Allied Sciences*, *12*(6), 6241–6246. https://doi.org/10.55522/jmpas.V12I6.5904

Al-Qaaneh, A. M., & Al-Ghamdi, F. H. (2021). Tocilizumab prescribing criteria for COVID-19 patients. *Human Vaccines & Immunotherapeutics*, *17*(4), 1128–1128. https://doi.org/10.1080/21645515.2020.1822137

Al-Qaaneh, A. M., Al-Ghamdi, F. H., AbdulAzeez, S., & Borgio, J. F. (2022). Safety of tocilizumab in COVID-19 patients and benefit of single-dose: The largest retrospective observational study. *Pharmaceutics*, *14*(3). https://doi.org/10.3390/pharmaceutics14030624

Allcock, S., Young, E. H., Holmes, M., Gurdasani, D., Dougan, G., Sandhu, M. S., Solomon, L., & Török, M. E. (2017). Antimicrobial resistance in human populations: Challenges and opportunities. *Global Health, Epidemiology and Genomics*, *2*, e4. https://doi.org/10.1017/gheg.2017.4

Arastehfar, A., Gabaldón, T., Garcia-Rubio, R., Jenks, J. D., Hoenigl, M., Salzer, H. J. F., Ilkit, M., Lass-Flörl, C., & Perlin, D. S. (2020). Drug-resistant fungi: An emerging challenge threatening our limited antifungal armamentarium. *Antibiotics (Basel)*, *9*(12). https://doi.org/10.3390/antibiotics9120877

Calam, C. T. (2008). *Process Development in Antibiotic Fermentations*. Cambridge University Press. https://books.google.jo/books?id=bxBwLAAACAAJ

Cars, O., & Nordberg, P. (2005). Antibiotic resistance-The faceless threat. *International Journal of Risk & Safety in Medicine*, *17*, 103–110.

CDC, (2024). Antimicrobial resistance. https://www.cdc.gov/antimicrobial-resistance/about/?CDC_AAref_Val=https://www.cdc.gov/drugresistance/about.html

Cui, X., Wang, L., Lü, Y., & Yue, C. (2022). Development and research progress of anti-drug resistant fungal drugs. *Journal of Infection and Public Health*, *15*(9), 986–1000. https://doi.org/10.1016/j.jiph.2022.08.004

Dadgostar, P. (2019). Antimicrobial resistance: Implications and costs. *Infection and Drug Resistance*, *12*, 3903–3910. https://doi.org/10.2147/idr.S234610

Ding, D., Wang, B., Zhang, X., Zhang, J., Zhang, H., Liu, X., Gao, Z., & Yu, Z. (2023). The spread of antibiotic resistance to humans and potential protection strategies. *Ecotoxicology and Environmental Safety*, *254*, 114734. https://doi.org/10.1016/j.ecoenv.2023.114734

European Centre for Disease Prevention and Control. (2020). Antimicrobial resistance in the EU/EEA (EARS-Net) – Annual Epidemiological Report 2019. https://www.ecdc.europa.eu/sites/default/files/documents/TQ-AC-20-001-EN-N_0.pdf

Fatima, Z., Kumari, P., Rehman, S., & Hameed, S. (2021). Virulence traits of Candida spp.: an overview. Recent Trends in Mycological Research: Volume 1: Agricultural and Medical Perspective, 439–455.

Fatima, Z., Purkait, D., Rehman, S., Rai, S., & Hameed, S. (2023). Multidrug resistance: a threat to antibiotic era. In Biological and Environmental Hazards, Risks, and Disasters (pp. 197–220). Elsevier.

Garvey, M., & Rowan, N. J. (2023). Pathogenic drug resistant fungi: A review of mitigation strategies. *International Journal of Molecular Sciences*, *24*(2), 1584. https://www.mdpi.com/1422-0067/24/2/1584

Gow, N. A. R., Johnson, C., Berman, J., Coste, A. T., Cuomo, C. A., Perlin, D. S., Bicanic, T., Harrison, T. S., Wiederhold, N., Bromley, M., Chiller, T., & Edgar, K. (2022). The importance of antimicrobial resistance in medical mycology. *Nature Communications*, *13*(1), 5352. https://doi.org/10.1038/s41467-022-32249-5

Iwata, K. (1992). Drug resistance in human pathogenic fungi. *European Journal of Epidemiology*, *8*(3), 407–421. https://doi.org/10.1007/bf00158576

Jubair, N., Rajagopal, M., Chinnappan, S., Abdullah, N. B., & Fatima, A. (2021). Review on the antibacterial mechanism of plant-derived compounds against Multidrug-Resistant Bacteria (MDR). *Evidence-Based Complementary and Alternative Medicine*, *2021*, 3663315. https://doi.org/10.1155/2021/3663315

Muteeb, G., Rehman, M. T., Shahwan, M., & Aatif, M. (2023). Origin of antibiotics and antibiotic resistance, and their impacts on drug development: A narrative review. *Pharmaceuticals (Basel)*, *16*(11). https://doi.org/10.3390/ph16111615

Nadgir, C. A., & Biswas, D. A. (2023). Antibiotic resistance and its impact on disease management. *Cureus*, *15*(4), e38251. https://doi.org/10.7759/cureus.38251

Nahvi, I., Farhan, M., Rehman, S., & Hameed, S. (2025). Nanomaterial-Based Vaccines. In Nanotechnology Based Microbicides and Immune Stimulators (pp. 37–53). Singapore: Springer Nature Singapore.

O'Neill, J. (2014). *Antimicrobial Resistance: Tackling a Crisis for the Health and Wealth of Nations*. Review on Antimicrobial Resistance. https://books.google.jo/books?id=b1EOkAEACAAJ

O'Neill, J., & Wellcome Trust. (2016). *Tackling Drug-Resistant Infections Globally: Final Report and Recommendations*. Review on Antimicrobial Resistance. https://books.google.jo/books?id=aa6lAQAACAAJ

Prasad, R., Shah, A., & Dhamgaye, S. (2014). Mechanisms of drug resistance in fungi and their significance in biofilms. In *Antibiofilm Agents* (Vol. 8, pp. 44–65). https://doi.org/10.1007/978-3-642-53833-9

Prestinaci, F., Pezzotti, P., & Pantosti, A. (2015). Antimicrobial resistance: A global multifaceted phenomenon. *Pathogens and Global Health*, *109*(7), 309–318. https://doi.org/10.1179/2047773215y.0000000030

Rao, G. G. (1998). Risk factors for the spread of antibiotic-resistant bacteria. *Drugs*, *55*(3), 323–330. https://doi.org/10.2165/00003495-199855030-00001

Rehman, S. (2023). A parallel and silent emerging pandemic: Antimicrobial resistance (AMR) amid COVID-19 pandemic. *Journal of Infection and Public Health*, *16*(4), 611–617.

Rehman, S., Albhishiri, G., Alsalem, Z., AlJameel, S. S., Al-Qaaneh, A., Shah, A. H., Akhtar, S., Hameed, S., & Jermy, B. R. (2024). Bionanocomposites comprising mesoporous metal organic framework (ZIF-8) phytofabricated with Allium sativum as alternative nanomaterials to combat antimicrobial drug resistance. *Bioprocess and Biosystems Engineering*. https://doi.org/10.1007/s00449-024-03027-y

Saba, I., Wani, K., Batoo, K. M., Rehman, S., & Hameed, S. (2025). Immunomodulatory Nanoparticles and Their Advancement in Health. In Nanotechnology Based Microbicides and Immune Stimulators (pp. 19–35). Singapore: Springer Nature Singapore.

Sabtu, N., Enoch, D. A., & Brown, N. M. (2015). Antibiotic resistance: What, why, where, when and how? *British Medical Bulletin*, *116*, 105–113. https://doi.org/10.1093/bmb/ldv041

Seneviratne, G., Kecskés, M., & Kennedy, I. (2008). Biofilmed biofertilisers: Novel inoculants for efficient nutrient use in plants. *ACIAR Proceedings*, *130*,126–130.

Silver, L. L., & Bostian, K. A. (1993). Discovery and development of new antibiotics: The problem of antibiotic resistance. *Antimicrobial Agents and Chemotherapy*, *37*(3), 377–383. https://doi.org/10.1128/aac.37.3.377

Smith, R. A., M'Ikanatha, N. M., & Read, A. F. (2015). Antibiotic resistance: A primer and call to action. *Health Communication*, *30*(3), 309–314. https://doi.org/10.1080/10410236.2014.943634

Urban-Chmiel, R., Marek, A., Stępień-Pyśniak, D., Wieczorek, K., Dec, M., Nowaczek, A., & Osek, J. (2022). Antibiotic resistance in bacteria-A review. *Antibiotics (Basel)*, *11*(8). https://doi.org/10.3390/antibiotics11081079

Ventola, C. L. (2015). The antibiotic resistance crisis: Part 1: Causes and threats. *Pharmacy and Therapeutics*, *40*(4), 277–283. https://www.ncbi.nlm.nih.gov/pmc/articles/PMC4378521/pdf/ptj4004277.pdf

Wilkins, A. S. (1996). Antibiotic resistance: Origins, evolution and spread. *Bioessays*, *18*(10), 847–848. https://onlinelibrary.wiley.com/doi/10.1002/bies.950181012

World Health Organization. (2014). *Antimicrobial Resistance: Global Report on Surveillance*. World Health Organization. https://iris.who.int/handle/10665/112642

Yudkin, J. (1953). Origin of acquired drug resistance in bacteria. *Nature*, *171*(4352), 541–546. https://doi.org/10.1038/171541a0

Zhang, J. F., Rehman, S., Alghamdi, T., Sheikh, F. A., Hassan, M. S., & Amna, T. (2022). Bionanotechnology approaches to combat biofilms and drug resistance. In Innovative approaches for nanobiotechnology in healthcare systems (pp. 230–248). IGI Global.

Zhu, Y., Huang, W. E., & Yang, Q. (2022). Clinical perspective of antimicrobial resistance in bacteria. *Infection and Drug Resistance*, *15*, 735–746. https://doi.org/10.2147/idr.S345574

12 Encapsulation and Delivery Strategies for Enhanced Viability and Efficiency of Probiotic Microorganisms

Luana Backes[1], Thayná Alves-Ferreira[1], Fabiane Cruz-Vieira[1], Rafael Caetano[1], Lucas de Oliveira Pereira[1], Priscila Vieira da Silva[1], Vitória Marina Abrantes[2], Edja Maria Melo de Brito Costa[3], Andréa Cristina Barbosa da Silva[3], and Diego Romário-Silva[1,2]

[1]Research Program in Integrated Dental Sciences, University of Cuiabá, Cuiabá, Brazil

[2]Department of Dentistry, Universidade Estadual da Paraíba - UEPB, Campina Grande, Paraíba, Brazil

[3]Department of Pharmacy, Universidade Estadual da Paraíba - UEPB, Campina Grande, Paraíba, Brazil

12.1 INTRODUCTION

Antimicrobial resistance has been a concern for researchers and healthcare professionals for decades. Currently, this challenge is considered a significant public health issue. In this context, alternative therapeutic approaches for treating infections have been encouraged. Natural products, which are sources of various bioactive molecules with different biological activities, have been considered promising for the development of new antimicrobials (Newman & Cragg, 2020). Thus, biomolecules produced by other microorganisms have been extensively studied for the development of antimicrobials against various types of pathogens (Schneider, 2021).

For a long time, probiotic bacteria that promote health in the human body have been included in the diet through fermented foods and specific preparations. This practice primarily aims to balance the gut microbiota and support proper digestive system function (Yu, Devine & Vernon, 2024). However, science has been rediscovering the potential of probiotic microorganisms for various applications due to their range of biological activities, including anti-inflammatory, antimicrobial, anti-allergic effects, and prevention of various diseases (Silva et al., 2020).

The most extensively studied probiotic microorganisms are *Lactobacillus*, which have been subdivided into 25 genera, encompassing various species with diverse biological activities of interest for human health (Fijan, 2023). However, current evidence indicates that research on the antimicrobial activity of probiotics is focused on identifying a new generation of strains with antimicrobial properties that benefit the composition and function of the human microbiota in different body sites. The pathogens most evaluated to identify these promising strains include pathogenic bacteria associated with systemic and nosocomial infections, such as *Klebsiella* spp., *Escherichia coli*, *Pseudomonas* spp., *Streptococcus* spp., *Bacillus* spp., and *Salmonella* spp. (Schifano et al., 2021; Sudan et al., 2021; Torres-Sánchez et al., 2021; Fijan et al., 2022; Tenea et al., 2022).

DOI: 10.1201/9781032676951-12

In vitro analyses, along with some *in vivo* trials, have shown promising results regarding antimicrobial activity against various human pathogens (Kim et al., 2021; Shahverdi et al., 2023). However, the widespread use of probiotics for infection treatment faces several challenges related to their delivery, viability, and bioavailability. In this context, various encapsulation systems and delivery formulations are being studied to optimize the use of these beneficial microorganisms. Therefore, this chapter discusses the main strategies and materials used for this purpose, addressing the current challenges and future perspectives on the subject.

12.2 CHALLENGES IN USING PROBIOTICS AS THERAPEUTIC APPROACHES

The human body presents various microenvironments, each with its own physical and biochemical characteristics that can limit the growth of microorganisms, including those of health interest such as probiotics (Wendel, 2022). The survival and viability of probiotics in the gastrointestinal tract, for example, represent one of the greatest challenges faced in administering these beneficial microorganisms (Peng et al., 2023).

Upon ingestion, probiotics must traverse the hostile environment of the stomach, where they encounter an acidic pH (1–4) and digestive enzymes that can compromise their viability. Gastric acidity is particularly detrimental, significantly reducing the number of viable probiotics that reach the intestine, where they exert their main beneficial effects. Figure 12.1 summarizes the main challenges faced by probiotics during their journey through the digestive system. The survival of probiotics during passage through the stomach is crucial for their therapeutic efficacy. Without adequate protection, most probiotic strains are destroyed before reaching the intestine, limiting their benefits to the host (da Silva et al., 2021).

The stability and viability of probiotics throughout the gastrointestinal tract are critical issues. The ability of probiotic strains to survive from ingestion to the large intestine, where they exert most of their beneficial effects, can be affected by factors such as gastric acidity, competition with resident microbiota, and environmental conditions along the digestive tract (Hill et al., 2014). Additionally, regulatory aspects add complexity to the clinical use of probiotics. The lack of standardization in product manufacturing and labeling can hinder informed choices by healthcare professionals and consumers, compromising the consistency in quality and efficacy of products available on the market. Despite these challenges, probiotics continue to be investigated as a promising therapeutic strategy for various clinical conditions, including gastrointestinal disorders, allergies, and even emerging neurological and psychiatric disorders (Slykerman et al., 2017).

To overcome the challenges faced by probiotics in the human body, it is essential to develop strategies that enhance their adhesion and persistence (Sanders et al., 2010). The adhesion of probiotics to the intestinal mucosa is crucial for their colonization and interaction with the host, which is fundamental for their beneficial effects. However, not all probiotic strains possess this ability effectively. Adhesion to the mucosa depends on factors such as the presence of specific receptors on the probiotic cell surface and compatibility with the host's intestinal mucosa (Ouwehand et al., 2003).

Probiotics must compete with the native intestinal microbiota and pathogens for colonization and available nutrients. An already established microbiota can limit the efficacy of probiotics by resisting the colonization of new microorganisms and creating a competitive environment that hinders the entry of new strains. Furthermore, the diversity and composition of the intestinal microbiota vary significantly among individuals (Syromyatnikov et al., 2022). Some probiotic strains may induce undesirable inflammatory responses, especially in susceptible individuals. Therefore, careful selection of probiotic strains with appropriate immunomodulatory properties is essential to minimize adverse effects and maximize therapeutic benefits (Al Bander et al., 2020).

One of the main challenges in the clinical use of probiotics is the variability in individual responses, as each person has a unique intestinal microbiota that may react differently to administered strains. Studies show that the efficacy of probiotics can vary significantly among individuals,

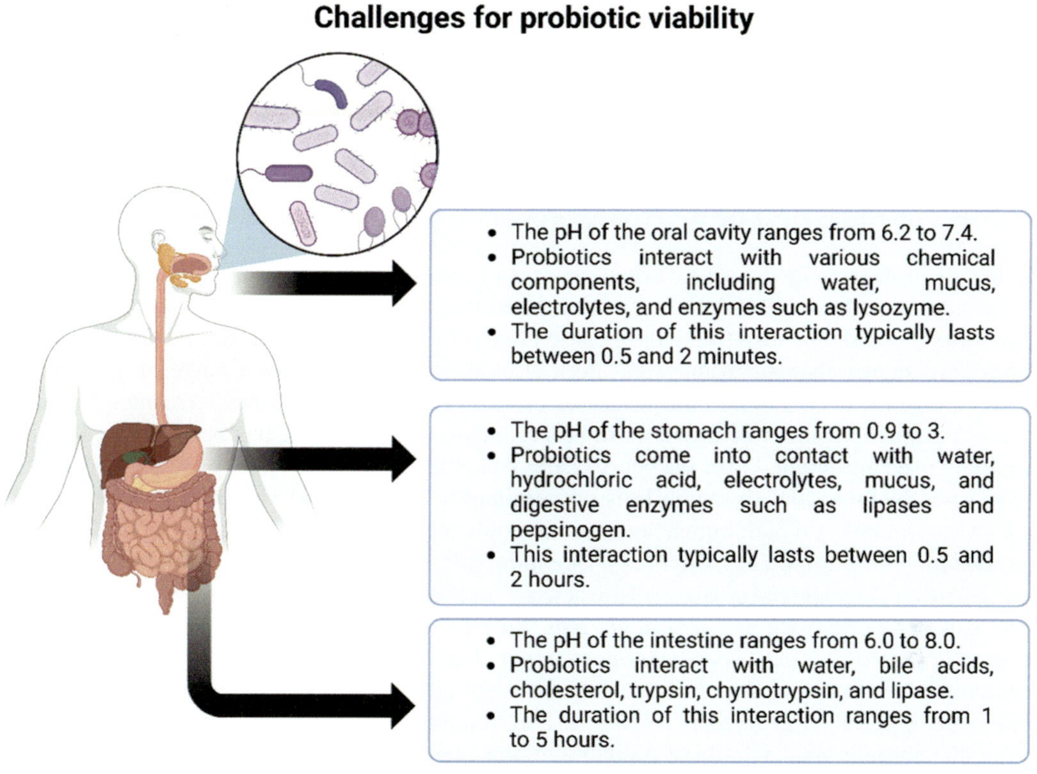

FIGURE 12.1 Schematic representation of the physicochemical challenges faced by probiotic microorganisms in the various microenvironments of the gastrointestinal tract. (Created with BioRender.com).

depending on factors such as the initial microbiota composition, underlying health conditions, and even diet (Hill et al., 2014). Additionally, there are concerns about the safety of probiotics, especially in more vulnerable populations. Patients with compromised immune systems, such as those undergoing immunosuppressive treatments or with autoimmune diseases, may have an increased risk of opportunistic infections associated with probiotic use (Goldenberg et al., 2017; Tom et al., 2022). Therefore, it is crucial that probiotic selection is rigorous and their administration is carefully monitored.

The variability of probiotic strains and the diversity of their metabolic products imply distinct biological properties that are often difficult to measure. A critical aspect is to better understand the mechanisms of action of probiotics. Although they are recognized for promoting the balance of the intestinal microbiota and enhancing the mucosal barrier, the molecular details and specific interactions with the host are still active research areas. Recent studies investigate how different probiotic strains can modulate local and systemic immune responses, influence intestinal gene expression, and interact with host metabolic pathways (Rooks & Garrett 2016; Wang et al., 2017). The diversity of probiotic strains is crucial for their clinical efficacy. Each strain may have distinct properties, such as the ability to produce bioactive metabolites, withstand adverse conditions in the gastrointestinal tract, and compete with potential pathogens. The selection of specific strains for particular clinical conditions is essential to ensure consistent therapeutic outcomes and should be carried out by a knowledgeable healthcare professional (Hill et al., 2014). In this context, the variability in the formulation of probiotic products available on the market can directly affect their clinical efficacy. Establishing clear guidelines for dosage, timing of administration, and treatment duration is essential to maximize the therapeutic benefits of probiotics for different health conditions (Kumar et al., 2010). International guidelines, such as those established by the World Health Organization and the Food and Agriculture Organization, are being developed to guide the industry and healthcare professionals in the appropriate selection and use of probiotics (FAO/WHO, 2002). Studies have indicated that certain probiotic strains can influence drug metabolism, altering their efficacy or toxicity. This is particularly relevant in patients requiring complex drug therapies, where probiotic–drug interactions may have significant clinical impacts (Fijan, 2014).

12.3 EVIDENCE OF THE ANTIMICROBIAL ACTIVITY OF PROBIOTIC MICROORGANISMS

The use of probiotics for treating various diseases began in the 19th century with observations of reduced populations of bifidobacteria in children with diarrhea, suggesting that oral intake of bifidobacteria could replenish this subpopulation of the microbiota and improve health. Since then, studies have shown changes in intestinal flora or skin microbiota during a range of diseases and suggested positive effects of certain probiotics (Wieërs et al., 2020). Various fermented foods can serve as promising sources of probiotic microorganisms, but raw milk has been one of the most commonly used and studied over the years (Zaib et al., 2024).

The microbial diversity in raw milk is predominantly composed of lactic acid bacteria, which are used as probiotics in both human and animal applications. Strains such as *Lactobacillus casei*, *Lactobacillus plantarum*, *Lactobacillus fermentum*, and *Lactobacillus paracasei* present in milk have shown varying degrees of antagonistic activity and have been capable of reducing the competitiveness of pathogens such as *Escherichia coli*, *Enterococcus faecalis*, *Salmonella Typhimurium*, *Salmonella enteritidis*, and *Listeria monocytogenes*. These properties are promising and have high potential for *in vivo* application or in combination with other strains in a multispecies probiotic for greater efficacy (Rúben et al., 2019). Table 12.1 summarizes findings from various studies demonstrating the antimicrobial activity of probiotic microorganisms against different human pathogens, including information on the types of pathogens tested, the probiotic strains used, and the observed effects. Additionally, it highlights the effectiveness of each probiotic in inhibiting or reducing the activity of the pathogens.

In vitro and *in vivo* studies assessed the antibacterial properties and safety of *Lactobacillus paracasei* strains. The results demonstrated that the probiotic exhibits antibacterial activity against *Escherichia coli* and *Staphylococcus aureus*, effectively inhibiting both Gram-positive and Gram-negative bacteria (Shahverdi et al., 2023). This activity is attributed to the production of antimicrobial compounds that lower the pH, compete for binding sites with pathogens, and stimulate immunomodulatory cells (Abotsi et al., 2011).

In addition, the probiotic *Bacillus subtilis* controls the colonization by *S. aureus* in humans, eliminating approximately 95% of the colonies without altering the microbiota. This approach offers several advantages over traditional pathogen decolonization strategies (Piewngam et al., 2023).

The antibiofilm and antifungal potential of the supernatants of *Lactobacillus acidophilus* and *Lactobacillus plantarum* against *Candida* spp. was evaluated through the analysis of *Candida* cell morphology. Inhibition of germ tube formation was observed in all tested *Candida. albicans* samples, with inhibitions of 68% and 53% for *C. albicans* and 69% and 59% for *C. krusei*. The inhibition was more pronounced with *L. acidophilus* and *L. plantarum*, due to their high capacity for coaggregation with *C. albicans* strains (Hashem & Hashem, 2021).

The antimicrobial effects of probiotics, both industrially produced and nonindustrially produced, were evaluated by immersing them for seven days in a solution containing *Staphylococcus aureus* for biofilm formation. Subsequently, the probiotics were plated on culture medium and divided into groups: *Lactobacillus casei*, *Lactobacillus brevis*, *Lactobacillus rhamnosus*, *Lactobacillus* from fermented milk, and *Yakult Batavo*, with the sixth group used as a negative control (Lemos et al., 2021). Quantitative and qualitative analyses were performed through colony-forming unit counts. The study demonstrated that noncommercial probiotic bacteria, such as *Lactobacillus casei* and *Lactobacillus rhamnosus*, as well as the commercially available fermented milk from the Batavo brand, showed satisfactory results in reducing colonization of mini-implants by *S. aureus* (Lemos et al., 2021). Despite the scientific evidence supporting the

TABLE 12.1
Summary of Findings from Various Studies on the Antimicrobial Activity of Probiotic Microorganisms Against Different Human Pathogens

Author and Year	Methodology	Microorganisms	Results and Conclusions
Teja et al. (2022)	*In vitro*	*Enterococcus faecalis* *Candida albicans*	• Antimicrobial activity of probiotics against the microorganisms. • Probiotics may effectively aid in endodontic treatment.
Janini and Ravindran (2022)	*In vivo*	*Streptococcus mutans*	• Reduction in plaque scores ($p<0.05$). • Decrease in oral microbiota. • Cariostatic effect.
Holm et al. (2023)	*In vitro*	*Aggregatibacter actinomycetemcomitans* *Streptococcus salivarius.*	• Reduction of *Aggregatibacter actinomycetemcomitans*. • Ineffective in reducing *Streptococcus salivarius*. • *Streptococcus salivarius* is effective against periodontal pathogens.
Shahverdi et al. (2023)	*In vitro* *In vivo*	*Escherichia coli* *Staphylococcus aureus, Shigella flexneri, Klebsiella pneumoniae, Yersinia enterocolitica subsp. Enterocolítica, Mycobacterium, Listeria monocytogenes*	• *L. paracasei* showed antibacterial activity against *Escherichia coli* and *Staphylococcus aureus*. • Adequate antibacterial properties.
Hashem and Hashem (2021)	*In vitro*	*Candida spp.*	• Inhibition of biofilm formation by 50%–72% with *L. plantarum* and 74%–85% with *L. acidophilus*.
	In vitro	*Staphylococcus aureus*	• *Lactobacillus casei* and *Lactobacillus rhamnosus* reduced *S. aureus* colonization.
Shaaban et al. (2022)	*In vitro*	*Enterococcus faecalis*	• *Lactobacilli* is more effective than calcium hydroxide paste.
Gobbi et al. (2020)	*In vitro*	*Streptococcus salivarius* M18 *Lactobacillus reuteri*	• Reduction in microorganism growth.
	In vitro	*Escherichia coli* *Enterococcus faecalis* *Salmonella typhimurium* *Salmonella enteritidis* *Listeria monocytogenes*	• Inhibition of pathogens.
Piewngam et al. (2023)	*In vivo*	*Staphylococcus aureus*	• Elimination of nearly 100% of *S. aureus*.

antimicrobial activity of lactobacilli and their potential against antimicrobial resistance, further research is needed to validate these activities *in vivo*. Most available evidence comes from laboratory experiments which, although promising, do not replicate the physiological conditions of the human body. Additionally, the main challenge in translating *in vitro* results to *in vivo* assays lies in the viability of microorganisms in different microenvironments within the body. Therefore, the development of encapsulation strategies and delivery formulations that enhance probiotic viability should be encouraged (Trush et al., 2020; Merenstein et al., 2023).

12.4 ENCAPSULATION TECHNIQUES FOR THE DEVELOPMENT OF DELIVERY FORMULATIONS

Due to the significant reduction in probiotic viability, encapsulation has been employed to enhance their resistance to the mentioned adverse conditions (Yao et al., 2020). However, although some mechanisms have been developed to overcome this barrier, probiotic viability remains a challenge, as the encapsulation process itself can compromise their stability. Factors such as overheating, mechanical pressures, and pH conditions during manufacturing can negatively affect probiotics (Yao et al., 2020; Singh et al., 2022).

With advancements in the field of polymers and drug delivery systems, microencapsulation technology has been proposed as an effective strategy to preserve probiotic viability. This approach can enhance their efficacy following oral administration. Recently, nanomaterials have demonstrated significant potential as encapsulating materials for this purpose. The integration of nanotechnology with microencapsulation techniques can optimize the controlled release of viable probiotic bacteria in the intestine, allowing these microorganisms to perform their biological activities *in vivo* with positive clinical outcomes (Razavi et al., 2021). Thus, biocompatible and biodegradable polymers with pH or enzyme sensitivity have shown great potential as probiotic carriers. These promising materials can offer various benefits, including (i) increased microorganism viability, (ii) extended storage time, (iii) protection in the acidic environment of the stomach, and (iv) targeted delivery to the desired site (Asgari et al., 2020).

Various encapsulating materials have been proposed, including whey protein isolate and dextran conjugates (Guo et al., 2022), polysaccharide–protein matrices with chitosan and alginate (Iqbal et al., 2019), cocoa powder combined with sodium alginate or fructooligosaccharides (Hossain et al., 2021), and alginate with Persian gum (Nami et al., 2020), among others that are being extensively investigated.

Although there is evidence of the application of probiotics for the microbiome balance in various parts of the body, their use in the digestive system remains the most extensively studied (Silva et al., 2020) and, therefore, will be the focus of this chapter. In this context, Figure 12.2 presents a schematic representation of an encapsulation system that protects probiotic microorganisms from the challenges of the gastrointestinal tract, ensuring their arrival in the intestine, where they exert their greatest benefits.

The polysaccharide–protein matrix encapsulation technique has shown effectiveness in improving the viability of *Bifidobacterium bifidum* during gastrointestinal transit and storage in foods. In the study, *Bifidobacterium bifidum* was encapsulated with sodium alginate (SA) and chitosan (CH) at various concentrations. The resulting microspheres were hardened in calcium chloride solution and coated with whey protein concentrate (WPC). The SA 6%–WPC 5% and CH 0.8%–WPC 5% combinations demonstrated the highest encapsulation efficiency, providing significantly greater cell viability under simulated stomach and intestinal conditions. This method allowed for controlled and sustained release of probiotic cells, promoting colonization at the target site and proving to be a promising approach for the targeted delivery of probiotics. However, the stability of calcium alginate microspheres may be compromised by calcium chelating agents and in highly acidic environments. To improve this stability, it is necessary to develop more robust coatings, possibly using polyelectrolyte polymers such as polylysine and proteins.

The effectiveness of chitosan coating on microcapsules containing *Lactobacillus plantarum* was investigated to assess its ability to maintain probiotic viability under heat and in simulated gastrointestinal juice. The study compared chitosan-coated microcapsules with those coated with Eudragit L100-55. The primary challenge encountered was the low viability of probiotic cells encapsulated with chitosan, while uncoated microcapsules and those coated with Eudragit maintained a high rate of viable cells. Flow cytometry indicated that chitosan released during the tests inhibited bacterial growth, explaining the drastic reduction in cell viability. Thus, although chitosan is effective in some applications, it may not be suitable for coating probiotics due to its bactericidal effect under adverse conditions. These results underscore the importance of selecting the appropriate coating to protect probiotics.

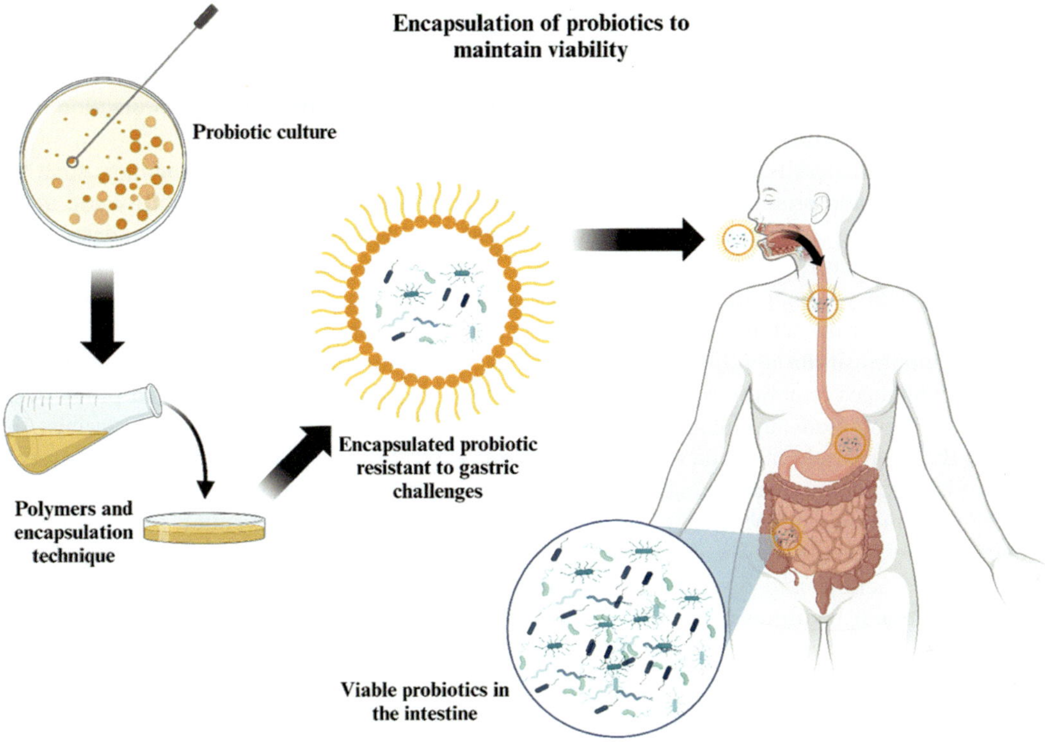

FIGURE 12.2 Schematic representation of a reverse micellar encapsulation system that protects probiotic microorganisms throughout the gastrointestinal tract, allowing their arrival in the intestine, where they exert their primary benefits (Created with BioRender.com).

The study by Hossain et al. (2021) assessed the impact of encapsulating probiotics with cocoa powder on viability during processing, storage, and gastrointestinal digestion in chocolates. The formulations combining cocoa powder with sodium alginate (A1) and sodium alginate with fructooligosaccharides (A2) demonstrated high encapsulation efficiency for seven probiotic strains, with rates of up to 93.40% and 95.36%, respectively. The encapsulated probiotics maintained viability above the recommended therapeutic level (10^7 CFU/g) for up to 180 days at 4°C and 120 days at 25°C. However, the study identified challenges in maintaining viable counts during processing at high temperatures, storage, and under severe gastrointestinal conditions.

Microencapsulation of *Lactococcus lactis* ABRIINW-N19 in orange juice was performed using a combination of alginate and Persian gum with prebiotics such as fructooligosaccharides (FOS) and inulin. The main challenges encountered were the stability of probiotic cells during prolonged storage and the high porosity of alginate in acidic environments. These challenges were addressed by combining alginate with Persian gum to form more stable hydrogels and adjusting the concentrations of the materials to ensure the controlled release of the cells under colonic conditions. The addition of FOS and inulin enhanced prebiotic activity and maintained cell viability. The study achieved a high encapsulation efficiency (98.4%) and a viability of 61% under simulated digestive conditions. The microencapsulated cells remained stable during 6 weeks of storage at 4°C, with the formulation containing 1.5% FOS and 2% inulin showing the highest survival rate (Nami et al., 2020).

Encapsulation of *Lactobacillus fermentum* K73 was analyzed using ionotropic gelation, with sodium alginate and whey protein as wall materials. Process optimization revealed that the combination of these components was effective, utilizing the same whey culture medium. This resulted in high bacterial viability, with an encapsulation yield of 94.8% and a final concentration of 9.261 $\log_{10}$ CFU/mL. The encapsulated spheres created an additional barrier that enhanced the bacteria's resistance to simulated gastrointestinal conditions. The study highlights the potential of these materials to meet the requirements for probiotics, providing an efficient technique for the protection and release of probiotics in functional food products.

A variety of biomaterials are used in probiotic encapsulation, with a focus on polymers such as alginate, chitosan, and proteins. These materials are selected for their biocompatibility and ability to form protective capsules. Alginate is widely used due to its low cost and safety, despite its high porosity and sensitivity to acidic conditions. Chitosan is valued for its antibacterial properties, although it may compromise probiotic viability under certain conditions. Proteins, such as those from whey, are chosen for their ability to form hydrogels that enhance probiotic stability during storage and transit through the gastrointestinal tract. Despite the advantages of each material, significant challenges remain in maintaining the viability and activity of encapsulated probiotics, necessitating ongoing advancements in the formulation and application of these biomaterials. In summary, the appropriate selection of material and encapsulation technique is crucial for preserving probiotic functionality and advancing the delivery of these microorganisms.

Despite promising evidence regarding the use of probiotics in the prevention and control of infections, there are still significant barriers to overcome, particularly concerning formulations, mechanisms of action, and dosage (Silva et al., 2020). Publications indexed in MEDLINE®/PubMed® (National Institutes of Health - NCBI) over the past 10 years highlight this reality. A comparison of the number of studies related to the search terms "probiotics," "antimicrobial probiotics," and "probiotics encapsulation" reveals a clear gap in research focused on strategies for maintaining probiotics within the body (Figure 12.3). Thus, it is crucial not only to demonstrate the promising biological activities of probiotics but also to develop strategies that ensure the replication of these activities *in vivo*. Despite advances in probiotic delivery systems, there is still no consensus on the best formulations and their ideal microenvironments within the body. Most studies focus on overcoming challenges within the gastrointestinal tract, which, while promising, limits the application of probiotics in the prevention and treatment of infections in other systems and tissues.

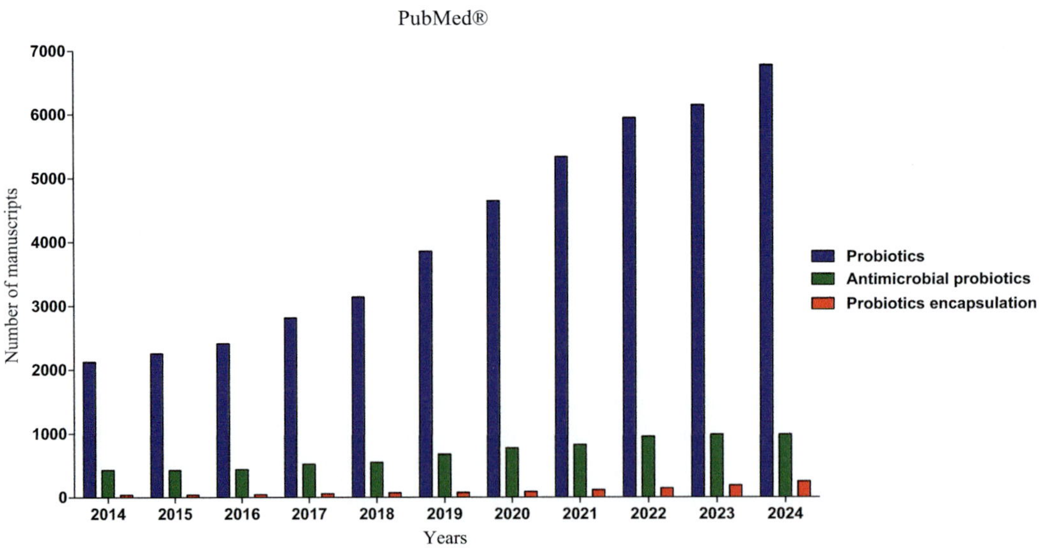

FIGURE 12.3 Compilation of indexed publications on PubMed®, highlighting the scientific gap related to encapsulation strategies and delivery formulations for probiotics.

12.5 FINAL CONSIDERATIONS AND FUTURE PERSPECTIVES

Scientific evidence from *in vitro* and preclinical *in vivo* studies supports the antimicrobial potential of probiotic strains against pathogenic microorganisms, primarily through the inhibition of microbes via metabolites produced by probiotics or direct competitive inhibition. However, clinical studies show that the benefits of these strains in treating infections, beyond their role in maintaining intestinal health, remain limited. Despite growing interest in discovering new probiotic strains, effective formulations for delivering these microorganisms to their targeted sites of action are still lacking. It is important to recognize that no single formulation strategy will be suitable for all microenvironments within the human body; various materials and techniques must be carefully chosen based on the specific application. Nevertheless, the biological properties of probiotics are promising, underscoring the need for further research to support their clinical use in addressing multidrug-resistant infections.

A particularly exciting future direction in medicine is the development of personalized probiotics, tailored to the unique microbiome profile of each individual. This approach aims to create customized probiotic therapies that consider the distinct characteristics of each microbiome, such as bacterial diversity and composition. Personalized probiotics hold the potential to enhance both the efficacy and safety of treatments. Given that the human microbiome influences immunity, digestion, mental health, and susceptibility to chronic diseases (Gloor et al., 2017), its personalization could improve the management of conditions such as gastrointestinal disorders, diabetes, obesity, autoimmune diseases, and psychiatric disorders, based on individual responses to specific bacterial strains (Zhao et al., 2021). However, implementing personalized probiotics requires robust regulatory frameworks, comprehensive clinical trials, and advancements in genetic sequencing and microbiome analysis (Sanders et al., 2019a,b). The future application of these individualized probiotics will rely on the integration of data science, technological innovations, and a deeper understanding of the complex interactions between human health and the microbiome.

12.6 CONCLUSION

The use of innovative encapsulation and delivery strategies for probiotics represents a crucial approach to fully harness the biological activities of these microorganisms, ensuring greater stability, viability, and efficacy in promoting human health. Modern encapsulation techniques, particularly those based on nanotechnology, address the challenges probiotics face in surviving the adverse conditions of the gastrointestinal tract and other microenvironments within the human body. These methods protect probiotics by enabling their controlled release at specific sites, thereby enhancing their therapeutic potential.

Additionally, advancements in delivery systems are continuously evolving, with increasing emphasis on personalized approaches tailored to the unique characteristics of each individual's microbiome. This personalization maximizes the functionality and bioavailability of probiotic strains, facilitating their clinical application in the prevention and treatment of gastrointestinal disorders, immune-related diseases, and other health conditions. However, the literature remains limited regarding the development of specific formulations for probiotic delivery. Thus, future research is needed to refine these strategies, incorporating cutting-edge materials and technologies designed to address the diverse requirements of various probiotic strains. Moreover, clinical trials will be essential to validate the long-term safety and efficacy of these advanced delivery systems.

REFERENCES

Abotsi W, Ainooson GK, Boakye-Gyasi E, Mensah AY, Woode E, Gbedema SY. Acute and sub-acute toxicity studies of the ethanolic extract of the aerial parts of Hilleria latifolia (Lam.) H. Walt (Phytolaccaceae) in Rodents. *West Afr J Pharm*. 2011;22(1):27–35.

Al Bander Z, Nitert MD, Mousa A, Naderpoor N. The gut microbiota and inflammation: An overview. *Int J Environ Res Public Health*. 2020 Oct 19;17(20):7618. doi: 10.3390/ijerph17207618.

Asgari S, Pourjavadi A, Licht TR, Boisen A, Ajalloueian F. Polymeric carriers for enhanced delivery of probiotics. *Adv Drug Deliv Rev*. 2020;161–162:1–21. doi: 10.1016/j.addr.2020.07.014.

FAO/WHO. Guidelines for the Evaluation of Probiotics in Food. Report of a Joint FAO/WHO Working Group on Drafting Guidelines for the Evaluation of Probiotics in Food. London, Ontario, Canada: World Health Organization and Food and Agriculture Organization of the United Nations; 2002.

Fijan S. Microorganisms with claimed probiotic properties: An overview of recent literature. *Int J Environ Res Public Health*. 2014;11(5):4745–4767.

Fijan S. Probiotics and their antimicrobial effect. *Microorganisms*. 2023 Feb 19;11(2):528. doi: 10.3390/microorganisms11020528.

Fijan S, Kocbek P, Steyer A, Vodičar PM, Strauss M. The antimicrobial effect of various single-strain and multi-strain probiotics, dietary supplements or other beneficial microbes against common clinical wound pathogens. *Microorganisms*. 2022 Dec 19;10(12):2518. doi: 10.3390/microorganisms10122518.

Gloor GB, et al. Microbiome analyses: From methodology to applications. *Nat Rev Microbiol*. 2017;15(8):452–466.

Gobbi E, Francesco MA, Piccinelli G, Caruso A, Bardellini E, Majorana A. In Vitro inhibitory effect of two commercial probiotics on chromogenic actinomycetes. *Eur Arch Paediatr Dent*. 2020;21:673–677. doi: 10.1007/s40368-020-00512-2.

Goldenberg JZ, Yap C, Lytvyn L, Lo CK, Beardsley J, Mertz D, Johnston BC. Probiotics for the prevention of Clostridium difficile-associated diarrhea in adults and children. *Cochrane Database Syst Rev*. 2017 Dec 19;12(12):CD006095. doi: 10.1002/14651858.CD006095.

Guo Q, Li S, Tang J, Chang S, Qiang L, Du G, Yue T, Yuan Y. Microencapsulation of Lactobacillus plantarum by spray drying: Protective effects during simulated food processing, gastrointestinal conditions, and in kefir. *Int J Biol Macromol*. 2022 Jan 1;194:539–545.

Hashem ZS, Hashem AS. An in vitro study on the antifungal and antibiofilm activities of probiotic bacteria against Candida species isolated from orthodontic appliances and dental caries. *Novel Res Microbiol J*. 2021;5(2):1176–1193.

Hill C, Guarner F, Reid G, et al. Expert consensus document. The International Scientific Association for Probiotics and Prebiotics consensus statement on the scope and appropriate use of the term probiotic. *Nat Rev Gastroenterol Hepatol*. 2014;11(8):506–514.

Holm WV, Carvalho R, Delanghe L, Eilers T, Zayed N, Mermans F, Bernaerts K, Boon N, Claes I, Lebeer S, Teughels W. Antimicrobial potential of known and novel probiotics on in vitro periodontitis biofilms. *NPJ Biofilms Microbiomes*. 2023;9(3):1–12. doi: 10.1038/s41522-023-00370-y.

Hossain MN, Ranadheera CS, Fang Z, Ajlouni S. Impact of encapsulating probiotics with cocoa powder on the viability of probiotics during chocolate processing, storage, and in vitro gastrointestinal digestion. *J Food Sci*. 2021 May;86(5):1629–1641. doi: 10.1111/1750-3841.15695.

Iqbal R, Zahoor T, Huma N, Jamil A, Ünlü G. In-vitro GIT tolerance of microencapsulated Bifidobacterium bifidum ATCC 35914 using polysaccharide-protein matrix. *Probiotics Antimicrob Proteins*. 2019 Sep;11(3):830–839. doi: 10.1007/s12602-017-9384-5.

Janini P, Ravindran V. Comparative evaluation of the antimicrobial effects of probiotic milk and probiotic powder on the salivary *Streptococcus mutans* counts and the plaque scores in children aged 3–6 years: A randomized controlled trial. *Dent Med Probl*. 2022;59(1):99–104. doi: 10.17219/dmp/139731

Kim H, Kim Y, Kang CH. In vivo confirmation of the antimicrobial effect of probiotic candidates against *Gardnerella vaginalis*. *Microorganisms*. 2021 Aug 9;9(8):1690. doi: 10.3390/microorganisms9081690.

Kumar M, Nagpal R, Kumar R, et al. Cancer-preventing attributes of probiotics: An update. *Int J Food Sci Nutr*. 2010;61(5):473–496.

Lemos IS, Jassé FFA, Sazuki SS, Alencar CM, Fujii DM, Zaniboni JF, Suzuki H, Segundo ASG. Antimicrobial activity of probiotics against oral pathogens around orthodontic mini-implants: An in vitro study. *Dental Press J Orthod*. 2021;26(5):1–19. doi: 10.1590/2177-6709.26.5.1-19.oar

Merenstein D, Pot B, Leyer G, Ouwehand AC, Preidis GA, Elkins CA, Hill C, Lewis ZT, Shane AL, Zmora N, Petrova MI, Collado MC, Morelli L, Montoya GA, Szajewska H, Tancredi DJ, Sanders ME. Emerging issues in probiotic safety: 2023 perspectives. *Gut Microbes*. 2023 Jan-Dec;15(1):2185034. doi: 10.1080/19490976.2023.2185034.

Nami Y, Lornezhad G, Kiani A, Abdullah N, Haghshenas B. Alginate-Persian Gum-Prebiotics microencapsulation impacts on the survival rate of Lactococcus lactis ABRIINW-N19 in orange juice. *LWT*. 2020; 124, 109190. doi: 10.1016/j.lwt.2020.109190

Newman DJ, Cragg GM. Natural products as sources of new drugs over the nearly four decades from 01/1981 to 09/2019. *J Nat Prod*. 2020 Mar 27;83(3):770–803. doi: 10.1021/acs.jnatprod.9b01285.

Ouwehand AC, et al. Adhesion to and interactions with intestinal mucosa of probiotic bacteria. *British J Nutr*. 2003;89(suppl. 1):S15–S20.

Peng P, Feng T, Yang X, Nie C, Yu L, Ding R, Zhou Q, Jiang X, Li P. Gastrointestinal microenvironment responsive nanoencapsulation of probiotics and drugs for synergistic therapy of intestinal diseases. *ACS Nano*. 2023 Aug 8;17(15):14718–14730. doi: 10.1021/acsnano.3c02646.

Piewngam P, Khongthong S, Roekngam N, Theapparat Y, Sunpaweravong S, Damrongsak F, Otto M. Probiotic for pathogen-specific Staphylococcus aureus decolonisation in Thailand: A phase 2, double-blind, randomised, placebo-controlled trial. *Lancet Microbe*. 2023;4(2). doi: 10.1016/S2666-5247(22)00263-3.

Razavi S, Janfaza S, Tasnim N, Gibson DL, Hoorfar M. Nanomaterial-based encapsulation for controlled gastrointestinal delivery of viable probiotic bacteria. *Nanoscale Adv*. 2021 Mar 30;3(10):2699–2709. doi: 10.1039/d0na00952k.

Rooks MG, Garrett WS. Gut microbiota, metabolites and host immunity. *Nat Rev Immunol*. 2016;16(6):341–352.

Sanders ME, et al. Probiotics and prebiotics in dietetics practice. *J Am Diet Assoc*. 2010; 110(1):104–114.

Sanders ME, et al. Probiotics and prebiotics in human health. *Ann Rev Food Sci Technol*. 2019a;10:93–125.

Sanders ME, Merenstein DJ, Reid G, et al. Probiotics and prebiotics in intestinal health and disease: From biology to the clinic. *Nat Rev Gastroenterol Hepatol*. 2019b;16(10):605–616.

Schifano E, Tomassini A, Preziosi A, Montes J, Aureli W, Mancini P, Miccheli A, Uccelletti D. Leuconostoc mesenteroides strains isolated from carrots show probiotic features. *Microorganisms*. 2021 Nov 4;9(11):2290. doi: 10.3390/microorganisms9112290.

Schneider YK. Bacterial natural product drug discovery for new antibiotics: Strategies for tackling the problem of antibiotic resistance by efficient bioprospecting. *Antibiotics (Basel)*. 2021 Jul 10;10(7):842. doi: 10.3390/antibiotics10070842.

Shaaban S, Hamad GM, Genena S, Meheissen MA, Moussa S. Evaluation of the antibacterial activity of Lactobacilli probiotics supernatants against Enterococcus faecalis (in vitro study). *BMC Oral Health*. 2022;22(407):1–11. doi: 10.1186/s12903-022-02434-5.

Shahverdi S, Barzegari AA, Bakhshayesh RV, Nami Y. In-Vitro and in-vivo antibacterial activity of potential probiotic Lactobacillus paracasei against Staphylococcus aureus and Escherichia coli. *Heliyon*. 2023;9(4): e14641. doi: 10.1016/j.heliyon.2023.e14641.

Silva D, Sardi J, Pitangui N, Roque S, Silva A, Rosalen P. Probiotics as an alternative antimicrobial therapy: Current reality and future directions. *J Funct Foods*. 2020;73:104080. doi: 10.1016/j.jff.2020.104080

da Silva NM, Tagliapietra BL, Flores VDA, Pereira Dos Santos Richards NS. In vitro test to evaluate survival in the gastrointestinal tract of commercial probiotics. *Curr Res Food Sci*. 2021 May 12;4:320–325. doi: 10.1016/j.crfs.2021.04.006.

Singh S, Gupta R, Chawla S, Gauba P, Singh M, Tiwari RK, Upadhyay S, Sharma S, Chanda S, Gaur S. Natural sources and encapsulating materials for probiotics delivery systems: Recent applications and challenges in functional food development. *Front Nutr*. 2022 Sep 21;9:971784. doi: 10.3389/fnut.2022.971784.

Slykerman RF, Hood F, Wickens K, et al. Effect of Lactobacillus rhamnosus HN001 in pregnancy on postpartum symptoms of depression and anxiety: A randomised double-blind placebo-controlled trial. *EBioMedicine*. 2017;24:159–165.

Sudan S, Flick R, Nong L, Li J. Potential probiotic Bacillus subtilis isolated from a novel niche exhibits broad range antibacterial activity and causes virulence and metabolic dysregulation in enterotoxic E. coli. *Microorganisms*. 2021 Jul 12;9(7):1483. doi: 10.3390/microorganisms9071483.

Syromyatnikov M, Nesterova E, Gladkikh M, Smirnova Y, Gryaznova M, Popov V. Characteristics of the gut bacterial composition in people of different nationalities and religions. *Microorganisms*. 2022 Sep 18;10(9):1866. doi: 10.3390/microorganisms10091866.

Teja GVC, Raju MRN, Reddy ULN, Satyanarayana UVV, Praneeth D, Maheswari K. An in vitro evaluation of the antimicrobial activity of probiotics against endodontic pathogens. *Cureus*. 2022;14(6). doi: 10.7759/cureus.26455.

Tenea GN, Gonzalez GL, Moreno JL. Probiotic characteristics and antimicrobial potential of a native Bacillus subtilis strain Fa17.2 rescued from wild Bromelia sp. flowers. *Microorganisms*. 2022 Apr 21;10(5):860. doi: 10.3390/microorganisms10050860.

Tom FS, Tucker KJ, McCracken CM, McGregor JC, Gore SJ. Infectious complications of probiotic use: A matched case-control study. *Infect Control Hosp Epidemiol*. 2022 Oct;43(10):1498–1500. doi: 10.1017/ice.2021.261.

Torres-Sánchez A, Pardo-Cacho J, López-Moreno A, Ruiz-Moreno Á, Cerk K, Aguilera M. Antimicrobial effects of potential probiotics of Bacillus spp. isolated from human microbiota: in vitro and in silico methods. *Microorganisms*. 2021 Jul 29;9(8):1615. doi: 10.3390/microorganisms9081615.

Trush EA, Poluektova EA, Beniashvilli AG, Shifrin OS, Poluektov YM, Ivashkin VT. The evolution of human probiotics: Challenges and prospects. Probiotics antimicrob proteins. 2020 Dec;12(4):1291–1299. doi: 10.1007/s12602-019-09628-4.

Wang Y, Wu Y, Wang Y, Xu H, Mei X, Yu D, et al. Antioxidant properties of probiotic bacteria. *Nutrients*. 2017;9(5):521.

Wendel U. Assessing viability and stress tolerance of probiotics-A review. *Front Microbiol*. 2022 Jan 27;12:818468. doi: 10.3389/fmicb.2021.818468.

Wieërs G, Belkhir L, Enaud R, Leclercq S, de Foy JM, Dequenne I, Timary P, Cani PD. How probiotics affect the microbiota. *Front Cell Infect Microbiol*. 2020;10:454. doi: 10.3389/fcimb.2020.00454.

Yao M, Xie J, Du H, Mcclements DJ, Xiao H, Li L. Progress in microencapsulation of probiotics: A review. *Compr Rev Food Sci Food Saf*. 2020; 19(2):857–874. doi:10.1111/1541-4337.12532.

Yu X, Devine DA, Vernon JJ. Manipulating the diseased oral microbiome: The power of probiotics and prebiotics. *J Oral Microbiol*. 2024 Jan 31;16(1):2307416. doi: 10.1080/20002297.2024.2307416.

Zaib S, Hayat A, Khan I. Probiotics and their beneficial health effects. *Mini Rev Med Chem*. 2024;24(1):110–125. doi: 10.2174/1389557523666230608163823.

Zhao L, et al. Gut microbiota in human health and disease. *Nat Rev Microbiol*. 2021;19(5):305–320.

13 Probiotics in Urinary Tract Infection

Potential Strategies to Address Antimicrobial Resistance

Shaista Hilal, Samreen Maqbool, Darakshan Bhat, and Nahida Tabassum
Department of Pharmaceutical Sciences, Pharmacology Division, University of Kashmir, Hazaratbal, Srinagar, Jammu and Kashmir, India

13.1 INTRODUCTION

Urinary tract infections (UTIs) are among the most common bacterial infections affecting individuals of all ages, with a higher prevalence in women due to anatomical factors. UTIs typically involve the invasion of the urinary tract by pathogenic microorganisms, leading to symptoms such as dysuria, frequent urination, and abdominal pain. The infections can range from simple bladder infections (cystitis) to more severe kidney infections (pyelonephritis), with the potential for serious complications if left untreated.

The conventional treatment for UTI primarily involves the use of antibiotics, which are effective in eliminating the causative bacteria. However, the frequent use of antibiotics has led to increasing concerns about antibiotic resistance, making some UTIs more difficult to treat. Additionally, the side effects associated with antibiotic use and the risk of recurrent infections have prompted the exploration of alternative treatment options. In recent years, probiotics have gained attention as a promising alternative or complementary therapy for UTIs.

Probiotics are live microorganisms that confer health benefits to the host by restoring and maintaining a healthy microbial balance. Studies suggest that probiotics, particularly those containing strains of Lactobacillus, may help prevent the recurrence of UTIs by inhibiting the growth of uropathogenic bacteria and enhancing the host's immune response. The growing interest in probiotics stems from their potential to provide a safer, natural, and more sustainable approach to UTI management compared to traditional antibiotics.

13.2 URINARY TRACT INFECTIONS

It is a bacterial infection affecting the urinary tract. Normally urine is sterile. It is usually free of bacteria, viruses, and fungi. An infection occurs when tiny microorganisms, usually bacteria from the digestive tract, cling to the opening of the urethra and begin to multiply.

When bacteria from the rectal area enter the urinary tract via the urethra to the bladder and multiply in the urine, an infection occurs. UTI is an infection that occurs in any part of the urinary system, which includes the kidneys, ureters, bladder, and urethra. Most infections involve the lower urinary tract, the bladder, and the urethra. UTIs are typically caused by bacteria, though fungi and viruses can also be responsible (Wagenlehner et al. 2020).

Most of infections arise from *E.coli* which normally lies in the colon. In addition, Mycoplasma and Chlamydia may be associated with sexually transmitted UTI (Komala et al. 2013).

DOI: 10.1201/9781032676951-13

13.2.1 Epidemiology

The epidemiology of UTIs reveals that they are among the most common bacterial infections globally, affecting more than 150 million people annually. In the United States, over 7 million individuals seek medical attention for UTIs each year, accounting for approximately 15% of antibiotic prescriptions. The peak incidence of uncomplicated UTIs is generally observed in sexually active women aged 18–39 years, with an estimated rate of 0.5 UTIs per person per year in this demographic. The burden of UTIs extends beyond individual health, as they represent a significant economic cost to healthcare systems, with consultations for UTIs constituting between 1% and 6% of all medical visits, leading to costs around 1.6 billion US dollars annually in the United States (Öztürk and Murt 2020).

These are the most common bacterial infections seen in childhood, affecting a significant proportion of pediatric hospital admissions. The epidemiology of UTIs in children reveals a higher incidence in girls (56%) compared to boys (44%), with the median age of affected children being 12 months. *Escherichia coli* is the predominant pathogen, responsible for 74% of UTIs, with a higher prevalence in girls (Daniel et al. 2023).

13.2.2 Pathophysiology of UTIs

UTIs occur when pathogenic microorganisms, primarily bacteria, colonize the urinary tract. These bacteria typically originate from the gastrointestinal tract and ascend through the urethra to the bladder, leading to cystitis. In more severe cases, the infection can ascend further to the kidneys, causing acute pyelonephritis.

The pathogenesis involves several key mechanisms. The initial step is the adherence of uropathogens to the uroepithelial cells, mediated by specific adhesins such as P-fimbriae and type 1 fimbriae. This attachment triggers an immune response, resulting in the release of cytokines and chemokines, which recruit neutrophils to the site of infection. Despite the host's immune defense, some bacteria can evade phagocytosis and persist within the bladder's superficial cells, forming intracellular bacterial communities, which protect them from the immune system and antibiotic treatment.

In patients with underlying risk factors such as congenital anomalies of the kidney and urinary tract, urinary stasis, or immunosuppression, the infection is more likely to become complicated, leading to a higher risk of recurrent infections and renal damage. The interplay between bacterial virulence factors and host susceptibility determines the severity and outcome of the infection (Walsh and Collyns 2020).

13.2.2.1 Types

UTIs are of different types. An infection limited to the urethra is called urethritis. If bacteria move to the bladder and multiply, bladder infection cystitis occurs. If the infection is not treated promptly, bacteria may then travel further up the ureters, multiplying and infecting the kidneys, which lead to pyelonephritis (Komala et al. 2013)

In adults, UTI is classified as cystitis (bladder infection) or pyelonephritis (kidney infection). For men and women, urethritis, a sexually transmitted infection, should also be considered in the differential diagnosis. The UTI spectrum includes acute and chronic bacterial prostatitis. Related infections in men include epididymitis, orchitis, and epididymo-orchitis (Bettcher et al. 2021). Other types include:

Uncomplicated Cystitis*:* A bladder infection in a healthy, nonpregnant, premenopausal female with a normal urinary tract.

Complicated Cystitis: A bladder infection associated with factors that increase the risk of serious complications or decrease the effectiveness of treatment. These factors include cystitis plus a foreign body (such as a catheter or urinary tract stone), recent instrumentation, urinary tract abnormalities, and vesico-ureteral reflux. Cystitis in men, pregnant women, and patients with renal transplant or other causes of an immunocompromised state, and UTIs caused by atypical organisms or multidrug-resistant bacteria.

Uncomplicated Pyelonephritis*:* A kidney infection that occurs in a healthy nonpregnant patient with a normal urinary tract.

Complicated Pyelonephritis: A kidney infection that occurs during pregnancy or in a patient with other complicating factors, typically requiring hospital admission.

Recurrent UTI: It can be defined as the type when more than two UTIs occur in 6 months or more than three UTIs in 12 months.

Asymptomatic bacteriuria (colonization): It occurs when more than 100,000 colony-forming units (CFU)/mL of bacteria are present in clean-catch urine culture, without signs or symptoms of illness.

13.2.3 Common Causative Agents

The ability of different uropathogens to successfully adhere to and colonize the epithelium of the lower urinary tract is related to their ability to express specific virulence factors (Govindarajan and Kandaswamy 2022). Most UTIs are caused by Gram-negative and Gram-positive bacteria residing in the colon, such as *Escherichia coli*, *Enterococcus faecalis*, *Proteus mirabilis*, and *Klebsiella pneumoniae* (Govindarajan and Kandaswamy 2022). Other causative agents include *Staphylococcus saprophyticus, Group B Streptococcus (GBS),* and *Pseudomonas aeruginosa* (Flores-Mireles et al. 2019) (Figure 13.1).

The organisms most commonly responsible for catheter-associated UTIs are *Escherichia coli, Proteus mirabilis, P. aeruginosa, Streptococcus faecalis* and *Staphylococcus aureus*, Besides *Klebsiella pneumoniae, Mycobacterium tuberculosis, Actinomycetes*, Nocardia, Candida, etc. can also cause UTI.

Escherichia coli: The microbial etiology of UTI is generally consistent with uropathogenic strains of *E. coli* being the primary cause, accounting for 70%–95% of outpatient cases. These strains possess virulence factors that aid their survival and evasion of host immune responses (Hooton 2012).

Klebsiella pneumonia: It has emerged as a major cause of healthcare-associated opportunistic infections, such as bacteremia, pneumonia, and UTIs (Effah et al. 2020). This pathogen, similar to Uropathogenic *Escherichia coli,* uses two types of fimbrial adhesin, type 1 fimbriae and type 3 fimbriae, for biofilm formation and bladder colonization (Schroll et al. 2010). These two types of fimbriae adhesins have different binding specificities. Type 1 fimbriae bind to mannose receptors in the urinary tract and promote bladder cell invasion (Govindarajan and Kandaswamy 2022). Type

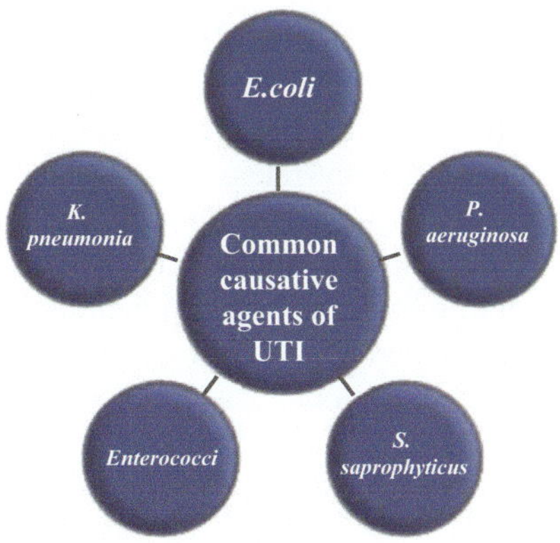

FIGURE 13.1 Some of the common causative agents of UTIs.

3 fimbriae, on the other hand, do not bind to mannose receptors but play an important role during biofilm aggregation on medical devices such as catheters (Govindarajan and Kandaswamy 2022).

Proteus mirabilis: It is the most commonly identified Gram-negative bacterium in UTIs, especially in catheterized patients or patients with urinary tract abnormalities (Jansen et al. 2004). It produces urease which hydrolyzes urea to carbon dioxide and ammonia. This increases the urine pH and generates calcium crystals and magnesium ammonium phosphate precipitates, which are incorporated into polysaccharide capsule, forming crystalline biofilms on the catheter (Jacobsen and Shirtliff 2011).

Enterococci: They cause several nosocomial infections, particularly surgical site/soft tissue infections, bloodstream infections, and UTIs (Khalil et al. 2022). These uropathogens do not contain pili and adhere to the host cell via their surface proteins, such as enterococcal surface proteins (Esp) and endocarditis- and biofilm-associated pilus (Ebp) (Singh et al. 2007). Catheter-associated urinary tract infections (CAUTI) caused by enterococci are due to the release of fibrinogen in the bladder after urinary catheterization. EbpA, which contains an N-terminal fibrinogen-binding domain, binds to fibrinogen deposited on the implanted catheter and promotes the formation of the biofilm responsible for *E. faecalis* CAUTI (Flores-Mireles et al. 2016). In addition, biofilm formation on the catheter helps bacteria evade the host immune system by masking antigenic determinants (Klein and Hultgren 2020).

Staphylococcus saprophyticus: It is a coagulase-negative microorganism responsible for UTIs, such as cystitis, in sexually active women (Sims et al. 2017). This Gram-positive bacterium is the second most common cause of community-acquired UTI, after *E. coli*. Bacterial adhesion to the bladder and ureter epithelium occurs through different types of adhesins, such as Aas, Uaf, and SdrI. These include cell-wall-associated proteins with hemagglutinic and adhesive properties, as well as surface-associated glycoproteins that facilitate bacterial adhesion to the host cell surface and promote bladder colonization (Govindarajan and Kandaswamy 2022). In case of *Proteus mirabilis* and *Klebsiella pneumoniae* but not in *E. coli*, *S. saprophyticus* urease is responsible for persistent bacterial colonization in the bladder and kidney (Flores-Mireles et al. 2019). This enzyme, which catalyzes the hydrolysis of urea into carbon dioxide and ammonia, causes an increase in urine pH and the production of carbonate precipitates (stone formation) in the urine (Sims et al. 2017).

Pseudomonas aeruginosa: Within the hospital setting, *P. aeruginosa* is the third most common cause of UTI (7%–10%) after *E. coli and E. faecalis* (Khalil et al. 2022). Patients with underlying conditions, such as urinary tract abnormalities or in dwelling urinary catheters, are more susceptible to UTI caused by *P. aeruginosa* (Gomila et al. 2018). The inherent multiple antibiotic resistance of this microorganism, combined with its ability to develop new resistance to multiple classes of antibiotics and to form biofilms, explains the high mortality and morbidity of UTIs caused by *P. aeruginosa* (Gomila et al. 2018).

This microorganism has the ability to form biofilms on catheters through the production of various components such as extracellular polysaccharides, elastase, exoenzyme S, and hemolytic phospholipase C. Elastase is a major virulence factor of *P. aeruginosa*. This enzyme, through its protease activity, induces tissue destruction by releasing nutrients essential for bacterial growth. Exoenzyme S is present in invasive strains of *P. aeruginosa* and acts on actin cytoskeleton rearrangement, affecting cell adhesion, morphology, and apoptosis in target host cells. Phospholipase C acts by hydrolyzing phosphatidylcholine from the host cell membrane, causing cellular damage and organ failure. All these factors are regulated by the quorum sensing system and are involved in the spread of UTI to the kidneys, resulting in pyelonephritis (Qin et al. 2022).

13.2.4 Symptoms

UTIs present with nonspecific symptoms, with fever being the most common symptom observed in 63.4% of cases. Acute pyelonephritis is associated with more severe clinical presentations, such as higher levels of inflammatory markers, compared to cystitis. Pain during urination, increased urgency and frequency, and cloudy or strong-smelling urine are the symptoms commonly observed in UTI.

Common urinary tract infection symptoms in

- Women include urge to urinate frequently, burning with urination, cloudy urine, strong unpleasant smell of urine, dark or bloody urine, pelvic pain, back pain, fever and chills.
- Males include urge to urinate frequently, burning with urination, cloudy urine, strong unpleasant smell of urine, dark or bloody urine, pelvic pain, back pain, rectal pain, penile, testicular and abdominal pain, and penile discharge.
- Children include urge to urinate frequently, burning with urination, cloudy urine, strong unpleasant smell of urine, and dark or bloody urine (Komala et al. 2013).

13.2.5 Diagnosis

The diagnostic methods currently available to detect pathogens in urine fall into two categories:

- Laboratory methods that require prior sample processing
- The point-of-care (POC) tools that perform an onsite direct analysis.

One of the most commonly used methods to detect a UTI is the urine examination by collecting a midstream urine sample or catheter specimens and subsequent microbiological evaluation (urine culture) for pathogen isolation, identification, and antimicrobial susceptibility testing (AST).

13.2.6 Risk Factors and Susceptibility

These include:

i. **Gender**: Women are more susceptible to UTIs due to their shorter urethra, which allows bacteria easier access to the bladder.
ii. **Sexual Activity**: Sexual intercourse can introduce bacteria into the urinary tract, increasing the risk of infection, particularly in women.
iii. **Use of Spermicides**: They can alter the normal flora of the vagina, making it easier for harmful bacteria to thrive and cause UTIs.
iv. **Menopause**: Decreased estrogen levels after menopause can lead to changes in the urinary tract, increasing the risk of infection.
v. **Age**: Older individuals, especially those over 65, are more susceptible to UTIs. This risk is particularly high for women and men aged 80 and older.
vi. **Incontinence**: Both urinary and fecal incontinence can increase the likelihood of UTIs.
vii. **Genetics**: A family history of UTIs can elevate an individual's risk, and genetic factors may also influence the severity of infections.
viii. **Medical Procedure**: Procedures involving the urinary tract, such as cystoscopy or surgery for pelvic organ prolapse, can increase the risk of UTIs. Prolonged use of urinary catheters can provide a direct pathway for bacteria to enter the urinary tract, leading to UTIs, especially in hospitalized patients.
ix. **Diet**: Dietary habits can influence susceptibility. Certain foods may alter the urinary environment, making it more conducive to bacterial growth.
x. **Disease State**:
    - Anatomical abnormalities in the urinary tract, such as urinary tract obstructions kidney or bladder stones can increase susceptibility by preventing the complete emptying of the bladder, allowing bacteria to grow (Ahmad et al. 2022).

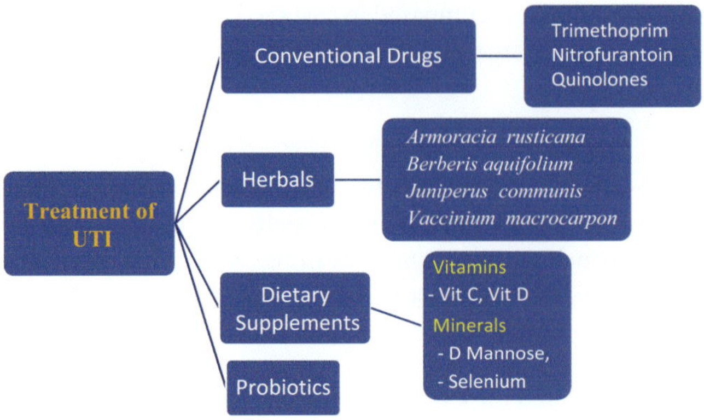

FIGURE 13.2 Treatment of UTI.

- **Diabetes Mellitus**: People with diabetes have higher risk of UTIs.
- **Immunocompromised Status**: Individuals with weakened immune systems, such as those who have undergone kidney or other solid organ transplants, are at a higher risk of UTIs. A study found that 28% of kidney transplant patients developed UTIs within 13 days of the procedure. While HIV infection is not thought to increase the risk of symptomatic UTIs, even in patients with low CD4 counts.

Understanding these factors can help in reducing the risk of developing UTIs through targeted dietary and lifestyle modifications (Kontiokari et al. 2004).

13.2.7 Treatment of UTI

Different ways of treating UTIs include conventional and herbal drugs, dietary modifications, etc. as given in Figure 13.2. Various regimens have been used to treat UTIs. For men, a quick antibiotic therapy for at least seven days is recommended. In the patients treated in the first line, the administration of trimethoprim and nitrofurantoin is currently used. Second-line antibiotics such as quinolones should be considered for patients with prostatitis (Lane and Takhar 2011). Empirical treatment involves the use of antibiotics for 3 days, which achieves a cure in 85%–90% of women.

Various antibiotic therapy regimens include:

- seven days of ciprofloxacin, co-amoxiclav, or cephalexin,
- 14 days of trimethoprim for men and nonpregnant women, and
- 7–10 days of cefalexin for pregnant women. Severe infections require hospital care and treatment with broad-spectrum parenteral antibiotics and admission for intravenous fluids.

13.3 ANTIBIOTIC RESISTANCE: A GROWING CONCERN

In the last decades, the extensive use of antibiotics has resulted in the emergence of antibiotic-resistant bacteria pathogens and led to the spread of antibiotic resistance. Bacterial resistance refers to a bacterium's ability to withstand or tolerate the effects of antibiotics. It is recognized by the World Health Organization (WHO) as a major health threat of the 21st century. It threatens the effective prevention and treatment of an ever-increasing range of infections (Edwin et al. 2017). Antibiotic resistance is a growing global problem in which antibiotics are no longer effective in the treatment of infectious diseases for which they were specifically designed. The WHO has now issued a warning that the world is "running out of antibiotics," escalating fears about global antibiotic resistance reaching new heights. The emergence of drug-resistant bacteria has posed a critical challenge to the treatment of clinical infectious diseases, resulting in a gradual increase in the frequency of nosocomial infections. Antibiotic resistance is on the rise in all parts of the world as bacterial infections are one of the leading causes of illness and mortality. New resistance mechanisms emerge regularly and spread internationally, putting the capacity to treat prevalent infectious diseases in jeopardy (David et al. 2022).

The irrational use of antibiotics across various sectors, including clinical care, agriculture, animal health, and the food system, has exacerbated this crisis. Global data indicate that deaths attributed to bacterial resistance exceeded 700,000 in 2019, with projections estimating a staggering 20 million deaths by 2050 if effective control measures are not implemented. Sub-Saharan Africa and South Asia currently experience the highest mortality rates globally, primarily due to their large populations, inadequate regulations to combat antibiotic resistance, and insufficient awareness among prescribers and the public regarding the implications of antibiotic resistance. Addressing these challenges is imperative to mitigate the adverse effects of antibiotic use (Yang et al. 2024).

Mortality rates associated with antibiotic-resistant infections are concerning, as new resistance mechanisms continue to emerge and spread rapidly. This dynamic escalation poses significant challenges to public health, making effective treatment increasingly difficult. Antibiotics are highly effective in treating bacterial UTIs by directly targeting and eliminating the pathogens responsible. They offer fast relief from UTI symptoms, often within 24 to 48 hours of initiation. Early treatment with antibiotics prevents the infection from spreading to the kidneys or causing sepsis, which are serious complications.

One of the major limitations is the increasing occurrence of antibiotic-resistant strains, making some UTIs difficult to treat. Antibiotics can cause side effects, such as gastrointestinal disturbances, allergic reactions, and yeast infections. Recurrent UTIs are common even after antibiotic treatment, necessitating repeated courses, which can exacerbate resistance and side effects. Antibiotics not only kill the pathogenic bacteria but also disrupt the beneficial microbiota in the gut and urogenital tract, potentially leading to dysbiosis and other health issues (Banerjee 2021).

Antibiotic resistance in UTIs is a pressing and growing concern, marked by bacteria's ability to develop mechanisms that make many standard antibiotics ineffective. This issue is particularly problematic in the treatment of UTIs, where prompt empirical therapy is often necessary before detailed diagnostic results are available. Recently, the rise of drug-resistant bacteria has presented significant challenges to treatments specifically designed to combat these infections. The WHO has emphasized the urgency of addressing this global health threat.

13.3.1 Mechanisms of Antibiotic Resistance

Enzyme Production: Many uropathogens, like *Pseudomonas aeruginosa*, produce enzymes such as beta-lactamases that degrade antibiotics before they can exert their effects. These enzymes can destroy the β-lactam ring, crucial for the action of β-lactam antibiotics like penicillin and cephalosporins.

Efflux Pumps and Membrane Permeability: Some bacteria increase the activity of efflux pumps, which actively remove antibiotics from the cell, while also reducing the permeability of their outer membranes to prevent antibiotic entry.

Biofilm Formation: Biofilms, protective layers formed by bacterial colonies, make infections particularly resistant to treatment as antibiotics struggle to penetrate these structures.

Horizontal Gene Transfer: Resistance genes can be transferred between bacteria through processes like conjugation, increasing the spread of resistant traits across different bacterial populations.

In recent years various approaches have been developed to overcome the problems associated with antibiotic resistance (Khameneh et al. 2019). Complementary and alternative medicine has been recognized as an effective approach for the treatment of antibiotic-resistant bacterial infection (Khameneh et al. 2019), which consists of a wide range of products such as natural compounds (*Arctostaphylos uva-ursi* (bearberry), *Juniperus communis* (juniper), *Vaccinium macrocarpon* (cranberry) (Fazly Bazzaz et al. 2021) dietary supplements as vitamins and minerals, and also probiotics (Baker et al. 2018).

D-mannose, a simple sugar found in many fruits including cranberries (Ala-Jaakkola et al. 2022), plays a crucial role in preventing UTIs by inhibiting the adhesion of *E. coli* to the urothelium, the lining of the urinary tract (Mydock-McGrane et al. 2017). *E. coli,* the most common causative agent of UTIs, uses a protein called FimH located on its type I fimbriae to bind to mannose-containing glycoproteins on the surface of the urothelial cells. This adhesion is a key step in the establishment of infection, as it allows the bacteria to colonize and invade the urinary tract (Sarshar et al. 2020).

When D-mannose is consumed, it is not metabolized by the human body and is excreted in the urine. In the urinary tract, D-mannose competes with the mannose residues on the urothelial cells for binding to the FimH adhesin on *E. coli*. The presence of excess D-mannose in the urine saturates the binding sites of FimH, preventing the bacteria from adhering to the urothelium. As a result, the bacteria are flushed out of the urinary tract during urination, reducing the risk of infection (Mydock-McGrane et al. 2017).

Cranberry Products: Cranberries have long been associated with UTI prevention, primarily due to their content of proanthocyanidins. These compounds are believed to inhibit the adhesion of *E. coli* to the urothelium by blocking the bacterial fimbriae that interact with the surface of the urinary tract cells. Although the efficacy of cranberry products can vary depending on their formulation and the concentration of active ingredients, they remain a popular preventive measure, particularly in populations prone to recurrent UTIs (Fazly Bazzaz et al. 2021).

Vaccination Strategies: Vaccines targeting uropathogens represent a forward-thinking approach to UTI prevention. Oral and sublingual vaccines, such as the MV140/Uromune, which contains a mixture of heat-killed uropathogens including *E. coli*, aim to stimulate the immune system at mucosal surfaces, where UTIs typically begin. By generating a targeted immune response, these vaccines can prevent bacterial colonization and subsequent infection. While vaccination requires a longer regimen to achieve efficacy, it holds promise as a long-term strategy for reducing antibiotic dependence in UTI management (Asadi Karam, Habibi and Bouzari 2019).

13.4 PROBIOTICS

The term probiotic is derived from Latin words "pro" (meaning "for") and "bios" (meaning life). The concept of probiotics was first introduced by Elie Metchnikoff, a Russian Nobel Prize winner in 1907. He proposed that the beneficial microorganisms in the digestive system could contribute to overall health, particularly in digestive disease. The WHO has defined probiotics as "useful living microorganisms that have a positive effect on the health and physiology of a person when taken in sufficient quantities" (Akgül and Karakan 2018).

Several bacteria belonging to the genera *Pediococcus, Lactococcus, Enterococcus, Streptococcus, Propionibacterium,* and *Bacillus* are considered potential microbes for probiotic status (Hamad et al. 2022). The frequently used strains belong to the group of *Bifidobacterium* and *Lactobacillus* that significantly affect health with various actions. They detoxify xenobiotics and environmental pollutants (Reid and Sobel 2004), bio-transform mycotoxins in foods (Hamad et al. 2022), synthesize vitamin K, riboflavin, and folate (Hamad et al. 2022), and ferment undigested fiber in the colon (Warman et al. 2022). Probiotics prevent pathogenic bacteria by restricting binding sites on mucosal epithelial cells and modulating the host immune response, thus improving intestinal barrier integrity (Fusco et al. 2023).

Probiotics have emerged as a promising complementary or alternative therapy for UTIs, particularly in women, including pregnant women and children (Palani et al. 2021). Traditionally, the urinary tract was believed to be sterile, but recent studies have shown that the urinary microbiome consists of various bacterial genera, *including Streptococcus, Escherichia, Enterococcus*, and *Citrobacter* (Perez-Carrasco et al. 2021). In healthy women, the vaginal microbiome is predominantly composed of Lactobacillus species, making them key candidates for probiotic formulations intended to prevent and treat UTIs (Akgül and Karakan 2018).

They can be consumed either by incorporating them into foods or drinks in the form of dairy or nondairy foodstuffs or as supplements (Fenster et al. 2019). Various fermented foods have active microbes genetically similar to the strains utilized as probiotics. It has been observed that fermented foods enhance the functional and nutritional aspects by transforming substrates and producing bioactive and bioavailable end-products (Marco et al. 2017). The approximate consumption of 109 CFU/day has been revealed as an effective dose (Hill et al. 2014). By keeping in view, the effective dosage, probiotics are being incorporated into many foods like beverages, ice cream, yogurt, bread, and many others by the food industry.

They help in maintaining intestinal wall integrity and prevent the adhesion of pathogenic bacteria to epithelial cells. Some Lactobacillus strains, including *Lactobacillus brevis* and *Lactobacillus salivarius*, have demonstrated the ability to adhere to vaginal epithelial cells, preventing the colonization of harmful bacteria (Gupta et al. 2017). While most studies on probiotics for UTIs have focused on oral administration, which is believed to enable the probiotics to colonize the vagina through the rectum, perineal area, and vagina, some research has explored the efficacy of intravaginal delivery or bladder instillation (Morelli et al. 2004).

13.4.1 Probiotics in Human Health, with Specific Implications for UTI Prevention and Management

Probiotics are increasingly recognized for their broad impact on human health, particularly in the prevention and management of various infections, including UTIs. The beneficial effects of probiotics extend beyond the gut, influencing systemic health, including the urogenital tract.

i. **Systemic Benefits of Probiotics**: Probiotics contribute to overall human health by maintaining a balanced microbiome across multiple body systems, including the gastrointestinal, urogenital, and respiratory tracts. This systemic balance is crucial in preventing infections like UTIs, which can occur when the normal flora is disrupted. Probiotics help by enhancing the body's natural defenses across these systems.
ii. **Modulation of the Immune System**: The immune-modulating properties of probiotics are well-documented. By interacting with the gut-associated lymphoid tissue, probiotics influence systemic immunity, which has implications for preventing infections like UTIs. Enhanced immune function can lead to a reduced incidence of UTIs, especially in populations with weakened immune systems, such as the elderly.
iii. **Impact on Urogenital Health**: The use of probiotics specifically targeting urogenital health has shown promise in both prevention and treatment of UTIs. *Lactobacillus* species, which are naturally dominant in a healthy vaginal microbiome, have been extensively studied for their role in preventing pathogenic bacteria from colonizing the urinary tract. Their ability to produce antimicrobial substances and compete with pathogens for adhesion sites on uroepithelial cells plays a key role in maintaining urogenital health.
iv. **Reduction of Antibiotic Use**: Probiotics offer a significant benefit in reducing the need for antibiotics, which are commonly used to treat UTIs. The widespread use of antibiotics has led to the emergence of resistant strains of bacteria, a major public health concern. Probiotics, by preventing the onset of UTIs, reduce the frequency of antibiotic use, thereby contributing to the mitigation of antibiotic resistance.
v. **Safety and Efficacy in Diverse Populations**: Probiotics are generally considered safe and have been used effectively in diverse populations, including pregnant women, the elderly, and children. Their role in UTI prevention is particularly beneficial in these groups, where the risks of antibiotic treatment may outweigh the benefits. Clinical trials have demonstrated the efficacy of probiotics in reducing the incidence of UTIs across various demographic groups, highlighting their broad applicability (Reid et al. 2001; Schwenger et al., 2015).

13.4.2 Key Probiotic Strains Used in UTI Management

Some Lactobacillus strains, including *Lactobacillus brevis* and *Lactobacillus salivarius*, have demonstrated the ability to adhere to vaginal epithelial cells, preventing the colonization of harmful bacteria (Gupta et al. 2017).

13.4.3 *Lactobacillus rhamnosus* GR-1 and *Lactobacillus reuteri* RC-14

These strains are among the most extensively studied for UTI prevention. They work by adhering to the uroepithelial cells, preventing the attachment and colonization of uropathogens like *E. coli*. They also produce biosurfactants and bacteriocins that inhibit the growth of harmful bacteria. Several studies have shown that these strains can reduce the recurrence of UTIs, particularly in women with a history of recurrent infections. They have been used in oral and vaginal formulations. These are generally considered safe for use in healthy individuals, with minimal side effects reported.

13.4.4 *Lactobacillus crispatus*

This strain is known for its ability to produce lactic acid, lowering the pH of the vaginal environment and creating unfavorable conditions for the growth of uropathogens. It also produces hydrogen peroxide, which has bactericidal properties. *Lactobacillus crispatus* has been shown to be effective in restoring the normal vaginal flora after antibiotic treatment, thus reducing the risk of subsequent UTIs. Widely used and safe, especially in vaginal applications.

13.4.5 *Lactobacillus acidophilus*

This strain helps maintain the balance of the vaginal microbiota and inhibits the growth of pathogenic bacteria by producing lactic acid and other antimicrobial substances. It is commonly found in probiotic supplements and has been associated with a reduction in the incidence of UTIs, especially when used as a prophylactic measure. It is considered safe with a long history of use in dietary supplements and fermented foods.

13.4.6 *Lactobacillus casei*

It is known for its role in boosting the immune system and enhancing the body's natural defenses against infections. It can also inhibit the growth of uropathogens through competitive exclusion. While less studied than other strains, it has shown promise in preliminary research for reducing UTI recurrence. It is generally safe, with few reported adverse effects.

13.4.7 *Lactobacillus plantarum*

This strain exhibits strong antibacterial properties against a range of pathogens, including those that cause UTIs. It produces lactic acid and other metabolites that inhibit the growth of uropathogens. Some studies suggest that *Lactobacillus plantarum* can help in maintaining urogenital health, although more research is needed to confirm its efficacy in UTI prevention. It is well-tolerated with a good safety profile.

13.4.8 *Bifidobacterium longum*

It is not as prominent as Lactobacillus species in UTI management. It contributes to gut and vaginal health, indirectly supporting urogenital health by promoting a balanced microbiota.

Preliminary studies suggest a role in reducing UTI risk by improving overall microbial balance. It is commonly used in probiotics with a well-established safety profile.

The selection of probiotic strains for UTI management should be tailored to the individual, considering factors such as the specific strain's mechanism of action, clinical evidence, and safety profile. While Lactobacillus strains are the most studied and commonly used, ongoing research is essential to optimize probiotic therapies for UTI prevention and treatment (Qasemi et al. 2023).

13.4.9 Mechanisms of Action

Probiotic bacteria, such as *L. rhamnosus*, interact with dendritic cells, leading to their activation and the stimulation of IgA production by B cells. Additionally, certain strains like *Lactobacillus plantarum* produce teichoic acid, which exhibits anti-inflammatory properties by enhancing IL10 production (Braat et al. 2004). Probiotics also produce antimicrobial substances, such as bacteriocins, H_2O_2, and lactic acid, and broad-spectrum antibiotics like reuterin, which inhibit the growth of uropathogens like *E. coli*, a major cause of UTIs (Gupta et al. 2017).

Strains of *Lactobacillus* and *Bifidobacterium* exhibit multiple mechanisms of action that contribute to the prevention and treatment of UTIs:

- **Acidic Environment Creation**: *Lactobacillus* strains can thrive in environments with a $pH \leq 4.5$, where they produce organic acids, like lactic acid, that lower the pH and inhibit the growth of uropathogens like *E. coli*. This acidic environment is hostile to many harmful bacteria, preventing their colonization.
- **Production of Antibacterial Substances**: Probiotics produce antimicrobial compounds such as bacteriocins and hydrogen peroxide, which directly inhibit or kill pathogenic bacteria in the urinary tract.
- **Inhibition of Pathogen Adhesion**: Probiotics can produce biosurfactants that reduce the ability of uropathogens to adhere to the uroepithelium. Additionally, certain strains can co-aggregate with uropathogens, blocking their attachment to the urinary tract lining and displacing those already attached.
- **Competitive Exclusion**: By colonizing the urogenital region, probiotics outcompete harmful bacteria for adhesion sites and nutrients, effectively preventing the growth and spread of pathogens.
- **Modulation of Immune Response**: Probiotics can influence the host's immune system, enhancing its ability to combat infections. This includes reducing the expression of virulence factors by pathogens and stimulating the host's defenses.
- **Biofilm Formation Reduction**: Probiotics can reduce the formation of biofilms by uropathogens, which are protective layers that bacteria form to shield themselves from the host's immune response and antibiotics.
- **Suppression of Pathogen Growth**: Certain strains of *Lactobacillus* and *Bifidobacterium* have been shown to suppress the growth of urinary pathogens, as observed with *L. rhamnosus* and *Bifidobacterium* species, which inhibit *E. coli*.
- **Vaginal Probiotic Therapy**: The administration of probiotics as vaginal suppositories can help prevent UTIs by establishing a dominant population of beneficial bacteria, which restricts the ascension of uropathogens into the bladder.

These mechanisms collectively help in reducing the incidence and recurrence of UTIs, enhancing recovery after an infection, and providing an adjunct treatment option alongside conventional therapies (Fazly Bazzaz et al. 2021).

13.5 IMMUNOMODULATORY EFFECTS OF PROBIOTICS IN RELATION TO UTIS

Probiotics strains like *Lactobacillus acidophilus* and *Bifidobacterium infantis* exhibit immunomodulatory properties that can be beneficial in treating and preventing UTIs. These probiotics help modulate the immune response by:

Cytokine Regulation: Probiotics reduce the production of pro-inflammatory cytokines such as IL-1β and TNF, which can be elevated during UTIs. They also increase anti-inflammatory cytokines like IL-22, which help in reducing inflammation and supporting the integrity of the mucosal barriers.

Enhancing Mucosal Immunity: Probiotics enhance the production of IgA, a crucial antibody in mucosal immunity. This helps in preventing the adherence and colonization of pathogenic bacteria in the urinary tract.

Tight Junction Integrity: By strengthening the tight junctions in the epithelial cells, probiotics help in maintaining the barrier function of the urinary tract, preventing the translocation of bacteria from the gut or vagina to the urinary tract.

Th1/Th2 Balance: Probiotics can modulate the Th1/Th2 balance, reducing Th1-driven inflammation that is often associated with autoimmune and inflammatory conditions, which can exacerbate UTI symptoms.

These immunomodulatory effects make probiotics a promising adjunct therapy in the management of UTIs, particularly in reducing recurrence and supporting overall urinary tract health (Sadaqat, M. H. 2024).

13.6 PRODUCTION OF ANTIMICROBIAL SUBSTANCES

Probiotics play a significant role in the production of antimicrobial substances, particularly in the prevention of UTIs. These microorganisms, particularly strains like *Lactobacillus* and *Bifidobacterium*, produce a variety of antimicrobial substances that inhibit the growth of pathogenic bacteria, thereby reducing the recurrence of UTIs. Probiotics are known to influence immune responses by enhancing the activity of immune cells like macrophages, T cells, and natural killer cells. They also help in the production of anti-inflammatory cytokines and in reducing the production of pro-inflammatory cytokines, which can be crucial during infections like UTIs. By modulating the immune system, probiotics may help in balancing the body's response to pathogenic bacteria in the urinary tract, potentially reducing the severity of the infection and promoting recovery (Raheem et al. 2021).

Key antimicrobial substances produced by probiotics include:

Organic Acids: These include organic acids, such as lactic acid and acetic acid, which lower the pH of their environment, making it inhospitable for many pathogens.

Antimicrobial Peptides (AMPs): Probiotics secrete AMPs, which are small proteins that can directly kill or inhibit the growth of pathogenic bacteria.

Bacteriocins: These are proteinaceous toxins produced by bacteria to inhibit the growth of similar or closely related bacterial strains. Bacteriocins have a broad spectrum of action against pathogenic bacteria, including those involved in UTIs.

Hydrogen Peroxide: Some probiotic strains, particularly *Lactobacillus* species, produce hydrogen peroxide, which has strong antimicrobial properties.

These substances contribute to the probiotic's ability to prevent pathogens from adhering to the urinary tract's epithelial cells, thereby reducing the likelihood of infection and recurrence. The production of these substances is crucial in maintaining a healthy microbiota and preventing the colonization of harmful bacteria that cause UTIs (Akgül and Karakan 2018).

The health benefits of probiotics are associated with preventing and reducing many diseases, i.e., allergic diseases, cancer, hypercholesterolemia, lactose intolerance, inflammatory bowel disease, diarrhea, and irritable bowel syndrome (Grom et al. 2020).

13.7 CONCLUSION

UTIs represent a significant healthcare concern due to their prevalence, impact on quality of life, and the challenges posed by antibiotic resistance. Traditional treatment approaches, while effective, are increasingly limited by the emergence of resistant bacterial strains and the potential for adverse effects. As a result, there is a growing interest in exploring alternative therapeutic options that are both effective and sustainable.

Probiotics have emerged as a promising alternative or complementary therapy for UTIs. The rationale for using probiotics in UTI management is rooted in their beneficial effects on the immune system. The evidence suggests that certain probiotic strains, particularly Lactobacillus, may reduce the risk of recurrent UTIs by promoting a healthy urogenital flora and inhibiting the colonization of harmful pathogens. While the research is still evolving, the potential of probiotics offering a natural and effective means of managing UTIs is an exciting development in the field.

Future research should focus on large-scale, well-designed clinical trials to better understand the efficacy of probiotics in preventing and treating UTIs. There is also a need for studies that explore the mechanisms by which probiotics exert their beneficial effects, as well as the identification of specific strains that are most effective for UTI management. Additionally, research into the optimal dosage, duration of treatment, and the long-term safety of probiotic use in different populations will be crucial for the integration of probiotics into standard UTI management protocols.

REFERENCES

Ahmad S, Ali F, Qureshi SA, Uzma B, Shakeela Q, Sabir MS, Farooq M. The evaluation of antibiotic susceptibility pattern and associated risk factors of UTI in tertiary care hospital of Peshawar. *Pak J Pharm Sci.* 2022, 35(3), 897–903.

Akgül T, Karakan T. The role of probiotics in women with recurrent urinary tract infections. *Turk J Urol.* 2018, 44(5), 377–383.

Ala-Jaakkola R, Laitila A, Ouwehand AC, Lehtoranta L. Role of D-mannose in urinary tract infections – A narrative review. *Nutr J.* 2022 Mar 22, 21(1), 18.

Asadi Karam MR, Habibi M, Bouzari S. Urinary tract infection: Pathogenicity, antibiotic resistance, and development of effective vaccines against uropathogenic Escherichia coli. *Mol Immunol.* 2019, 108, 56–67.

Baker S, Thomson N, Weill FX, Holt KE. Genomic insights into the emergence and spread of antimicrobial-resistant bacterial pathogens. *Science.* 2018, May 18, 360(6390), 733–738.

Banerjee S. Recommendation and limitation of probiotics supplements. *Curr Trends Pharm Pharm Chem.* 2021, 3(3), 19–22.

Bettcher CM, Campbell E, Petty LA, et al. *Urinary Tract Infection.* Ann Arbor: Michigan Medicine University of Michigan, 2021 May.

Braat H, van den Brande J, van Tol E, Hommes D, Peppelenbosch M, van Deventer S. Lactobacillus rhamnosus induces peripheral hyporesponsiveness in stimulated CD4+ T cells via modulation of dendritic cell function. *Am J Clin Nutr.* 2004 Dec, 80(6), 1618–1625.

Daniel M, Szymanik-Grzelak H, Sierdziński J, Podsiadły E, Kowalewska-Młot M, Pańczyk-Tomaszewska M. Epidemiology and risk factors of UTIs in children—A single-center observation. *J Pers Med.* 2023, 13(1), 138.

David N, Malachy U, Clement A, Mushtak A, Chinedu I, Victor C, Morteza S. Antibiotic resistance: The challenges and some emerging strategies for tackling a global menace. *J Clin Lab Anal.* 2022, 36. doi:10.1002/jcla.24655.

Edwin D, Meena D, Deeksha NA. Antibiotic resistance. *Int J Health Sci Pharm (IJHSP).* 2017, 1(1), E1–E3.

Effah CY, Sun T, Liu S, Wu Y. Klebsiella pneumoniae: An increasing threat to public health. *Ann Clin Microbiol Antimicrob.* 2020 Jan 9, 19(1), 1.

Fazly Bazzaz BS, Darvishi Fork S, Ahmadi R, Khameneh B. Deep insights into urinary tract infections and effective natural remedies. *Afr J Urol*. 2021, 27, 1–13.

Fenster K, Freeburg B, Hollard C, Wong C, Rønhave Laursen R, Ouwehand AC. The production and delivery of probiotics: A review of a practical approach. *Microorganisms*. 2019 Mar 17, 7(3), 83.

Flores-Mireles A, Hreha TN, Hunstad DA. Pathophysiology, treatment, and prevention of catheter-associated urinary tract infection. *Top Spinal Cord Inj Rehabil*. 2019 Summer, 25(3), 228–240.

Flores-Mireles AL, Walker JN, Bauman TM, Potretzke AM, Schreiber HL, Park AM, Pinkner JS, Caparon MG, Hultgren SJ, Desai A. Fibrinogen release and deposition on urinary catheters placed during urological procedures. *J Urol*. 2016, 196, 416–421.

Fusco V, Wu R, Zhang W, Zhai Q. Editorial: Role of probiotics and probiotics' metabolites in food and intestine. *Front Microbiol*. 2023, 14, 1183550.

Gomila A, Carratala J, Eliakim-Raz N, et al. Risk factors and prognosis of complicated urinary tract infections caused by Pseudomonas aeruginosa in hospitalized patients: A retrospective multicenter cohort study. *Infect Drug Resist*. 2018, 11, 2571–2581.

Govindarajan DK, Kandaswamy K. Virulence factors of uropathogens and their role in host pathogen interactions. *The Cell Surface*. 2022, 8, 100075.

Grom LC, Coutinho NM, Guimarães JD, Balthazar CF, Silva R, Rocha RD, Freitas MQ, Duarte MC, Pimentel TC, Esmerino EA, Silva MC, Cruz AG. Probiotic dairy foods and postprandial glycemia: A mini-review. *Trends Food Sci Technol*. 2020, 101, 165–171.

Gupta V, Nag D, Garg P. Recurrent urinary tract infections in women: How promising is the use of probiotics? *Indian J Med Microbiol*. 2017 Jul–Sep, 35(3), 347–354.

Hamad G, Rabee O, Michael E, Taha M, Fify A, Reham A, Sabrien O, Jose ML, Sarah A.. Detection and inhibition of Clostridium botulinum in some Egyptian fish products by probiotics cell-free supernatants as bio-preservation agents. *LWT*. 2022, 163. 113603.

Hooton TM. Uncomplicated urinary tract infection. *N Engl J Med*. 2012, 366, 1028–1037.

Jacobsen SM, Shirtliff ME. Proteus mirabilis bioflims and catheter-associated urinary tract infections. *Virulence*. 2011, 2, 460–465.

Jansen AM, Lockatell V, Johnson DE, Mobley HL. Mannose-resistant Proteus-like fimbriae are produced by most Proteus mirabilis strains infecting the urinary tract, dictate the in vivo localization of bacteria, and contribute to biofilm formation. *Infect. Immun*. 2004, 72, 7294–7305.

Karam AMR, Bouzari SHM. Urinary tract infection: Pathogenicity, antibiotic resistance and development of effective vaccines against uropathogenic Escherichia coli. *Mol Immunol*. 2019 Apr, 108, 56–67.

Khalil MA, Alorabi JA, Al-Otaibi LM, Ali SS, Elsilk SE. Antibiotic resistance and biofilm formation in *Enterococcus* spp. isolated from urinary tract infections. *Pathogens*. 2022 Dec 25, 12(1), 34.

Khameneh B, Iranshahy M, Soheili V, Fazly Bazzaz BS. Review on plant antimicrobials: A mechanistic viewpoint. *Antimicrob Resist Infect Control*. 2019 Jul 16, 8, 118.

Klein RD, Hultgren SJ. Urinary tract infections: Microbial pathogenesis, host-pathogen interactions and new treatment strategies. *Nat. Rev. Microbiol*. 2020, 18, 211–226.

Komala M, Bhowmik D, Kumar KP. Urinary tract infection: Causes, symptoms, diagnosis and it's management. *J Chem Pharm Sci*. 2013, 6, 22–28.

Kontiokari T, Nuutinen M, Uhari M. Dietary factors affecting susceptibility to urinary tract infection. *Pediatr Nephrol*. 2004, 19, 378–383.

Lane DR, Takhar S. (2011). Diagnosis and management of urinary tract infection and pyelonephritis. *Emerg Med Clin North Am*. 29(3), 539–552.

Marco ML, Heeney D, Binda S, Cifelli CJ, Cotter PD, Foligné B, Gänzle M, Kort R, Pasin G, Pihlanto A, Smid EJ, Hutkin R. Health benefits of fermented foods: Microbiota and beyond. *Curr Opin Biotechnol*. 2017, 44, 94–102.

Morelli L, Zonenenschain D, Del Piano M, Cognein P. Utilization of the intestinal tract as a delivery system for urogenital probiotics. *J Clin Gastroenterol*. 2004 Jul, 38(6 Suppl), S107–S110.

Mydock-McGrane LK, Hannan TJ, Janetka JW. Rational design strategies for FimH antagonists: New drugs on the horizon for urinary tract infection and Crohn's disease. *Expert Opin Drug Discov*. 2017 Jul, 12(7), 711–731.

Öztürk R, Murt A. Epidemiology of urological infections: A global burden. *World J Urol*. 2020, 38, 2669–2679.

Palani Kumar MK, Halami PM, Serva Peddha M. Effect of *Lactobacillus fermentum* MCC2760-based probiotic curd on hypercholesterolemic C57BL6 mice. *ACS Omega*. 2021 Mar 12, 6(11), 7701–7710.

Perez-Carrasco V, Soriano-Lerma A, Soriano M, Gutiérrez-Fernández J, Garcia-Salcedo JA. Urinary microbiome: Yin and Yang of the urinary tract. *Front Cell Infect Microbiol*. 2021 May 18, 11, 617002.

Qasemi A, Lagzian M, Rahimi F, Majd FK, Bayat Z. The power of probiotics to combat urinary tract infections: A comprehensive review. *Res Biotechnol Environ Sci.*, 2023, 2(1), 1–11.

Qin S, Xiao W, Zhou C, Pu Q, Deng X, Lan L, Liang H, Song X, Wu M. Pseudomonasaeruginosa: Pathogenesis, virulence factors, antibiotic resistance, interaction with host, technology advances and emerging therapeutics. *Signal Transduct. Target. Ther.* 2022, 7, 199.

Raheem A, Liang L, Zhang G, Cui S. Modulatory effects of probiotics during pathogenic infections with emphasis on immune regulation. *Front Immunol.* 2021, 12, 616713.

Reid G. et al. Probiotics for urogenital health. *Nutr Res Rev.* 2001, 14(1), 27–34.

Reid G, Sobel JD. Probiotics in the treatment of urinary tract infections. *Clin Microbiol Rev.* 2004, 17(1), 132–144.

Sadaqat MH. Antibacterial, antiviral, antifungal, and immunomodulatory properties of probiotics. *Afghanistan J Basic Med Sci.* 2024, 1(2), 111–120.

Sarshar M, Behzadi P, Ambrosi C, Zagaglia C, Palamara AT, Scribano D. FimH and anti-adhesive therapeutics: A disarming strategy against uropathogens. *Antibiotics (Basel).* 2020 Jul 10, 9(7), 397.

Schroll C, Barken KB, Krogfelt KA, Struve C. Role of type 1 and type 3 fimbriae in Klebsiella pneumoniae biofilm formation. *BMC Microbiol.* 2010, 10, 179.

Schwenger EM, Tejani AM, Loewen PS. Probiotics for preventing urinary tract infections in adults and children. *Cochrane Database Syst Rev.* 2015 Dec 23;2015(12), CD008772. doi: 10.1002/14651858.CD008772.pub2. PMID: 26695595; PMCID: PMC8720415.

Sims M, Mariyanovski V, McLeroth P, Akers W, Lee YC, Brown ML, Du J, Pedley A, Kartsonis NA, Paschke A. Prospective, randomized, double-blind, Phase 2 dose-ranging study comparing efficacy and safety of imipenem/cilastatin plus relebactam with imipenem/cilastatin alone in patients with complicated urinary tract infections. *J Antimicrob Chemother.* 2017, 72, 2616–2626.

Singh KV, Nallapareddy SR, Murray BE. Importance of the ebp (endocarditis- and biofilm-associated pilus) locus in the pathogenesis of Enterococcus faecalis ascending urinary tract infection. *J Infect Dis.* 2007, 195, 1671–1677.

Wagenlehner FM, Bjerklund Johansen TE, Cai T, Koves B, Kranz J, Pilatz A, & Tandogdu Z. Epidemiology, definition and treatment of complicated urinary tract infections. *Nat Rev Urol.* 2020, 17(10), 586–600.

Walsh C, Collyns T. Pathophysiology of urinary tract infections. *Surgery (Oxford).* 2020, 38(4), 191–196.

Warman DJ, Jia H, Kato H. The potential roles of probiotics, resistant starch, and resistant proteins in ameliorating inflammation during aging (inflammaging). *Nutrients.* 2022 Feb 10, 14(4), 747.

Yang S, Qiao J, Zhang M, Kwok L, Matijašić B, Zhang H, Zhang W. Prevention and treatment of antibiotics-associated adverse effects through the use of probiotics: A review. *J Adv Res.* 2024, 4, S2090-1232(24)00230-3.

14 Oral Health and Antimicrobial Resistance

Probiotic Intervention for Dental Care

Darakshan Bhat, Samreen Maqbool, Shaista Hilal, Suhail Maqbool, and Nahida Tabassum
Department of Pharmaceutical Sciences, Pharmacology Division, University of Kashmir, Hazaratbal, Srinagar, Jammu and Kashmir, India

14.1 INTRODUCTION

Oral diseases are among the most common diseases around the world that people usually suffer from during their lifetime. Tooth decay is a multifactorial disease, and the composition of oral microbiota is a critical factor in its development. Also, *Streptococcus mutans* is considered as the most important caries-causing species. It is expected that probiotics, as they adjust the intestinal microbiota and reduce the number of pathogenic bacteria in the human intestine, can exert their health-giving effects, especially the anti-pathogenic effect, in the oral cavity, which is part of the human gastrointestinal tract (Rad et al. 2023).

More than 700 species of microorganisms inhabit the oral cavity (Palmer 2014). These include mostly prokaryotes and some eukaryotes. The total collection of microorganisms residing in the oral cavity has been often referred to as the oral microbiota, oral microflora or the oral microbiome (Dewhirst et al. 2010). These microorganisms nourish from saliva and the gingival crevicular fluid enabling the maintenance of oral ecosystems. The overall general health and dietary habits of the host greatly influence the microbiome in the oral cavity. The wide range of pH, nutrient availability, shedding and non-shedding surfaces, salivary and crevicular fluids are the influencing elements of the oral microenvironment. While most of these resident microorganisms appear to be harmless, there are some capable of oral infections, including caries (tooth decay), gingivitis and periodontitis (gum infections), and endodontic (root canal) infections. Deranging the oral bacterial homeostasis by any process of weakened immunity or any external factors, such as steroid therapy, can cause opportunistic infections, such as candidiasis, actinomycosis, and mucositis (Julkunen et al. 2017). Adverse diets can also change the composition of the oral microbiome leading to the onset of caries, erosion of the enamel and periodontal disease. Once infected, the treatment options are very painful, expensive, and time-consuming. Given the extent of the problem, oral diseases are a serious public health issue. They cause considerable loss of productivity on account of the pain, impairment of function, and reduced quality of life apart from financial burden to the host. Oral diseases are the fourth most expensive disease to treat in developed countries (Listl et al. 2015). Probiotics are emerging as a powerful prophylactic in oral care. Bacteriotherapy, in the form of probiotic bacteria with an inhibitory effect on oral pathogens, is a promising and cost-effective concept that has been developed recently (Çaglar et al. 2008).

DOI: 10.1201/9781032676951-14

14.2 ORAL DISEASES

Oral diseases are among the most common diseases found in humans. Despite their high social and economic burdens, these receive little attention in many countries and remain a neglected area of international health (FDI 2015). Oral health is defined as "a state of being free from mouth and facial pain, oral and throat cancer, oral infections and sores, periodontal (gum) disease, tooth decay, tooth loss and other diseases and disorders that limit an individual's capacity in biting, chewing, smiling, speaking and psychosocial wellbeing"(WHO 2012). The organisms mostly responsible for oral diseases are Gram-positive (*Streptococcus, Enterococcus, Micrococcus, Peptococcus, Peptosteptococcus, Lactobacillus, Corynebacterium, Actinomyces, Arachnia, Rothia, Eubacterium, Propioni bacterium, Bifidobacterium, Bacillus, and Clostridium*) and Gram-negative (*Nisseria, Veillonela, Campylobacter, Eikenella, Actinobacillus, Capnocytophaga, Heamophilus, Simonsiella, Bactertoides, Fusobacterium, Porphyromonas, Provotella, Laptotrichia,* and *Wolinella*), and *some eukaryotes* (*protozoa, yeasts, mycoplasma, candida,* and *spirochetes*) (Dewhirst et al. 2010).

14.3 COMMON DENTAL DISEASES AND CONDITIONS

Common dental problems include bad breath, periodontal (gum) disease, mouth sores, oral cancers, tooth decay, tooth wear, tooth loss, temporomandibular disorders, wisdom tooth issues, and unattractive smiles.

Oral cancer is the eighth most common cancer in the world. Prevalence of oral cancers is relatively higher in men, older people, and people of lower education and income levels, with tobacco and smoking as major risk factors. These can develop on the lip, tongue, oral mucosa, floor of mouth, base of tongue, and oropharynx.

Nearly all adults have tooth decay, and severe periodontal disease occurs in 15%–20% of middle-aged adults. Tooth decay is the destruction of tooth structure caused by acid-forming bacteria in the presence of sugars. Untreated caries in permanent teeth remains the most prevalent oral condition worldwide, affecting more than 2.4 billion people (Kassebaum et al. 2015). Furthermore, the burden of untreated caries has shifted from children to adults, as three peaks in prevalence have been observed: at ages 6, 25, and 70 years. Increasing occurrences of untreated caries have been attributed to population growth and longevity and a significant decrease in tooth loss. Signs and symptoms of tooth decay include (i) toothache, (ii) tooth sensitivity, (iii) mild to sharp pain when eating or drinking something sweet, hot, or cold, (iv) visible holes or pits in teeth, (v) brown, black, or white staining on any surface of a tooth, and (vi) pain when biting down. Periodontal disease is an inflammatory disease that affects the soft and hard tissues supporting the teeth. In the early stage (gingivitis), the gums become red and swollen due to bacterial inflammation. In the more advanced stages (periodontitis), bone loss and gum recession occur, and the teeth may become mobile and fall out. The signs and symptoms of periodontal disease include (i) bleeding gums when brushing and flossing, (ii) red, swollen, or tender gums, (iii) persistent bad breath, (iv) pus between teeth and gums, (v) shaky, mobile, or separating teeth, and (vi) changes in the way teeth fit when biting. Tooth decay and periodontal diseases are the foremost reasons for tooth loss and acute orofacial pain (Yap 2015).

The World Health Organization (WHO) estimates that about half of people suffer from oral disease. About 2.4 billion people have permanent tooth decay, and 532 million children are also effected by primary teeth decay (Kassebaum et al. 2017).

Some of the common oral health problems are as follows.

14.3.1 Caries

Dental caries, or cavities or tooth decay, are a localized breakdown of tooth enamel caused by bacterial activity in dental plague. This damage can progress to inflammation and pulp teeth, leading to symptoms like toothache and sensitivity. Globally, caries prevalence ranges from 49% to

83%. *Streptococcus mutans* is the key cariogenic bacterium due to its ability to produce a sticky substance called mutan, aiding its attachment in plague (Krzyściak et al. 2014). This mutan, from which the bacteria derive its name, facilitates bacterial establishment in the oral biofilm. These bioflims thrive in acidic environments, created by the metabolism of sugars like sucrose, fructose, and glucose. This acidity favors the growth of other acid-producing bacteria, such as *Actinomyces spp.*, contributing to enamel demineralization and erosion (Neel et al. 2016).

14.3.2 Plague

Dental plaque, a complex biofilm of microorganisms, is the primary cause of dental caries. It provides a favorable environment for these bacteria to thrive and protects them from both antimicrobial agents and the body's defenses. Persistent plague buildup leads to tartar formation, which can cause gum disease, tooth decay, and even tooth loss.

14.3.3 Gingivitis

Gingivitis is an inflammatory condition of the gums, triggered by the accumulation of bacterial plague and tartar. *Porphyromonas gingivalis*, a key pathogen, colonizes the gum line, invading epithelial cells and causing tissue damage. This invasion contributes to the development of chronic inflammatory diseases. GI (Gingival Index) and BOP (bleeding on probing) serve as indicators of inflammation and are used to assess the severity of gingivitis. Gingipain proteases, enzymes produced by these bacteria contribute to the virulence by disrupting the host's complement system, a crucial part of the immune response (Slaney et al. 2006).

14.3.4 Periodontitis

Periodontal diseases are prevalent conditions that cause inflammation and damage to both soft and hard tissues surrounding teeth. These diseases are triggered by components within the plague that forms on tooth surfaces. The infection can spread beyond the gums (gingivitis) to deeper structures, leading to the destruction of the periodontal ligament and bone. The WHO estimates that 15%–20% of middle-aged (35–44 years) adults experience tooth loss due to severe gum diseases. It has been reported that severe periodontitis can cause tooth mobility, potentially leading to tooth loss (Llambés et al. 2015). Periodontal bacteria can enter bloodstream, causing systemic issues like pregnancy complications, diabetes, and hospital-acquired pneumonia. Common bacteria associated with chronic perionditis include *P. gingivalis, Treponema denticola, Tannerella forsythia, Aggregatibacter actinomycetemcomitans, Bacteroides sp., Actinomyces sp., Staphylococcus intermedius, and Candida albicans*. A recent study by Arjunan et al. (2018) has highlighted the role of *P. gingivalis* in activating anti-apoptotic pathway thereby contributing to immune suppression and promoting the growth of cancer cells.

14.3.5 Halitosis

In the absence of oxygen, certain Gram-negative bacteria (*Fusobacterium nucleatum, P. gingivalis, Prevotella intermedia, and T. denticola*) breakdown food proteins, producing sulfur-containing compounds (hydrogen sulfide, methyl mercaptan, and dimethyl sulfide). These compounds contribute to both periodontal disease and bad breath (halitosis) (Persson et al. 1990). These volatile substances, originating from the back of the mouth or exhaled air, include sulfur-containing gases such as hydrogen sulfide, methanethiol, and dimethyl sulfide. They may also contain traces of other gases, such as indole, skatole, putrescine, cadaverine, and acetone. These compounds primarily cause bad breath, while an imbalance in the oral microbiome further contributes to unpleasant odor (Porter and Scully 2006). Bad breath can have significant social and psychological impacts on individuals.

14.3.6 *CANDIDA ALBICANS*

Candida species, including *C. albicans*, are commonly found in the human oral cavity. While typically harmless, candida can become pathogenic under certain conditions, leading to oral candidiasis. *C. albicans* is the primary cause of fungal infections among candida species. Factors, such as weakened immune systems due to chronic illnesses and the use of immunosuppressive medications (e.g., chemotherapy), have contributed to an increase in oral candidiasis cases in recent decades (Farah et al. 2010).

14.4 ANTIBIOTIC RESISTANCE AND ORAL HEALTH

Dentistry accounts for approximately 3%–11% of all antibiotic prescriptions worldwide. The human oral cavity is a huge reservoir of more than 750 microbial species. Studies suggest that the normal human flora may act as a reservoir for antibiotic-resistant genes, which, under the right conditions, can be transferred to pathogenic species. Oral microbial biofilm formation, such as dental plaque, which initiates periodontal diseases and dental caries, is the major reason for the antimicrobial therapy failure and rise in resistance against the action of antimicrobials (Kumari et al. 2020). Microbial biofilms are formed by colonization of various unicellular organisms to form a community that is attached to any solid surface and adherence of other microorganisms within a polymeric matrix made up of exo-polysaccharide thereby forming a complex community (Smith2005). These biofilms initiate changes within microorganisms thereby exhibiting characteristics different from their planktonic counterparts, their susceptibility to various antibiotic agents, that includes responses against immunological reactions, their interaction with the host tissues, or their physiological properties. Biofilm bacteria embedded with a matrix of extracellular polysaccharides are more than 10,000 times less sensitive to antimicrobials than free-floating planktonic bacteria. Biofilm creates a protected environment for the growth and survival of microbes in a hostile environment (Kanwar et al. 2017).

The oral or dental biofilm that colonizes on the surfaces of the tooth is a complex three-dimensional structure that is formed of around 100–300 cell layers and is a potential reservoir for the oral pathogenic microorganisms (Sbordone and Bortolaia2003). A number of Gram-negative bacteria involved in periodontal infections such as *P. gingivalis, A. actinomycetemcomitans, F. nucleatum, P. intermedia, B. forsythus, S. intermedius, and T. denticola*. These members are responsible for increased periodontal pocket formation as well as tooth loss (Chung et al. 2014). Biofilms formed by bacteria protect themselves from antimicrobial agents and host immune responses as; (i) it protects against oxidative stress, (ii) effluxes out antibiotics, (iii) provides protective measures by its polysaccharide matrix and stops easy diffusion of antibiotics (Kaur et al. 2015). The sequential events that occur during the formation of plaque on dental surfaces are as follows: (i) A pellicle is formed on the tooth surface from proteins and glycoproteins present in saliva. (ii) Bacterial molecules are adsorbed by the pellicle formed on the tooth surface through weak physical forces. (iii) Reversible addition of oral bacteria to the tooth surface through binding between specific molecules present on the surface of the microbial cell with the complementary receptor present in the formed pellicle over the tooth surface. These constitute the primary colonizers particularly *Streptococci* sps and *Actinomycetes* sps. (iv) Subsequently, other microorganisms colonize to form secondary colonizers by coadhesion with primary colonizers through cell surface adhesion to the receptors present in primary colonizing organisms. This creates a rise in the microbial diversity in the plaque biofilm. (v) Bacterial proliferation takes place by multiplication resulting in the rise of biomass and formation of biofilm matrix by synthesis of exo-polymers. (vi) Plaque growth and maturation takes place and all together they behave as a complex organism. There is growth and colonization of anaerobic organisms. Due to the release of exo-polymers by the colonized microbes, the penetration of the antibiotic agents is restricted into the 3-d structured plaque biofilm and also novel properties are exhibited by the slowly growing microorganisms on the biofilm surface (Melo and Perussi 2013).

14.4.1 Mechanism of Drug Resistance by Biofilms

Several groups of antimicrobial therapeutic agents (e.g., β lactams, metronidazole, tetracyclines, and quinolones) are practiced for the treatment of oral infections in addition to dentifrices and oral rinses (chlorhexidine, triclosan, and fluoride). In course of time, oral microbiota develop resistance following many different mechanisms such as the following:

- Production of hydrolytic enzymes to inactivate antibiotics (Smith 2005).
- Effluxing out the antibiotics taken up by the bacteria (Smith 2005).
- Bringing conformational changes in drug targets by causing mutation in genes responsible for (Melo and Perussi 2013).
- Differential expression of porin proteins in biofilm cells reduced the accession of antibiotics resulting in antibiotic resistance in Gram-negative bacteria (Kouidhi et al. 2015).
- The formation of biofilm is controlled by quorum sensing of molecules which coordinates the gene expression to regulate the virulence factor (Wu et al. 2015).
- Bacteria present in biofilm adapt themselves to changed physiological conditions in response to exposure to antimicrobial agents and reduce their growth rate.
- Formation of exopolysaccharide biofilm matrix by bacteria retards the permeability of antibiotics following different mechanisms, for example: electrostatic interaction, absorption, enzymatic neutralization, and hydrophobicity (Mah and Toole 2001).

14.5 PROBIOTICS

Probiotics were discovered by scholars as early as 1908, and since then the field of studying the health effects of probiotics on the host has gradually developed. In 2013, The International Scientific Association of Probiotics and Prebiotics (ISAPP) defined probiotics as "live microorganisms that, when administered in adequate amounts, confer a health benefit on the host". Probiotics are commonly used by humans to maintain their overall well-being. Although their effectiveness in promoting gastrointestinal health is well known, research has also shown that probiotics can be effective in preventing and treating various oral diseases, such as dental caries, oral mucositis, and halitosis (Mercenier et al. 2003).

In dentistry, probiotics were first introduced by Meurman and colleagues, who found that *Lacticaseibacillus rhamnosus* GG ATCC 53103 could colonize the human mouth. With further research, probiotics have been found to have a remarkable ability to prevent dental caries (Meurman et al. 1994).

14.5.1 Probiotic Bacteria for Oral Health

To be able to exert its properties in the oral cavity, it is essential for probiotic organism to resist the oral environmental conditions and defense mechanisms, to be able to adhere to saliva-coated surfaces, to colonize and grow in the mouth, and to inhibit oral pathogens. Generally, there is scarce evidence that probiotics permanently resides in the human body and in the mouth, in particular (Dave et al. 2013). Oral probiotics are those bacteria that are intended to work in the mouth to alter the oral biofilm to be more healthful. A basic prerequisite to be an oral probiotics is the ability to bond and inhabitant over the oral mucosal surfaces.

Some of the oral probiotic strains are as follows (Saraf et al. 2010):

- Lactic acid-producing bacteria (LAB): Lactobacillus, Bifidobacterium, and Streptococcus
- Non-lactic acid-producing bacterial species: Bacillus and propionibacterium
- Non-pathogenic yeasts: Saccharomyces
- Non-spore forming and non-flagellated rod or coccobacilli.

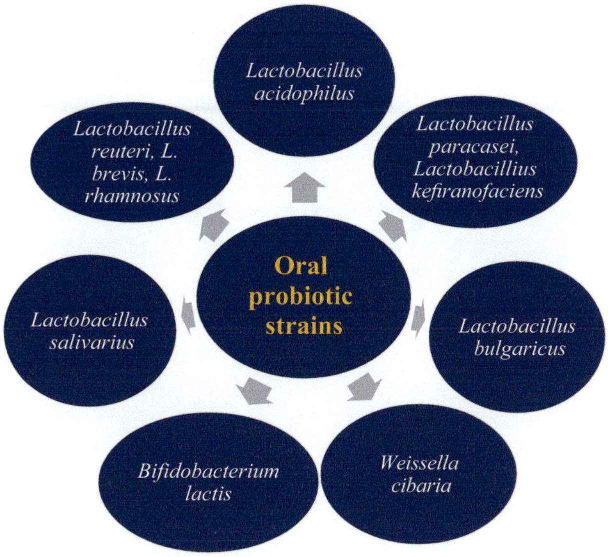

FIGURE 14.1 Various oral probiotic strains.

The most commonly used strains belong to the genera *Lactobacillus* and *Bifidobacterium* that are commonly found in the oral cavity (Stamatova and Meurman 2009). Researchers have proved that lactobacilli strains maintain oro microbiological balance (Ramachandran et al. 2013). The lactobacilli show good survival in saliva and *Weissella cibaria* adheres to epithelial cells. Evidence indicates that these probiotics consumed in food products can colonize the oral cavity as shown in Figure 14.1 (Bonifait et al. 2009).

14.5.2 Probiotic Sources

Probiotics are available to consumers mainly in the form of dietary supplements and foods. Yogurt and other fermented foods, soybeans, and asparagus are the main sources of naturally occurring probiotic bacteria in the diet and are considered as ideal vehicles. The fresher the yogurt, the more viable bacteria it contains. These dairy sources of probiotics are rich in calcium and thus may prevent demineralization of teeth. Other carriers for probiotics may include kombucha, miso, kefir, biodrink, etc. (Mallikarjuna et al. 2013).

14.5.3 Mechanism of Action

In oral health, the possible mechanisms of action, as illustrated in Figure 14.2, include the following:

14.5.4 Production of Active Metabolites

Bacteriocins first identified as a cationic antibacterial peptide produced by the ribosome, bacteriocins are classified into four classes, with most probiotics focusing on Classes I and II. Nisin, a Type A bacteriocin in Class I, is well known as a small positively charged protein (2–5 kDa) that causes target cells to form membrane pores, while Class II bacteriocins kill bacteria by increasing membrane permeability and leaking target bacterial contents. In addition to the aforementioned mechanisms, bacteriocins can also inhibit the synthesis of film and cell wall, exert the activities of DNase and RNase, and regulate microbiota (Luo et al. 2024).

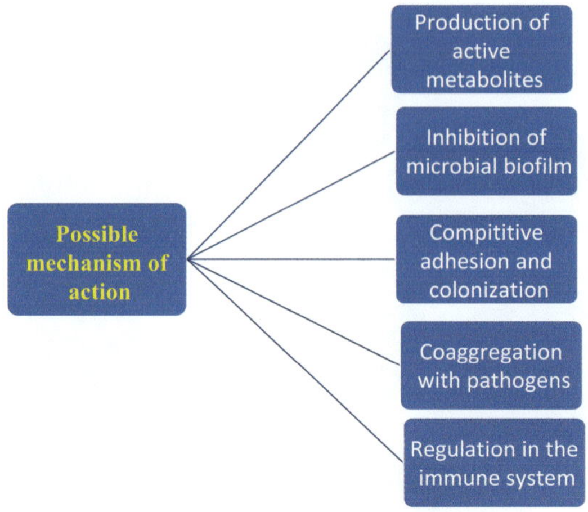

FIGURE 14.2 Possible mechanism of action of probiotics.

Proteins, carbohydrates, and lipids make up the majority of biosurfactants, which are amphiphilic compounds with both hydrophobic and hydrophilic groups that are created by microbial metabolism (Satpute et al. 2016). Because of their distinct anti-adhesion and antibacterial qualities, lactobacillus usually produces surfactin-type biosurfactants, which are protein-rich and have the ability to significantly inhibit pathogen adhesion (Satpute et al. 2016). For instance, BS produced by *L. reuteri* DSM17938, *L. acidophilus* DDS-1, *L. rhamnosus* ATCC 53103, and *L. paracasei* B21060 inhibited *S. mutans* and *Streptococcus oralis* adhesion and biofilm formation on titanium surfaces in a dose-dependent manner (Ciandrini et al. 2016).

It is well known that organic acids, including lactic acid and butyric acid, which are created by *Lactobacillus* in the human gastrointestinal system and other body regions, are healthy. To a certain extent, oral pathogenic microbes may be bacteriostatically affected by these organic acids. To have antibacterial properties, several probiotic species, such as *Bifadobacterium bifadum*, *Lactobacillus johnsonii*, *Lactobacillus crispatus*, and *Lactobacillus jensenii*, create hydrogen peroxide. The pathogenic bacteria die as a result of hydrogen peroxide's action on their epithelium.

14.5.5 Inhibition of Cariogenic Microbial Biofilm

Dental caries is frequently caused by biofilm. One important characteristic of probiotics is their capacity to prevent or stop the growth of biofilms and pathogenic microorganisms in the oral cavity. Some strains, such as *S. oralis 89a, Limosilactobacillus fermentum TCUESC01, L. acidophilus 4A, and Bifobacterium longum subsp. longum*, have frequently been reported to have anti-biofilm activity (Barzegari et al. 2020). For example, *Lacticaseibacillus casei ATCC 393, L. reuteri ATCC 23272, L. plantarum ATCC 14917,* and *Ligilactobacillus salivarius* may suppress the biofilms of *S. mutans* by down regulating genes such as gtf B, gtf C, and gtf D in *S. mutans* (Wasfi et al. 2018).

14.5.6 Competitive Adhesion and Colonization

One of the key characteristics of probiotics contributing to their health effects is their capacity for outcompeting oral pathogens concerning adhesion and colonization (Lin et al. 2018). For example, *L. reuteri* LR6 displayed the most substantial adhesion capabilities among eight tested probiotic strains, which corresponded to a higher ability for inhibiting the adherence of pathogens to Caco-2cells (Singh et al. 2017). Enhanced colonization efficacy by *S. salivarius* M18 results in stronger anti-caries activity as evidenced by a reduction in plaque scores and *S. mutans* levels (Burton et al. 2013). *Levilactobacillus brevis* KCCM 202399 inhibit the adherence of *S. mutans* KCTC5458 by reducing the self-aggregation, cell surface hydrophobicity, and EPS production of *S. mutans* (Ha Kim et al. 2022). Probiotics can reduce pathogen adhesion even without direct contact. Saliva treated with probiotics has shown to reduce the adhesion of *S. mutans* to hydroxyapatite surfaces (a model for enamel) (Haukioja et al. 2008). Further studies showed that the above salivary membrane lacked two proteins: salivary lectin gp340, the primary receptor for *S. mutans* in the salivary membrane, and salivary peroxidase, an innate defense factor found in human saliva (Tenovuo 1998).

14.5.7 Coaggregation with Pathogens

Coaggregation is among the favorable qualities of probiotics as it allows them to establish a barrier that impedes pathogen colonization. According to a study, *S. mutans* invitro116 showed particular coaggregation with 6 of the 624 lactic acid bacteria. These species were identified as *L. paracasei and L. rhamnosus*. This coaggregation process was found to be extremely resistant to both protease and heat, and it is not affected by saliva or lectins (Luo et al. 2024).

14.5.8 Regulation of the Immune System

Probiotics have the ability to directly affect pathogenic microbes or biofilm, but they can also activate and modify the host's immune system (Sliepen et al. 2009). According to clinical research, individuals with severe early childhood caries who consume *L. paracasei* SD1 daily or triweekly had significantly higher salivary levels of human neutrophil peptides 1–3 with wide bactericidal action and lower levels of *S. mutans*, which may halt the progression of caries (Wattanarat et al. 2021). Moreover, salivary immunoglobulin A levels have shown rise after 6 months of consuming milk containing *L. paracasei* SD1, and this rise is positively connected with a load of *L. paracasei* (Pahumunto et al. 2019).

14.6 PROBIOTICS IN ORAL HYGIENE

14.6.1 Probiotics in Dental Caries

Probiotics have shown promise in combating dental caries. Research suggests that certain strains, such as *Lactobacillius kefiranofaciens* DD2, from kefir can effectively inhibit oral pathogens like *S. mutans* and *Streptococcus sobrinus* (Jeong et al. 2018). This inhibition extends to bioflim formation by down regulating genes involved in carbohydrate metabolism, adhesion, and other regulatory mechanisms. Additionally, probiotics like *L. rhamnosus* GG exhibit strong adhesion properties and produce antimicrobial substances, including organic acids and bacteriocins, which contribute to inhibiting Streptococcus (Piqué et al. 2019). Other beneficial strains include *Streptococcus dentisani*, which has been isolated from caries-free individuals and demonstrates the ability to inhibit oral pathogens through bacteriocin production and pH buffering of oral cavity (López-López et al. 2017). *Lactococcus lactis* can form biofilms similar to dental plaque, inhibiting the growth of oral bacteria like *S. oralis, Veillonella dispar, Actinomyces naeslundii*, and *S. sobrinus* (Huang et al. 2011). In a study by Lin et al. (2017), four probiotic strains of Lactobacilli have demonstrated strong inhibitory effects on *S. mutans*, isolated from children with active caries, achieving inhibition rates of 70%–90%. These probiotic strains (*L. casei Shirota, L. casei LC01, L. plantarum STIII*, and *L. paracasei LPC37*) also significantly reduced the levels of *S. mutans, Streptococcus sanguinis*, and other bacteria in mixed biofilms compared to the control group.

14.6.2 Probiotics and Dental Plague

Probiotics can help prevent plague formation through several mechanisms. They neutralize free electrons, regulate the immune system, and control mucosal permeability. Research has shown that strains of *Streptococcus thermophilus and L. lactis* can attach to surfaces coated with saliva, similar to how plague forms on teeth (Comelli et al. 2002). *L. lactis* NCC2211 can create bioflims that resemble dental plague and can regulate the growth of *S. sobrinus*, a bacterium that contributes to tooth decay (Chug et al. 2020).

14.6.3 Probiotics and Gingivitis

Probiotic strains help in managing gingivitis by maintaining a balanced oral microbiome. Acidogenicbacteria, such as Lactobacilli, Streptococci, and Bifidobacterium, produce antimicrobial compounds that inhibit harmful pathogens. These mechanisms include coaggregation, production of toxic byproducts, and competition for substrates. Research has shown that nisin, a commercially available antimicrobial agent can effectively reduce plague buildup and gingivitis in dogs as compared to chlorhexidine and placebo treatments (Howell et al. 1993).

14.6.4 Probiotocs and Periodontitis

Probiotics can help manage periodontitis by maintaining a healthy balance of bacteria in the mouth. Many studies have shown that certain strains of *Lactobacillus* can suppress the growth of harmful bacteria associated with periodontal disease. Many latest studies showed application of beneficial bacteria, as an adjunct to scaling and root planning, can inhibit the recolonization of periodontopathogen and in overall pocket depth reduction and clinical attachment gain. Probiotics adhere to dental tissues as a part of the biofilm (or plaque) and compete with the growth of cariogenic bacteria or periodontal pathogens (Penala et al. 2016). A study by Allaker and Stephen (2017) suggests that probiotics could be useful in the improvement of oral health, mostly in subjects at a high risk of periodontal disease. *L. ruteri, L. brevis, L. rhamnosus, and L. salivarius* WB2I inhibit periodontitis-causing pathogens. *L. brevisis* is effective in curing chronic periodontitis, GI, and plaque index. They also inhibit the release of nitric oxide, consequently reducing the release of salivary prostaglandins (PGE2), thus diminishing the activation of salivary matrix metalloproteases (MMPs) (Kuka et al. 2019).

14.6.5 Probiotics and Halitosis

Reduction in halitosis was observed after gargling with rinse containing *W. cibaria*. It showed marked reduction in the levels of hydrogen sulfide (H_2S) and Methanethiol ($2CH_3SH$) by approximately 48.2% and 59.4%, respectively. *W. cibaria* secretes hydrogen peroxide and bacteriocins that act against Gram-positive bacteria and works in coaggregation with *Fusobacterium nucleatem* (Bonifait et al. 2009). It adheres to host epithelial cell wiping away the pathogenic bacteria.

14.6.6 Probiotics in *Candida albicans*

A reduction in the prevalence of *C. albicans* in the elderly after consumption of probiotic cheese-containing *L. rhamnosus* strains GG and LC705 and *Propionibacterium freudenreichii* ssp. shermanii JS has been demonstrated. Lactobacilli probiotics inhibits the growth of *C. albicans* possibly due to the low pH milieu produced by the lactobacilli. Candida-infected mice fed with *L. acidophilus* exhibited accelerated clearance of *C. albicans* from the mouth. Also *Lactobacillus fermentum* appears to be promising. Hasslöf et al. (2010) have found that two *L. plantarum* strains and *L. reuteri* ATCC 55730 displayed the strongest inhibition on *C. albicans*. *Bifidobacterium animalis* reduces the incidence and severity of mucosal candidiasis (Gupta and Garg 2009).

14.6.7 Probiotics and Orthodontics

The presence of fixed orthodontic appliances in the mouth can allow microorganisms to accumulate, causing enamel demineralization that produces white spot lesions. Studies have shown that short-term consumption of fruit yogurt-containing bifidobacteria alters the levels of salivary mutans streptotocci and lactobacilli in patients with fixed appliances. In one study, 200 g once daily fruit yogurt-containing *B. animalis* subsp lactis N 173010 significantly reduced salivary mutans streptococci counts in 2 weeks. No effect was produced on lactobacilli counts (Cildir et al.2009). A study showed that the consumption of probiotic curd (Active plus) and the topical application of probiotic toothpaste (GD) caused significant decreases in the *S. mutans* levels in the plaque around the brackets of orthodontic patients (Jose et al. 2013).

14.6.8 Probiotics and Voice Prosthesis

Turkish yogurt containing *S. thermophilus* and *Lactobacillus bulgaricus* effectively abolished the biofilm formation on indwelling voice prostheses. Patients in the Netherlands consuming buttermilk containing *Lactococcus cremoris, Lactococcus lactis spp.* which produces antimycotics and other substances prolonged the lifetime of indwelling voice prostheses. Lactobacilli have long been known for their capacity to interfere with the adhesion of uropathogens to epithelial cells and catheter materials, while *S. thermophilus* can effectively compete with yeasts in their adhesion to substratum surfaces, like silicone rubber (Chopra and Mathur 2013).

14.6.9 Xerostomia

Evidence suggests that probiotics can also efficiently lessen the risk xerostomia (Hatakka and Ahola 2007).

14.6.10 Cancer

Probiotics represent an emerging option for cancer prevention. Probiotics can interfere at various stages of cancer process such as interfering with chromosomal and DNA damage (Nagraj et al. 2012) They may also detoxify carcinogens such as β-glucuronidase, for example *L. Bulgaricus and S. thermophilus* in colon preventing colon cancers. In a recent study, probiotic *L. salivarius* REN isolated from centralians has showed highly potent anti-genotoxicity in rat oral cancer models in an initial assay. High dose of *L. salivarius* REN has effectively suppressed 4-nitroqunoline-1-oxide (4NQO)-induced oral carcinogenesis in initial and post-initial stage and the inhibition was in a dose-dependent manner. It has also decreased neoplasm incidence (65% to 0%). Probiotic acts by protecting DNA against oxidative damage and down regulating Cyclooxygenase (COX)-2 expressions. It has also significantly decreased the expression of proliferating cell nuclear antigen (PCNA), and induced apoptosis in a dose-dependent manner. The findings of this study suggest that probiotics may act as potential agents for oral cancer prevention (Zhang et al. 2013).

14.7 MISCELLANEOUS

Few strains of lactic acid bacteria, such as *S. thermophilus*, *Lactobacillus bulgaricus*, and other lactobacilli in fermented milk products, can alleviate symptoms of lactose intolerance by providing bacterial lactase to the intestine and stomach and favoring calcium absorption, thus minimizing osteoporosis. Studies have indicated a decrease in the numbers of lactobacilli, preceding the development of an allergy. Probiotics have been shown to reduce the incidence of childhood eczema by half. They may exert a beneficial effect on allergic reaction by improving mucosal barrier function.

In addition, probiotics consumption by young children may beneficially affect immune system development. Probiotics such as *Lactobacillus* GG may be helpful in alleviating some of the symptoms of food allergies such as those associated with milk protein (Singh et al. 2011).

14.8 CONCLUSION

The increasing prevalence of antibiotic-resistant bacteria poses a significant threat to global health, including oral health. Indiscriminate use of antibiotics has led to the emergence of multi-drug-resistant oral pathogens, compromising the effectiveness of conventional treatments. Probiotics, live microorganisms, offer promising alternative approach to combat oral diseases and reduce reliance on antibiotics. By restoring the balance of the oral microbiome, probiotics can inhibit pathogenic bacteria, reduce inflammation, and enhance overall oral health. While promising, further research is imperative to elucidate the optimal probiotic strains, dosages, and delivery methods for various oral conditions. A comprehensive approach, combining prudent antibiotic use, improved oral hygiene practices, and the strategic use of probiotics, is essential to mitigate the impact of antibiotic resistance and maintain oral health.

REFERENCES

Allaker, R. P., Stephen, A. S. (2017). Use of probiotics and oral health. *Current Oral Health Reports*, 4, 309–318. doi: 10.1007/s40496-017-0159-6. Epub 2017 Oct 19. PMID: 29201598; PMCID: PMC5688201.

Arjunan, P., Meghil, M. M., Pi, W., Xu, J., Lang, L., El-Awady, A., et al. (2018). Oral pathobiont activates anti-apoptotic pathway, promoting both immune suppression and oncogenic cell proliferation. *Scientific Reports*, 8(1), 16607.

Barzegari, A. et al.(2020). The battle of probiotics and their derivatives against biofilms. *Infection and Drug Resistance*, 13, 659–672.

Bonifait, I., Chandad, F., Grenier, D. (2009). Probiotics for oral health: Myth or reality?*Journal/Canadian Dental Association. Journal de l'Association Dentaire Canadienne*, 75, 585–590.

Burton, J. P. et al. (2013). Influence of the probiotic *Streptococcus salivarius* strain M18 on indices of dental health in children: A randomized double-blind, placebo-controlled trial. *Journal of Medical Microbiology*. 62, 875–884.

Çaglar, E., Kuscu, O. O., Kuvvetli, S. S., Cildir, S. K., Sandalli, N., Twetman, S. (2008). Short-term effect of ice-creamcontaining Bifidobacterium lactis Bb-12 on the number of salivary mutans streptococci and lactobacilli. *Acta Odontologica Scandinavica*, 66(3), 154–158.

Chopra, R., Mathur, S.(2013). Probiotics in dentistry: A boon or sham. *Dental Research Journal (Isfahan)*, 10, 302–306.

Chug, P., Dutt, R., Sharma, A., Bhagat, N., Dhar, M. S.(2020). A critical appraisal of the effects of probiotics on oral health. *Journal of Functional Foods*, 70, 103985.

Chung, J. H., Kim, Y. K., Kim, K. H., et al.(2014). Synthesis, characterization, biocompatibility of hydroxyapatite natural polymers nanocomposites for dentistry applications. *Artificial Cells, Nanomedicine, and Biotechnology*, 11, 18.

Ciandrini, E. et al.(2016). Characterization of biosurfactants produced by Lactobacillusspp.and their activity against oral streptococci biofilm. *Applied Microbiology and Biotechnology*, 100, 6767–6777.

Cildir, S. K, Germec, D., Sandalli, N., Ozdemir, F. I., Arun, T., Twetman, S.et al. (2009). Reduction of salivary mutans streptococci in orthodontic patients during daily consumption of yoghurt containing probiotic bacteria. *European Orthodontic Society*, 31, 407–411.

Comelli, E. M., Guggenheim, B., Stingele, F., Neeser, J. (2002). Selection of dairy bacterial strains as probiotics for oral health. *European Journal of Oral Sciences*, 110, 218–224.

Dave, H.D., Shah, C.S., Shah, M., Deshpande, N.(2013). Probiotics in periodontics. Good for bad: A review. *Research and Reviews: Journal of Dental Sciences*, 1, 7–12.

Dewhirst, F.E.,Chen, T.,Izard, J.,Paster, B.J.,Tanner, A.C.,Yu, W.H.,etal.(2010).Thehuman oral microbiome. *Journal of Bacteriology*, 192(19), 5002–5017.

Farah, C., Lynch, N., & McCullough, M. (2010). Oral fungal infections: An update for the general practitioner. *Australian Dental Journal*, 55, 48–54.

FDI World Dental Federation. 2015. The challenge of oral diseases, 2nd Ed. Brighton, UK: Myraid Editions.

GuptaV, Garg R. (2009). Probiotics. *Indian Journal of Medical Microbiology*, 27, 202–209. doi: 10.4103/0255-0857.53201. PMID: 19584499.

Ha Kim, J., Jang, H. J., Lee, N.-K., Paik, H.-D.(2022). Antibacterial and antibiofilm effect of cell-free supernatant of Lactobacillus brevis KCCM 202399 isolated from Korean fermented food against Streptococcus mutans KCTC 5458. *Journal of Microbiology and Biotechnology*, 32, 56–63.

Hasslöf, P., Hedberg, M., Twetman, S., Stecksén-Blicks, C. (2010). Growth inhibition of oral mutans streptococci and candida by commercial probiotic lactobacilli--an in vitro study. *BMC Oral Health*, 10, 18. doi: 10.1186/1472-6831-10-18. PMID: 20598145; PMCID: PMC2908555.

Hatakka, K., Ahola, A. J.(2007). Probiotics reduce the prevalence of oral candida in the elderly-a randomized control trial. *Journal of Dental Research*, 86, 125–130.

Haukioja, A., Loimaranta, V., Tenovuo, J.(2008). Probiotic bacteria affect the composition of salivary pellicle and streptococcal adhesion in vitro. *Oral Microbiology and Immunology*, 23, 336–343.

Howell, T. H., Fiorellini, J. P., Blackburn, P., Projan, S. J., Harpe, J.,Williams, R. C. (1993). The effect of a mouthrinse based on nisin, a bacteriocin, on developing plaque and gingivitis in beagle dogs. *Journal of Clinical Periodontology*, 20, 335–339.

Huang, R., Li, M., Gregory, R. L. (2011). Bacterial interactions in dental biofilm. *Virulence*, 2(5), 435–444.

Jeong, D., Kim, D. H., Song, K. Y., Seo, K. H. (2018). Antimicrobial and anti-biofilm activities of Lactobacillus kefiranofaciens DD2 against oral pathogens. *Journal of Oral Microbiology*, 10(1), 1472985.

Jose, E.J., Padmanabhan, S., Chitaranjan, B.A.(2013). Systemic consumption of probiotic curd and use of probiotic toothpaste to reduce Streptococcus mutans in plaque around orthodontic brackets. *American Journal of Orthodontics and Dentofacial Orthopedics*, 144, 67–72.

Julkunen, A., Heikkinen, A. M., Söder, B., Söder, P.Ö., Toppila-Salmi, S., Meurman, J. H. (2017). Autoimmune diseases and oral health: 30-year follow-up of a Swedish cohort. *Dentistry Journal*, 6(1), 1.

Kassebaum, N. J., Bernabé, E., Dahiya, M., Bhandari, B., Murray, C. J. L., Marcenes, W. (2015). Global burden of untreated caries: A systematic review and metaregression. *Journal of Dental Research,* 94(5), 650–658. doi: 10.1177/0022034515573272.

Kassebaum, N. J., Smith, A. G., Bernabé, E., Fleming, T. D., Reynolds, A. E., Vos, T., et al. (2017). Global, regional, and national prevalence, incidence, and disability-adjusted life years for oral conditions for 195 countries, 1990–2015: a systematic analysis for the global burden of diseases, injuries, and risk factors. *Journal of Dental Research*, 96(4), 380–387. doi: 10.1177/0022034517693566. PMID: 28792274; PMCID: PMC5912207.

Kanwar, I., Sah, A. K., Suresh, P. K. (2017). Biofilm-mediated antibiotic-resistant oral bacterial infections: Mechanism and combat strategies. *Current Pharmaceutical Design*, 23,2084–2095.

Kaur, N., Sahni, P., Singhvi, A., Hans, M. K., Ahluwalia, A. S. (2015). Screening the drug resistance property among aerobic pathogenic microorganisms of dental caries in north-western Indian population: A preliminary study.*Journal of Clinical and Diagnostic Research*, 9, Zc05–Zc08.

Kouidhi, B., Qurashi, Y. M., Chaieb, K.(2015). Drug resistance of bacterial dental biofilm and the potential use of natural compounds as alternative for prevention and treatment. *Microbial Pathogenesis*, 80, 3949.

Krzyściak, W., Jurczak, A., Kościelniak, D., Bystrowska, B., Skalniak, A. (2014). The virulence of Streptococcus mutans and the ability to form biofilms. *European Journal of Clinical Microbiology & Infectious Diseases: Official Publication of the European Society of Clinical Microbiology*, 33(4), 499–515.

Kuka, G. I., Gursoy, H., Alturfan, E. E., Ustundag, U. V., Kuru, B. (2019). Evaluation of nitric oxide levels in chronic periodontitis patients treated with initial periodontal therapy and probiotic food supplements: A double blind, randomized controlled clinical trial. *Biotechnology & Biotechnological Equipment*, 33(1), 974–979.

Kumari, K. S., Dash, P., Subudhi, E., Pattnaik, S., Singh, N. R., Jena, D., Mishra, L.(2020). Antimicrobial resistance: A dentists' prospective. *Indian Journal of Forensic Medicine & Toxicology*, 14(4), 8456–8462.

Lin, X., Chen, X. I., Tu, Y., Wang, S., Chen, H. (2017). Effect of probiotic lactobacilli on the growth of Streptococcus mutans and multispecies biofilms isolated from children with active caries. *Medical Science Monitor: International Medical Journal of Experimental and Clinical Research*, 23, 4175. doi: 10.12659/msm.902237. PMID: 28851857; PMCID: PMC5589056.

Lin, T.-H., Lin, C.-H., Pan, T.-M. (2018). Theimplicationofprobioticsinthe prevention of dental caries. *Applied Microbiology and Biotechnology*, 102, 577–586.

Listl, S., Galloway, J., Mossey, P. A., Marcenes, W. (2015). Global economic impact of dental diseases. *Journal of Dental Research*, 94(10), 1355–1361.

Llambés, F., Arias-Herrera, S., Caffesse, R. (2015). Relationship between diabetes and periodontal infection. *World Journal of Diabetes*, 6(7), 927–935.

López-López, A., Camelo-Castillo, A., Ferrer, M. D., Simon-Soro, Á., Mira, A. (2017). Health-associated niche inhabitants as oral probiotics: The case of streptococcus dentisani. *Frontiers in Microbiology*, 8, 379.

Luo, S., Wei, S., Luo, X., Yang, Q., Wong, K., Cheung, P. C. K., Zhang, B.(2024). How probiotics, prebiotics, synbiotics, and postbiotics prevent dental caries: An oral microbiota perspective. *NPJ Biofilms and Microbiomes*, 10, 14.

Mah, T. F. C., Toole, G. A.(2001). Mechanisms of biofilm resistance to antimicrobial agents. *Trends in Microbiology*, 9, 34–39.

Mallikarjuna, K., Gupta, S., Singh, S., Dadarya, B., Dausage, P., Gupta, P.(2013). Probiotics in dentistry: Review of the current status. *International Journal of Contemporary Dentistry*, 4, 66–75.

Melo, W. C. M. A., Perussi, J. R. (2013). Strategies to overcome biofilm resistance. In Méndez-Vilas A (Ed.) *Microbial Pathogens and Strategies for Combating Them: Science, Technology and Education*, pp. 179–187.Badajoz, Spain: Formatex Research Center.

Mercenier, A., Pavan, S., Pot, B. (2003). Probiotics as biotherapeutic agents: Present knowledge and future prospects. *Current Pharmaceutical Design*, 9, 175.

Meurman, J. H., Antila, H., Salminen, S. (1994). Recovery of Lactobacillus strain GG (ATCC 53103) from saliva of healthy volunteers after consumption of yoghurt prepared with the bacterium. *Microbial Ecology in Health and Disease*, 7, 295–298.

Nagraj, T., Ravi, B., Sankara, S. N., Madhu, K. (2012). Probiotics and oral health. *Journal ofIndian Academy of Oral Medicine and Radiology*, 24, 146–148.

Neel, E. A. A., Aljabo, A., Strange, A., Ibrahim, S., Coathup, M., Young, A. M., Bozec, L., Mudera, V. (2016). Demineralization–remineralization dynamics in teeth and bone. *International Journal of Nanomedicine*, 11, 4743–4763.

Pahumunto,N.,Sophatha,B.,Piwat,S., Teanpaisan,R.(2019). Increasing salivary IgA and reducing Streptococcus mutans by probiotic Lactobacillus paracasei SD1: A double-blind, randomized, controlled study. *Journal of Dental Sciences*, 14, 178–184.

Palmer, R. J. (2014). Composition and development of oral bacterial communities. *Periodontology*, 64(1), 20–39.

Penala, S., Kalakonda, B., Pathakota, K. R., Jayakumar, A., Koppolu, P., Lakshmi, B. V., et al. (2016). Efficacy of local use of probiotics as an adjunct to scaling and root planing in chronic periodontitis and halitosis: A randomized controlled trial. *Journal of Research in Pharmacy Practice*, 5(2), 86–93.

Persson, S., Edlund, M., Claesson, R., Carlsson, J. (1990). The formation of hydrogen sulfide and methyl mercaptan by oral bacteria. *Oral Microbiology and Immunology*, 5, 195–201.

Piqué, N., Berlanga, M., Miñana-Galbis, D. (2019). Health benefits of heat-killed (tyndallized) probiotics: An overview. *International Journal of Molecular Sciences*, 20(10), 2534.

Porter, S. R., Scully, C. (2006). Oral malodour (halitosis). *BMJ (Clinical Research Edition)*, 333(7569), 632–635.

Rad, A.H., Pourjafar, H., Mirzakhani, E. (2023). A comprehensive review of the application of probiotics and postbiotics in oral health. *Frontiers in Cellular and Infection Microbiology*, 13, 1120995.

Ramachandran, S., Vijaybala, S., Dhinesh Raj, K. S.(2013). Probiotics the promising future-A review. *The SoutheastAsian Journal of Case Report and Review*, 2, 98–105.

Saraf, K., Shashikanth, M. C., Priy, T., Sultana, N., Chaitanya, N. C.(2010). Probiotics--do they have a role in medicine and dentistry?*Journal of the Association of Physicians of India*, 58, 488–490.

Satpute, S. K. et al.(2016). Biosurfactant/s from lactobacilli species: Properties, challenges and potential biomedical applications. *Journal of Basic Microbiology*, 56, 1140–1158.

Sbordone, L., Bortolaia, C. (2003).Oral microbial biofilms and plaque related diseases: Microbial communities and their role in the shift from oral health to disease. *Clinical Oral Investigations*, 7, 181–188.

Singh, K., Kallali, B., Kumar, A., Thaker, V.(2011). Probiotics: A review. *Asian Pacific Journal of Tropical Biomedicine*, 21, S287–S290.

Singh, T. P., Kaur, G., Kapila, S., Malik, R. K. (2017). Antagonistic activity of Lactobacillus reuteri strains on the adhesion characteristics of selected pathogens. *Frontiers in Microbiology*, 8, 486.

Slaney, J. M., Gallagher, A., Aduse-Opoku, J., Pell, K., Curtis, MA. (2006). Mechanisms of resistance of Porphyromonas gingivalis to killing by serum complement. *Infection and Immunity*, 74(9), 5352–5361.

Sliepen, I. et al. (2009). Microbial interactions influence inflammatory host cell responses. *Journal of Dental Research*, 88, 1026–1030.

Smith, A. W. (2005). Biofilms and antibiotic therapy: Is there a role for combating bacterial resistance by the use of novel drug delivery systems. *Advanced Drug Delivery Reviews*, 57, 1539–1550.

Stamatova, I., Meurman, H. J. (2009). Probiotics: Health benefits in the mouth. *American Journal of Dentistry*, 22, 329–338.

Tenovuo, J. (1998). Antimicrobial function of human saliva-how important is it for oral health? *Acta Odontologica Scandinavica*, 56, 250–256.

Wasfi, R., Abd El-Rahman, O. A., Zafer, M. M., Ashour, H. M. (2018). Probiotic Lactobacillus sp. inhibit growth, biofilm formation and gene expression of caries-inducing Streptococcus mutans. *Journal of Cellular and Molecular Medicine*, 22, 1972–1983.

Wattanarat, O. et al. (2021). Significant elevation of salivary human neutrophil peptides 1-3 levels by probiotic milk in preschool children with severe early childhood caries: A randomized controlled trial. *Clinical Oral Investigations*, 25, 2891–2903.

World Health Organization. Oral health fact sheet. 2012. Available from: http://www.who.int/ mediacentre/factsheets/fs318/en/.

Wu, H., Moser, C., Wang, H. Z., et al.(2015). Strategies for combating bacterial biofilm infections. *International Journal of Oral Science*, 7, 1–7. doi: 10.1038/ijos.2014.65. PMID: 25504208; PMCID: PMC4817533.

Yap, T., McCullough M. (2015). Oral medicine and the ageing population. *Australian Dental Journal*, 60(S1), 44–53. doi: 10.1111/adj.12283.

Zhang, M., Wang, F., Jiang, L., Liu, R., Zhang, L., Lei, X. et al. (2013). Lactobacillus salivarius REN inhibit rat oral cancer induced by 4-nitroquioline 1-oxide. *Cancer Prevention Research (Philadelphia)*, 6, 686–694.

15 Clinical Trials and Evidence-Based Research on Probiotics and Antimicrobial Resistance

Agnieszka Rudzka[1], *Dorota Zielińska*[2], *Katarzyna Neffe-Skocińska*[2], *Barbara Sionek*[2], *Aleksandra Szydłowska*[2], *and Danuta Kołożyn Krajewska*[1,2]

[1]Faculty of Science and Technology, Department of Dietetics and Food Studies, Jan Dlugosz University in Czestochowa, Częstochowa, Poland

[2]Faculty of Human Nutrition, Department of Gastronomic Technology and Food Hygiene, Institute of Human Nutrition Sciences, Warsaw University of Life Sciences, Nowoursynowska Warsaw, Poland

15.1 INTRODUCTION

Many well-known microbial pathogens developed antimicrobial resistance (AMR) in response to antibiotic exposure or were inherently antibiotic-resistant. In cases where these pathogens cause the infection, probiotics may offer an alternative, or most often, an adjuvant treatment to antibiotics. Probiotics are thought to use multiple mechanisms to prevent or alleviate infections. They reduce the ability of pathogens to colonize the human intestine through competition as well as the production of antimicrobial substances. Further, probiotics improve the barrier properties of the intestinal epithelium and stimulate the immune system (Kainz et al., 2020; Ouwehand et al., 2016; Silva et al., 2020). Clinical trials have hence explored the efficacy of probiotics in eradication or prevention of colonization by AMR pathogens and treatment of the symptoms of diseases.

The major body of work that looked at the effect of probiotics on the status of AMR microorganisms in humans was focused on bacteria, although AMR is also associated with other pathogens such as fungi, viruses, and protozoa (World Health Organization, 2023) Illnesses caused or associated with AMR bacteria are responsible for millions of deaths per year worldwide (Antimicrobial Resistance Collaborators, 2022). A list of AMR bacteria that are regarded as the most threatening to human health was updated and published by the World Health Organization in May (2024). Out of these, pathogens that are a common cause of infections in immunocompromised and antibiotic-exposed hospital patients and non-nosocomial, but especially difficult to eradicate *Helicobacter pylori* were a subject of research to clinical trials that used probiotics as a test treatment against infections or for decolonization of AMR bacteria in humans (Rahman et al., 2024).

In this chapter, clinical trials and meta-analyses that used probiotics to treat or prevent infections and colonization with AMR bacterial pathogens were reviewed. First, this chapter introduces the potential for the use of probiotics in the alleviation of symptoms of diseases that may be caused by these pathogens, such as the infections of digestive, respiratory, and urinary tracts, as well as infections of wounds. Then, it reports the results of clinical trials that researched the effects of probiotics on the status of AMR bacteria associated with mentioned infections in humans.

DOI: 10.1201/9781032676951-15

15.2 INFECTIONS CAUSED BY AMR PATHOGENS AND THE EFFECTIVENESS OF PROBIOTICS IN THEIR TREATMENT

15.2.1 Infections of the Digestive System

Probiotics have been proven useful in the treatment of several ailments of the digestive system such as diarrhea, stomach and intestine infections, and many others. Among them, the anti-*H. pylori* activity has been broadly reported.

Helicobacter pylori is a bacterium, commonly residing in a human digestive tract of about a half of the global population (Li et al., 2023). Although most of the carriers do not experience any symptoms, the presence of this pathogen increases the risk of the development of gastrointestinal disorders such as chronic gastritis, peptic ulcer, and gastric adenocarcinoma. *H. pylori* is also involved in the development of other extra-gastric disorders such as mucosa-associated lymphoid tissue lymphoma, idiopathic thrombocytopenic purpura, deficiencies of iron and B12 vitamin (Malfertheiner et al., 2014).

The international medical guidelines strongly recommend treatment for *H. pylori* eradication as the triple therapy using a proton pump inhibitor or ranitidine bismuth citrate, combined with clarithromycin and amoxicillin or metronidazole (Lee et al., 2022). However, during the last decades, the eradication rates decreased from more than 90% (at the beginning of 1990), to less than 70% due to *H. pylori* resistance to key antibiotics, mainly clarithromycin, but also metronidazole and levofloxacin. Most of WHO regions had high pooled prevalence of primary and secondary resistance to metronidazole, clarithromycin, and levofloxacin (>15%). The most common antibiotic resistance in all regions was metronidazole. The prevalence of metronidazole resistance has been estimated to be from 38.9% to 75.8% for Europe and Africa, respectively (Aumpan et al., 2023).

A significant improvement in *H. pylori* eradication rates with probiotics supplementation when used alongside standard therapy was demonstrated in many systematic reviews and meta-analyses (Jiang et al., 2023; Wang et al., 2023; Z. Yang et al., 2024). The crucial benefits of probiotic therapy in *H. pylori* case are improved tolerability by preventing the occurrence of treatment and related side effects. This improvement is important in considering the prevailing issues with the standard triple or quadruple regimens, which confront with reduced effectiveness due to rising antibiotic resistance. Probiotics supplementation, by providing a means of modulating the gastrointestinal microbiota, may offer a new approach to combating such resistance. Furthermore, their capacity to alleviate common antibiotic-associated side effects, such as diarrhea, nausea, and vomiting, may improve patient adherence to therapy (Goderska et al., 2018).

Probiotics could defend against *H. pylori* via several mechanisms, including (1) increasing acidity, (2) production of antimicrobial agents (short-chain fatty acids and bacteriocins), (3) competition for adhesion, and (4) production of anti-inflammatory cytokines. Moreover, probiotics could play an important role in the eradication of mucosal pathogens, by increasing the expression of MUC2 and MUC3 genes and thus the secretion of mucin (3).

There is a scientific report published by Nature in 2024, an umbrella review of systematic reviews with meta-analyses that described the data included eligible randomized controlled trials with the aim of determining whether probiotic supplementation can improve *H. pylori* eradication and reduce side effects during therapy. Overall, 28 unique meta-analyses based on 534 RCTs were included. The results suggests that probiotics supplementation with pooled probiotic strains was significantly associated with improved eradication rates and reduced risk of total side effects compared with standard therapy alone. Similar results were obtained when single-strained or multi-strained preparation of probiotics supplementation were implemented. It was however claimed that *Bifidobacterium* spp. showed the highest potential for eradication (Z. Yang et al., 2024). Based on the data found in recent meta-analysis, probiotics cannot be considered as an alternative to anti-*H. pylori* treatment, but their use in association with standard therapy could increase eradication rates as well as decrease the adverse effects of infection.

15.2.2 Respiratory Infections

Klebsiella pneumoniae is a common hospital pathogen involved in severe respiratory infections such as pneumoniae. Pneumonia is a leading cause of morbidity and mortality among older people. The incidence of pneumonia is more common among institutionalized countries and increases rapidly with age. Moreover, pneumonia is the most common clinical cause of sepsis (Gonzalez-Ferrer et al., 2021). It has been estimated that approximately 3%–8% of all hospital-acquired bacterial infections are caused by *K. pneumoniae* (Y. Tang et al., 2024).

Klebsiella spp. cause more frequent infections compared to other gram-negative opportunistic pathogens, however the reasons why are unclear. Possibilities include the bacteria's ability to withstand starvation, naturally resist antibiotics, outcompete other bacteria, readily exchange DNA with other members of the human microbiome, and acquire mobile genetic elements encoding a wide range of antibiotic resistance and virulence-enhancing genes (Feizabadi et al., 2007). *Klebsiella* spp. can produce extended spectrum beta-lactamases (ESBL), enzymes which can hydrolyze penicillins as well as cephalosporins and monobactam. The majority of these enzymes are also resistant to other antibiotics such as fluroquinolones and aminoglycosides (Ramazanzadeh, 2010).

Klebsiella pneumoniae infections are being recognized as a major health threat due to the increasing AMR therefore limiting the efficiency of therapies. An alternative strategy to improve immune response could be the application of probiotics. Recently, the gut-lung axis concept is gaining ground, indicating that gut microbiome and its alteration (dysbiosis) may alter the immune responses to lung infection. (Mindt & DiGiandomenico, 2022).

Many studies suggested that probiotics can serve as preventive factors in respiratory tract infections. Some authors reported the prophylactic effect for ventilator-associated pneumonia, as well as delayed respiratory tract infection, when probiotics were used (Fan et al., 2019). A network meta-analysis concluded that probiotic strains of: *Bifidobacterium longum* + *Lactobacillus bulgaricus* + *Streptococcus thermophilus* have higher efficacy in decreasing the incidence of respiratory infections than other probiotics (Fan et al., 2019). Other systematic reviews and meta-analyses have demonstrated that probiotics played a positive role in preventing ventilator-associated pneumonia and are associated with a reduction in its incidence, as well as the shortened intensive care unit stay. However, no significant effects were observed on mortality and occurrence of diarrhea (Darbandi et al., 2023; Song et al., 2021; Zhao et al., 2021).

The recent umbrella systematic review and meta-analysis evaluated the quality of 30 meta-analyses and nine clinical outcomes (Anvarifard et al., 2024). It was concluded that probiotics significantly decreased ventilator-associated pneumonia incidence, nosocomial infections, intensive care unit length of stay, hospital length of stay, ICU mortality, hospital mortality, mechanical ventilation duration, duration of antibiotic use, and diarrhea. The obtained results indicate that probiotic administration could be considered an adjunct therapy for critically ill patients; however, no specific probiotic treatment regimen can be recommended due to the diverse probiotics used in the included meta-analyses.

15.2.3 Urinary Tract Infections, Included Female Diseases

Urinary tract infection (UTI) is one of the most common effects of bacterial infections requiring medical intervention and antibiotic therapy. The clinical course of UTI may be atypical and complicated. With the exception of infancy, UTI affect mainly females, with the highest incidence in the 14–35 age groups. The prevalence of UTIs increases with age (Medina & Castillo-Pino, 2019). In 30%–44% of women, after the first incident of this disease, a relapse usually occurs within 3–12 months. Due to various forms of UTI, about 15% of all available antibiotics are prescribed (Chu & Lowder, 2018).

The most common etiological cause of UTI are gram-negative intestinal bacilli from the *Enterobacteriaceae* family. Common uropathogenic strains include *Escherichia coli*, *Enterococcus faecalis*, and *Klebsiella* species. Infections with the above-mentioned bacterial species cause the following clinical symptoms: painful urination, dysuria, hematuria, urinary urgency, burning micturition, frequent urination, nausea, and vomiting (Shaheen et al., 2019). According to available scientific literature, uropathogenic *E. coli* (UPEC) is the dominant infectious agent of both UTI, uncomplicated and complicated. This bacterium when resides in the intestine is generally harmless, but once it reaches the opening of the vagina during washing or after sexual intercourse, then it may colonize the urinary bladder to cause infection (Shaheen et al., 2019). On the other hand, in complicated infections, more frequent pathogens are *Enterococcus* spp. and yeast *Candida* spp. (Medina & Castillo-Pino, 2019). The etiology of UTI and the main causes have not changed over the years and included comorbidities, impaired autonomic nervous system function, menopause, female sexual activity, or in older men, reduced antibacterial activity of prostate secretions (Shaheen et al., 2019).

The etiological cause of recurrent bladder infections is repeated antibiotic therapy, often with poorly selected drugs, which contributes to the acquisition of resistance mechanisms by uropathogenic bacterial strains and dysbiotic changes in the natural microbiota of the digestive and urogenital systems, especially in women.

Currently, despite the development of medicine, effective treatment of UTI using known drugs with bactericidal and bacteriostatic effects is difficult. Moreover, this phenomenon concerns not only hospital strains but also non-hospital ones. The reason for this situation is the frequency of infections and the acquisition of many mechanisms of resistance to antibiotics by pathogenic microorganisms. One of the most dangerous resistance mechanisms is the ability of many gram-negative Enterobacteriaceae rods to produce ESBLs. This causes resistance to all beta-lactam antibiotics except carbapenems. In individual cases, uropathogenic strains may be sensitive to third-generation cephalosporins and penicillins with beta-lactamase inhibitors. Additionally, in recent years, Gram-negative rods producing various carbapenemases (CP) have appeared and spread throughout the world, the most dangerous of which are *K. pneumoniae* CP, and New-Delhi metallocarbapenemase, also OXA-48 (oxacillinase). They determine resistance to all antibiotics from the beta-lactam group (penicillins, cephalosporins, monobactams, and carbapenems), and are not inhibited by beta-lactamase inhibitors (Khoshnood et al., 2017).

The condition of the vaginal microbiota directly correlates with infections of the female urinary tract. The vaginal ecosystem is the main component protecting against colonization of the vaginal epithelium by undesirable microorganisms, including atypical ones, and consequently protecting against the development of infections and inflammations. The qualitative and quantitative composition of the vaginal microbiota is not constant and is subject to fluctuations depending on a number of factors: past diseases (diabetes, anemia), medications used (antibiotic therapy, hormonal treatment), infectious factors (fungal, bacterial, viral, or parasitic infections). It also depends on age. In a sexually mature woman, the dominant microorganism in the vagina is *Lactobacillus* spp. This system ensures a balanced vaginal environment and acts as a barrier against pathogenic microorganisms. Due to their strong adherence properties, *Lactobacillus* spp. bacteria adhere tightly to the vaginal epithelium and cover its surface, making it difficult for pathogenic microorganisms to colonize it (Liu & Yi, 2022; Mei & Li, 2022). Their antagonistic effect results from competition for nutrients, the ability to produce bacteriocins, biofilm, or hydrogen peroxide. The results of the conducted studies suggest that hydrogen peroxide-producing strains may play a key role in maintaining homeostasis in the vagina, and thus prevent infections of the entire urogenital system in women. In addition to the factors mentioned above, the presence of *Lactobacillus* spp. is directly correlated with the level of female sex hormones (Boris & Barbés, 2000).

An alternative to antibiotic therapy is probiotic therapy, including dietary supplementation with appropriate strains of *Lactobacillus* spp. or the use of vaginal drugs containing selected probiotic bacteria with proven beneficial effects on the urogenital system of women. Probiotics play an important role in maintaining the stability of vaginal microenvironment, improving immune defense, and blocking growth of pathogenic microorganisms. Research so far has focused on *Lactobacillus* spp. in gynecological diseases such as human papilloma virus (HPV) infection, bacterial vaginosis (BV), Genitourinary Syndrome of Menopause (GSM), and prevention of female UTI (Mei & Li, 2022). The available systematic reviews showed that the new solutions for a non-antibiotic management of UTI, including the application of probiotics, cranberry products, the phytodrug Canephron N, non-steroidal anti-inflammatory drugs (NSAID), D-mannose, estrogens, vitamins, and immunotherapy (Wawrysiuk et al., 2019). An interesting solution is to design drugs or supplements that combine beneficial probiotic strains with a factor that supports the multiplication of *Latobacillus* spp. bacteria (e.g., the female hormone estradiol) or a factor that limits the ability of pathogenic bacteria to adhere to the walls of the urinary bladder (D-mannose sugar) or vaginal mucosa. Changes in the vaginal microbiota with reduction of protective *Lactobacillus* spp. colonies are associated with an increased risk of UTI. Probiotic strains indicated in the literature as having beneficial effects in the treatment or support of UTI treatment include *L. rhamnosus* GR-1, *L. reuteri* RC-14 (Beerepoot et al., 2012; Wawrysiuk et al., 2019), or *L. crispatus* strains (Stapleton et al., 2011; Wawrysiuk et al., 2019). The use of either oral or intravaginal probiotics to restore the natural vaginal microbiota seems to be a promising approach to reducing antibiotic consumption and to decreasing antimicrobial resistance (Wawrysiuk et al., 2019).

15.2.4 Infections of Wounds, Including Surgical and Chronically Infected Wounds

Wound healing is a complex process, in which there are many factors that delay treatment. The most effective wound management strategy is to prevent infections, promote healing, and prevent excess scarring. Surgical site infections are currently one of the most frequent types of nosocomial infections. On the other hand, chronic wounds require the most attention, as they are colonized by bacteria, most of which adversely affect the rate of their healing. Chronically infected wounds, such as venous or arterial ulcers, diabetic foot ulcers, pressure sores, and non-healing surgical wounds, are more difficult to heal. These types of skin lesions and their infections have a significant impact on the patients' quality of life and represent a significant cause of morbidity and mortality (Baker et al., 2016; Howard et al., 2004). The rate of bacterial colonization of a wound depends on their numbers, species diversity, microbial virulence, and the host immune response. *Staphylococcus aureus*, *Pseudomonas aeruginosa*, *Enterococcus faecalis*, *Acinetobacter baumannii*, *Escherichia coli*, *K. pneumoniae*, *Enterobacter* spp., *Peptostreptococcus* spp. are the most common causes of wound infections, especially due to the ability to form biofilms (Howard et al., 2004; Scalise et al., 2015). Bacteria that colonize chronic wounds form a biofilm that helps them to survive body's defenses and antibiotics.

Routine methods of treating infected wounds, whether postoperative or chronic, involve rinsing skin lesions with saline solutions, debridement of necrotic tissue, and then applying antiseptics and antibiotics to the wounds. This type of procedure does not always bring beneficial effects, because in addition to pathogenic microorganisms, beneficial microbiota is removed from the affected skin area. An important direction is the search for alternatives to effective wound treatment for chronic lesions, for example, diabetic wounds, where as a result of the disease, the physiological skin microbiota occurs in a state of dysbiosis. Conventionally used antibiotic therapy *per os* in the treatment of wounds additionally burdens the patient's digestive system (Nakamura & Daya, 2007). Patients on broad-spectrum antibiotics may develop "overgrowth" of pathogenic microorganisms leading to disease, such as *Candida albicans* vaginitis or oropharyngitis and *Clostridium difficile* pseudomembranous enterocolitis (Howard et al., 2004).

The beneficial effect of probiotics on the digestive system and the balance of intestinal microbiota is a well-known and widely described research area. Beneficial effects of probiotics have also been demonstrated in *per os* use of it in dental medicine, for women in urogenital infections, and in the respiratory tract. A new direction and very promising area of the use of probiotics is the prevention and treatment of skin diseases. Subsequently, studies on the antagonistic activity of probiotics against pathogens of infected wounds are also presented in the literature (Fijan et al., 2019).

Some populations are at increased risk of developing wounds that are difficult to heal due to the illness and adverse changes in the skin microbiome. Such is the case with diabetes. The composition of the skin microbiome of healthy and diabetic people differs. Studies show that blood glucose levels are correlated with the qualitative and quantitative composition of the skin microbiome of both animals and humans. Diabetics have more delicate skin, often in an inflamed state. It has been shown that an increase in blood glucose was associated with a 40-fold increase in the number of bacteria living on the skin, and they were much less diverse than in healthy animals. Comprehensive studies of the skin microbiome of diabetic mice showed an increased number of bacteria from the genera *Staphylococcus, Aerococcus* and *Weissela*. *Clostridium* spp. and *Streptococcus* spp. were missing (Lukic et al., 2017; Patel et al., 2022). Thus, understanding the physiology and composition of the multi-species bacterial biofilm and skin microbiome of a sick person could help designing solutions for the treatment of chronic wounds in predisposed populations. Solutions such as therapeutic ointments containing selected strains of missing, safe microorganisms could present a good solution in the treatment of chronic and postoperative wounds. Recent literature suggests that probiotics may also have a role to play in enhancing wound healing and ameliorating surgical infections (Lukic et al., 2017).

15.3 EFFICACY OF PROBIOTIC THERAPIES AGAINST COLONIZATION OF HUMANS WITH AMR BACTERIA

A major part of clinical trials on the effectiveness of colonization prevention or decolonization of AMR bacteria by means of probiotics, focused on *C. difficile*, multidrug-resistant (MDR) Enterobacterales including ESBL and CP producers, *H. pylori*, and vancomycin-resistant enterococci (Rahman et al., 2024). Infections caused by these bacterial pathogens, with exception of *H. pylori* are typical for hospital setting and besides gastrointestinal (GI) tract may inflict respiratory infections, UTIs and infections of wounds as discussed in previous sections of this chapter.

Clinical trials that assessed the efficacy of probiotics against particular bacterial AMR pathogens may be divided into four groups, trials that used probiotics:

1. as a sole treatment for eradication of AMR bacteria;
2. as a sole means to prevent colonization with AMR bacteria;
3. next to antibiotics for eradication of AMR bacteria;
4. for preventing colonization with AMR bacteria during antibiotic therapy.

The results of these trials along with drugs that studied microorganisms are typically resistant to are summarized in Table 15.1.

Several clinical trials looked at the possibility of eradicating AMR bacteria with probiotics only. Such studies were possible in cases where no active infection has yet developed but participants were colonized by AMR bacteria that are known opportunistic pathogens. These clinical trials included the vancomycin-resistant enterococci (Doron et al., 2015; Rubin et al., 2022), ESBL Enterobacteriaceae (Ljungquist et al., 2020), norfloxacin-resistant *E. coli* (Tannock et al., 2011), and methicillin-resistant *S. aureus* (Eggers et al., 2018; Warrack et al., 2013). Neither of the studies demonstrated that probiotics helped to eradicate mentioned bacteria in treated patients compared to respective control groups that received placebo or no treatment at all.

Not many studies used probiotics as a sole preventive measure against the colonization of GI tract with AMR microorganisms. In one such study, Dall (2019) looked at the effect of *L. rhamnosus* GG administered to Danish travelers who were visiting India on the status of ESBL and CP Enterobacteriaceae. High proportion of the participants in both control and test groups were colonized after visiting India with ESBL Enterobecteriaceae (83% and 81%, respectively) showing no effect from probiotics. The study failed to demonstrate any effects on CP Enterobacteriaceae status since these bacteria were not detected in any sample. In contrast, another study that used

TABLE 15.1
Results of Clinical Trials that Tested Probiotics against Specific Antimicrobial-Resistant Bacteria and Antibiotics that These Bacteria Are Commonly Resistant to

			Odds Ratio for Colonization with AMR Bacteria in People who Were Given Probiotics (Values below 1 Favor Probiotics)				
Microorganism	Commonly Resistant to	Strains Studied in Clinical Trials	As a Sole Treatment for Eradication	With Antibiotics for Eradication	To Prevent Colonization (No Antibiotic Treatment)	To Prevent Colonization during Antibiotic Therapy	Probiotic Strains Applied in Studies
Helicobacter pylori	Clarithromycin, Metronidazole, Levofloxacin, Tetracycline, Amoxicillin, Rifampicin (Fernández-Caso et al., 2023)			1.10[11], 0.72[12], 0.66[13], 0.57*[14] 0.47[15]			*S. boulardii*[12], non-viable *L. reuteri*[11] at least two of the following list: *S. boulardii, B. subtilis, S. faecium, E. faecium, L. acidophilus* LA-5, *L. plantarum, B. lactis* BB-12[13–15]
Clostridioides difficile	Ampicillin, amoxicillin, cephalosporins, clindamycin, fluoroquinolones and many other antibiotics but less commonly (Leffler & Lamont, 2015)			1.17[16], 0.29[17]	0.29*[21]	2.38[19], 0.31[23], 0.30*[24], 0.29[25], 0.10*[26]	Mixture of 4 strains[16], *L. plantarum* 299v[17,26], Spores of nontoxigenic *C. difficile*M3[21], *L. casei* sp. *paracasei* CNCM I-1518[25], *S. boulardii*[24], *L. rhamnosus* G[G19,23]
Enterobacterales	Penicillins, cephalosporins, Fluoroquinolones carbapenems, aminoglycosides[1] and tetracyclines[2]	MDR			0.84[22]	1.55[19]	*L. rhamnosus* GG[19,22]
		ESBL	0.36[5]			5.64[23]	Mixture of 8 strains[5], *L. rhamnosus* GG[23]
		Norfloxacin-resistant *E. coli*	2.82[6]				*E. coli* Nissle 1917[6]
		E. coli ESBL				2.84[27]	Mixture of lactobacilli and bifidobacteria[27]
		CP *K. pneumoniae*		0.10*[18]			Commercial probiotic no name or strains given[18]
		ESBL *K. pneumoniae*				2.04[27]	Mixture of lactobacilli and bifidobacteria[27]

(Continued)

TABLE 15.1 (*Continued*)
Results of Clinical Trials that Tested Probiotics against Specific Antimicrobial-Resistant Bacteria and Antibiotics that These Bacteria Are Commonly Resistant to

Microorganism	Commonly Resistant to	Strains Studied in Clinical Trials	Odds Ratio for Colonization with AMR Bacteria in People who Were Given Probiotics (Values below 1 Favor Probiotics)				Probiotic Strains Applied in Studies
			As a Sole Treatment for Eradication	With Antibiotics for Eradication	To Prevent Colonization (No Antibiotic Treatment)	To Prevent Colonization during Antibiotic Therapy	
Staphylococcus aureus	Methicillin, vancomycin[3]	MRSA	1.11[7], 0.92[8]				*L. rhamnosus* HN001[7,8]
Enterococcus	Vancomycin, ampicillin[4]	MDR				1.40[4]	Mixture of 8 strains[4]
		VRE	N/A[9], 0.94[10]			2.25[23], 0.63[19]. 0.19*[28], 0.01*[29]	*L. rhamnosus* G[G9,10,19,23,28,29]
Pesudomonas aeruginosa	Penicillins, carbapenems, aminoglycosides, fluoroquinolones[1]	MDR				0.91[23]	*L. rhamnosus* GG[23], mixture of lactobacilli and bifidobacteria[27]
		Not specified				0.41[27]	
		Ceftazidime-resistant				0.10[30]	*L. casei rhamnosus*[30]
Multiple strains of AMR bacteria	-			1.44[19], 1.33[20]		1.00[31], 0.75[19], 0.63[23], 0.51*[27]	*L. rhamnosus* GG[19,23], *L. paracasei* ssp. *paracasei* F19[31], More than 1 probiotic of genus *Lactobacillus* or *Bifidobacterium*[20,27]

*, Statistical significance of the effect showed in the study, AMR, antimicrobial resistance; MDR, multidrug resistant; ESBL, extended spectrum beta-lactamases producing; CP, carbapenemase producing; VRE, vancomycin resistant; MRSA, methicillin-resistant *S. aureus*, numbers in superscripts denote references in a following order: [1]Iredell et al. (2016), [2]Robilotti and Deresinski (2014), [3]Rungelrath and DeLeo (2021), [4]de Regt et al. (2010), [5]Ljungquist et al. (2020), [6]Tannock et al. (2011), [7]Warrack et al. (2013), [8]Eggers et al. (2018), [9]Doron et al. (2015), [10]Rubin et al. (2022), [11]Yang et al. (2021), [12]He et al. (2022), [13]Tang et al. (2021), [14]Viazis et al. (2022), [15]Oh et al. (2016), [16]Barker et al. (2017), [17]Wullt et al. (2003), [18]Nouvenne et al. (2015), [19]Rauseo et al. (2022), [20]Salomão et al. (2016), [21]Gerding et al. (2015), [22]Dall et al. (2019), [23]Kwon et al. (2015), [24]McFarland et al. (1994), [25]Alberda et al. (2018), [26]Klarin et al. (2008), [27]Wieërs et al. (2020), [28]Szachta et al. (2011), [29]Manley et al. (2007), [30]Forestier et al. (2008), [31]Sullivan et al. (2004).

only probiotics against AMR microorganism colonization showed that spores of nontoxigenic *C. difficile* M3 prevented the recurrence of *C. difficile* infection in a significant number of hospitalized patients (Gerding et al., 2015). This effect could be partly attributed to the used probiotic strain and the nature of infection. Given that *C. difficile* infection is preceded by the decline of the abundance of Clostridiales in the human intestinal microbiota (Vincent et al., 2013), administering probiotic from this family could hamper the colonization of toxigenic *C. difficile*. The validity of this hypothesis was confirmed in several studies (Gerding et al., 2015).

Many clinical trials looked at employing probiotics in support of antibiotics for the eradication of AMR bacteria. Most research in this group was devoted to infection caused by *H. pylori* (see Table 15.1). Only one study with a high number of participants (664) showed a significant effect of multistrain probiotic (*L. acidophilus* LA-5, *L. plantarum*, *B. lactis* BB-12, and *S. boulardii*) in enhancing the eradication of *H. pylori* among quadruple non-bismuth therapy-treated patients (Viazis et al., 2022). Other studies on *H. pylori* were underpowered (He et al., 2022; Oh et al., 2016; B. Tang et al., 2021) or did not show any effect probably due to the use of non-viable probiotics (*L. reuteri* DSM17648) (C. Yang et al., 2021). In this group of clinical trials, no significant effect was achieved for *C. difficile* (Barker et al., 2017; Wullt et al., 2003) and where authors looked at the eradication of multiple AMR bacteria (Rauseo et al., 2022; Salomão et al., 2016). However, Nouvenne et al. (2015) demonstrated very good efficacy of applying a commercial probiotic against CP *K. pneumoniae* (OR=0.10).

Most of the studies where probiotic efficacy was tested against AMR bacteria aimed to prevent colonization in antibiotic-treated patients, whose illness was not directly attributed to the resistome. Supplementation of these patients with *L. plantarum* 299v and *S. boulardii* decreased significantly the odds of contracting *C. difficile* (OR=0.10 and 0.30, respectively) (Klarin et al., 2008; McFarland et al., 1994), whereas *L. rhamnosus* GG protected patients from colonization by vancomycin-resistant *Enterococcus* (Manley et al., 2007; Szachta et al., 2011) (OR=0.19 and 0.01 in two independent studies). Furthermore, the mixture of probiotics containing *L. acidophilus* NCFM, *L. paracasei* Lpc-37, *B. lactis* Bl-04, and *B. lactis* Bi-07 was effective against a range of AMR bacteria (OR=0.51, including ESBL and AmpC β-lactamase producers, as well as non-fermenting bacteria, including *P. aeruginosa* and other members of gensu *Pseudomonas*) (Wieërs et al., 2020). On the other hand, a number of studies showed that probiotis were not effective against colonization of antibiotic-treated patients with AMR microorganisms (see Table 15.1).

Overall, although meta-analysis of available data reported that including probiotics in the range of mentioned treatments altogether gives significantly lower odds of colonization with AMR bacteria (OR=0.59 at 95% confidence interval of 0.43–0.81) (Rahman et al., 2024), differences in the effects from various probiotic strains against a variety of pathogens reported by a range of available studies are very prominent. Therefore, further research that would allow the optimization of probiotic treatments against AMR pathogens is required.

15.4 CONCLUSIONS

Clinical trials indicate that probiotics may provide protection against diseases caused by AMR pathogens either by the prevention of contracting or alleviation of symptoms of the infections. Undoubtedly, one major mechanism that supports such action of probiotics is their apparent ability to influence the resistome of a human. Available clinical trials reported very variable results as to the effectiveness of probiotics in the prevention of colonization or eradication of AMR bacteria. The research in this area should be continued and include more studies with a number of participants that would provide a sufficient statistical power.

Overall, evidence gathered by clinical trials supports the use of probiotics against AMR infections. Further research is needed to optimize the effects achieved especially in relation to most effective probiotic strains and their combinations.

REFERENCES

Alberda, C., Marcushamer, S., Hewer, T., Journault, N., & Kutsogiannis, D. (2018). Feasibility of a Lactobacillus casei drink in the intensive care unit for prevention of antibiotic associated diarrhea and Clostridium difficile. *Nutrients*, *10*(5), 539. https://doi.org/10.3390/nu10050539

Antimicrobial Resistance Collaborators. (2022). Global burden of bacterial antimicrobial resistance in 2019: A systematic analysis. *The Lancet*, *399*(10325), 629–655. https://doi.org/10.1016/S0140-6736(21)02724-0

Anvarifard, P., Anbari, M., Ghalichi, F., Ghoreishi, Z., & Zarezadeh, M. (2024). The effectiveness of probiotics as an adjunct therapy in patients under mechanical ventilation: An umbrella systematic review and meta-analysis. *Food & Function*, *15*(11), 5737–5751. https://doi.org/10.1039/d3fo04653b

Aumpan, N., Mahachai, V., & Vilaichone, R.-K. (2023). Management of Helicobacter pylori infection. *JGH Open: An Open Access Journal of Gastroenterology and Hepatology*, *7*(1), 3–15. https://doi.org/10.1002/jgh3.12843

Baker, A. W., Dicks, K. V., Durkin, M. J., Weber, D. J., Lewis, S. S., Moehring, R. W., Chen, L. F., Sexton, D. J., & Anderson, D. J. (2016). Epidemiology of surgical site infection in a community hospital network. *Infection Control and Hospital Epidemiology*, *37*(5), 519–526. https://doi.org/10.1017/ice.2016.13

Barker, A. K., Duster, M., Valentine, S., Hess, T., Archbald-Pannone, L., Guerrant, R., & Safdar, N. (2017). A randomized controlled trial of probiotics for Clostridium difficile infection in adults (PICO). *Journal of Antimicrobial Chemotherapy*, *72*(11), 3177–3180. https://doi.org/10.1093/jac/dkx254

Beerepoot, M. A. J., ter Riet, G., Nys, S., van der Wal, W. M., de Borgie, C. A. J. M., de Reijke, T. M., Prins, J. M., Koeijers, J., Verbon, A., Stobberingh, E., & Geerlings, S. E. (2012). Lactobacilli vs antibiotics to prevent urinary tract infections: A randomized, double-blind, noninferiority trial in postmenopausal women. *Archives of Internal Medicine*, *172*(9), 704–712. https://doi.org/10.1001/archinternmed.2012.777

Boris, S., & Barbés, C. (2000). Role played by lactobacilli in controlling the population of vaginal pathogens. *Microbes and Infection*, *2*(5), 543–546. https://doi.org/10.1016/s1286-4579(00)00313-0

Chu, C. M., & Lowder, J. L. (2018). Diagnosis and treatment of urinary tract infections across age groups. *American Journal of Obstetrics and Gynecology*, *219*(1), 40–51. https://doi.org/10.1016/j.ajog.2017.12.231

Dall, L. B., Lausch, K. R., Gedebjerg, A., Fuursted, K., Storgaard, M., & Larsen, C. S. (2019). Do probiotics prevent colonization with multi-resistant Enterobacteriaceae during travel? A randomized controlled trial. *Travel Medicine and Infectious Disease*, *27*, 81–86. https://doi.org/10.1016/j.tmaid.2018.11.013

Darbandi, A., Banar, M., Koupaei, M., Afifirad, R., Asadollahi, P., Bafandeh, E., Rasooli, I., Emamie, A., Navidifar, T., & Owlia, P. (2023). Clinical efficacy of probiotics in prevention of infectious diseases among hospitalized patients in ICU and non-ICU wards in clinical randomized trials: A systematic review. *Health Science Reports*, *6*(8), e1469. https://doi.org/10.1002/hsr2.1469

de Regt, M. J. A., Willems, R. J. L., Hené, R. J., Siersema, P. D., Verhaar, H. J. J., Hopmans, T. E. M., & Bonten, M. J. M. (2010). Effects of probiotics on acquisition and spread of multiresistant Enterococci. *Antimicrobial Agents and Chemotherapy*, *54*(7), 2801–2805. https://doi.org/10.1128/aac.01765-09

Doron, S., Hibberd, P. L., Goldin, B., Thorpe, C., McDermott, L., & Snydman, D. R. (2015). Effect of Lactobacillus rhamnosus GG administration on vancomycin-resistant enterococcus colonization in adults with comorbidities. *Antimicrobial Agents and Chemotherapy*, *59*(8), 4593–4599. https://doi.org/10.1128/AAC.00300-15

Eggers, S., Barker, A. K., Valentine, S., Hess, T., Duster, M., & Safdar, N. (2018). Effect of Lactobacillus rhamnosus HN001 on carriage of Staphylococcus aureus: Results of the impact of probiotics for reducing infections in veterans (IMPROVE) study. *BMC Infectious Diseases*, *18*(1), 129. https://doi.org/10.1186/s12879-018-3028-6

Fan, Q.-L., Yu, X.-M., Liu, Q.-X., Yang, W., Chang, Q., & Zhang, Y.-P. (2019). Synbiotics for prevention of ventilator-associated pneumonia: A probiotics strain-specific network meta-analysis. *The Journal of International Medical Research*, *47*(11), 5349–5374. https://doi.org/10.1177/0300060519876753

Feizabadi, M. M., Etemadi, G., Rahmati, M., Mohammadi-Yeganeh, S., Shabanpoor, S., & Asadi, S. (2007). Antibiotic resistance patterns and genetic analysis of Klebsiella pneumoniae isolates from the respiratory tract. *TANAFFOS (Respiration)*, *6*(3(summer)), 20–25.

Fernández-Caso, B., Miqueleiz, A., & Alarcón, T. (2023). Whole genome sequencing for studying Helicobacter pylori antimicrobial resistance. *Antibiotics*, *12*(7), 7. https://doi.org/10.3390/antibiotics12071135

Fijan, S., Frauwallner, A., Langerholc, T., Krebs, B., Ter Haar Née Younes, J. A., Heschl, A., Mičetić Turk, D., & Rogelj, I. (2019). Efficacy of using probiotics with antagonistic activity against pathogens of wound infections: An Integrative Review of Literature. *BioMed Research International*, *2019*, 7585486. https://doi.org/10.1155/2019/7585486

Forestier, C., Guelon, D., Cluytens, V., Gillart, T., Sirot, J., & De Champs, C. (2008). Oral probiotic and prevention of Pseudomonas aeruginosa infections: A randomized, double-blind, placebo-controlled pilot study in intensive care unit patients. *Critical Care (London, England)*, *12*(3), R69. https://doi.org/10.1186/cc6907

Gerding, D. N., Meyer, T., Lee, C., Cohen, S. H., Murthy, U. K., Poirier, A., Van Schooneveld, T. C., Pardi, D. S., Ramos, A., Barron, M. A., Chen, H., & Villano, S. (2015). Administration of spores of nontoxigenic Clostridium difficile strain M3 for prevention of recurrent C. difficile infection: A randomized clinical trial. *JAMA*, *313*(17), 1719–1727. https://doi.org/10.1001/jama.2015.3725

Goderska, K., Agudo Pena, S., & Alarcon, T. (2018). Helicobacter pylori treatment: Antibiotics or probiotics. *Applied Microbiology and Biotechnology*, *102*(1), 1–7. https://doi.org/10.1007/s00253-017-8535-7

Gonzalez-Ferrer, S., Peñaloza, H. F., Budnick, J. A., Bain, W. G., Nordstrom, H. R., Lee, J. S., & Van Tyne, D. (2021). Finding order in the chaos: Outstanding questions in Klebsiella pneumoniae pathogenesis. *Infection and Immunity*, *89*(4), e00693–20. https://doi.org/10.1128/iai.00693-20

He, C., Kong, F., Chai, X., Zou, C., Zhu, X., & Zhao, D. (2022). Effect of probiotic-assisted eradication of cagA+/vacA s1m1 Helicobacter pylori on intestinal flora. *BioMed Research International*, *2022*, 8607671. https://doi.org/10.1155/2022/8607671

Howard, J. C., Reid, G., & Gan, B. S. (2004). Probiotics in surgical wound infections: Current status. *Clinical and Investigative Medicine. Medecine Clinique Et Experimentale*, *27*(5), 274–281.

Iredell, J., Brown, J., & Tagg, K. (2016). Antibiotic resistance in Enterobacteriaceae: Mechanisms and clinical implications. *BMJ (Clinical Research Ed.)*, *352*, h6420. https://doi.org/10.1136/bmj.h6420

Jiang, X., Xu, C., Liu, B., Chen, P., Xu, Q., & Zhang, L. (2023). Efficacy and safety of bifidobacterium quadruple viable tablets in the treatment of Helicobacter pylori-infected peptic ulcer or gastritis patients: A systematic review and meta-analysis. *BMC Infectious Diseases*, *23*(1), 313. https://doi.org/10.1186/s12879-023-08211-1

Kainz, K., Bauer, M. A., Madeo, F., & Carmona-Gutierrez, D. (2020). Fungal infections in humans: The silent crisis. *Microbial Cell*, *7*(6), 143–145. https://doi.org/10.15698/mic2020.06.718

Khoshnood, S., Heidary, M., Mirnejad, R., Bahramian, A., Sedighi, M., & Mirzaei, H. (2017). Drug-resistant gram-negative uropathogens: A review. *Biomedicine & Pharmacotherapy = Biomedecine & Pharmacotherapie*, *94*, 982–994. https://doi.org/10.1016/j.biopha.2017.08.006

Klarin, B., Wullt, M., Palmquist, I., Molin, G., Larsson, A., & Jeppsson, B. (2008). Lactobacillus plantarum 299v reduces colonisation of Clostridium difficile in critically ill patients treated with antibiotics. *Acta Anaesthesiologica Scandinavica*, *52*(8), 1096–1102. https://doi.org/10.1111/j.1399-6576.2008.01748.x

Kwon, J. H., Bommarito, K. M., Reske, K. A., Seiler, S. M., Hink, T., Babcock, H. M., Kollef, M. H., Fraser, V. J., Burnham, C.-A. D., & Dubberke, E. R. (2015). Randomized controlled trial to determine the impact of probiotic administration on colonization with multidrug-resistant organisms in critically ill patients. *Infection Control and Hospital Epidemiology*, *36*(12), 1451–1454. https://doi.org/10.1017/ice.2015.195

Lee, Y.-C., Dore, M. P., & Graham, D. Y. (2022). Diagnosis and treatment of Helicobacter pylori infection. *Annual Review of Medicine*, *73*, 183–195. https://doi.org/10.1146/annurev-med-042220-020814

Leffler, D. A., & Lamont, J. T. (2015). Clostridium difficile Infection. *New England Journal of Medicine*, *372*(16), 1539–1548. https://doi.org/10.1056/NEJMra1403772

Li, Y., Choi, H., Leung, K., Jiang, F., Graham, D. Y., & Leung, W. K. (2023). Global prevalence of Helicobacter pylori infection between 1980 and 2022: A systematic review and meta-analysis. *The Lancet Gastroenterology & Hepatology*, *8*(6), 553–564. https://doi.org/10.1016/S2468-1253(23)00070-5

Liu, H.-F., & Yi, N. (2022). A systematic review and meta-analysis on the efficacy of probiotics for bacterial vaginosis. *European Review for Medical and Pharmacological Sciences*, *26*(1), 90–98. https://doi.org/10.26355/eurrev_202201_27752

Ljungquist, O., Kampmann, C., Resman, F., Riesbeck, K., & Tham, J. (2020). Probiotics for intestinal decolonization of ESBL-producing Enterobacteriaceae: A randomized, placebo-controlled clinical trial. *Clinical Microbiology and Infection: The Official Publication of the European Society of Clinical Microbiology and Infectious Diseases*, *26*(4), 456–462. https://doi.org/10.1016/j.cmi.2019.08.019

Lukic, J., Chen, V., Strahinic, I., Begovic, J., Lev-Tov, H., Davis, S. C., Tomic-Canic, M., & Pastar, I. (2017). Probiotics or pro-healers the role of beneficial bacteria in tissue repair. *Wound Repair and Regeneration: Official Publication of the Wound Healing Society [and] the European Tissue Repair Society*, *25*(6), 912–922. https://doi.org/10.1111/wrr.12607

Malfertheiner, P., Link, A., & Selgrad, M. (2014). Helicobacter pylori: Perspectives and time trends. *Nature Reviews. Gastroenterology & Hepatology*, *11*(10), 628–638. https://doi.org/10.1038/nrgastro.2014.99

Manley, K. J., Fraenkel, M. B., Mayall, B. C., & Power, D. A. (2007). Probiotic treatment of vancomycin-resistant enterococci: A randomised controlled trial. *The Medical Journal of Australia*, *186*(9), 454–457. https://doi.org/10.5694/j.1326-5377.2007.tb00995.x

McFarland, L. V., Surawicz, C. M., Greenberg, R. N., Fekety, R., Elmer, G. W., Moyer, K. A., Melcher, S. A., Bowen, K. E., Cox, J. L., Noorani, Z., Harrington, G., Rubin, M., & Greenwald, D. (1994). A randomized placebo-controlled trial of Saccharomyces boulardii in combination with standard antibiotics for Clostridium difficile disease. *JAMA*, *271*(24), 1913–1918. https://doi.org/10.1001/jama.1994.03510480037031

Medina, M., & Castillo-Pino, E. (2019). An introduction to the epidemiology and burden of urinary tract infections. *Therapeutic Advances in Urology*, *11*, 1756287219832172. https://doi.org/10.1177/1756287219832172

Mei, Z., & Li, D. (2022). The role of probiotics in vaginal health. *Frontiers in Cellular and Infection Microbiology*, *12*, 963868. https://doi.org/10.3389/fcimb.2022.963868

Mindt, B. C., & DiGiandomenico, A. (2022). Microbiome modulation as a novel strategy to treat and prevent respiratory infections. *Antibiotics*, *11*(4), 474. https://doi.org/10.3390/antibiotics11040474

Nakamura, Y., & Daya, M. (2007). Use of appropriate antimicrobials in wound management. *Emergency Medicine Clinics of North America*, *25*(1), 159–176. https://doi.org/10.1016/j.emc.2007.01.007

Nouvenne, A., Ticinesi, A., & Meschi, T. (2015). Carbapenemase-producing Klebsiella pneumoniae in elderly frail patients admitted to medical wards. *Italian Journal of Medicine*, *9*(2), 2. https://doi.org/10.4081/itjm.2014.476

Oh, B., Kim, B.-S., Kim, J. W., Kim, J. S., Koh, S.-J., Kim, B. G., Lee, K. L., & Chun, J. (2016). The effect of probiotics on gut microbiota during the Helicobacter pylori eradication: Randomized controlled trial. *Helicobacter*, *21*(3), 165–174. https://doi.org/10.1111/hel.12270

Ouwehand, A. C., Forssten, S., Hibberd, A. A., Lyra, A., & Stahl, B. (2016). Probiotic approach to prevent antibiotic resistance. *Annals of Medicine*, *48*(4), 246–255. https://doi.org/10.3109/07853890.2016.1161232

Patel, B. K., Patel, K. H., Huang, R. Y., Lee, C. N., & Moochhala, S. M. (2022). The gut-skin microbiota axis and its role in diabetic wound healing: A review based on current literature. *International Journal of Molecular Sciences*, *23*(4), 2375. https://doi.org/10.3390/ijms23042375

Rahman, M. N., Barua, N., Tin, M. C. F., Dharmaratne, P., Wong, S. H., & Ip, M. (2024). The use of probiotics and prebiotics in decolonizing pathogenic bacteria from the gut; a systematic review and meta-analysis of clinical outcomes. *Gut Microbes*, *16*(1), 2356279. https://doi.org/10.1080/19490976.2024.2356279

Ramazanzadeh, R. (2010). Prevalence and characterization of extended-spectrum beta-lactamase production in clinical isolates of Klebsiella spp. *African Journal of Microbiology Research*, *4*(13), 1359–1362.

Rauseo, A. M., Hink, T., Reske, K. A., Seiler, S. M., Bommarito, K. M., Fraser, V. J., Burnham, C.-A. D., Dubberke, E. R., & Program, for the C. P. E. (2022). A randomized controlled trial of Lactobacillus rhamnosus GG on antimicrobial-resistant organism colonization. *Infection Control & Hospital Epidemiology*, *43*(2), 167–173. https://doi.org/10.1017/ice.2021.94

Robilotti, E., & Deresinski, S. (2014). Carbapenemase-producing Klebsiella pneumoniae. *F1000Prime Reports*, *6*, 80. https://doi.org/10.12703/P6-80

Rubin, I. M. C., Mollerup, S., Broholm, C., Knudsen, S. B., Baker, A., Helms, M., Holm, M. K. A., Kallemose, T., Westh, H., Dahl Knudsen, J., Pinholt, M., & Petersen, A. M. (2022). No effect of Lactobacillus rhamnosus GG on eradication of colonization by vancomycin-resistant Enterococcus faecium or microbiome diversity in hospitalized adult patients. *Microbiology Spectrum*, *10*(3), e02348–21. https://doi.org/10.1128/spectrum.02348-21

Rungelrath, V., & DeLeo, F. R. (2021). Staphylococcus aureus, antibiotic resistance, and the interaction with human neutrophils. *Antioxidants & Redox Signaling*, *34*(6), 452–470. https://doi.org/10.1089/ars.2020.8127

Salomão, M. C. C., Heluany-Filho, M. A., Menegueti, M. G., Kraker, M. E. A. D., Martinez, R., & Bellissimo-Rodrigues, F. (2016). A randomized clinical trial on the effectiveness of a symbiotic product to decolonize patients harboring multidrug-resistant Gram-negative bacilli. *Revista Da Sociedade Brasileira de Medicina Tropical*, *49*, 559–566. https://doi.org/10.1590/0037-8682-0233-2016

Scalise, A., Bianchi, A., Tartaglione, C., Bolletta, E., Pierangeli, M., Torresetti, M., Marazzi, M., & Di Benedetto, G. (2015). Microenvironment and microbiology of skin wounds: The role of bacterial biofilms and related factors. *Seminars in Vascular Surgery*, *28*(3–4), 151–159. https://doi.org/10.1053/j.semvascsurg.2016.01.003

Shaheen, G., Akram, M., Jabeen, F., Ali Shah, S. M., Munir, N., Daniyal, M., Riaz, M., Tahir, I. M., Ghauri, A. O., Sultana, S., Zainab, R., & Khan, M. (2019). Therapeutic potential of medicinal plants for the management of urinary tract infection: A systematic review. *Clinical and Experimental Pharmacology and Physiology*, *46*(7), 613–624. https://doi.org/10.1111/1440-1681.13092

Silva, D. R., Sardi, J. de C. O., Pitangui, N. de S., Roque, S. M., Silva, A. C. B. da, & Rosalen, P. L. (2020). Probiotics as an alternative antimicrobial therapy: Current reality and future directions. *Journal of Functional Foods*, *73*, 104080. https://doi.org/10.1016/j.jff.2020.104080

Song, H., Hu, W., Zhou, X., Tao, J., Zhang, S., Su, X., & Wu, W. (2021). Clinical benefits from administering probiotics to mechanical ventilated patients in intensive care unit: A PRISMA-guided meta-analysis. *Frontiers in Nutrition*, *8*, 798827. https://doi.org/10.3389/fnut.2021.798827

Stapleton, A. E., Au-Yeung, M., Hooton, T. M., Fredricks, D. N., Roberts, P. L., Czaja, C. A., Yarova-Yarovaya, Y., Fiedler, T., Cox, M., & Stamm, W. E. (2011). Randomized, placebo-controlled phase 2 trial of a Lactobacillus crispatus probiotic given intravaginally for prevention of recurrent urinary tract infection. *Clinical Infectious Diseases: An Official Publication of the Infectious Diseases Society of America*,*52*(10), 1212–1217. https://doi.org/10.1093/cid/cir183

Sullivan, Å., Johansson, A., Svenungsson, B., & Nord, C. E. (2004). Effect of Lactobacillus F19 on the emergence of antibiotic-resistant microorganisms in the intestinal microflora. *Journal of Antimicrobial Chemotherapy*, *54*(4), 791–797. https://doi.org/10.1093/jac/dkh406

Szachta, P., Ignyś, I., & Cichy, W. (2011). An evaluation of the ability of the probiotic strain Lactobacillus rhamnosus GG to eliminate the gastrointestinal carrier state of vancomycin-resistant enterococci in colonized children. *Journal of Clinical Gastroenterology*, *45*(10), 872–877. https://doi.org/10.1097/MCG.0b013e318227439f

Tang, B., Tang, L., Huang, C., Tian, C., Chen, L., He, Z., Yang, G., Zuo, L., Zhao, G., Liu, E., Wang, S., Lin, H., He, J., & Yang, S. (2021). The effect of probiotics supplementation on gut microbiota after Helicobacter pylori Eradication: A multicenter randomized controlled trial. *Infectious Diseases and Therapy*, *10*(1), 317–333. https://doi.org/10.1007/s40121-020-00372-9

Tang, Y., Chen, L., Yang, J., Zhang, S., Jin, J., & Wei, Y. (2024). Gut microbes improve prognosis of Klebsiella pneumoniae pulmonary infection through the lung-gut axis. *Frontiers in Cellular and Infection Microbiology*, *14*, 1392376. https://doi.org/10.3389/fcimb.2024.1392376

Tannock, G. W., Tiong, I. S., Priest, P., Munro, K., Taylor, C., Richardson, A., & Schultz, M. (2011). Testing probiotic strain Escherichia coli Nissle 1917 (Mutaflor) for its ability to reduce carriage of multidrug-resistant E. coli by elderly residents in long-term care facilities. *Journal of Medical Microbiology*, *60*(Pt 3), 366–370. https://doi.org/10.1099/jmm.0.025874-0

Viazis, N., Argyriou, K., Kotzampassi, K., Christodoulou, D. K., Apostolopoulos, P., Georgopoulos, S. D., Liatsos, C., Giouleme, O., Koustenis, K., Veretanos, C., Stogiannou, D., Moutzoukis, M., Poutakidis, C., Mylonas, I. I., Tseti, I., & Mantzaris, G. J. (2022). A four-probiotics regimen combined with a standard Helicobacter pylori-eradication treatment reduces side effects and increases eradication rates. *Nutrients*, *14*(3), 3. https://doi.org/10.3390/nu14030632

Vincent, C., Stephens, D. A., Loo, V. G., Edens, T. J., Behr, M. A., Dewar, K., & Manges, A. R. (2013). Reductions in intestinal Clostridiales precede the development of nosocomial Clostridium difficile infection. *Microbiome*, *1*(1), 18. https://doi.org/10.1186/2049-2618-1-18

Wang, Y., Wang, X., Cao, X.-Y., Zhu, H.-L., & Miao, L. (2023). Comparative effectiveness of different probiotics supplements for triple helicobacter pylori eradication: A network meta-analysis. *Frontiers in Cellular and Infection Microbiology*, *13*, 1120789. https://doi.org/10.3389/fcimb.2023.1120789

Warrack, S. R., Panjikar, P., Duster, M., & Safdar, N. (2013). Probiotics for the reduction of colonization with MRSA. *American Journal of Infection Control*, *41*(6), S135–S136. https://doi.org/10.1016/j.ajic.2013.03.269

Wawrysiuk, S., Naber, K., Rechberger, T., & Miotla, P. (2019). Prevention and treatment of uncomplicated lower urinary tract infections in the era of increasing antimicrobial resistance-non-antibiotic approaches: A systemic review. *Archives of Gynecology and Obstetrics*, *300*(4), 821–828. https://doi.org/10.1007/s00404-019-05256-z

Wieërs, G., Verbelen, V., Van Den Driessche, M., Melnik, E., Vanheule, G., Marot, J.-C., & Cani, P. D. (2020). Do probiotics during in-hospital antibiotic treatment prevent colonization of gut microbiota with multi-drug-resistant bacteria? A randomized placebo-controlled trial comparing saccharomyces to a mixture of lactobacillus, bifidobacterium, and saccharomyces.*Frontiers in Public Health*, *8*, 578089. https://doi.org/10.3389/fpubh.2020.578089

World HealthOrganization. (2023). *Antimicrobial Resistance*. World Health Organization. https://www.who.int/news-room/fact-sheets/detail/antimicrobial-resistance, accessed 2024 Aug 1.

World HealthOrganization. (2024). *WHO Bacterial Priority Pathogens List, 2024: Bacterial Pathogens of Public Health Importance to Guide Research, Development and Strategies to Prevent and Control Antimicrobial Resistance*. World Health Organization. https://www.who.int/publications/i/item/9789240093461, accessed 2024 Jun 30.

Wullt, M., Hagslätt, M.-L. J., & Odenholt, I. (2003). Lactobacillus plantarum 299v for the treatment of recurrent Clostridium difficile-associated diarrhoea: A double-blind, placebo-controlled trial. *Scandinavian Journal of Infectious Diseases*, *35*(6–7), 365–367. https://doi.org/10.1080/00365540310010985

Yang, C., Liang, L., Lv, P., Liu, L., Wang, S., Wang, Z., & Chen, Y. (2021). Effects of non-viable Lactobacillus reuteri combining with 14-day standard triple therapy on Helicobacter pylori eradication: A randomized double-blind placebo-controlled trial. *Helicobacter*, *26*(6), e12856. https://doi.org/10.1111/hel.12856

Yang, Z., Zhou, Y., Han, Z., He, K., Zhang, Y., Wu, D., & Chen, H. (2024). The effects of probiotics supplementation on Helicobacter pylori standard treatment: An umbrella review of systematic reviews with meta-analyses. *Scientific Reports*, *14*(1), 10069. https://doi.org/10.1038/s41598-024-59399-4

Zhao, J., Li, L.-Q., Chen, C.-Y., Zhang, G.-S., Cui, W., & Tian, B.-P. (2021). Do probiotics help prevent ventilator-associated pneumonia in critically ill patients? A systematic review with meta-analysis. *ERJ Open Research*, *7*(1), 00302–02020. https://doi.org/10.1183/23120541.00302-2020

16 Safety Considerations and Regulatory Framework for Probiotics in Antimicrobial Resistance

Barbara Sionek[1], *Aleksandra Szydłowska*[1], *Katarzyna Neffe-Skocińska*[1], *Agnieszka Rudzka*[2], *Dorota Zielińska*[1], *and Danuta Kołożyn-Krajewska*[1,2]

[1]Faculty of Human Nutrition, Institute of Human Nutrition Sciences, Department of Gastronomic Technology and Food Hygiene, Warsaw University of Life Sciences (WULS), Warsaw, Poland

[2]Faculty of Science and Technology, Department of Dietetics and Food Studies, Jan Dlugosz University in Czestochowa, Częstochowa, Poland

ABBREVIATIONS

LAB	lactic acid bacteria
WHO	World Health Organization
FAO	Food and Agriculture Organization
DL	deep learning
EFSA	European Food Safety Authority
GRAS	Generally Recognized as Safe
QPS	Qualified Presumption of Safety
NGPs	next-generation probiotics
HMP	Human Microbiome Project
FDA	Food and Drug Administration
LBPs	Live Biotherapeutic Products
EMA	European Medicines Agency
AHQR	Agency for Healthcare Research and Quality
ESLP	European Scientific League for Probiotics
WGS	whole genome sequencing
SNP	single nucleotide polymorphism
PATRIC	Pathosystems Resource Integration Center
AR	antibiotic resistance
CDC	Centers for Disease Control and Prevention
CARD	Comprehensive Antibiotic-Resistance Database
MLST	multilocus sequence typing
RGI	Resistome Gene Identifier
NDARO	National Database of Antibiotic Resistant Organisms
MEGARes	Microbial Ecology Group
AI	artificial intelligence
ML	Machine learning

DOI: 10.1201/9781032676951-16

AAB	acetic acid bacteria
AMR	antimicrobial resistance
ARGs	antimicrobial resistance genes
TU	taxonomic unit
MIC	minimum inhibitory concentration
FEEDAP	Panel on Additives and Products or Substances used in Animal Feed
HIAs	highly important antimicrobials
CIAs	critically important antimicrobials
WAAW	World Awareness Week
NF	Novel Food
RCT	randomized controlled trials

16.1 INTRODUCTION

The assessment of the safety of probiotics is a complex issue. It should consider the identification of the microorganism used, the method of administration, the level of exposure, the purpose of use, and the consumer's health status. The list of side effects caused by probiotics is still expanding. The side effects are strain-specific. Therefore, molecular typing, including whole genome sequences, should be used for probiotic assessment. It is of particular significance because the list of "good microorganisms" extends. There is a lack of standardized safety protocols and widely recognized standards governing the use of probiotics. Probiotics are classified as foods or dietary supplements (in the USA and Europe), as natural health products (in Canada), or as foods for specific health uses (in Japan). A 2002 FAO/WHO document identified four categories of probiotic adverse reactions: systemic infections, adverse metabolic effects, immune overstimulation in susceptible individuals, and gene transmission. Current regulations and laws regarding the use of probiotics are not uniform. In North America, the principle of long-term, safe use of the microorganism, the so-called GRAS (Generally Recognized as Safe) status, is vital. In the European Union, the European Food Safety Authority (EFSA) is responsible for regulating, evaluating, and monitoring the safety of probiotics. There is a list of microorganisms found to be safe and approved for use (Qualified Presumption of Safety – QPS). In North America and Europe, regulations are restrictive, and the approval of a microorganism requires numerous and expensive scientific studies. This chapter will discuss safety considerations of probiotics, issues related to antibiotic resistance (AR), rules, and recommendations.

16.2 SAFETY CONSIDERATION OF PROBIOTICS

The positive effect of probiotics on human health has been known for many years, and their numerous benefits are widely described in the literature. They are effectively used both to prevent illness in healthy individuals and to support treatment in sick patients, including those with chronic conditions. The consumption of probiotics, whether in the form of ready-to-use preparations, traditional food products, or food additives is gaining more and more popularity. The rapid development of the functional food market, including products containing live colonies of probiotic microorganisms, responds to the growing needs of health-conscious consumers seeking healthy, traditional, minimally processed, and tasty foods. Due to various positive effects, probiotics are also incorporated in animal foods (Borase et al., 2022). The increased interest in probiotics and their common applicability raises concerns about the safety. In contrast to the well-known health benefits of probiotics, safety issues seem to be under-reported or even neglected. The post-marketing follow-up and surveillance of probiotic food products seem to be insufficient. Many studies and reports of probiotics are focused mainly on health benefits and do not include adverse event reporting (Sniffen et al., 2018). In 2018, Bafeta et al. analyzed 384 randomized trials with probiotics. In 98% of them, there was no definition of adverse events. In 28% of studies, no adverse events were reported, and in 37%, safety reports were omitted (Bafeta et al., 2018). This may result from the absence of standardized safety protocols and widely recognized standards for probiotic applications. Even in clinical trials, there are inconsistencies in adverse events reporting (Doron and Snydman, 2015). The adverse effects of probiotics, although relatively rare, may carry potential of health risks. This is particularly important in vulnerable populations, that is, people with chronic diseases, newborns, pregnant women, elderly people, and immunocompromised patients. The safety of probiotics largely depends on who consumes them. They are well tolerated by a group of healthy consumers, and the side effects are mild and do not threaten life and health. Paradoxically, the group of the most seriously ill patients may benefit the most from the use of probiotics, but they are also most at risk of undesirable effects, including sepsis (Sionek and Kołożyn-Krajewska, 2019). In the case of groups with increased risks, side effects may occasionally be severe. Probiotics can cause systemic and local infections, including opportunistic infections. They can be responsible for allergic reactions, adverse immune effects, and metabolic disorders. One of the important safety concerns is the spread of antimicrobial resistance (AMR; Liu et al., 2024). The resistant determinants can be both acquired by probiotics and transferred further to commensals or pathogenic microorganisms. Probiotics are commonly co-administered with drugs, which might influence the drug's pharmacokinetics/pharmacodynamics. Therefore, the bioavailability and efficacy of various drugs can be changed, which can have clinical significance (Purdel et al., 2023). Another concern of probiotic safety is the degradation of the gut mucosal defense barrier. Some probiotic *Bifidobacterium* strains are suspected to have mucolytic properties (Splichalova et al., 2020). Translocation of gut bacteria through impaired intestinal mucus is possible. Therefore, the strains harboring mucolytic activity can be categorized as virulent (Oba et al., 2022).

Traditionally probiotic safety is mainly based on the principle of a documented history of traditional, safe use. The conventional source of probiotics is the human gastrointestinal tract. Recently, there has been a growing interest in probiotic microorganisms derived from unconventional sources. They come from various environments, such as fermented and unfermented food, breast milk, soil, and the digestive tract of animals (Zielińska and Kolożyn-Krajewska, 2018). Moreover, advances in microbiology and molecular biology, the genetic sequencing methods opened new perspectives for discovering new microorganisms, which are candidates to become probiotics. This group of new microorganisms with known human health benefits was called the next-generation probiotics (NGPs) (O'Toole et al., 2017). The human microbiome is still an unexplored ecosystem. It is estimated that until now, approximately 80% of microbes in the human intestine are unknown (Eckburg et al., 2005). To investigate the human microbiome, research projects were designed, such as The Human Microbiome Project (HMP), which creates "resources that would enable the comprehensive characterization of the human microbiome and analysis of its role in human health and disease" (NIH Human Microbiome Project, 2024). Extending the family of "good microorganisms without a history of safe use," new challenges are created. In this regard, the most important issue is the safety. The probiotics from unconventional sources, including NGPs as well as genetically modified microorganisms, do not fulfill the authorities' criteria of conventional probiotics and are not allowed for human application as food, a food supplement, or a drug. Therefore, the strain-specific double-blind, randomized trials in the appropriate population should be designed to provide convincing data ensuring their safety (Sionek et al., 2023a). In 2012, the US Food and Drug Administration (FDA) created a novel category of microorganisms called Live Biotherapeutic Products (LBPs). This type of product is defined as "a biological product that (i) contains live organisms, such as bacteria; (ii) is applicable to the prevention, treatment, or cure of a disease or condition of human beings; and (iii) is not a vaccine". The FDA recognized LBPs as "medicinal products containing live microorganisms (bacteria or yeasts) for human use" (FDA Early Clinical Trials with Live Biotherapeutic Products 2016, Cordaillat-Simmons et al., 2020). Since 2019, the European Pharmacopoeia has also recognized the category of products intended to prevent or treat diseases, including LBPs, as medicinal products (European Pharmacopoeia 3053E, 2019). Therefore, LBPs require registration similar to newly developed drugs: in the USA according to the FDA rules, and in Europe according to the European Medicines Agency (EMA) (European Medicines Agency).

Randomized clinical trials are a core source of information and evidence on the safety of probiotics. However, available tools also include meta-analyses and clinical case reports. In 2011, the Agency for Healthcare Research and Quality (AHRQ) published an independent, expert report, Safety of Probiotics to Reduce Risk and Prevent or Treat Disease, sponsored by U.S. government agencies. The authors analyzed 11,977 publications, from which 622 studies were selected. A total of 235 studies found probiotics well tolerated, and 387 reported adverse effects. The use of probiotics was not linked to a higher incidence of adverse effects. The conclusion was that the available information was insufficient to answer questions about the safety of probiotics (Hempel et al., 2011). Doron and Snydman (2015) reviewed the safety of probiotics in 24 studies, including various vulnerable populations, such as children, pregnant women, the elderly, patients with irritable bowel syndrome, and immunocompromised patients. The authors reported only two cases of adverse events related to probiotics (Doron and Snydman, 2015).

To confer health benefits, probiotics must be administered in sufficiently large quantities of live microorganisms. Additionally, dead microorganisms and their fragments are also supplied. The food products or preparations used contain not only one strain but commonly a mixture of strains. There is an increase in commercially available probiotic products on the market. Unfortunately, the provided information on probiotic content, its viability, and stability throughout shelf life, purity, and contaminations is commonly insufficient and inaccurate (Kolaček et al., 2017). Moreover, there are no standards for probiotic manufacturing processes, transportation, and marketing. However, products for medicinal purposes probably have better quality than food products. Considering variations between products and even batches, recognized manufacturers should be preferred (Szajewska, 2016). The composition of the final probiotic product, including contaminations, has implications on efficacy and safety profile. Therefore, molecular typing methods appear to be very useful tools for the assignment of species and strain identity (Celandroni et al., 2019). The key issues are strain identification and correct labeling of commercialized probiotic products. The label should include strain specification. The label data and quality control should be strictly regulated by low and thereafter supervised (Nataraj et al., 2023). Misinformation or deviations on commercialized probiotic product labels can have a negative impact on consumers and health professionals' credibility and confidence (Sanders et al., 2010). The initiative of product certification can overcome those problems. The independent, accredited, and audited laboratories should perform detailed qualitative and quantitative microbiological analyses. Therefore, a certificate can be issued, confirming the label information such as strain specificity, viability, and sufficient quantity until declared shelf life (Warzée et al., 2021). The European Scientific League for Probiotics (ESLP) is an international Scientific League for Preparations based on live or inanimate *Lactobacilli, Bifidobacteria* preparations – and/or their components and metabolites – for Prebiotics and for LPB's – Live Biotherapeutic Products. The list of certified probiotic products is available online (Table 16.1). ESLP has created a quality label to validate quality products and to differentiate them from cowboy products. The ESLP quality label is reserved for proprietary medicinal products distributed in pharmacies and is assigned on spontaneous application. The ESLP label does not provide in any way a quality scale between labeled products, nor does it disqualify products that do not wish to participate in the trials proposed for granting the label. It has no health claim value. By way of its Scientific Committee, composed of neutral experts and scientists recognized in the field, the ESLP helps to promote scientific research.

According to Merenstein et al. (2023), whole genome sequencing (WGS) enables the determination of virulence, toxin antibiotic-resistance genes, tracking the strain during production, and the investigation and etiology of suspected infections. The genome can be interrogated for any genes of concern, including toxigenicity, pathogenicity, or AR (Merenstein et al., 2023). Two approaches are currently used to analyze WGS: single nucleotide polymorphism (SNP) and gene-by-gene-based methods. In the SNP method, the comparison of nucleotide differences between the target genome and reference genome is analyzed. In the gene-by-gene-based method, alleles are assigned through multiple comparisons between genomes (Sharma et al., 2020). Probiotics are living microorganisms,

TABLE 16.1
ESLP List of Certified Probiotic Products 2024, according to ESLP (probiotics-scientific-league.org) (ESLP-European Scientific League for Probiotics)

Bacterial Strain	Name of Commercial Product
Lactobacillus reuteri DSM 17938	Protectis Adult, Protectis Easy Drops, Protectis Junior
Lacticaseibacillus rhamnosus LA801	Lactibiane ATB
Bifidobacterium animalis ssp. *lactis* HNO19	Bactiol Easy
Lactobacillus acidophilus NCFM, *B. animalis* ssp. *lactis* Bi-07	Bactiol Plus
Lactobacillus acidophilus CUL21, *L. acidophilus* CUL60 *B. bifidum* CUL20, *B. animalis* ssp. *lactis* CUL34	Probiomax
Lactobacillus helveticus LA102, *Bifidobacterium longum* LA101, *Lactococcus lactis* LA103, *Streptococcus thermophilus* LA104	Lactibiane reference
Propionibacterium freudenreichii (2 specific strains)	Securil
Lacticaseibacillus rhamnosus GG, *B. animalis* ssp. *lactis* Bi-07	Bactiol Junior, Bactiol HMO fucose
Lacticaseibacillus rhamnosus GG, *B. animalis* ssp. *lactis* Bb12	Bactiol Natal

and variations through mutations, deletions, and insertion, as well as gene transfer, are possible, which can change their virulence properties. Therefore, regular strain typing should be considered to predict virulence, susceptibility of antibiotic-resistance genes, and, finally, to confirm genome stability (Nataraj et al., 2023, Kojic et al., 2020, Fatahi-Bafghi et al., 2022). The data show that strain virulence elements are very useful tools in identifying threats of bacterial pathogens (Boyte et al., 2025). The Pathosystems Resource Integration Center (PATRIC) is the bacterial bioinformatics resource center, freely available online, designed to support the biomedical research community's work on bacterial infectious diseases by integrating vital pathogen information with rich data and analysis tools (PATRIC–Database Commons, 2017).

The overall safety of the probiotics use can be attributed to the risks related to the product and to the risks related to the consumer or the patient. According to Rouanet et al. (2020), among risks related to the product, there are (i) isolation/selection issues (origin of the strain, isolation process, and handling), (ii) the strain characteristics (i.e., virulence profile, gene transfer, biogenic amines production drug, and hormonally related metabolism), (iii) preparation process and final product characteristic (i.e., impurities, route of administration). The risks related to the consumer or patients include age, gender, vulnerable populations, microbiome composition, nature of the target, dosage and route of administration, concomitant medications, diet, and lifestyle (Rouanet et al., 2020). This diversity and interdependence indicate that probiotics should be tested and assessed in different target populations, states, and age groups. The safety of new probiotic microorganisms cannot be supported by a long history of safe use as its conventional counterparts. Therefore, laboratory, animal studies, and, finally, expensive long-term randomized trials should be designed and conducted. Regarding public health authority, strict rules and demands are proper and very important, but on the other hand, they can suppress and limit the introduction of novel probiotic strains for human applications.

16.3 ANTIMICROBIAL RESISTANCE OF PROBIOTICS

Probiotics are commonly used around the world because of their potential health benefits. They are recommended as dietary supplements to prevent healthy people from contracting an illness and are used in population groups with acute and chronic diseases. They are also recommended, along with antibiotic therapy. Due to their positive impact on the human microbiome, probiotics can prevent or even support the treatment of diseases. Moreover, probiotic microorganisms are naturally presented in various traditionally preserved foods or added as starter cultures or additives (Sionek et al., 2023b). Functional food with probiotics gained common consumer acceptance and seems to become the future trend (Sionek et al., 2023a). The fermentation is the oldest method of effective food preservation. It allows for consuming raw food (vegetables and fruits) or minimally processed food (meat, dairy, and fish) containing various beneficial microorganisms. Probiotics need to be provided in high amounts, at least 10^6 CFU, to deliver health benefits (Probiotics in Food, Report of a Joint FAO/WHO 2020). However, the use of probiotic microorganisms also carries potential risks. Food can be the source of antibiotic-resistance genes as well as resistant pathogens and resistant non-pathogens (Wolfe, 2023).

The AMR of probiotics can be congenital or acquired. Microorganisms can attain AMR genes through vertical (intrinsic resistance) or horizontal transfer (acquired resistance) (Rouanet et al., 2020). Therefore, the risks of developing or transferring AR by probiotics to different microorganisms should be considered. The most important mechanisms of AR include genes primarily located in transposing elements (plasmids and transposons), mechanisms of antibiotic clearance (efflux pump), and factors affecting antibiotics or antibiotic targets (e.g., antibiotic-inactivating enzymes and mutations in antibiotic target domains) (Aghamohammad and Rohani, 2023, Byarugaba, 2010). Although probiotic harboring antibiotic-resistant genes can be considered as potential reservoirs of multidrug resistance, the evidence of such gene transmissions from probiotics to the pathogen is scarce (Liu et al., 2024). The transmission risk increases when resistance genes are located in mobile elements, while intrinsic resistance (acquired resistance) presents a very low potential of spreading (Roe et al., 2022). Once the bacteria acquire AR, they can further reproduce, and the transmission risk of the resistance genes to surrounding microbiota increases (Das et al., 2019). In 1998, the transfer of AR genes *in vivo* from probiotic bacteria *Lactobacillus reuteri* to *Enterococcus faecalis* was reported by Morelli et al. (1988). The transfer of AR genes is also possible from non-viable cells. According to Merenstein et al. (2023), when considering the safety of AR genes in the genome of probiotics, several issues should be considered:

i. are the genes associated with genes that provide a likely mechanism for horizontal transfer;
ii. are the genes capable of functional expression if transferred to a naïve host;
iii. is the resistance phenotype for the antibiotic typical for the species (intrinsic resistance);
iv. is the resistance phenotype so widespread that the probiotic would not substantively contribute to its presence among environmental microbes; and
v. is the resistance to a clinically relevant antibiotic (Merenstein et al., 2023).

The safety considerations are particularly relevant in sensitive populations such as chronically ill, immunocompromised patients, premature infants, neonatal, or patients hospitalized in intensive care units. In this regard, probiotic safety requires in-depth research and clinical randomized trials to confirm efficacy and safety. In those vulnerable populations, the potential health benefits should be confronted with the possible side effects of probiotics (Yelin et al., 2019).

The problem of AR is a global issue. Widespread and even overuse of antibiotics and their improper use in medicine, veterinary medicine, and food production (animal growth promoters, prophylaxis, and treatment) are the main causes of AR (Phillips et al., 2004). Antibiotics in agriculture are common mainly due to response to increased world food production demands. They are present in subinhibitory concentrations in food, and therefore, microorganisms can develop adaptive responses leading to AR (Okaiyeto et al., 2024). Although misuse of antibiotics poses a great threat to global food safety and human health, the issue of AR is still underestimated. It is projected that by the year 2050, agricultural contamination with antibiotic-resistant bacteria will be responsible for 10 million annual deaths around the world (Okaiyeto et al., 2024).

Bacteria display multiple adaptive strategies and mechanisms to combat environmental stress as well to resist antibiotics (Li and Nair, 2012, Sionek et al., 2024). Increased resistance, selection, gene transfer, acquisition, and activation of antibiotic-resistance genes evolved some bacteria into antibiotic-resistant strains (Yang et al., 2024). The result is an antibiotic failure in eliminating bacterial infection. According to the CDC report (2019), more than 2.8 million antibiotic-resistant infections occur in the USA each year, and more than 35,000 deaths are reported (Centers for Disease Control and Prevention (U.S.), 2019). In the European Union, annually, 33,000 people die as a result of AR infections (Dadgostar, 2019). The World Health Organization considers AMR as one of the top global public threats. WHO estimates that bacterial AR was directly responsible for 1.27million global deaths in 2019 and contributed to 4.95 million deaths (WHO, 2023). In regard to public health and consumer safety, probiotic safety should be closely monitored. Risk assessments and WGS should be considered. To monitor antibiotic resistance, bioinformatic databases of resistance genes were created. The Comprehensive Antibiotic-Resistance Database (CARD) is a database of resistance genes, their products, and associated phenotypes, providing a dynamic view of antibiotic-resistant isolates being analyzed worldwide (CARD, 2024). This project collects pathogen identification, multilocus sequence typing (MLST), AMR gene list, date, and geographical region for genome sequences submitted to Resistome Gene Identifier (RGI) (A bioinformatic database of resistance genes). Another database is the National Database of Antibiotic-Resistant Organisms (NDARO) is a collaborative, cross-agency, centralized hub for researchers to access AR data to facilitate real-time surveillance of pathogenic organisms. It was created to increase standardization, to make AR-related data more widely available, to make more effective use of bacterial genomic data, and to assist researchers and public health officials in allowing researchers to identify bacterial genomes with AR genes (National Database of Antibiotic Resistant Organisms (NDARO)). The MEGARes have been designed for the computational analysis of large-scale sequencing data (on the order of terabytes) in a fast and accurate way for statistical analyses on count-based data and the construction of sequence classifiers. MEGARes (Microbial Ecology Group) incorporate previously published resistance gene sequences for antimicrobial drugs, biocides, and metals. Sources include Antibacterial Biocide and Metal Resistance Genes (BacMet), CARD, ResFinder and PointFinder, AMRFinderPlus (MEGARes – Microbial Ecology Group. https://www.meglab.org/). The ProbResist database catalogs published reports of probiotic bacteria from various sources that have been experimentally proven to be resistant to one or more antibiotics databases. It presents (i) the breakdown of the type of probiotic bacteria most frequently reported to be resistant to antibiotics as a pie chart, (ii) the abundance of AR reported in the scientific literature for probiotic bacteria as a bar graph, (iii) the antibiotic classes, generations, and mode of actions as tables and (iv) the frequency of reports of probiotics from different regions of countries that are resistant to antibiotics as heat map (Dou et al., 2022). The significance of AR gene databases seems to be particularly useful for the safety assessment of next-generation probiotics as well as for genetically modified probiotics (O'Toole et al., 2017, Cao et al., 2020). These groups of microorganisms harbor a huge potential for human health benefits but also carry the risks of adverse effects, including AR. Therefore, NGPs are not recognized as microorganisms intended for human use to reach the market as food, a food supplement, or a drug (Sionek et al., 2023a).

The probiotic impact on AR depends on the characteristics of populations. In the study of Montassier et al. (2021), the gut metagenomes analysis of healthy antibiotics-naïve humans who received an 11-probiotic-strain preparation resulted in a decrease in the number of antibiotic-resistance genes. In the cohort of humans, a course of antibiotics resulted in an increase of strains carrying vancomycin-resistance genes but not resistance genes encoded by the probiotic strains (Montassier et al., 2021). The widespread use of metagenomics tests and studies provides a variety of data. Antibiotic-resistance genes have been identified in probiotic *Lactobacillus* strains present in various food products (cheeses, fermented meats, and vegetables). Widely used LAB poses resistance to commonly used antibiotics: tetracyclines, penicillins, chloramphenicol, and macrolides. Acetic acid bacteria strains were shown to be resistant to some antibiotics (gentamicin, ampicillin, chloramphenicol, ciprofloxacin, erythromycin, and trimethoprim), but the potential of transmission of AR genes seems to be limited (Cepec and Trček, 2022). *Lactobacillus* showed an intrinsic resistance to vancomycin and aminoglycosides. This is important because vancomycin is commonly used for *C. difficile* infection (Goldstein et al., 2015). On the other side, survival of LAB in the presence of vancomycin can avoid antibiotic-associated diarrhea and support the treatment of *C. difficile* infection. The intrinsic resistance of *Lactobacillus* is rather not-transferable and, therefore, can be considered safe. In a study by Mehdi Fatahi-Bafghi et al. (2022), complete bacterial genomes of 126 probiotic bacteria were analyzed for antibiotic-resistance genes. It was shown that probiotic bacteria harbor various antibiotic-resistance genes and some of them are transferable. In *Bifidobacterium* and *Lactobacillus*, the prevalence of the tetracycline resistance gene *tet*W was found to be more prevalent than other antibiotic-resistance genes (Fatahi-Bafghi et al., 2022).

Current methods of AR gene identification are based on genome mapping and alignment-based sequence similarities. In this regard, there are concerns about false-negative and false-positive results. Moreover, traditional methods are unable to identify novel AR gene sequences and cannot distinguish between chromosomal or plasmid sequences. In recent years, research has been carried out with emerging artificial intelligence (AI) methods as a tool to identify and classify AR genes (Olatunji et al., 2024). Although AI, machine learning (ML), and deep learning (DL) in genomics are at an early stage, those approaches are promising and can support the detection of AR genes and their significance (NHGRI, 2022). Do probiotics and other beneficial microorganisms pose a real risk of AR? The question needs more systematic studies, including metagenomic surveys. According to the EFSA panels (2012), when a bacterial strain demonstrates higher resistance to a specific antimicrobial than the other strains of the same taxonomical unit, the presence of acquired resistance is indicated, and additional information is needed on the genetic basis of the AMR. Any bacterial strain carrying an acquired resistance to antimicrobial that is shown to be due to the acquisition of genetic determinant presents the greatest potential for horizontal spread and should not be used as a feed additive (EPSA, 2012). A long history of safe use of fermented foods is an argument, which in the past gained authority's acceptance (i.e., GRAS and QPS status). However, in a changed world with antibiotic-contaminated food and globally increasing AR, the question of probiotic safety should be reopened.

16.4 RECOMMENDATION OF SCIENTIFIC ASSOCIATIONS AND COMMITTEES FOR PROBIOTICS IN ANTIMICROBIAL RESISTANCE

The progressive aging process of societies, harmful habits or lifestyle or threats such as "pandemics" motivate to look for new ways to support health (Report Top Ten Trends, Innova Market Insights). The healthy lifestyle promoted in recent years is manifested by an increased demand for food products with the highest nutritional value and specific health-promoting properties. For this reason, consumers are increasingly turning to functional foods, which not only provide nutritional value but also have scientifically proven benefits for one or more bodily functions. Probiotics and probiotic foods fit into the above trend. The value of many foods on the market can be often enhanced by the addition of probiotics. It's no surprise that these foods are seen as a better choice by consumers, who view them as beneficial to health (Zielińska et al., 2018).

Probiotic microorganisms are defined as "live microorganisms that, when administered in adequate amounts, confer a health benefit on the host". This definition is inclusive of a broad range of microbes and applications, while capturing the essence of probiotics (microbial, viable, and beneficial to health) (Hill et al., 2014).

Strains of lactic acid bacteria, which are part of our gut microbiota, are frequently recognized probiotics. The two most common probiotic bacteria are *Lactobacillus* and *Bifidobacterium*, while there are other significant and common strains of *Lactococcus, Streptococcus, Enterococcus, Propionibacterium*, and *Saccharomyces* yeasts. Lactobacilli and Bifidobacteria were the first probiotics in terms of technology and application in the food sector (Elshaghabee et al., 2017). In addition, several writers emphasized the significance of determining probiotic traits in groups of microorganisms, such as acetic acid bacteria (AAB) (Neffe-Skocińska et al., 2024, Hata et al., 2023).

Probiotic and potentially probiotic microorganisms are more and more often a component of the daily diet of healthy people. They are used in food technology, pharmaceutical industry, or as dietary supplements. The purpose of their use is primarily to ensure health and prevent diseases, especially within the digestive system. Probiotics have been used to prevent diseases and improve the health of groups of people chronically ill and susceptible to various diseases, such as newborns, pregnant women, or the elderly. Due to their widespread use, safety research is crucial. They are based mainly on the principle of documented history of traditional, safe use (Zielińska et al., 2018). The criteria developed by EFSA, FAO, and WHO determine the characteristics of a probiotic microorganism, considering the safety aspect in addition to functionality or technological properties.

There is a common view that the use of probiotics is safe and well tolerated, which is why in the European Union countries probiotics have the status of a "QPS". However, it should be noted that there are circumstances in which probiotics should not be applied, due to a risk of sepsis development. This applies to hospitalized people, patients in intensive care units, patients with inserted puncture or catheter, patients receiving immunosuppressive or anticancer therapy. It is also contraindicated to give probiotics to premature babies and people suffering from pancreatitis (Dudek-Wicher et al., 2020). Most consumers pay more attention to the declared functions of probiotics by manufacturers, but ignore the fact, that fermented foods may pose unique AMR risks compared to other commonly consumed foods. Most resistant microbes in ferments are non-pathogenic, meaning that a direct threat to human health is unlikely. But even the presence of genetically labile antimicrobial resistance genes (ARGs) in beneficial fermentation microbes could be problematic (Li et al., 2020).

LAB have numerous applications in the production of fermented foods (Rzepkowska et al., 2014), have high adaptability, and produce many resistance mechanisms. One of the most dangerous is the transfer of resistance genes located on the moving elements of DNA by conjugation to other bacteria. This is associated with the risk of dynamic development of AR of microorganisms present in the human gastrointestinal tract (Courvalin, 2006). The vertical transfer of genes from probiotic bacteria to commensals and pathogens in the gut indicates that the safety of probiotics is becoming an emerging concern (Toomey et al., 2009). Some researchers (Tóth et al., 2021, Selvin et al., 2020, Zarzecka et al., 2020) report that the problem of AMR concerns both starter and probiotic cultures.

The microbial cells in ferments have the chance to interact with different elements of the human microbiome, and many fermented foods are eaten raw. ARGs in fermented food microbiomes may be more dense than those in other raw foods (vegetables, fruits, etc.) since the densities of microbial cells in many ferments are high per gram of food consumed (Tóth et al., 2021, Selvinet al., 2020). Therefore, strains used as starter cultures or probiotics must be tested for AR, especially the ability to transfer these genes. In the case of many strains with probiotic properties, the presence of determinants of resistance to mobile genetic elements is found, and their ability to transfer genes through horizontal gene transfer to other bacteria, including pathogenic ones, is also confirmed (Chajęcka-Wierzchowska and Zadernowska, 2019).

The emergence of AMR in the aspect of probiotics has become a global concern (Selvinet al., 2020, WHO, 2023). Therefore, the member states of the European Union initiate actions leading to the reduction of excessive use of antibiotics and chemotherapeutics in humans and animals (EFSA, 2024). Many of the advancements in contemporary medicine are jeopardized by AMR. It increases the danger of certain medical procedures and treatments, including cancer chemotherapy, cesarean sections, and surgery, and makes infections more difficult to treat (WHO, 2023).

The agency responsible in the European Union for the regulation, assessment, and monitoring of food safety, food supplements, and therefore most probiotic products is the EFSA. The safety of probiotics draws from studies on their production methods and microbial characterization.

There is no uniform, universally accepted system for regulating the safety of probiotics. According to FAO/WHO, microorganisms that have been used safely in food for years have acquired the status of organisms safe for health human (GRAS). Rules for the gradual evaluation of probiotics have been introduced:

1. Strain identification.
2. *In vitro* studies in the evaluation of probiotic properties.
3. *In vivo* studies in animals and humans to determine safety and efficacy.

Probiotics are classified as food or dietary supplements (in the USA and Europe), as natural health products (in Canada) or as food for specific health uses (in Japan). In most countries, dietary supplements are subject to far less stringent regulatory requirements than medicines. As a result, they are not held to the same level of control and monitoring, both before and after entering the market. In contrast, medicines must meet strict safety regulations throughout their development and distribution. In the United States, dietary supplements are also prohibited from containing ingredients regulated as new drugs or biologics, unless those ingredients were approved prior to the introduction of relevant regulations (U.S. Food and Drug Administration: New Dietary Ingredients in Dietary Supplements – Background for Industry, U.S. FDA).

Detailed guidelines on the safety of probiotics in food were developed by the FAO/WHO Expert Group in 2002. Due to the fact that probiotic features are strain-dependent, the microorganism should be identified first by phenotypic and genotypic methods. The document entitled "Guidelines for the Evaluation of Probiotics in Food" contains indications regarding the study of bacterial functionality; it also emphasizes the need to assess the safety of the tested strains (Probiotics in Food, Report of a Joint FAO/WHO, 2020).

Currently, in Europe, the basic document on the safety of probiotics is the QPS list. Four factors are taken into consideration when determining a QPS status: (i) the taxonomic unit (TU) for which QPS is sought (also known as "taxonomic identification"); (ii) whether there is enough pertinent information about the proposed TU to draw conclusions about human/animal exposure through food or feed (also known as the "body of knowledge"); (iii) whether the proposed TU contains known "safety concerns"; and, finally, (iv) the intended end use (also known as the "intended use"). A "qualification" to rule out a danger linked to a TU may be created and added if one is found and tested at the strain or product level. The EFSA unit assigned to the application dossier assesses the qualifications of the microbial strain being examined (Allende et al., 2017).

Lack of acquired genes for all bacterial TUs, coding for resistance to important antimicrobials for people and animals is a generic criterion; if the relevant yeasts are to be employed as viable organisms in the food or feed chains, the absence of antimycotic resistance should be demonstrated. The term "for production purpose only" refers to a product's lack of live cells from the producing organism and can also be used to describe food and feed items made using microbial biomass (Koutsoumani et al., 2020).

The QPS status is assessed by the EFSA Panel on Biological Hazards. Since 2014, the assessment of microorganisms recommended for inclusion has been carried out twice a year, and every 3 years, the assessment of microorganisms previously entered on the list, that is, the QPS status is determined. The absence of a microorganism on the QPS list does not mean that it is dangerous to health, nor does it exclude the possibility of using products in the European Union (the principle of a long history of use and the absence of side effects is enough). It only means that the microorganism has not been evaluated by EFSA. To date, EFSA has rejected all submitted health claims for probiotics (Allende et al., 2017).

Microbial feed additives should not promote the spread of AMR in any other way or add to the pool of AMR genes already present in the gut bacterial community. The antibiotics that are taken into consideration include those that are pertinent to their usage in both humans and animals (also known as highly important antimicrobials, or HIAs) or critically important antimicrobials (CIAs) (WHO, 2016).

According to EFSA (2018), the estimation of antimicrobial susceptibility of bacteria intended for use as viable cells in feed additives and used as production organisms includes: (i) phenotypic testing based on the determination of a minimum inhibitory concentration (MIC) for a selected group of antimicrobials and (ii) search of the WGS for the presence of known AMR genes (EFSA, 2018).

The scheme proposed by the Panel on Additives and Products or Substances used in Animal Feed (FEEDAP Panel) for the AMR assessment of a bacterial strain used as a feed additive is shown in Figure 16.1. For the purpose of distinguishing resistant from susceptible strains, the FEEDAP Panel has defined microbiological cut-off values based on published data. On this basis, strains can be categorized as follows:

1. susceptible when its growth is inhibited at a concentration of a specific antimicrobial equal to or lower than the established cut-off value ($S \leq \times$ mg/L),
2. resistant when it is able to grow at a concentration of a specific antimicrobial higher than the established cut-off value ($R > \times$ mg/L).

The detection of the MIC above the cut-off values proposed by the FEEDAP Panel for one or more antimicrobials requires further investigation using genomic data to determine the nature of the resistance (EFSA, 2012):

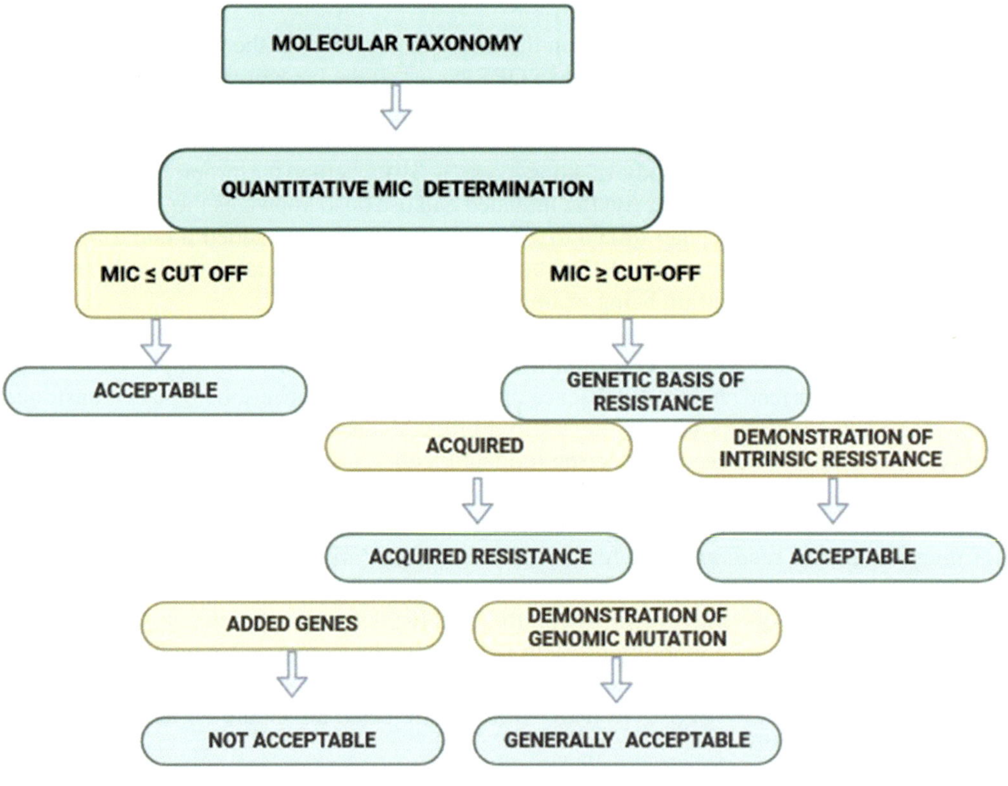

FIGURE 16.1 The scheme proposed by the Panel on Additives and Products or Substances used in Animal Feed (FEEDAP Panel) for the AMR assessment of a bacterial strain used as a feed additive. (Adopted from EFSA, 2012). Created with BioRender.com.

- If no known AMR gene is identified that can be linked to the phenotype, no further studies are required.
- If the phenotypic resistance can be directly related to the presence of a known AMR gene, this is considered as a hazard.

In all cases, if the genetic analysis reveals AMR genes for antimicrobials considered to be CIAs or HIAs, the MIC values should be determined and compared with values in the literature:

- If MIC≤(reference values), the likelihood of the AMR gene to become active should be assessed (e.g., based on sequence comparison with active genes).
- If MIC>(reference values), this is considered as a hazard (EFSA, 2018).

The FEEDAP Panel general recommendations on AMR of bacterial strains are shown in Table 16.2.

According to the EFSA, microorganisms with transferable genes for AR should be used less frequently or not at all in feed, fermented foods, and probiotic foods intended for human use. For usage in food and supplements, only probiotic strains that do not promote the spread of functional AR should be chosen (EFSA, 2007; Ouwehand et al., 2016). For food products containing multiple microbial strains, the viability should be separately assessed for each strain (Rychen et al., 2018).

The regulations regarding the occurrence of AMR in probiotics differ between regions of the world. The requirements relating to the issue – Regulatory Framework for Probiotics in Antimicrobial Resistance, the proposed solutions depending on the country are shown in Table 16.3.

An estimated 700,000–1 million fatalities globally are attributed to AMR each year; estimates suggest that this number could rise to 10 million by 2050.

Due to the seriousness of the AMR problem, EFSA has proposed several interactive communication tools addressing this issue (EFSA, News). One of them is a data visualization tool that presents resistance levels in humans, animals, and food of animal origin in the individual countries (EFSA, Interactive communication tools). It is important to note that a number of initiatives are run to bring attention to the issue of "AMR". The "AMR World Awareness Week" (WAAW) is an international campaign aimed at increasing knowledge of AMR and encouraging One Health stakeholders to adopt best practices that will slow the emergence and spread of drug resistance. The WAW is organized yearly from November 18 to 24 (World AMR Awareness Week, 2024).

TABLE 16.2
The FEEDAP Panel Classification of Microorganisms Used in Terms of AMR Combined with the Possibility of Their Use asFeed Additives (Rychen et al., 2018)

Serial No.	Bacterial Strain	Application as a Feed Additive
1.	Any bacterial strain carrying an intrinsic resistance to antimicrobial(s) presents a minimal potential for horizontal spread	May be used.
2.	Any bacterial strain carrying an acquired resistance to antimicrobial(s) that is shown to be due to chromosomal mutation(s) presents a low potential for horizontal spread	Generally may be used.
3.	Any bacterial strain carrying an acquired resistance to antimicrobial(s) that is shown to be due to the acquisition of genetic determinant(s) presents the greatest potential for horizontal spread.	Should not be used.
4.	The absence of information on the genetic nature of a demonstrated resistance.	Should not be used.

TABLE 16.3
Global Regulatory Framework Regarding the Occurrence of AMR of Probiotics in Food Supplements and Food Products

	Establishing Safety Factor	
Country	**Antimicrobial Resistance (AMR)**	**Regulatory**
European Union	Antimicrobial resistance risk assessment is required for all; determination of intrinsicor acquired resistance and potential transmissible antimicrobial resistance genes is required	**QPS** – https://zenodo.org/records/12793170 **Novel Foods:**https://efsa.onlinelibrary.wiley.com/doi/10.2903/j.efsa.2016.4594
USA	Antimicrobial resistance should be reported	**For Food Products**: https://www.federalregister.gov/documents/2016/08/17/2016-19164/substances-generally-recognized-as-safe **For Dietary Supplements**: https://www.federalregister.gov/documents/2016/10/04/2016-23931/dietary-supplements-new-dietary-ingredient-notifications-and-related-issues-revised-draft-guidance
Canada	The phenotypic and genotypic investigation/explanation of resistance; Demonstration of lack of horizontal antimicrobial resistance transfer ability	**Food**: https://www.canada.ca/en/health-canada/services/food-nutrition/food-labeling/health-claims/accepted-claims-about-nature-probiotic-microorganisms-food.html
Brasil	Resistance to antimicrobials must be investigated	**Food Supplements**: http://antigo.anvisa.gov.br/documents/10181/5809185/IN_76_2020_COMP.pdf/28f071f8-8079-4671-a1ec-b47c8dd30917
India	The risk of transferable antimicrobial resistance must be investigated	**Foods with added probiotic ingredients (Schedule VII):**https://fssai.gov.in/upload/uploadfiles/files/Compendium_Nutra_29_09_2021.pdf

Source: Adapted from Roe et al. (2022).

16.5 LEGAL REGULATIONS AND RECOMMENDATIONS FOR PROBIOTICS IN ANTIMICROBIAL RESISTANCE

Due to the widespread use of probiotics, their safety is crucial. It is known that probiotics must be identified at the strain level and safe for use in humans. Food cultures (FCs), which are microorganisms used directly in food production, are considered in the EU as food ingredients with a very long history of use in a wide variety of food products. FC, like other food ingredients, must meet the requirements set out in EU Regulation No. 178/2002 (Regulation (EC) No. 178/2002, 2002) including Article 14, point 1: "Food shall not be placed on the market if it is unsafe".

Separate regulations apply to Novel Food (NF). Novelty, in this context, means food not consumed to a significant extent in the Community before 14 May 1997. If this is the case, then the food must comply with the updated novel foods Regulation (EU Regulation 2015/2283) (Regulation (EU) 2015/2283, 2015) "Novel foods and food ingredients covered by this Regulation are those that are not currently used for human consumption. This Regulation applies to foods and food ingredients belonging to the following categories: (…) food consisting of, isolated from or produced from microorganisms, fungi or algae; (…)." Before being placed on the market, NF must be subject to a risk price conducted by EFSA and a risk management assessment by the European Commission and undergo the procedure of issuing the necessary permits.

Gene modification of a microorganism used as FC is also considered NF, but in this particular case, the NF notification procedure is waived with regard to the regulations on genetically modified food and must comply with EU guidelines and EU law (Directive 2001/18/EC, 2001, Regulation (EC) No 1830/2003, 2003) and the compatible legal act with detailed rules for the implementation of Regulation (EC) No 1829/2003 of the European Parliament and of the Council as regards the application for the authorization of new genetically modified food and feed, the notification of existing products and adventitious or technically unavoidable presence of genetically modified material which has benefited from a favorable risk evaluation (Commission Regulation (EC) No 641/2004 of 6 April 2004).

The QPS list is EFSA's rapid risk assessment tool that is used by EFSA panels to assess products containing microorganisms requiring pre-market authorization (e.g., feed additive cultures, cell factories producing enzymes/additives/vitamins, new microorganisms, plant protection). It covers all risk assessments for microorganisms for use by humans, animals, and the environment – "from farm to fork". In the EFSA, Biological Hazards (BIOHAZ) panel assesses the safety of biological agents, taking into account:

- definition of a taxonomic unit (determining group identity),
- knowledge base,
- possible safety concerns (pathogenicity),
- for some species, final use.

The QPS list includes species presented to EFSA or proposals submitted by interested parties during public consultations. The last published update of the list includes 110 species. Microorganisms that are not on the QPS list are not necessarily considered hazardous. There are more reasons why they are not on the QPS list, for example, the species has not been assessed by EFSA or is not assessed as safe regardless of substrate, condition and use and then undergoes a full safety assessment for the specific use. FCs with a long history of safe use in food are considered traditional food ingredients and are legally permitted for use in food in the EU without pre-market authorization (Koutsoumani et al., 2020).

In order to fill the gap left by the regulations, EFFCA (European Food and Feed Cultures Association) proposed the definition of FC as "safe live bacteria, yeasts or molds used in food production, which are themselves a characteristic food ingredient. FC preparations are preparations consisting of concentrates (>10^8 colony-forming units/g or ml) of one or more live and active species and/or strains of microorganisms, including unavoidable components of the medium carried over from fermentation and components that are necessary for their survival, storage and facilitating their use in the food production process, and in some cases they are standardized in small numbers with carriers".

In line with the suggestion of FAO and WHO in 2006, The European Parliament and the Council published EC Regulation No. 1924/2006 (Regulation (EC) No 1924/2006, 2014) on declarations of nutrition and health related to food. This document applies to everyone's nutrition and health claims for all types of food intended for consumers in final form, including probiotic products sold with health claims within the European Union.

The EFSA has developed a scientific and technical guide for applications to obtain approval for health claims. This office may consent to the issuance of a health claim appearing on a product label if the product has an adequate amount of scientific evidence to support its effects. To determine whether there is a relationship between a substance (such as a probiotic) and the desired one outcome (such as maintaining a healthy digestive system), it is important to investigate the following criteria:

- temporal relationship,
- strength of the relationship,
- dose-response,
- duplication of arrangements,
- biological credibility,

- considering alternative explanations,
- cessation of exposure,
- consistency of results with other knowledge.

Well-designed randomized controlled trials (RCT), best evaluated through a review and meta-analysis process, provide evidence about causation. However, such tests are not always available for a given food group (probiotic foods), so there is insufficient evidence confirming the health benefits resulting from their use (Hill et al., 2014). Due to the need to provide sufficient scientific evidence confirming the health benefits of using probiotic food, no probiotic claim was deemed sufficiently substantiated for food in the European Union. EFSA has not yet agreed to make claims about "probiotics" on any products available on the EU market.

The only country in the EU that has published a list of probiotics and a guide on their use is Italy (Zielińska et al., 2018). In Italy, there has been a tradition of using "good bacteria" for over 30 years, administered as dietary supplements or food ingredients that support the intestinal microflora. Therefore, the Italian Ministry of Health has regulated the use of bacteria in food and confirmed the use of the word probiotic in relation to food and dietary supplements provided:

- occurrence of a minimum of $1x10^9$ CFU of live bacterial cells administered daily,
- full genetic characterization of the probiotic strain,
- documented history of safe use on the Italian market (Hill et al., 2014).

The regulations regarding probiotic food in Denmark, in addition to the regulations described above in force throughout the European Union, are also subject to the Danish Veterinary and Food Authority (Danish Statute on Food Additives /2000, Order No. 1247 on Food Additives). By law, the manufacturer is obliged to determine the technological function of the culture used. The goal of a positive culture should be defined. If the culture is added for biopreservation purposes, information on the mode of action of the culture and the active biopreservation ingredient should be provided. In addition, the manufacturer must specify the type of food in which the culture will be used and how the culture will be stored, indicating possible culture carriers (Wright, 2005).

According to Danish requirements, detailed information must be provided regarding the culture to be added to the food product, that is:

- characteristics of culture,
- purity of the culture in the form of test results for the presence of other microorganisms,
- information whether the culture is capable of creating toxic or antibiotic compounds,
- information on microbiological/toxicological tests that enable assessment of whether a given culture can be used appropriately,
- all results must be available for each culture as signed certificates of analysis, not just as reference values, and must document the purity of the individual cultures.

The safety of traditionally used LAB has never been questioned, however, the used isolates of intestinal bacteria such as *Bifidobacterium, Lactobacillus*, and *Enterococcus*, offered to consumers in large numbers as probiotics, have raised safety concerns. The definition of probiotics covers issues related to the use of probiotics in the diet of healthy people, but is insufficient when probiotics are recommended for specific diseases. So far, no universally accepted recommendations for the use of probiotics have been developed. In the group of sick people, there is a more frequent use, even abuse, of antibiotics, which adversely affects the commensal intestinal microflora. In the case of high-risk groups, side effects may occasionally be severe. Paradoxically, the group of the most seriously ill patients may benefit the most from the use of probiotics, but they are also most at risk of side effects, including sepsis. Therefore, further, multi-disciplinary research conducted by various independent research centers is needed. Additionally, the safety of probiotics should be confirmed in the sick and be based on a research that considered large group of volunteers and a long observation period.

According to the WHO/FAO working group, probiotics may theoretically be responsible for four types of side effects:

1. systemic infections,
2. unfavorable metabolic effects,
3. excessive immune stimulation in sensitive,
4. gene transmission.

Almost all patients with reported infections caused by probiotic or probiotic-like microorganisms had comorbidities that predisposed them to infection, in particular abnormal heart valves in cases of endocarditis and the presence of a catheter in cases of sepsis. No reports have described sepsis associated with the use of probiotics in healthy individuals. Other opportunistic risk factors such as old age, pregnancy, immunodeficiency, or digestive changes have not been identified as risk factors for probiotic infections. In almost all cases, the source of the infecting organisms appears to be human commensal flora rather than food. It is also suggested (end confirmed in animal studies) that the main risk factors for probiotic sepsis are: impaired immunity, including a weakened state or malignant tumor, and premature babies (Kołożyn-Krajewska et al., 2011).

One theoretical concern about the safety of probiotics is that some have been designed or selected to adhere well to the intestinal mucosa, which is thought to be important for their mechanism of action. Adhesion to the intestinal mucosa may also increase bacterial translocation and virulence. Therefore, the strongest probiotics may have increased pathogenicity. The association between mucosal adhesion and pathogenicity in *Lactobacillus* spp. is supported by the finding that blood cultures isolate *Lactobacillus* spp. adhere to intestinal mucus in greater numbers than isolates from human feces or dairy products. It is estimated that the cases of infections caused by LAB represent the following percentages: in the case of *Enterococcus* 5%–15%, *Lactobacillus* 0.1%, *Leuconastoc* 0.01% of cases of infectious bacteremia (Kołożyn-Krajewska et al., 2011, Ojha et al., 2023).

Enterococci are considered major nosocomial pathogens that cause bacteremia, endocarditis, urinary tract infections, and others. Enterococci are opportunistic pathogens and usually cause infections in patients with severe underlying disease or immunocompromised. *E. fecalis* predominates among enterococci isolated from human infections (over 80%), while *E. faecium* is associated with most other infections (Kołożyn-Krajewska et al., 2011, Bourgin et al., 2021).

Lactobacillus sp. strains exhibit undesirable metabolic effects, causing changes in the activity of bile salt hydrolase, which may affect absorption processes in the gastrointestinal tract. Moreover, the production of lactic acid by bacteria may cause lactic acidosis in sensitive patients, for example, those with short bowel syndrome. Carbohydrates not absorbed in the intestine are available to bacteria that produce excessive amounts of lactic acid. The neurotoxic effect of D-lactic acid causes neurological complications, such as behavioral disorders, speech disorders, and in extreme cases leads to impaired consciousness and encephalopathic coma. In the case of *L. lactis*, the ability to produce biogenic amines, such as cadaverine and putrescine, has been demonstrated, which have a harmful effect on the nervous and circulatory systems (Kołożyn-Krajewska et al., 2011, Bourgin et al., 2021).

A small number of studies have shown that selected probiotic strains have the ability to increase the immune response. This is beneficial in cases of decreased immune response, infection, or vaccination. However, this would not be desirable in the case of allergies, where a reduction in the allergy-causing immune response is required. No reports of an increase in allergic reactions or weakening of immune functions after the use of probiotics have been published so far (Kołożyn-Krajewska et al., 2011, Ojha et al., 2023).

Antibiotic-resistant plasmids can be conjugatively transferred to other strains, species, and even genera, including potential human or animal pathogens. This raises the question of whether genetic exchange can occur between food microbes and human gut microbiota or pathogens. Among LAB,

AR has been reported in strains isolated from pork and beef, and especially from the gastrointestinal tract of humans and animals. Many LAB species such as *L. rhamnosus, L. casei, L. paracasei*, many other lactobacilli, pediococci, and leuonostobacteria are intrinsically resistant to vancomycin due to the special structure of their cell wall peptidoglycan precursors. Although enterococci and the abovementioned bacteria apparently share a similar phenotype of vancomycin resistance, their safety implications are radically different due to the different mechanism of vancomycin resistance (Kołożyn-Krajewska et al., 2011, Ojha et al., 2023).

It is believed that enterococcal strains may become resistant when genes are transferred into the strain. Therefore, transferred resistance to vancomycin among enterococcal strains is a serious clinical problem, also due to the possibility of transferring resistance to staphylococcal strains. Therefore, the safety and long-term impact of using *E. faecium* strains as probiotics, particularly in relation to AR, require thorough evaluation.

Currently, the status of probiotic-based products is fraught with confusion as different regulatory agencies, in different countries, define and categorize probiotics differently, and therefore there is much variation among regulators, manufacturers, and consumers regarding claims made about probiotic-containing foods. This seems to require further research into the concept, properties, and rational use of probiotics for healthy and sick people.

16.6 CONCLUSION

The increased interest in probiotic usage, especially in the form of functional food, raises concerns about their safety. Although the occurrence of probiotic's adverse effects is rare, the possible health risks should be considered and surveilled. The safety issues are of key importance especially in vulnerable populations. Moreover, probiotics may harbor antibiotic resistance. Nevertheless, the evidence of AMR gene transmissions from probiotics to the pathogen is scarce. Molecular typing, including whole genome sequences, offers new possibilities and therefore, should be used for probiotic assessment.

Effective risk management, harmonization of international guidelines, and responsible approach to their application will ensure the maximization of benefits while minimizing risks. A big challenge here is the need for global cooperation in combating AMR. Probiotic regulation should be in line with international strategies to combat AMR. It is also crucial to raise awareness among consumers and healthcare professionals about the safe use of probiotics.

REFERENCES

Aghamohammad, S., & Rohani, M. (2022). Antibiotic resistance and the alternatives to conventional antibiotics: The role of probiotics and microbiota in combating antimicrobial resistance. *Microbiological Research*, *267*, 127275. https://doi.org/10.1016/j.micres.2022.127275

Allende, A., Bolton, D., Chemaly, M., Davies, R., Escámez, P. S. F., Girones, R., Herman, L., Koutsoumanis, K., Lindqvist, R., Nørrung, B., Ricci, A., Robertson, L., Ru, G., Sanaa, M., Simmons, M., Skandamis, P., Snary, E., Speybroeck, N., Ter Kuile, B., Threlfall, J., Wahlström, H. (2017). Update of the list of QPS-recommended biological agents intentionally added to food or feed as notified to EFSA 5: Suitability of taxonomic units notified to EFSA untilSeptember 2016. *EFSA Journal*, *15*(3), 4663. https://doi.org/10.2903/j.efsa.2017.4663

Bafeta, A., Koh, M., Riveros, C., & Ravaud, P. (2018). Harms reporting in randomized controlled trials of inter ventions aimed at modifying microbiota: A systematic review. *Annals of Internal Medicine*, *169*(4), 40–247.

Borase, H., Dwivedi, M. K., Krishnamurthy, R., & Patil, S.(2022). Chapter 30- Probiotics: Health safety considerations. In: *Probiotics in the Prevention and Management of Human Diseases* (pp. 449–463); Dwivedi, M. K., Amaresan, N., Sankaranarayanan, A., Kemp E. H., (Eds.); Academic Press. https://www.researchgate.net/profile/Hiteshkumar-Patel-5/publication/357225636_Dwivedi-PPD-1632906/links/61c216d98bb2010184294907/Dwivedi-PPD-1632906.pdf

Bourgin, M., Kriaa, A., Mkaouar, H., Mariaule, V., Jablaoui, A., Maguin, E., & Rhimi, M. (2021). Bile salt hydrolases: At the crossroads of microbiota and human health. *Microorganisms*, *9*(6), 1122. https://doi.org/10.3390/microorganisms9061122

Boyte, M. -E., Akhtar, N., Koshy, B., & Roe, A. L. (2025). A review of probiotic ingredient safety supporting monograph development conducted by the United States Pharmacopeia (USP). *Journal of Dietary Supplements*, *22*(1), 123–161. https://doi.org/10.1080/19390211.2024.231448.8

Byarugaba, D.K. (2009). Mechanisms of antimicrobial resistance. In Springer *eBooks* (pp. 15–26). https://doi.org/10.1007/978-0-387-89370-9_2

CARD. (2024). A bioinformatic database of resistance genes, their products and associated phenotypes. https://card.mcmaster.ca/. Accessed 2024 Jul 28.

Cao, C., Wang, J., Liu, Y., Kwok, L., Zhang, H., & Zhang, W. (2020). Adaptation of Lactobacillus plantarum to ampicillin involves mechanisms that maintain protein homeostasis. *mSystems*, *5*(1). https://doi.org/10.1128/msystems.00853-19

Celandroni, F., Vecchione, A., Cara, A., Mazzantini, D., Lupetti, A., & Ghelardi, E. (2019). Identification of Bacillus species: Implication on the quality of probiotic formulations. *PLoS One*, *14*(5), e0217021. https://doi.org/10.1371/journal.pone.0217021

Centers for Disease Control and Prevention (U.S.) (2019).Antibiotic resistance threats in the United States. US Department of Health and Human Services. https://ndc.services.cdc.gov/wp-content/uploads/Antibiotic-Resistance-Threats-in-the-United-States-2019.pdf. Accessed 2024 Jul 27.

Cepec, E., & Trček, J. (2022). Antimicrobial resistance of acetobacter and komagataeibacter species originating from vinegars. *International Journal of Environmental Research and Public Health*, *19*(1), 463. https://doi.org/10.3390/ijerph19010463

Chajęcka-Wierzchowska, W, & ZadernowskaA. (2019). Bakterie fermentacji mlekowej w tym szczepy probiotyczne jako rezerwuar genów oporności na antybiotyki. Lactic acid bacteria including probiotic strains as a reservoir of antibiotic resistance genes. *Żywność, Nauka Technologia, Jakość/Food Science Technology Quality*. *26*, 22–35. doi:10.15193/zntj/2019/120/29

Commission Regulation (EC) No 641/2004 of 6 April 2004on detailed rules for the implementation of Regulation (EC) No 1829/2003 of the European Parliament and of the Council as regards the application for the authorisation of new genetically modified food and feed, the notification of existing products and adventitious or technically unavoidable presence of genetically modified material which has benefited from a favourable risk evaluation (Text with EEA relevance).

Cordaillat-Simmons, M., Rouanet, A., & Pot, B. (2020). Live biotherapeutic products: The importance of a defined regulatory framework. *Experimental & Molecular Medicine*, *52*(9), 1397–1406. https://doi.org/10.1038/s12276-020-0437-6

Courvalin, P. (2006) Antibiotic resistance: The pros and cons of probiotics. *Digestive and Liver Disease 38*(2), 261–265. https://doi.org/10.1016/s1590-8658(07)60006-1

Dadgostar, P. (2019). Antimicrobial resistance: Implications and costs. *Infection and Drug Resistance*, *12*, 3903–3910. https://doi.org/10.2147/IDR.S234610

Danish Statute on Food Additives/2000, OrderNo. 1247 on Food Additives. https://leap.unep.org/en/countries/dk/national-legislation/order-no-1247-food-additives. Accessed 2024 Jul 28.

Das, D. J., Shankar, A., Johnson, J. B., & Thomas, S. (2019). Critical insights into antibiotic resistance transferability in probiotic Lactobacillus. *Nutrition*, *69*, 110567. https://doi.org/10.1016/j.nut.2019.110567

Directive 2001/18/EC of the European Parliament and of the Council of12 March 2001on the deliberate release into the environment of genetically modified organisms and repealing Council Directive 90/220/EEC – Commission Declaration. Official Journal of the European Communities, Current consolidated version: 27/03/2021.

Doron, S., & Snydman, D. R. (2015). Risk and safety of probiotics. *Clinical Infectious Diseases*, *60*(suppl_2), S129–S134. https://doi.org/10.1093/cid/civ085

Dou, W., Abdalla, H. B., Chen, X., Sun, C., Chen, X., Tian, Q., Wang, J., Zhou, W., Chi, W., Zhou, X., Ye, H., Bi, C., Tian, X., Yang, Y., & Wong, A. (2022). ProbResist: A database for drug-resistant probiotic bacteria. *Database*, *2022*. https://doi.org/10.1093/database/baac064

Dudek-Wicher, R., Junka, A., Paleczny, J., & Bartoszewicz, M. (2020). Clinical trials of probiotic strains in selected disease entities. *International Journal of Microbiology*, 2020, 8854119. https://doi.org/10.1155/2020/8854119

Eckburg, P. B., Bik, E. M., Bernstein, C. N., Purdom, E., Dethlefsen, L., Sargent, M., Gill, S. R., Nelson, K. E., & Relman, D. A. (2005). Diversity of the human intestinal microbial flora. *Science*, *308*(5728), 1635–1638. https://doi.org/10.1126/science.1110591

EPSA. (2012). Guidance on the assessment of bacterial susceptibility to antimicrobials of human and veterinary importance.*EPSA Journal*, *10*(6), 2740. https://doi.org/10.2903/j.efsa.2012.2740

EFSA. (2018). Guidance on the characterisation of microorganisms used as feed additives or as production organisms. *EFSA Journal*, *16*(3), e05206. https://doi.org/10.2903/j.efsa.2018.5206

EFSA. (2024). The European Union summary report on antimicrobial resistance in zoonotic and indicator bacteria from humans, animals and food in 2021–2022. *EFSA Journal*, *22*(2), e8583. https://doi.org/10.2903/j.efsa.2024.8583

EFSA BIOHAZPanel (EFSA Panel on Biological Hazards), Koutsoumanis, K., Allende, A., Alvarez-Ordóñez, A., Bolton, D., Bover-Cid, S., Chemaly, M., Davies, R., De Cesare, A., Hilbert, F., Lindqvist, R., Nauta, M., Peixe, L., Ru, G., Simmons, M., Skandamis, P., Suffredini, E., Sandro Cocconcelli, P., Fernández Escámez, P. S., Prieto Maradona, M., Querol, A., Evaristo Suarez, J., Sundh, I., Vlak, J., Barizzone, F., Correia, S.,& Herman, L.2020. Scientific opinion on the update of the list of QPS-recommended biological agents intentionally added to food or feed as notified to EFSA (2017–2019). *EFSA Journal*, *18*(2), 5966. https://doi.org/10.2903/j.efsa.2020.5966

EFSA. Interactive communication tools. https://multimedia.efsa.europa.eu/dataviz-2022/index.htm. Accessed 2024 Jun 10.

EFSA. News. https://www.efsa.europa.eu/en/news/continued-efforts-needed-fight-antimicrobial-resistance-amr-humans-and-animals. Accessed 2024 Jul 10.

Elshaghabee, F. M. F., Rokana, N., Gulhane, R. D., Sharma, C., & Panwar, H. (2017). Bacillus as potential probiotics: Status, concerns, and future perspectives. *Frontiers in Microbiology*, *8*, 1490. https://doi.org/10.3389/fmicb.2017.01490

ESLP—European Scientific League for Probiotics. www.probioleague.org. Accessed 2024 Jul 15.

European Committeeon Antimicrobial Susceptibility Testing. Breakpoints tables for interpretation of MICs and zones diameters. Version8.1, 2018. https://www.eucast.org. Accessed 2024 Jun 10.

European Food SafetyAuthority (EFSA). (2007). Introduction of a Qualified Presumption of Safety (QPS) approach for assessment of selected microorganisms referred to EFSA—Opinion of the Scientific Committee. *EFSA Journal*, 5, 587.

European Pharmacopoeia 3053E (2019). General Monograph on Live Biotherapeutic Products Published.pdf; EDQM.https://www.edqm.eu/en/-/live-biotherapeutic-products-lbps-unprecedented-quality-requirements-by-the-ph.-eur.-commission-1. Accessed 2024 Aug 28.

Fatahi-Bafghi, M., Naseri, S., & Alizehi, A. (2022). Genome analysis of probiotic bacteria for antibiotic resistance genes. *Antonie Van Leeuwenhoek*, *115*(3), 375–389. https://doi.org/10.1007/s10482-021-01703-7

FDA Early Clinical Trials with Live Biotherapeutic Products: Chemistry, Manufacturing, and Control Information. (2016). https://www.fda.gov/downloads/BiologicsBloodVaccines/GuidanceCompliance RegulatoryInformation/Guidances/General/UCM292704.pdf. Accessed 2024 Aug 20.

Goldstein, E. J. C., Tyrrell, K. L., & Citron, D. M. (2015). Lactobacillus species: Taxonomic complexity and controversial susceptibilities. *Clinical Infectious Diseases*, *60*(suppl_2), S98–S107. https://doi.org/10.1093/cid/civ072

Hata, N. N. Y., Surek, M., Sartori, D., Serrato, R. V., & Spinosa, W. A. (2023). Role of acetic acid bacteria in food and beverages. *Food Technology and Biotechnology*, *61*(1), 85–103. https://doi.org/10.17113/ftb.61.01.23.7811

Hempel, S., Newberry, S., Ruelaz, A., Wang, Z., Miles, J. N. V., Suttorp M.J., Johnsen, B., Shanman, R., Slusser W., Fu, N., Smith, A., Roth, B., Polak, J., Motala, A., Perry, T., & Shekelle. P. G. (2011). Safety of probiotics to reduce risk and prevent or treat disease. *Evidence Report Technology Assessment, 200*, 1–645. https://www.ahrq.gov/downloads/pub/evidence/pdf/probiotics/probiotics.pdf

Hill, C., Guarner, F., Reid, G., Gibson, G. R., Merenstein, D. J., Pot, B., Morelli, L., Canani, R. B., Flint, H. J., Salminen, S., Calder, P. C., & Sanders, M. E. (2014). The International Scientific Association for Probiotics and Prebiotics consensus statement on the scope and appropriate use of the term probiotic. *Nature Reviews Gastroenterology & Hepatology*, *11*(8), 506–514. https://doi.org/10.1038/nrgastro.2014.66

Kojic, M., Jovcic, B., Miljkovic, M., Novovic, K., Begovic, J., & Studholme, D. J. (2020). Large-scale chromosome flip-flop reversible inversion mediates phenotypic switching of expression of antibiotic resistance in lactococci. *Microbiological Research*, *241*, 126583. https://doi.org/10.1016/j.micres.2020.126583

Kolaček, S., Hojsak, I., Canani, R.B., Guarino, A., Indrio, F., OrelR., et al. ESPGHAN working group for probiotics and prebiotics. (2017). Commercial probiotic products: A call for improved quality control. A position paper by the ESPGHAN working group for probiotics and prebiotics. *Journal of Pediatric Gastroenterology and Nutrition, 65*(1), 117–24. https://doi.org/10.1097/mpg.0000000000001603

Kołożyn-Krajewska, D., Dolatowski, Z. J., & Zielińska, D. (2011). Risk assessment of probiotic use particularly in meat products – A review. *Fleischwirtschaft*, *4*, 61–68.
Li, T., Teng, D., Mao, R., Hao, Y., Wang, X., & Wang, J. (2020). A critical review of antibiotic resistance in probiotic bacteria. *Food Research International*, *136*, 109571. https://doi.org/10.1016/j.foodres.2020.109571
Li, Z., & Nair, S. K. (2012). Quorum sensing: How bacteria can coordinate activity and synchronize their response to external signals?*Protein Science*, *21*(10), 1403–1417. https://doi.org/10.1002/pro.2132
Liu, X., Zhao, H., & Wong, A. (2024). Accounting for the health risk of probiotics. *Heliyon*, *10*(6), e27908. https://doi.org/10.1016/j.heliyon.2024.e27908
Medicines | European Medicines Agency (EMA) (europa.eu); https://www.ema.europa.eu/en/homepage. Accessed 2024 Aug 24.
MEGARes – Microbial Ecology Group (MEG). Research addressing the issues of microbial ecology in animal, public, and environmental health. https://www.meglab.org/. Accessed 2024 Jul 28.
Merenstein, D., Pot, B., Leyer, G., Ouwehand, A. C., Preidis, G. A., Elkins, C. A., Hill, C., Lewis, Z. T., Shane, A. L., Zmora, N., Petrova, M. I., Collado, M. C., Morelli, L., Montoya, G. A., Szajewska, H., Tancredi, D. J., & Sanders, M. E. (2023). Emerging issues in probiotic safety: 2023 perspectives. *Gut Microbes*, *15*(1). https://doi.org/10.1080/19490976.2023.2185034
Montassier, E., Valdés-Mas, R., Batard, E., Zmora, N., Dori-Bachash, M., Suez, J., & Elinav, E. (2021). Probiotics impact the antibiotic resistance gene reservoir along the human GI tract in a person-specific and antibiotic-dependent manner. *Nature Microbiology*, *6*(8), 1043–1054. https://doi.org/10.1038/s41564-021-00920-0
Morelli, L., Sarra, P. G., & Bottazzi, V. (1988). In vivo transfer of pAMβ1 from Lactobacillus reuteri to Enterococcus faecalis. *Journal of Applied Bacteriology*, *65*(5), 371–375. https://doi.org/10.1111/j.1365-2672.1988.tb01905.x
Nataraj, B. H., Behare, P. V., Yadav, H., & Srivastava, A. K. (2023). Emerging pre-clinical safety assessments for potential probiotic strains: A review. *Critical Reviews in Food Science and Nutrition*, *64*(23), 8155–8183. https://doi.org/10.1080/10408398.2023.2197066
National Database of Antibiotic Resistant Organisms (NDARO). Pathogen Detection – NCBI (nih.gov). https://www.ncbi.nlm.nih.gov/pathogens/antimicrobial-resistance/. Accessed 2024 Jul 28.
Neffe-Skocińska, K., Długosz, E., Szulc-Dąbrowska, L., & Zielińska, D. (2024). Novel Gluconobacter oxydans strains selected from Kombucha with potential postbiotic activity. *Applied Microbiology and Biotechnology*, *108*(1). https://doi.org/10.1007/s00253-023-12915-4
NHGRI.(2022). Artificial intelligence, machine learning and genomics. National Human Genome Research Institute, Bethesda, MD. https://www.genome.gov/about-genomics/educational-resources/fact-sheets/artificial-intelligence-machine-learning-and-genomics. Accessed 2024 Aug 2.
NIH Human Microbiome Project—About the Human Microbiome (hmpdacc.org). https://commonfund.nih.gov/hmp/publications. Accessed 2024 Aug 20.
O'Toole, P. W., Marchesi, J. R., & Hill, C. (2017). Next-generation probiotics: the spectrum from probiotics to live biotherapeutics. *Nature Microbiology*, *2*(5). https://doi.org/10.1038/nmicrobiol.2017.57
Oba, S., Yildirim, T., & Karataş, Ş. M. (2022). Probiotics safety aspect of functional foods. Journal of Culinary Science & Technology, 1–30. https://doi.org/10.1080/15428052.2022.2135156
Ojha, A. K., Shah, N. P., Mishra, V., Emanuel, N., & Taneja, N. K. (2023). Prevalence of antibiotic resistance in lactic acid bacteria isolated from traditional fermented Indian food products. *Food Science and Biotechnology*, *32*(14), 2131–2143. https://doi.org/10.1007/s10068-023-01305-1
Okaiyeto, S. A., Sutar, P. P., Chen, C., Ni, J., Wang, J., Mujumdar, A. S., Zhang, J., Xu, M., Fang, X., Zhang, C., & Xiao, H. (2024). Antibiotic resistant bacteria in food systems: Current status, resistance mechanisms, and mitigation strategies. *Agriculture Communications*, *2*(1), 100027. https://doi.org/10.1016/j.agrcom.2024.100027
Olatunji, I., Bardaji, D. K. R., Miranda, R. R., Savka, M. A., & Hudson, A. O. (2024). Artificial intelligence tools for the identification of antibiotic resistance genes. *Frontiers in Microbiology*, *15*. https://doi.org/10.3389/fmicb.2024.1437602
Ouwehand, A. C., Forssten, S., Hibberd, A. A., Lyra, A., & Stahl, B. (2016). Probiotic approach to prevent antibiotic resistance. *Annals of Medicine*, *48*(4), 246–255. https://doi.org/10.3109/07853890.2016.1161232
PATRIC –Database Commons (cncb.ac.cn). (2017). https://www.patricbrc.org/. Accessed 2024 Jul 21.
Phillips, I.,et al. (2004). Does the use of antibiotics in food animals pose a risk to human health? A critical review of published data. *Journal of Antimicrobial Chemotherapy*, *53*(1), 28–52. https://doi.org/10.1093/jac/dkg483

Probioticsin Food. (2002). Health and nutritional properties and guidelines for evaluation. Report of a Joint FAO/WHO Working Group on Drafting Guidelines for the Evaluation of Probiotics in Food.https://openknowledge.fao.org/server/api/core/bitstreams/382476b3-4d54-4175-803f-2f26f3526256/content. Accessed 2024 Jul 20.

Purdel, C., Ungurianu, A., Adam-Dima, I., & Margină, D. (2023). Exploring the potential impact of probiotic use on drug metabolism and efficacy. *Biomedicine & Pharmacotherapy*, *161*, 114468. https://doi.org/10.1016/j.biopha.2023.114468

Regulation (EC) No 178/2002 of the European Parliament and of the Council of 28 January2002 laying down the general principles and requirements of food law, establishing the European Food Safety Authority and laying down procedures in matters of food safety. Official Journal of the European Communities1.2.2002. Accessed 2024 Jul 10.

Regulation (EC) No 1830/2003 of the European Parliament and of the Council of 22September 2003 concerning the traceability and labelling of genetically modified organisms and the traceability of food and feed products produced from genetically modified organisms and amending Directive 2001/18/EC. Official Journal of the European Communities22.9.2003. Accessed 2024 Jul 10.

Regulation (EC) No 1924/2006 of the European Parliament and of the Council of 20December 2006 on nutrition and health claims made on foods. Official Journal of the European Communities13.12.2014. https://eur-lex.europa.eu/legal-content/en/ALL/?uri=CELEX%3A32006R1924. Accessed 2024 Jul 10.

Regulation (EU) No 2015/2283 of the European Parliament and of the Council of 25 November 2015on novel foods, amending Regulation (EU) No 1169/2011 of the European Parliament and of the Council and repealing Regulation (EC) No 258/97 of the European Parliament and of the Council and Commission Regulation (EC) No 1852/2001. Official Journal of the European Union11.12.2015. Accessed 2024 Jul 10.

Report Top Ten Trends, Innova Market Insights. https://www.innovamarketinsights.com/trends/global-health-trends/. Accessed 2024 May 20.

Roe, A. L., Boyte, M., Elkins, C. A., Goldman, V. S., Heimbach, J., Madden, E., Oketch-Rabah, H., Sanders, M. E., Sirois, J., & Smith, A. (2022). Considerations for determining safety of probiotics: A USP perspective. *Regulatory Toxicology and Pharmacology*, *136*, 105266. https://doi.org/10.1016/j.yrtph.2022.105266

Rouanet, A., Bolca, S., Bru, A., Claes, I., Cvejic, H., Girgis, H., Harper, A., Lavergne, S. N., Mathys, S., Pane, M., Pot, B., Shortt, C., Alkema, W., Bezulowsky, C., Blanquet-Diot, S., Chassard, C., Claus, S. P., Hadida, B., Hemmingsen, C.,... Cordaillat-Simmons, M. (2020). Live Biotherapeutic Products, a road map for safety assessment. *Frontiers in Medicine*, *7*. https://doi.org/10.3389/fmed.2020.00237

Rychen, G., Aquilina, G., Azimonti, G., Bampidis, V., De Lourdes Bastos, M., Bories, G., Chesson, A., Cocconcelli, P. S., Flachowsky, G., Gropp, J., Kolar, B., Kouba, M., López-Alonso, M., Puente, S. L., Mantovani, A., Mayo, B., Ramos, F., Saarela, M., Villa, R. E.,... Galobart, J. (2018). Guidance on the characterisation of microorganisms used as feed additives or as production organisms. *EFSA Journal*, *16*(3). https://doi.org/10.2903/j.efsa.2018.5206

Rzepkowska, A., Zielińska, D., & Kołożyn-Krajewska, D. (2014). Antybiotykooporność bakterii z rodzaju Lactobacillus pochodzących z żywności, jako kryterium stawiane probiotykom/ Antibiotic resistance to bacteria of Lactobacillus genus isolated from food, as a criterion for probiotic. *Zeszyty Problemowe Postępów Nauk Rolniczych*, *578*, 99–110.

Sanders, M. E., Akkermans, L. M., Haller, D., Hammerman, C., Heimbach, J. T., Hörmannsperger, G., & Huys, G. (2010). Safety assessment of probiotics for human use. *Gut Microbes*, *1*(3), 164–185. https://doi.org/10.4161/gmic.1.3.12127

Selvin, J., Maity, D., Sajayan, A., & Kiran, G. S. (2020). Revealing antibiotic resistance in therapeutic and dietary probiotic supplements. *Journal of Global Antimicrobial Resistance, 22*, 202–205. https://doi.org/10.1016/j.jgar.2020.02.007

Sharma, A., Lee, S., & Park, Y. (2020). Molecular typing tools for identifying and characterizing lactic acid bacteria: A review. *Food Science and Biotechnology*, *29*(10), 1301–1318. https://doi.org/10.1007/s10068-020-00802-x

Sionek, B., & Kołożyn-Krajewska, D. (2019). Safety in use of probiotics by humans. *Żywność Nauka Technologia Jakość/Food Science Technology Quality, 26*(3), 5–21.

Sionek, B., Szydłowska, A., Küçükgöz, K., & Kołożyn-Krajewska, D. (2023b). Traditional and new microorganisms in lactic acid fermentation of food. *Fermentation*, *9*(12), 1019. https://doi.org/10.3390/fermentation9121019

Sionek, B., Szydłowska, A., Trząskowska, M., & Kołożyn-Krajewska, D. (2024). The impact of physicochemical conditions on lactic acid bacteria survival in food products. *Fermentation*, *10*(6), 298. https://doi.org/10.3390/fermentation10060298

Sionek, B., Szydłowska, A., Zielińska, D., Neffe-Skocińska, K., & Kołożyn-Krajewska, D. (2023a). Beneficial bacteria isolated from food in relation to the next generation of probiotics. *Microorganisms*, *11*(7), 1714. https://doi.org/10.3390/microorganisms11071714

Sniffen, J. C., McFarland, L. V., Evans, C. T., & Goldstein, E. J. C. (2018). Choosing an appropriate probiotic product for your patient: An evidence-based practical guide. *PLoS One*, *13*(12), e0209205. https://doi.org/10.1371/journal.pone.0209205

Splichalova, A., Pechar, R., Killer, J., Splichalova, Z., Bunesova, V. N., Vlkova, E. & SplichalI. (2020). Colonization of germ-free piglets with mucinolytic and non-mucinolytic bifidobac terium boum strains isolated from the intestine of wild boar and their interference with Salmonella typhimurium. *Microorganisms*, *8*(12):2002. https://doi.org/10.3390/microorganisms8122002

Szajewska, H. (2016). Quality of Probiotic Products. *Journal of Pediatric Gastroenterology and Nutrition*, *63*(1S), S43. https://doi.org/10.1097/01.mpg.0000489605.04419.55

Toomey, N., Monaghan, Á., Fanning, S., & Bolton, D. (2009). Transfer of antibiotic resistance marker genes between lactic acid bacteria in model rumen and plant environments. *Applied and Environmental Microbiology*, *75*(10), 3146–3152. https://doi.org/10.1128/aem.02471-08

Tóth, A. G., Csabai, I., Judge, M.F., Maróti, G., Becsei, Á., Spisák, S., & Solymosi, N. (2021). Mobile antimicrobial resistance genes in probiotics. *Antibiotics*,*10*, 1287.

U.S. Food and Drug Aministration: New dietary ingredients in dietary supplements – Background for industry. US FDA.https://www.fda.gov/food/new-dietary-ingredients-ndi-notification-process/new-dietary-ingredients-dietary-supplements-background-industry. Accessed 2024 May 20.

Warzée, J., Elli, M., Fall, A., Cattivelli, D., & François, J. (2021). Supranational assessment of the quality of probiotics: Collaborative initiative between independent accredited testing laboratories. *Microorganisms*, *9*(7), 1456. https://doi.org/10.3390/microorganisms9071456

WHO (World Health Organisation).(2016). Critically important antimicrobials for human medicine –4th rev. WHO Press. https://www.who.int/publications/i/item/9789241595742. Accessed 2024 Jul 19.

WHO. (2023). Antimicrobial resistance. https://www.who.int/news-room/fact-sheets/detail/antimicrobial-resistance. Accessed 2024 Jul 28.

Wolfe, B. E. (2023). Are fermented foods an overlooked reservoir of antimicrobial resistance?*Current Opinion in Food Science*, *51*, 101018. https://doi.org/10.1016/j.cofs.2023.101018

World AMR Awareness Week 2024. (2024). https://www.who.int/campaigns/world-amr-awareness-week/2024. Accessed 2024 Jun 22.

Wright, A. (2005). Regulating the safety of probiotics – The European approach. *Current Pharmaceutical Design*, *11*(1),17–23.

Yang, S., Qiao, J., Zhang, M., Kwok, L., Matijašić, B. B., Zhang, H., & Zhang, W. (2024). Prevention and treatment of antibiotics-associated adverse effects through the use of probiotics: A review. *Journal of Advanced Research*. https://doi.org/10.1016/j.jare.2024.06.006

Yelin, I., Flett, K. B., Merakou, C., Mehrotra, P., Stam, J., Snesrud, E., Hinkle, M., Lesho, E., McGann, P., McAdam, A. J., Sandora, T. J., Kishony, R., & Priebe, G. P. (2019). Genomic and epidemiological evidence of bacterial transmission from probiotic capsule to blood in ICU patients. *Nature Medicine*, *25*(11), 1728–1732. https://doi.org/10.1038/s41591-019-0626-9

Zarzecka, U., Zadernowska, A., & Chajęcka-Wierzchowska, W. (2020). Starter cultures as a reservoir of antibiotic resistant microorganisms. *LWT*, *127*, 109424. https://doi.org/10.1016/j.lwt.2020.109424

Zielińska, D., & Kolożyn-Krajewska, D. (2018). Food-origin lactic acid bacteria may exhibit probiotic properties: Review. *BioMed Research International*, 2018, 1–15. https://doi.org/10.1155/2018/5063185

Zielińska, D., Sionek, B., & Kołożyn-Krajewska, D. (2018). Safety of probiotics. In: *Diet, Microbiome and Health*(pp. 131–161); Mihai Grumezescu, A., Holban, A. M., (Eds.); Academic Press, Cambridge, MA

17 Future Directions and Potential Applications of Probiotics in Addressing Antimicrobial Resistance

Mohmmad Younus Wani[1], *Manzoor Ahmad Malik*[2], *Bilal Ahmad Najar*[3], *and Aijaz Ahmad*[4]

[1]Department of Chemistry, College of Science, University of Jeddah, Jeddah, Saudi Arabia

[2]Department of Chemistry, University of Kashmir, Srinagar, Jammu and Kashmir, India

[3]Department of Paediatrics, Government Medical College, Anantnag, Jammu and Kashmir, India

[4]Division of Pulmonary, Allergy, and Critical Care Medicine, Department of Medicine, University of Pittsburgh Medical Center, Pittsburgh, USA

17.1 INTRODUCTION

Antimicrobial resistance (AMR) stands as one of the most pressing global health crises of the 21st century, with profound implications for public health, clinical practice, and the broader healthcare system (Salam et al., 2023; Prestinaci et al., 2015). The rise of resistant strains of bacteria, fungi, and parasites threatens the efficacy of commonly used antibiotics, rendering once-treatable infections increasingly difficult, if not impossible, to manage. AMR not only prolongs hospital stays and complicates surgical procedures, but it also significantly increases the morbidity and mortality rates associated with infections caused by multidrug-resistant (MDR) pathogens. Organisms such as *Staphylococcus aureus*, *Escherichia coli* (*E. coli*), *Klebsiella pneumoniae*, *Pseudomonas aeruginosa*, and *Acinetobacter baumannii* have emerged as key contributors to this growing crisis, posing severe challenges to clinicians worldwide (De Oliveira et al., 2020).

The traditional approach to combating AMR has primarily focused on the development of new antibiotics, yet this strategy has struggled to keep pace with the rapid emergence of resistant strains (Aslam et al., 2018). Moreover, the widespread and often inappropriate use of antibiotics in both human and veterinary medicine, as well as the agricultural sector, has accelerated the development and spread of resistant pathogens (Muteeb et al., 2023). Efforts to curb this issue through stringent antibiotic stewardship, while essential, have been insufficient without the introduction of novel therapeutic approaches that can effectively combat AMR.

In this context, probiotics, which are live microorganisms that confer health benefits to the host when administered in adequate amounts, have garnered significant attention as a promising alternative or adjunct to conventional antibiotics (Ji et al., 2020, Maftei et al., 2024). Traditionally recognized for their ability to support gut health, enhance immunity, and maintain a balanced microbiota, probiotics are now being investigated for their potential to mitigate the impact of AMR. The mechanisms by which probiotics could contribute to the fight against AMR extend beyond their well-established roles in gastrointestinal health (Latif et al., 2023, Plaza-Diaz et al., 2019). Recent

 DOI: 10.1201/9781032676951-17

research suggests that probiotics may exert antimicrobial effects through a variety of mechanisms, including the inhibition of pathogenic bacteria, the modulation of the immune response, and the enhancement of the host's natural resistance mechanisms (Petrariu et al., 2024).

This chapter aims to explore the current landscape of probiotic research in the context of AMR, highlighting the emerging understanding of how probiotics can be utilized to combat resistant pathogens. It will delve into the diverse therapeutic applications of probiotics, ranging from their use in preventing infections in high-risk populations, such as hospital patients and immunocompromised individuals, to their potential role in restoring microbial balance in the gut following antibiotic therapy. Additionally, this chapter will discuss the challenges and limitations associated with the widespread use of probiotics, including issues related to strain specificity, dosage, and regulatory standards. As we look toward the future, the potential for probiotics to play a critical role in addressing AMR will depend on ongoing research to elucidate their precise mechanisms of action, optimize their therapeutic use, and overcome the barriers that currently limit their widespread application. This chapter will also highlight the need for collaborative efforts between microbiologists, clinicians, and regulatory agencies to ensure the safe and effective use of probiotics in combating AMR, and it will propose potential directions for future research that could further harness the power of probiotics as a vital tool in the global fight against AMR.

17.2 THE PROMISE OF PROBIOTICS IN COMBATING AMR

Probiotics offer a multifaceted and promising approach to tackling AMR, with several advantages over traditional antibiotic therapies (Maftei et al., 2024). One of the key benefits is their ability to act through multiple mechanisms that complement those of conventional antibiotics (Plaza-Diaz et al., 2019). Probiotics can produce antimicrobial peptides (AMPs) and organic acids that inhibit the growth of pathogenic bacteria, providing an alternative to broad-spectrum antibiotics that often disrupt beneficial microbiota (Ali et al., 2023, Rabetafika et al., 2023). This antimicrobial action can help manage infections without promoting resistance in the same way antibiotics do, which can inadvertently select for resistant strains. In addition to their direct antimicrobial effects, probiotics enhance the host's immune response. Studies show that probiotics can stimulate immune cells such as macrophages and neutrophils, increasing their activity and facilitating cytokine production that helps clear infections (Mazziotta et al., 2023). Unlike antibiotics, which target a broad range of bacteria and may disturb the gut microbiota, probiotics work in a more localized and selective manner. They inhibit pathogenic organisms while preserving the balance of the microbiome, thereby reducing the risk of opportunistic infections and secondary infections (Skoufou et al., 2024). This aspect is especially important when considering the growing problem of *Clostridium difficile* infections following antibiotic therapy, a condition where probiotics have demonstrated significant promise (Patangia et al., 2022). Another important role of probiotics in combating AMR is their ability to prevent the colonization of MDR pathogens. By competing for nutrients and adhesion sites, probiotics can limit the entry and proliferation of resistant strains in the gastrointestinal tract and other mucosal surfaces. Probiotics have been shown to reduce the colonization of pathogens such as *Salmonella* and *Clostridium difficile*, making them valuable in preventing gastrointestinal infections and reducing the need for antibiotics (Milner et al., 2021). This competitive exclusion mechanism presents a promising alternative to systemic antibiotic use, which often leads to the selection of resistant strains.

Recent advancements also highlight the potential of engineered probiotics as delivery systems for therapeutic agents, including AMPs, antibodies, and even vaccines. These innovations suggest that probiotics could not only prevent infections but also directly combat AMR by delivering targeted therapeutic agents in a controlled and localized manner (Mugwanda et al., 2023). This approach is a significant step toward the development of more precise and effective treatments for resistant infections.

17.3 ADVANTAGES OF TRADITIONAL ANTIBIOTIC APPROACHES

Probiotics represent a significant shift away from the traditional approach to fighting infections, which has primarily focused on the development of new antibiotics. While antibiotics remain a cornerstone of modern medicine, their overuse and misuse have accelerated the development of resistance, rendering many previously effective drugs obsolete. In contrast, probiotics offer a more sustainable solution, with less likelihood of promoting resistance. They work by maintaining or restoring microbial balance without disrupting the normal microbiota as antibiotics often do, thereby avoiding the promotion of resistant strains (Skoufou et al., 2024).

Probiotics are generally considered safe, well-tolerated, and non-toxic, with minimal side effects reported in clinical trials, making them particularly suitable for long-term use in vulnerable populations, including the elderly, children, and immunocompromised individuals. This contrasts with the adverse effects associated with prolonged antibiotic use, such as gastrointestinal disturbances, allergic reactions, and systemic toxicity (Patangia et al., 2022). Furthermore, probiotics can be especially effective in mitigating the detrimental effects of antibiotics, such as dysbiosis which is the imbalance of gut bacteria, by replenishing beneficial bacteria and supporting the restoration of a healthy microbiome (Milner et al., 2021). This feature is crucial in preventing secondary infections like *Clostridium difficile*-associated diarrhea or *Candida* overgrowth, which are common consequences of antibiotic treatment. Thus, the dual role of probiotics which is promoting immune health, while preserving microbiome integrity presents a viable complement to traditional antibiotic therapy, offering a sustainable strategy to combat infections and reduce AMR in the long term. Their use could be instrumental in addressing the growing challenges posed by resistant pathogens in clinical settings.

17.4 MECHANISMS OF ACTION OF PROBIOTICS

17.4.1 Competitive Exclusion and Colonization Resistance

One of the primary mechanisms by which probiotics combat pathogens is through competitive exclusion. This strategy involves probiotics outcompeting harmful microorganisms for critical resources, such as nutrients and adhesion sites on the host's mucosal surfaces. By occupying these ecological niches, probiotics effectively prevent the colonization of pathogenic microorganisms, thus reducing the risk of infection. This mechanism is particularly vital in the gastrointestinal tract, where a dense microbial community creates a highly competitive environment.

Probiotic strains such as *Lactobacillus* and *Bifidobacterium* have been extensively studied for their ability to inhibit pathogen growth by competitive exclusion (Latif et al., 2023). For example, recent studies have demonstrated that *Lactobacillus* species can displace *Clostridium difficile*, a major cause of antibiotic-associated diarrhea (AAD; van Zyl et al., 2020). Similarly, *Bifidobacterium* strains are known to inhibit the growth of *E. coli*, a common pathogen responsible for urinary tract infections and gastrointestinal disorders (Katouli, 2010). These findings suggest that probiotics play a critical role in maintaining microbial equilibrium, especially in the context of antibiotic disruptions.

In addition to direct competition for nutrients and attachment sites, probiotics produce a variety of antimicrobial substances that contribute to pathogen exclusion. These include bacteriocins, organic acids like lactic acid, and hydrogen peroxide, all of which inhibit the growth of harmful microorganisms (Silva et al., 2020). The production of lactic acid by *Lactobacillus* spp. is particularly noteworthy, as it lowers the pH in the gut, creating an acidic environment that is hostile to many pathogens (De Filippis et al., 2020). Furthermore, probiotics have been shown to enhance the host's immune system by stimulating the mucosal immune response and promoting the secretion of AMPs, further assisting in the exclusion of pathogens (Zhou et al., 2019).

Recent research has expanded our understanding of how a balanced gut microbiota mediates colonization resistance. Disruptions in microbiota diversity, often caused by antibiotics, illness, or stress, can lead to an imbalance that favors pathogenic organisms. However, the administration of probiotics can help restore microbial equilibrium. Studies have shown that *Lactobacillus* and *Bifidobacterium* strains, when administered during or after antibiotic treatment, contribute to the re-establishment of gut flora balance, thus reducing the risk of opportunistic infections, such as *C. difficile* (Yang et al., 2024). Probiotics can also help prevent the growth of MDR pathogens by maintaining a healthy microbial community, which functions as a barrier against pathogen colonization. Moreover, the role of probiotics in high-risk populations, such as those with compromised immune systems or undergoing chemotherapy, has gained attention. Probiotics, particularly *Lactobacillus* and *Bifidobacterium*, have been shown to reduce the incidence of gastrointestinal infections, including those caused by MDR pathogens (Van Rossum et al., 2024). These findings are consistent with the growing body of evidence supporting the use of probiotics as a strategy to prevent infections and maintain gut health, especially in vulnerable individuals.

17.4.2 Production of Antimicrobial Substances

Probiotics are known for their ability to produce bioactive metabolites that exert antimicrobial effects, providing a key mechanism of protection against pathogenic microorganisms. These metabolites include lactic acid, hydrogen peroxide, bacteriocins, and short-chain fatty acids (SCFAs), all of which contribute to creating an environment that inhibits the growth of harmful bacteria and disrupts pathogen biofilms. Lactic acid, produced by many Lactobacillus species, lowers the pH of the surrounding environment, which creates unfavorable conditions for many pathogens, including *E. coli* and *Streptococcus pneumoniae* (Papadimitriou et al., 2016). Hydrogen peroxide, another product of *Lactobacillus* spp., also contributes to antimicrobial activity by generating oxidative stress, which can damage bacterial cells (Vieco-Saiz et al., 2019). In addition to these acids, bacteriocins, which are proteinaceous AMPs are produced by various probiotic strains, including *Lactobacillus acidophilus* and *Lactobacillus plantarum.* These peptides target and inhibit the growth of MDR pathogens such as *E. coli* and *Salmonella* (Papadimitriou et al., 2016; Vieco-Saiz et al., 2019).

SCFAs, such as butyrate and acetate, also play an important role in maintaining gut homeostasis and exhibiting antimicrobial activity. These SCFAs have been shown to disrupt pathogen biofilm formation, thereby enhancing the efficacy of the host's immune system in clearing infections (Fusco et al., 2023a). Recent studies highlight that SCFAs not only lower the pH of the gut but also directly interfere with the metabolic processes of pathogens (Fusco et al., 2023b). The production of antimicrobial substances by probiotics is a multifaceted approach that not only directly inhibits the growth of pathogens but also enhances the overall health of the microbiome, supporting the host's defense mechanisms.

17.4.3 Immune Modulation

Probiotics have been shown to interact with the host immune system, enhancing both innate and adaptive immunity to bolster resistance against infections, including those caused by MDR pathogens (Zheng et al., 2020, Raheem et al., 2021). One of the primary ways probiotics achieve this is by stimulating the production of AMPs such as defensins, which directly inhibit or kill pathogenic microbes. Recent studies confirm that probiotic strains like *Lactobacillus rhamnosus* can enhance AMP production, strengthening gut epithelial defenses and reducing pathogen load (Fusco et al., 2023b). Additionally, probiotics can activate immune cells, such as phagocytes, natural killer (NK) cells, and T lymphocytes, all crucial for immune defense (Aziz and Bonavida, 2016). For instance, *Lactobacillus* and *Bifidobacterium* strains have been found to enhance the activity of NK cells and the phagocytic capacity of macrophages, thereby increasing the host's ability to detect and destroy pathogens.

The gut-associated lymphoid tissue (GALT) is also a key target of probiotic-mediated immune modulation (Feria et al., 2017). Probiotics can promote the development of regulatory T cells (Tregs) and other adaptive immune responses within the GALT, which not only strengthens gut immunity but also contributes to systemic immune defenses, reducing susceptibility to infections beyond the gastrointestinal tract.

17.4.4 Biofilm Disruption and Inhibition

Biofilm formation is a critical factor in the persistence of infections caused by antibiotic-resistant bacteria, as biofilms act as protective barriers, making pathogens more resistant to antibiotics and immune defenses (Mirghani et al., 2022). Recent studies indicate that probiotics can effectively disrupt biofilm formation by producing enzymes such as proteases, lipases, and glycosidases, which degrade the biofilm matrix (Tomé et al., 2023). For instance, *Lactobacillus acidophilus* and *Lactobacillus rhamnosus* have been found to produce extracellular proteolytic enzymes that break down biofilm structures, reducing bacterial adherence and biofilm integrity. Probiotics also interfere with quorum sensing (QS), a communication mechanism crucial for biofilm development and maintenance (Salman et al., 2023). By producing signaling molecules that inhibit QS pathways, probiotics such as *Lactobacillus* and *Bifidobacterium* species disrupt the coordinated behaviors necessary for biofilm stability. Research has shown that *Lactobacillus reuteri* inhibits QS-related gene expression in *Pseudomonas aeruginosa*, a common biofilm-forming pathogen, thereby reducing its biofilm-forming ability. Furthermore, combining probiotics with antibiotics has shown promise in combating biofilm-associated infections (Hawas et al., 2022). Probiotics can enhance the susceptibility of biofilm-forming bacteria to antibiotics, possibly due to biofilm matrix disruption and inhibition of QS pathways, which weakens biofilm integrity. This combination has been particularly effective against MDR organisms, highlighting the potential of probiotics as an adjunct therapy.

17.5 POTENTIAL APPLICATIONS OF PROBIOTICS IN CLINICAL SETTINGS

17.5.1 Probiotics as Adjuncts to Antibiotics

Probiotics have gained significant attention in clinical settings for their role as adjuncts to antibiotic therapy. Although antibiotics are essential for treating severe infections, their widespread use often results in disruptions to the gut microbiota, leading to undesirable side effects like AAD, dysbiosis, and overgrowth of resistant pathogens. In this context, probiotics can help restore balance to the gut microbiota and alleviate some of these adverse effects.

Recent studies have provided robust evidence supporting the use of probiotics alongside antibiotics (Meroni et al., 2021). A meta-analysis concluded that the administration of probiotics during antibiotic therapy significantly reduced the incidence of AAD, particularly in hospitalized patients or those undergoing intensive antibiotic regimens (Goodman et al., 2021). In another study, the combination of probiotics, including strains such as *Lactobacillus* and *Bifidobacterium*, was shown to maintain microbiota diversity, protect against opportunistic infections, and promote the recovery of the gut's beneficial bacterial populations during and after antibiotic treatment (Goodman et al., 2021). Furthermore, the use of probiotics in conjunction with antibiotics can also aid in improving gut health by enhancing the intestinal barrier function and modulating immune responses. Clinical trials have demonstrated that probiotics help restore gut barrier integrity, reduce intestinal inflammation, and modulate cytokine levels in patients receiving antibiotics (Chandrasekaran et al., 2024). Notably, probiotics such as *Saccharomyces boulardii* have been particularly effective in preventing gut permeability and inflammation, providing a dual benefit of both preventing and treating AAD. In addition to their effects on gastrointestinal disturbances, probiotics have shown promise in mitigating the risk of opportunistic infections and enhancing immune function. Studies have suggested that probiotics can suppress the growth of pathogens like *Clostridium difficile* and *Candida* species, which may proliferate in the gut due to antibiotic-induced dysbiosis (Ehrhardt et al., 2016). Probiotic supplementation has also been linked to increased levels of secretory immunoglobulin A (SIgA), a key component of the gut's immune defense, helping to strengthen the body's resistance to infections (Johnston et al., 2007). While probiotics may not replace antibiotics in treating severe infections, they play a crucial role as adjuncts by mitigating the side effects of antibiotic therapy. Their ability to reduce the incidence of AAD, restore gut health, and prevent the overgrowth of harmful microorganisms makes them a valuable tool in clinical practice, especially for vulnerable patient groups.

17.5.2 Prevention and Treatment of Antibiotic-Associated Diarrhea (AAD)

AAD is a common side effect of broad-spectrum antibiotic therapy. It occurs when antibiotics disrupt the normal gut microbiota, leading to the overgrowth of opportunistic pathogens like *Clostridium difficile*, which is known to cause more severe complications such as pseudomembranous colitis. The condition is characterized by loose stools, abdominal cramping, and fever, and can lead to prolonged hospital stays and increased healthcare costs. Probiotics, including *Saccharomyces boulardii*, *Lactobacillus spp.*, and *Bifidobacterium spp.*, have been widely investigated for their role in both the prevention and treatment of AAD. *Saccharomyces boulardii*, a yeast-based probiotic, has garnered significant attention for its effectiveness in reducing the incidence and recurrence of *C. difficile* infections (Tung et al., 2009). A study demonstrated that *S. boulardii* could significantly reduce the recurrence of *C. difficile* infections in patients who had previously been treated with antibiotics (Hopkins and Wilson, 2018). This probiotic is thought to work by enhancing the integrity of the intestinal mucosa and competing with pathogenic bacteria for available nutrients and binding sites, thus preventing overgrowth of harmful pathogens. Furthermore, several studies have investigated the role of *Lactobacillus* strains in mitigating the harmful effects of antibiotics on gut health. *Lactobacillus rhamnosus* GG and *Lactobacillus acidophilus* have significantly reduced the incidence of AAD and improved gastrointestinal function during and after antibiotic therapy (Song et al., 2010). These probiotics help restore the balance of the gut microbiome, reducing inflammation and promoting beneficial microbial communities, thus preventing, or ameliorating symptoms of AAD. Additionally, probiotics, such as *Bifidobacterium spp.,* have demonstrated potential in mitigating antibiotic-induced disruptions in the gut (Wong et al., 2019). Studies highlight the beneficial effects of *Bifidobacterium* on intestinal health, showing that supplementation with this probiotic can reduce the incidence and severity of AAD while enhancing gut microbial diversity. Overall, the use of probiotics as adjuncts to antibiotic therapy has gained attention as an effective approach to prevent and manage AAD. Their capacity to restore intestinal balance, strengthen the gut barrier, and regulate immune responses presents a valuable therapeutic strategy in clinical practice.

17.5.3 Probiotics in Respiratory Infections

Infections of the upper and lower respiratory tract, often caused by antibiotic-resistant pathogens, are becoming increasingly difficult to treat, particularly in vulnerable populations such as children and the elderly. Probiotics can enhance the body's natural defense mechanisms, particularly by modulating the mucosal immune system. They have been shown to stimulate the production of immunoglobulins (IgA) and AMPs, which are crucial for preventing infections at mucosal surfaces like the respiratory tract. Probiotics may also compete with harmful pathogens for adhesion sites, thereby reducing pathogen colonization and promoting a healthy balance of microbial communities in the upper respiratory tract. Recent studies have underscored the potential of specific probiotic strains in reducing the incidence and severity of respiratory infections. Strains such as Lactobacillus and Bifidobacterium have been shown to lower the frequency of respiratory infections in vulnerable populations, including children and the elderly (Du et al., 2022, Shahbazi et al., 2020). In particular, *Lactobacillus rhamnosus* and *Bifidobacterium lactis* have been found to decrease the occurrence of upper respiratory tract infections while enhancing immune responses in older adults, a group at heightened risk for respiratory complications. These probiotics not only bolster immune function but also help preserve the integrity of the respiratory mucosa, a critical barrier against pathogen invasion.

Moreover, probiotics have shown promise in mitigating the effects of respiratory infections in populations with chronic respiratory conditions. *Lactobacillus* spp. and *Bifidobacterium* spp. reduced the duration and severity of symptoms in patients suffering from chronic obstructive pulmonary disease exacerbations, demonstrating the therapeutic potential of probiotics in managing both acute and chronic respiratory conditions (Du et al., 2022). In addition to supporting the immune system, probiotics have been studied for their role in enhancing the production of mucosal AMPs. These peptides, such as defensins and cathelicidins, play a crucial role in directly killing or inhibiting the growth of pathogenic microorganisms (Zong et al., 2020). By promoting the secretion of these peptides, probiotics may further reduce the susceptibility to respiratory infections. While more clinical studies are needed to fully understand their efficacy and mechanisms of action, current evidence suggests that probiotics can play an important role in boosting immune responses and preventing infections, especially in vulnerable populations.

17.5.4 Probiotics in Urinary Tract Infections (UTIs)

Urinary tract infections (UTIs) remain a prevalent and challenging clinical problem, particularly when caused by antibiotic-resistant organisms such as *E. coli*, *Klebsiella* spp., and other uropathogens. Prolonged or recurrent UTIs, often requiring frequent courses of antibiotics, increase the risk of AMR. Recent evidence supports the role of probiotics as an adjunctive or alternative therapy in preventing and managing UTIs, especially recurrent ones, by reducing the need for antibiotics and maintaining microbial balance within the urinary tract (Schwenger et al., 2015, Qasemi et al., 2023). *Lactobacillus* strains, including *Lactobacillus rhamnosus* and *Lactobacillus reuteri*, have been widely studied for their ability to modulate the urogenital microbiota. These probiotics are believed to protect against UTIs by adhering to the uroepithelial cells, thereby preventing the adhesion of uropathogenic *E. coli*, the most common causative agent of UTIs (Vagios et al., 2020). By enhancing the local immune response and promoting a favorable microbiota, probiotics can prevent the overgrowth of pathogenic bacteria.

Probiotics also offer protective effects in specific high-risk populations, including pregnant women, postmenopausal women, and individuals with indwelling urinary catheters, who are more susceptible to UTIs. Clinical trials have found that oral or vaginal administration of *Lactobacillus* spp. significantly reduced the incidence of UTIs in these groups, providing a safe and effective prophylactic approach (Ng et al., 2018). Moreover, *Lactobacillus* spp. can help reduce the colonization of *E. coli* in the urinary tract, offering a promising strategy for preventing UTIs in individuals at risk of recurrent infections (Ng et al., 2018). A meta-analysis further supported the use of probiotics in preventing recurrent UTIs, showing that supplementation with *Lactobacillus* strains significantly lowered the risk of infection (Gupta et al., 2024). Recent evidence also points to the potential role of other probiotic strains, such as *Bifidobacterium* spp., in managing UTIs. These probiotics may offer additional benefits in improving gut and urinary tract health, as well as in reducing the frequency of UTIs by modulating the gut microbiota, which is thought to influence the urogenital tract (Sihra et al., 2018). The potential of probiotics to reduce the reliance on antibiotics, thus helping mitigate the development of AMR, makes them a valuable tool in modern healthcare.

17.6 EMERGING TRENDS AND FUTURE DIRECTIONS

17.6.1 Development of Engineered Probiotics

The field of engineered probiotics has gained significant attention in recent years, driven by advances in genetic engineering techniques. These probiotics are designed to deliver targeted therapies, such as AMPs or other therapeutic molecules, directly to infection sites, offering a novel and efficient strategy for combating MDR pathogens (Mugwanda et al., 2023). One key innovation is the incorporation of AMPs, like defensins, into probiotic strains. These peptides have antimicrobial properties and can target resistant bacteria, potentially overcoming the limitations of conventional antibiotics.

The development of engineered probiotics is not limited to AMPs alone. Other therapeutic molecules, such as enzymes that degrade bacterial biofilms or immune-modulatory compounds, are being incorporated to enhance the treatment of infections (Bisht et al., 2024). For example, *Lactobacillus* and *Bifidobacterium* strains have been genetically modified to produce AMPs, improving their potential to combat infections such as *Clostridium difficile* and *E. coli* (Aggarwal et al., 2020, Ma et al., 2022). These engineered probiotics can be tailored to provide a site-specific delivery of these molecules, enhancing the therapeutic effect while reducing the side effects commonly associated with systemic treatments. Additionally, synthetic biology techniques enable the fine-tuning of probiotic strains, ensuring that they can not only deliver therapeutic agents but also modulate the host's immune response. For instance, probiotics can be engineered to stimulate the production of specific cytokines or enhance mucosal immunity, providing an added layer of protection against infections (Kang et al., 2020). This promising area of research may offer new ways to treat chronic or recurrent infections, especially those caused by MDR pathogens, and could play a pivotal role in the future of personalized medicine.

17.6.2 Personalized Probiotic Therapy

The growing understanding of the human microbiome has led to the concept of personalized probiotic therapy, which aims to tailor treatments based on an individual's unique microbial composition (Kashyap et al., 2017). Since microbiomes vary widely between individuals, personalized probiotics could offer more targeted and effective treatments for infections, particularly those caused by AMR pathogens. Advances in metagenomic analysis and microbiome profiling allow researchers to identify the most suitable probiotic strains for each person, enhancing the efficacy of probiotic interventions. This personalized approach can also optimize the balance of the gut microbiota, preventing dysbiosis while strengthening the body's defense mechanisms. Studies have shown that tailored probiotic therapies can improve outcomes in patients with recurrent infections, such as UTIs and respiratory tract infections, by restoring microbial diversity and promoting the growth of beneficial bacteria (Gulliver et al., 2022). Furthermore, personalized probiotics may contribute to a more sustainable management of AMR by reducing the need for broad-spectrum antibiotics. By specifically addressing the underlying microbial imbalances that contribute to infection recurrence, this approach holds significant promise for reducing reliance on antibiotics, thus limiting the development of resistant strains.

17.6.3 Probiotic-Driven Vaccine Development

Probiotics have the potential to play a pivotal role in enhancing vaccine efficacy, particularly for immunocompromised individuals or populations in resource-limited settings (Kazemifard et al., 2022). These probiotics can stimulate mucosal immunity, improving the body's overall response to vaccines, and providing enhanced protection against infections, including those caused by drug-resistant pathogens. Studies have shown that probiotics can enhance both humoral and cellular immunity, which are essential for effective vaccination responses (Xu et al., 2024).

An exciting avenue of research is the use of probiotic strains as vehicles for vaccine delivery. By engineering probiotics to express specific antigens, they can serve as a live platform for vaccine delivery, potentially reducing the need for traditional injection-based vaccines, which can be less feasible in underdeveloped regions. Probiotic-based delivery systems could also improve the stability and efficacy of vaccines, making them more accessible and practical for use in low-resource settings (Xu et al., 2024). For example, studies have explored the use of *Lactobacillus* and

Bifidobacterium strains as carriers for various vaccines, demonstrating their ability to elicit a robust immune response without adverse effects (Abavisani et al., 2024). These probiotics not only aid in the immune response but may also enhance vaccine longevity by maintaining intestinal homeostasis, which is critical for sustained immunity. As the field of probiotic-driven vaccine development advances, it could revolutionize the way we approach infectious disease prevention, particularly in populations at high risk of developing infections caused by resistant pathogens.

17.6.4 Probiotics in Veterinary Medicine

AMR is a growing concern not only in human health but also in veterinary medicine (Salam et al., 2023). The widespread use of antibiotics in livestock farming has significantly contributed to the emergence of resistant bacterial strains, which can then be transmitted to humans through the food supply, posing serious health risks. In this context, probiotics offer a promising alternative to reduce the reliance on antibiotics in veterinary practices. Probiotics have been increasingly recognized for their ability to improve gut health and enhance the immune system in animals, helping to control infections without the need for antibiotics. They are particularly beneficial in reducing the incidence of gastrointestinal and systemic infections in livestock, including pathogens like *Salmonella*, *E. coli*, and *Clostridium perfringens*, which are often responsible for outbreaks in farm animals (Chaucheyras-Durand et al., 2010). By promoting a healthy gut microbiota, probiotics can prevent infections, reduce pathogen load, and improve overall animal health, which ultimately helps in reducing the need for antibiotic treatments.

Probiotics have been successfully used in poultry and swine industries to improve growth performance and reduce infection rates. Studies have demonstrated that probiotics can inhibit the colonization of pathogenic bacteria by enhancing competitive exclusion mechanisms, thereby reducing the incidence of diseases that require antibiotic interventions (Halder et al., 2024). In cattle, the administration of probiotics has been shown to improve rumen health, digestion, and absorption of nutrients, further contributing to the reduction of antibiotic usage. The use of probiotics in veterinary medicine also plays a critical role in addressing the issue of AMR transmission from animals to humans. By reducing antibiotic usage in animals, probiotics can limit the selective pressure that drives the development of resistant strains. This strategy could potentially minimize the risk of AMR being transmitted via the food chain, thereby benefiting both animal and human health.

17.7 CHALLENGES AND BARRIERS TO CLINICAL IMPLEMENTATION

17.7.1 Strain-Specific Efficacy and Standardization

One of the major challenges in the clinical use of probiotics is the strain-specific nature of their efficacy. Probiotics are not a homogenous group of organisms, and different strains of the same species can exhibit significant variations in their ability to exert antimicrobial effects or promote health benefits. This variation complicates the process of identifying and standardizing probiotics for specific therapeutic uses. While certain strains, such as *Lactobacillus rhamnosus* GG, *Saccharomyces boulardii*, and *Bifidobacterium lactis*, have been studied extensively for their beneficial effects in various clinical settings, not all strains have proven effective in combating specific pathogens or conditions (Liu et al., 2023).

The lack of standardization across probiotic formulations further complicates clinical applications. Variations in probiotic dosages, viability, and composition can lead to inconsistent results across clinical trials and studies, making it difficult to compare findings or create universally applicable treatment protocols (Baral et al., 2021). For example, the same probiotic strain might show different therapeutic outcomes depending on its formulation, storage conditions, or administration method. Additionally, the complex interactions between probiotics and the human microbiome where the microbiota of each individual differs significantly can contribute to variability in probiotic efficacy. This highlights the need for more rigorous clinical trials and standardized protocols to assess and optimize the use of probiotics in clinical settings (Baral et al., 2021). Recent advancements in microbiome research have the potential to overcome these challenges by enabling the development of more tailored probiotic therapies. However, further research is needed to identify the most beneficial strains for specific health conditions and to establish clear guidelines for their use in clinical practice. This research will play a crucial role in ensuring that probiotics can be used safely and effectively as adjuncts to conventional therapies, thereby unlocking their full potential as therapeutic tools.

17.7.2 Regulatory and Safety Issues

The safety and regulatory issues surrounding probiotics as therapeutic agents are important considerations for their clinical application, particularly in the context of combating AMR (Iannitti and Palmieri, 2010). While probiotics are generally regarded as safe, there have been instances of adverse effects, particularly in immunocompromised individuals, such as infections from probiotic strains in patients with weakened immune systems. Thus, the need for robust safety guidelines and regulatory frameworks is critical (Merenstein et al., 2023, Borase et al., 2022). Regulatory agencies, including the U.S. Food and Drug Administration (FDA) and the European Food Safety Authority (EFSA), play a key role in establishing standards for probiotic use, especially concerning the quality, purity, and clinical applications of probiotic strains. These agencies aim to ensure that probiotics used in clinical settings meet specific safety criteria to mitigate the risk of adverse outcomes, especially as the clinical use of probiotics grows in the fight against AMR. For instance, EFSA has introduced guidelines for health claims made by probiotics, while the FDA has worked on categorizing probiotics based on their intended use to ensure their efficacy and safety. Furthermore, significant challenges exist in defining clear, standardized protocols for the strain-specific effectiveness of probiotics, which complicates the regulation process. Comprehensive studies and clear labeling practices are necessary to guide clinical use and prevent misuse or misrepresentation of probiotic products.

17.8 CONCLUSION

Probiotics offer a promising and innovative strategy in the fight against AMR. By utilizing mechanisms such as competitive exclusion, AMP production, immune modulation, and biofilm disruption, probiotics serve as an effective complement to traditional antibiotics. Despite their potential, challenges such as strain-specific efficacy, safety concerns, and regulatory hurdles remain significant barriers to their widespread clinical use. These issues must be addressed to integrate probiotics fully into medical practice for combating AMR. Additionally, ongoing research into engineered probiotics, personalized therapies, and their potential role in vaccine development holds great promise for advancing this field. As research progresses, optimizing probiotic formulations and their therapeutic applications will be essential to meet the challenges posed by drug-resistant infections.

REFERENCES

Abavisani M, Ebadpour N, Khoshrou A, Sahebkar A. Boosting vaccine effectiveness: The groundbreaking role of probiotics. *J Agri Food Res*. 2024;16:101189. https://doi.org/10.1016/j.jafr.2024.101189

Aggarwal N, Breedon AME, Davis CM, Hwang IY, Chang MW. Engineering probiotics for therapeutic applications: Recent examples and translational outlook. *Curr Opin Biotechnol*. 2020 Oct;65:171–179. https://doi.org/10.1016/j.copbio.2020.02.016

Ali MS, Lee EB, Hsu WH, Suk K, Sayem SAJ, Ullah HMA, Lee SJ, Park SC. Probiotics and postbiotics as an alternative to antibiotics: An emphasis on pigs. *Pathogens*. 2023 Jun 26;12(7):874. https://doi.org/10.3390/pathogens12070874

Aslam B, Wang W, Arshad MI, Khurshid M, Muzammil S, Rasool MH, Nisar MA, Alvi RF, Aslam MA, Qamar MU, Salamat MKF, Baloch Z. Antibiotic resistance: A rundown of a global crisis. *Infect Drug Resist.* 2018 Oct 10;11:1645–1658. https://doi.org/10.2147/IDR.S173867

Aziz N, Bonavida B. Activation of natural killer cells by probiotics. *For Immunopathol Dis Therap.* 2016;7 (1–2):41–55. https://doi.org/10.1615/ForumImmunDisTher.2016017095

Baral KC, Bajracharya R, Lee SH, Han HK. Advancements in the pharmaceutical applications of probiotics: Dosage forms and formulation technology. *Int J Nanomedicine.* 2021;16:7535–7556. https://doi.org/10.2147/IJN.S337427

Bisht V, Das B, Hussain A. et al. Understanding of probiotic origin antimicrobial peptides: A sustainable approach ensuring food safety. *NPJ Sci Food.* 2024;8:67. https://doi.org/10.1038/s41538-024-00304-8

Borase H, Dwivedi MK, Krishnamurthy R, Patil S. Probiotics: Health safety considerations. In Mitesh Kumar Dwivedi, N. Amaresan, A. Sankaranarayanan, E. Helen Kemp (Eds.), *Probiotics in the Prevention and Management of Human Diseases*, Academic Press, 2022, pp. 449–463, ISBN 9780128237335; https://doi.org/10.1016/B978-0-12-823733-5.00020-9

Chandrasekaran P, Weiskirchen S, Weiskirchen R. Effects of probiotics on gut microbiota: An overview. *Int J Mol Sci.* 2024; 25(11):6022. https://doi.org/10.3390/ijms25116022

Chaucheyras-Durand F, Durand H. Probiotics in animal nutrition and health. *Benef Microbes.* 2010 Mar;1(1): 3–9. https://doi.org/10.3920/BM2008.1002

De Filippis F, Pasolli E, Ercolini D. The food-gut axis: Lactic acid bacteria and their link to food, the gut microbiome and human health. *FEMS Microbiol Rev.* 2020 Jul 1;44(4):454–489. https://doi.org/10.1093/femsre/fuaa015

De Oliveira DMP, Forde BM, Kidd TJ, Harris PNA, Schembri MA, Beatson SA, Paterson DL, Walker MJ. Antimicrobial resistance in ESKAPE pathogens. *Clin Microbiol Rev.* 2020 May 13;33(3):e00181–19. https://doi.org/10.1128/CMR.00181-19

Du T, Lei A, Zhang N, Zhu C. The beneficial role of probiotic lactobacillus in respiratory diseases. *Front Immunol.* 2022;13:908010. https://doi.org/10.3389/fimmu.2022.908010

Ehrhardt S, Guo N, Hinz R, Schoppen S, May J, Reiser M, Schroeder MP, Schmiedel S, Keuchel M, Reisinger EC, Langeheinecke A, de Weerth A, Schuchmann M, Schaberg T, Ligges S, Eveslage M, Hagen RM, Burchard GD, Lohse AW. Saccharomyces boulardii to prevent antibiotic-associated diarrhea: A randomized, double-masked, placebo-controlled trial. *Open Forum Infect Dis.* 2016 Jan 29;3(1):ofw011. https://doi.org/10.1093/ofid/ofw011

Feria MG, Taborda NA, Hernandez JC, Rugeles MT. Effects of prebiotics and probiotics on gastrointestinal tract lymphoid tissue in HIV infected patients. *Revista médica de Chile.* 2017;145(2):219–229. https://doi.org/10.4067/S0034-98872017000200010

Fusco A, Savio V, Cimini D, D'Ambrosio S, Chiaromonte A, Schiraldi C, Donnarumma G. In vitro evaluation of the most active probiotic strains able to improve the intestinal barrier functions and to prevent inflammatory diseases of the gastrointestinal system. *Biomedicines.* 2023a Mar 12;11(3):865. https://doi.org/10.3390/biomedicines11030865

Fusco W, Lorenzo MB, Cintoni M, Porcari S, Rinninella E, Kaitsas F, Lener E, Mele MC, Gasbarrini A, Collado MC, Cammarota G, Ianiro G. Short-chain fatty-acid-producing bacteria: Key components of the human gut microbiota. *Nutrients.* 2023b May 6;15(9):2211. https://doi.org/10.3390/nu15092211

Goodman C, Keating G, Georgousopoulou E, Hespe C, Levett K, Probiotics for the prevention of antibiotic-associated diarrhoea: A systematic review and meta-analysis. *BMJ Open.* 2021;11:e043054. https://doi.org/10.1136/bmjopen-2020-043054

Gulliver EL, Young RB, Chonwerawong M, D'Adamo GL, Thomason T, Widdop JT, Rutten EL, Rossetto Marcelino V, Bryant RV, Costello SP, O'Brien CL, Hold GL, Giles EM, Forster SC. Review article: The future of microbiome-based therapeutics. *Aliment Pharmacol Ther.* 2022 Jul;56(2):192–208. https://doi.org/10.1111/apt.17049

Gupta V, Mastromarino P, Garg R. Effectiveness of prophylactic oral and/or vaginal probiotic supplementation in the prevention of recurrent urinary tract infections: A randomized, double-blind, placebo-controlled trial. *Clin Infect Dis.* 2024 May 15;78(5):1154–1161. https://doi.org/10.1093/cid/ciad766

Halder N, Sunder J, De AK. et al. Probiotics in poultry: A comprehensive review. *J Basic Appl Zoology.* 2024;85:23. https://doi.org/10.1186/s41936-024-00379-5

Hawas S, Verderosa AD, Totsika M. Combination therapies for biofilm inhibition and eradication: A comparative review of laboratory and preclinical studies. *Front. Cell. Infect. Microbiol.* 2022;12:850030. https://doi.org/10.3389/fcimb.2022.850030

Hopkins RJ, Wilson RB. Treatment of recurrent Clostridium difficile colitis: A narrative review. *Gastroenterol Rep (Oxf).* 2018 Feb;6(1):21–28. https://doi.org/10.1093/gastro/gox041

Iannitti T, Palmieri B. Therapeutical use of probiotic formulations in clinical practice. *Clin Nutr*. 2010 Dec;29(6):701–25. https://doi.org/10.1016/j.clnu.2010.05.004

Ji J, Jin W, Liu SJ, Jiao Z, Li X. Probiotics, prebiotics, and postbiotics in health and disease. *MedComm (*2020). 2023 Nov 4;4(6):e420. https://doi.org/10.1002/mco2.420

Johnston BC, Supina AL, Ospina M, Vohra S. Probiotics for the prevention of pediatric antibiotic-associated diarrhea. *Cochrane Database Syst Rev*. 2007 Apr 18;(2):CD004827. https://doi.org/10.1002/14651858.CD004827.pub2

Kang M, Choe D, Kim K, Cho BK, Cho S. Synthetic biology approaches in the development of engineered therapeutic microbes. *Int J Mol Sci*. 2020 Nov 19;21(22):8744. https://doi.org/10.3390/ijms21228744

Kashyap PC, Chia N, Nelson H, Segal E, Elinav E. Microbiome at the frontier of personalized medicine. *Mayo Clin Proc*. 2017 Dec;92(12):1855–1864. https://doi.org/10.1016/j.mayocp.2017

Katouli M. Population structure of gut Escherichia coli and its role in development of extra-intestinal infections. *Iran J Microbiol*. 2010 Jun;2(2):59–72.

Kazemifard N, Dehkohneh A, Baradaran Ghavami S. Probiotics and probiotic-based vaccines: A novel approach for improving vaccine efficacy. *Front Med (Lausanne)*. 2022 Oct 13;9:940454. https://doi.org/10.3389/fmed.2022.940454

Latif A, Shehzad A, Niazi S, Zahid A, Ashraf W, Iqbal MW, Rehman A, Riaz T, Aadil RM, Khan IM, Özogul F, Rocha JM, Esatbeyoglu T, Korma SA. Probiotics: Mechanism of action, health benefits and their application in food industries. *Front Microbiol*. 2023 Aug 17;14:1216674. https://doi.org/10.3389/fmicb.2023.1216674

Liu C, Ma N, Feng Y, Zhou M, Li H, Zhang X, Ma X. From probiotics to postbiotics: Concepts and applications. *Anim Res One Health*. 2023;1(1):92–114. https://doi.org/10.1002/aro2.7

Ma J, Lyu Y, Liu X, Jia X, Cui F, Wu X, Deng S, Yue C, Engineered probiotics. *Microb Cell Fact*. 2022;21:72. https://doi.org/10.1186/s12934-022-01799-0

Maftei N-M, Raileanu CR, Balta AA, Ambrose L, Boev M, Marin DB, Lisa EL. The potential impact of probiotics on human health: An update on their health-promoting properties. *Microorganisms*. 2024; 12(2):234. https://doi.org/10.3390/microorganisms12020234

Mazziotta C, Tognon M, Martini F, Torreggiani E, Rotondo JC. Probiotics mechanism of action on immune cells and beneficial effects on human health. *Cells*. 2023 Jan 2;12(1):184. https://doi.org/10.3390/cells12010184

Merenstein D, Pot B, Leyer G, Ouwehand AC, Preidis GA, Elkins CA, Hill C, Lewis ZT, Shane AL, Zmora N, Petrova MI, Collado MC, Morelli L, Montoya GA, Szajewska H, Tancredi DJ, Sanders ME. Emerging issues in probiotic safety: 2023 perspectives. *Gut Microbes*. 2023 Jan–Dec;15(1):2185034. https://doi.org/10.1080/19490976.2023.2185034

Meroni G, Panelli S, Zuccotti G, Bandi C, Drago L, Pistone D. Probiotics as therapeutic tools against pathogenic biofilms: Have we found the perfect weapon? *Microbiol Res*. 2021; 12(4):916–937. https://doi.org/10.3390/microbiolres12040068

Milner E, Stevens B, An M, Lam V, Ainsworth M, Dihle P, Stearns J, Dombrowski A, Rego D and Segars K. Utilizing probiotics for the prevention and treatment of gastrointestinal diseases. *Front. Microbiol*. 2021;12:689958. https://doi.org/10.3389/fmicb.2021.689958

Mirghani R, Saba T, Khaliq H, Mitchell J, Do L, Chambi L, Diaz K, Kennedy T, Alkassab K, Huynh T, Elmi M, Martinez J, Sawan S, Rijal G. Biofilms: Formation, drug resistance and alternatives to conventional approaches. *AIMS Microbiol*. 2022 Jul 4;8(3):239–277. https://doi.org/10.3934/microbiol.2022019

Mugwanda K, Hamese S, Van Zyl WF, Prinsloo E, Du Plessis M, Dicks LMT, Thimiri Govinda Raj DB. Recent advances in genetic tools for engineering probiotic lactic acid bacteria. *Biosci Rep*. 2023 Jan 31;43(1):BSR20211299. https://doi.org/10.1042/BSR20211299

Muteeb G, Rehman MT, Shahwan M, Aatif M. Origin of antibiotics and antibiotic resistance, and their impacts on drug development: A narrative review. *Pharmaceuticals (Basel)*. 2023 Nov 15;16(11):1615. https://doi.org/10.3390/ph16111615

Ng QX, Peters C, Venkatanarayanan N, Goh YY, Ho CY, Yeo WS. Use of Lactobacillus spp. to prevent recurrent urinary tract infections in females. *Med. Hypotheses*. 2018;114:49–54. https://doi.org/10.1016/j.mehy.2018.03.001

Papadimitriou K, Alegría Á, Bron PA, de Angelis M, Gobbetti M, Kleerebezem M, Lemos JA, Linares DM, Ross P, Stanton C, Turroni F, van Sinderen D, Varmanen P, Ventura M, Zúñiga M, Tsakalidou E, Kok J. Stress physiology of lactic acid bacteria. *Microbiol Mol Biol Rev*. 2016 Jul 27;80(3):837–90. https://doi.org/10.1128/MMBR.00076-15

Patangia DV, Anthony Ryan C, Dempsey E, Paul Ross R, Stanton C. Impact of antibiotics on the human microbiome and consequences for host health. *Microbiologyopen*. 2022 Feb;11(1):e1260. https://doi.org/10.1002/mbo3.1260

Petrariu O-A, Barbu IC, Niculescu A-G, Constantin M, Grigore GA, Cristian R-E, Mihaescu G, Vrancianu CO. Role of probiotics in managing various human diseases, from oral pathology to cancer and gastrointestinal diseases. *Front Microbiol.* 2024;14:1296447. https://doi.org/10.3389/fmicb.2023.1296447

Plaza-Diaz J, Ruiz-Ojeda FJ, Gil-Campos M, Gil A. Mechanisms of Action of Probiotics. Adv Nutr. 2019 Jan 1;10(suppl_1):S49–S66. https://doi.org/10.1093/advances/nmy063

Prestinaci F, Pezzotti P, Pantosti A. Antimicrobial resistance: A global multifaceted phenomenon. *Pathog Glob Health.* 2015;109(7):309–318. https://doi.org/10.1179/2047773215Y.0000000030

Qasemi A, Lagzian M, Rahimi F, Majd FK, Bayat Z. The power of probiotics to combat urinary tract infections: A comprehensive review. *Res Biotechnol Environ Sci.* 2023;2(1):1–11. https://doi.org/10.58803/RBES.2023.2.1.01

Rabetafika HN, Razafindralambo A, Ebenso B, Razafindralambo HL. Probiotics as antibiotic alternatives for human and animal applications. *Encyclopedia.* 2023; 3(2):561–581. https://doi.org/10.3390/encyclopedia3020040

Raheem A, Liang L, Zhang G, Cui S. Modulatory effects of probiotics during pathogenic infections with emphasis on immune regulation. *Front Immunol.* 2021;12:616713. https://doi.org/10.3389/fimmu.2021.616713

Salam MA, Al-Amin MY, Salam MT, Pawar JS, Akhter N, Rabaan AA, Alqumber MAA. Antimicrobial resistance: A growing serious threat for global public health. *Healthcare (Basel).* 2023 Jul 5;11(13):1946. https://doi.org/10.3390/healthcare11131946

Salman MK, Abuqwider J, Mauriello G. Anti-quorum sensing activity of probiotics: The mechanism and role in food and gut health. *Microorganisms.* 2023 Mar 20;11(3):793. https://doi.org/10.3390/microorganisms11030793

Schwenger EM, Tejani AM, Loewen PS. Probiotics for preventing urinary tract infections in adults and children. *Cochrane Database Syst Rev.* 2015 Dec 23;2015(12):CD008772.

Shahbazi R, Yasavoli-Sharahi H, Alsadi N, Ismail N, Matar C. Probiotics in treatment of viral respiratory infections and neuroinflammatory disorders. *Molecules.* 2020 Oct 22;25(21):4891. https://doi.org/10.3390/molecules25214891

Sihra N, Goodman A, Zakri R. et al. Nonantibiotic prevention and management of recurrent urinary tract infection. *Nat Rev Urol.* 2018;15:750–776. https://doi.org/10.1038/s41585-018-0106-x

Silva DR, Sardi JD, de Souza Pitangui N, Roque SM, da Silva AC, Rosalen PL. Probiotics as an alternative antimicrobial therapy: Current reality and future directions. *J Funct Foods.* 73;2020:104080. https://doi.org/10.1016/j.jff.2020.104080

Skoufou M, Tsigalou C, Vradelis S, Bezirtzoglou E. The networked interaction between probiotics and intestine in health and disease: A promising success story. *Microorganisms.* 2024;12(1):194. https://doi.org/10.3390/microorganisms12010194

Song HJ, Kim JY, Jung SA, Kim SE, Park HS, Jeong Y, Hong SP, Cheon JH, Kim WH, Kim HJ, Ye BD, Yang SK, Kim SW, Shin SJ, Kim HS, Sung JK, Kim EY. Effect of probiotic lactobacillus (Lacidofil® Cap) for the prevention of antibiotic-associated diarrhea: A prospective, randomized, double-blind, multicenter study. *J Korean Med Sci.* 2010 Dec;25(12):1784–1791. https://doi.org/10.3346/jkms.2010.25.12.1784

Tomé AR, Carvalho FM, Teixeira-Santos R, Burmølle M, Mergulhão FJM, Gomes LC. Use of probiotics to control biofilm formation in food industries. *Antibiotics.* 2023;12(4):754. https://doi.org/10.3390/antibiotics12040754

Tung JM, Dolovich LR, Lee CH. Prevention of Clostridium difficile infection with Saccharomyces boulardii: A systematic review. *Can J Gastroenterol.* 2009;23(12):817–821.

Vagios S, Hesham H, Mitchell C. Understanding the potential of lactobacilli in recurrent UTI prevention, *Microbial Pathogenesis.* 2020;148:104544. https://doi.org/10.1016/j.micpath.2020.104544

Van Rossum T, Haiß A, Knoll RL, et al. Bifidobacterium and Lactobacillus probiotics and Gut dysbiosis in preterm infants: The PRIMAL randomized clinical trial. *JAMA Pediatr.* 2024;178(10):985–995. https://doi.org/10.1001/jamapediatrics.2024.2626

van Zyl WF, Deane SM, Dicks LMT. Molecular insights into probiotic mechanisms of action employed against intestinal pathogenic bacteria. *Gut Microbes.* 2020 Nov 9;12(1):1831339. https://doi.org/10.1080/19490976.2020.1831339

Vieco-Saiz N, Belguesmia Y, Raspoet R, Auclair E, Gancel F, Kempf I, Drider D. Benefits and inputs from lactic acid bacteria and their bacteriocins as alternatives to antibiotic growth promoters during food-animal production. *Front. Microbiol.* 2019;10:57. https://doi.org/10.3389/fmicb.2019.00057

Wong CB, Odamaki T, Xiao JZ. Beneficial effects of Bifidobacterium longum subsp. longum BB536 on human health: Modulation of gut microbiome as the principal action. *J Funct. Foods*, 54, 2019, 506–519. https://doi.org/10.1016/j.jff.2019.02.002

Xu C, Aqib AI, Fatima M, Muneer S, Zaheer T, Peng S, Ibrahim EH, Li K. Deciphering the potential of probiotics in vaccines. *Vaccines (Basel)*. 2024 Jun 26;12(7):711. https://doi.org/10.3390/vaccines12070711

Yang S, Qiao J, Zhang M, Kwok LY, Matijašić BB, Zhang H, Zhang W. Prevention and treatment of antibiotics-associated adverse effects through the use of probiotics: A review. *J Adv Res*. 2024. https://doi.org/10.1016/j.jare.2024.06.006

Zheng D, Liwinski T, Elinav E. Interaction between microbiota and immunity in health and disease. *Cell Res*. 2020;30:492–506. https://doi.org/10.1038/s41422-020-0332-7

Zhou X, Liu H, Zhang J, Mu J, Zalan Z, Hegyi F, et al. Protective effect of Lactobacillus fermentum CQPC04 on dextran sulfate sodium-induced colitis in mice is associated with modulation of the nuclear factor-κBsignaling pathway. *J Dairy Sci*. 2019;102:9570–9585. https://doi.org/10.3168/jds.2019-16840

Zong X, Fu J, Xu B, Wang Y, Jin M. Interplay between gut microbiota and antimicrobial peptides. *Anim Nutr*. 2020;6(4):389–396. https://doi.org/10.1016/j.aninu.2020.09.002

Index

For Product Safety Concerns and Information please contact our
EU representative GPSR@taylorandfrancis.com Taylor & Francis
Verlag GmbH, Kaufingerstraße 24, 80331 München, Germany